W0246248

IUTAM SYMPOSIUM ON MECHANICS OF GRANULAR AND POROUS MATERIALS

SOLID MECHANICS AND ITS APPLICATIONS
Volume 53

Series Editor: **G.M.L. GLADWELL**
Solid Mechanics Division, Faculty of Engineering
University of Waterloo
Waterloo, Ontario, Canada N2L 3G1

Aims and Scope of the Series

The fundamental questions arising in mechanics are: *Why?*, *How?*, and *How much?*
The aim of this series is to provide lucid accounts written by authoritative research-ers giving vision and insight in answering these questions on the subject of mechanics as it relates to solids.

The scope of the series covers the entire spectrum of solid mechanics. Thus it includes the foundation of mechanics; variational formulations; computational mechanics; statics, kinematics and dynamics of rigid and elastic bodies; vibrations of solids and structures; dynamical systems and chaos; the theories of elasticity, plasticity and viscoelasticity; composite materials; rods, beams, shells and membranes; structural control and stability; soils, rocks and geomechanics; fracture; tribology; experimental mechanics; biomechanics and machine design.

The median level of presentation is the first year graduate student. Some texts are monographs defining the current state of the field; others are accessible to final year undergraduates; but essentially the emphasis is on readability and clarity.

For a list of related mechanics titles, see final pages.

IUTAM Symposium on

Mechanics of Granular and Porous Materials

Proceedings of the IUTAM Symposium
held in Cambridge, U.K.,
15–17 July 1996

Edited by

N. A. FLECK
*Department of Engineering,
University of Cambridge, U.K.*

and

A. C. F. COCKS
*Department of Engineering,
University of Leicester, U.K.*

SPRINGER-SCIENCE+BUSINESS MEDIA, B.V.

A C.I.P. Catalogue record for this book is available from the Library of Congress

ISBN 978-94-010-6324-1   ISBN 978-94-011-5520-5 (eBook)
DOI 10.1007/978-94-011-5520-5

Printed on acid-free paper

All Rights Reserved
© 1997 Springer Science+Business Media Dordrecht
Originally published by Kluwer Academic Publishers in 1997
No part of the material protected by this copyright notice may be reproduced or
utilized in any form or by any means, electronic or mechanical,
including photocopying, recording or by any information storage and
retrieval system, without written permission from the copyright owner.

CONTENTS

Contact Mechanics and Computational Methods

Granular Flow: Theory and Applications

vii

Instabilities

Applications

Preface

This volume constitutes the Proceedings of the IUTAM Symposium on Mechanics of Granular and Porous Materials, held in Cambridge from 15th to 17th July 1996. The objectives were:

1. To review existing experimental results and practical phenomena on the flow and compaction of particulate media;

2. To review the current state of constitutive models, and their implementation for predicting the macroscopic response.

3. Identification of the shortcomings of existing models and procedures in understanding practical phenomena.

The Symposium brought together the research communities of solid mechanics, materials science, geomechanics, chemical engineering and mathematics to review current knowledge of the flow and compaction of granular and porous media. The meeting emphasised the development and use of constitutive laws to model practical processes such as mixing, drainage and drying, compaction of metal and ceramic powders and soils, and instabilities associated with these processes. A common theme was to develop constitutive models from an understanding of the underlying physical mechanisms of deformation and fracture.

It was particularly rewarding to find that the separate research communities came together during the meeting and came to a consensus as to the main mechanisms of deformation and failure of particulate and porous solids.

The Symposium consisted of forty lectures, all of which were invited and accorded equal weight in the programme. In addition, two poster sessions allowed for a further twenty presentations. Only the content of the lectures is reflected in this volume. All the papers contained herein have been reviewed to the standard of leading scientific journals.

The smooth running of the Symposium owes much to the initiative and organisational skills of Jo Ladbrooke and Roxana Saadatnejad.

Norman A. Fleck
Alan C. F. Cocks

Cambridge, England
December 1996

International Scientific Committee

R. de Boer (Germany)
I. F. Collins (New Zealand)
O. Coussy (France)
N. A. Fleck (UK)
P. Germain (France)
R. M. McMeeking (USA)
A. Needleman (USA)
M. Satake (Japan)
S. B. Savage (Canada)

Local Organising Committee

M. F. Ashby
M. D. Bolton
J. Bridgwater
A. C. F. Cocks *(Secretary)*
N. A. Fleck *(Chairman)*
R. M. McMeeking
M. Thiercelin
J. R. Willis

Sponsors of the IUTAM Symposium on Mechanics of Granular and Porous Materials

International Union of Theoretical and Applied Mechanics (IUTAM)
US Office of Naval Research
Kluwer Academic Publishers BV
Schlumberger Cambridge Research, Ltd.
The Royal Society, London
Cambridge Centre for Micromechanics

Remembering

(John) Mark Duva

(1955 - 1995)

Mark grew up in Michigan, attended a Jesuit high school, but dropped out after the 11th grade to seek something more fulfilling.

In the late 60's, a number of communes sprang up around the country including several in Virginia. Getting back to basics and living with an extended family of like-minded idealists appealed to Mark so he applied and was accepted into the Springtree Community, a commune which still exists some 20 miles south of Charlottesville. But *Reality Bites*, as the title of a popular movie reminds us. At Springtree, instead of weighty philosophical discussions of the good life, there were, each week, compulsory "group therapy" sessions. "Here I was," Mark once told me, "a naive, 17 year-old kid, forced, week after week, to listen to squabbling lovers talking about the most intimate details of their on-again-off-again relationships." Still, as was his wont, Mark pulled something valuable from this emotional cauldron: he learned to be a carpenter. And he did it in a way that came to be characteristic - through community service, specifically, through the Charlottesville Home Improvement Project, an enterprise devoted to repairing and building homes for poor people.

In 1978 Mark started taking classes at the local community college and the next year he got married and transferred to the School of Engineering & Applied Sciences at the University of Virginia (U.Va.).

I first encountered Mark in 1980 in an undergraduate math class I was teaching. Not long after the semester began, Mark, who rarely said anything in class, came up to me and said, politely, "You know, I think it would have made much more sense if you had explained that last concept this way." And when it became apparent that I could expect these commentaries regularly, I knew that I was going to have to start spending a lot more time preparing my lectures. That episode revealed another aspect of Mark's academic personality: gentle, unpretentious, and absolutely committed to getting to the essence of things.

In the Engineering School, all fourth-year students have to write a thesis. Mark had gotten interested in certain apparent paradoxes in Einstein's Theory of Relativity and asked me to be his advisor. What an intellectually rewarding experience this was for me, for not only did Mark teach me new technical facts, he taught me the importance of style - that technical writing can be fluid, engrossing, informative, and convincing. For Mark believed that *how* he wrote was as important as *what* he wrote about.

The summer after he was graduated from U.Va. (with the second highest grade-point average in his class) he worked for me on a research project. We published a joint paper on our work and when it came time to order re-prints from the publisher, I, in my youthful enthusiasm, said "Mark, we've written a classic, I'm going to order 500 copies." Just before he died, I joked with Mark about that paper and how we had probably gotten all of 5 requests for reprints since we wrote it.

When he got his bachelor's degree in 1981, Mark was already sure that he wanted to teach at U.Va. but, much as I wanted to keep him on as a graduate student, I told him that it would be better for several reasons if he got his Ph.D. elsewhere and then return a conquering hero. And what better graduate school than Harvard? Mark's reputation preceded him; I remember him recalling with relish that the first words spoken to him by Bernie Budiansky were, "Duva, I hear you're a carpenter. That's great because I just bought an air-conditioning unit which needs to be built-in into my living room wall." Let this be a lesson to us all: good hands are sometimes more valued than a good mind.

And return to U.Va. Mark did. Was his road to academic success a superhighway? Not by a long shot. For one thing, he got off on the wrong foot with the Dean who had put Mark on his own research contract only to find out that Mark had his own ideas about research and wasn't about to make foolish claims or put his name on anything he didn't fully understand or didn't think was up to snuff. On top of this, Mark's first teaching ratings were - to put it mildly - terrible; but terrible for the best reasons: Mark cared deeply about the academic mission of the University and refused to cheat his students by feeding them intellectual junk food. He would spend hours preparing his lectures and crafting tests which stretched his students, making them think in new and therefore difficult ways, but in ways that he hoped would ultimately make them reflect back and say, "Yes, I got something special and valuable from that class."

Mark was a thinker but he was a doer as well. He felt that the academic life should be more than just teaching and research, much as he loved those activities. He felt that one has a duty not only to one's academic community, but to one's social community as well. And so, within the Engineering School, Mark voluntarily chaired several important but time-consuming committees on undergraduate studies and he initiated a symposium in which the top student theses were presented to and awarded prizes by a panel of faculty and industry representatives. Outside the University, Mark taught in the Charlottesville literacy program and devoted parts of his weekends to teaching in the Saturday Academy, an academic enrichment program organized for youngsters in the community by the Dean of African-American Affairs at the University.

Let me close with an anecdote that epitomizes Mark's never failing concern for students. I was in France, on sabbatical, the year Mark died, but we kept in touch by e-mail until nearly the end. With others, Mark and I were working on a big proposal to the National Science Foundation on a revolutionary curriculum in structural mechanics. In preparation, everyone read student exit interviews to try to get student's perspectives. I and another senior professor on the project agreed that the picture we got was pretty discouraging - a lot of bellyaching by a bunch of pretty privileged kids we concluded. But Mark turned our thinking around. "Look below the surface," he said, "and you'll see the same message over and over: 'we want faculty to spend time with us. We want to feel that they are personally concerned with our education and our particular goals.'" This was a wonderful illustration of Mark's philosophy: universities are first and foremost educational institutions and our students deserve nothing less than the best that we, as faculty, can offer them.

So if I were to draw one overarching impression from Mark's life it would be that it represented thought into action, but action always tempered by compassion, humility, and the belief that to pull the best from others, you must first demand it of yourself.

J. G. Simmonds
University of Virginia
January 1997

THE VISCOPLASTIC COMPACTION OF POWDERS

N.A. FLECK
Cambridge University Engineering Department
Trumpington Street, Cambridge, CB2 1PZ, UK

B. STORÅKERS
Department of Solid Mechanics, Royal Institute of Technology
S-100 44 Stockholm, Sweden

R.M. McMEEKING
Department of Mechanical and Environmental Engineering
University of California, Santa Barbara, CA, 93106, USA

Abstract. A model based on the affine motion of spherical particles of uniform size is developed for the early stage of compaction. Newly available results for the interaction of two viscoplastic spheres are used along with elements from previous models to develop a definitive statement of the results for arbitrary deformations. A prescription is given for computing the macroscopic stress in general, and pressure versus density relations are presented for both isostatic and closed die compaction.

1. Introduction

Good progress has been made on the modelling of the plastic and creep response of powder aggregates (Fischmeister *et al.*, 1978; Helle *et al.*, 1985; Arzt, 1982). During the initial stage of compaction (designated Stage I), the microstructure consists of individual particles connected by discrete contacts. Fleck, Kuhn and McMeeking (1992) have calculated the macroscopic multi-axial response for an isotropic distribution of contacts around each particle for plastic flow, and Kuhn and McMeeking (1992) have handled the case for creep. An initial attempt to deal with the development of anisotropy and finite cohesive strength of the contacts has been introduced by Fleck (1995). The case of cold and hot isostatic compaction has been analysed in a self-consistent way by Larsson *et al.* (1996).

2. Kinematics

The powder aggregate is considered to consist of spherical particles of uniform size in an initial state of random dense packing, with an initial relative density $D_o = 0.64$. Arzt (1982) approximated this state by a radial distribution function for particle centres adjacent to a reference sphere:

1

N. A. Fleck and A. C. F. Cocks (eds.), IUTAM Symposium on Mechanics of Granular and Porous Materials, 1–10.
© 1997 *Kluwer Academic Publishers.*

$$G(\hat{r}) = Z + \Gamma\left(\frac{\hat{r}}{2R} - 1\right) \qquad , \qquad \hat{r} \geq 2R \qquad (1)$$

where G is the number of particle centres within the radius $\hat{r}$ from the centre of the reference particle, Z and Γ are constants and R is the radius of the powder particles. The best estimates for the constants are $Z = 7.3$ and $\Gamma = 15.5$. Thus, on average each particle has 7.3 contacts with neighbours in the initial state.

We assume affine deformation in which each particle moves at the velocity prescribed for its centre. Incompatibilities in this process are accommodated by local distortions (e.g. flattening) where contact is made with neighbours. The aggregate is assumed to experience a macroscopically homogeneous deformation which therefore prescribes the motion of the particles' centres. Consider the deformation state for a reference sphere centred at a fixed origin, and a neighbouring sphere, as shown in Fig. 1.

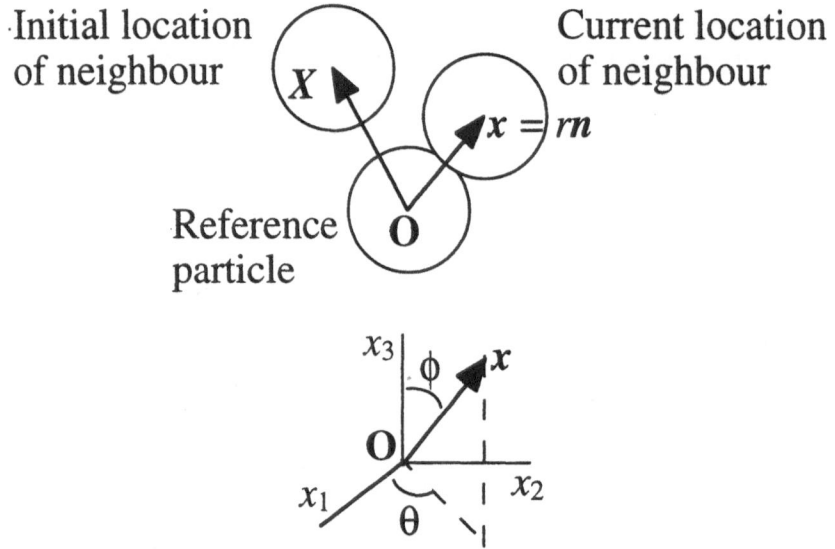

Figure 1. Relative displacement of a neighbouring particle with respect to a reference particle under a deformation gradient F.

The neighbouring particle with its centre originally at position X, a distance $\hat{r}$ from the reference particle centre, moves under a macroscopic deformation gradient F to a new location with its centre at x, such that

$$x = F \cdot X \ . \qquad (2)$$

Now introduce a unit normal n and a particle spacing r in the deformed configuration, such that $x = rn$. The stretch ratio λ can be stated as

$$\frac{r}{\hat{r}} = \lambda(n) = \frac{1}{\sqrt{n \cdot \left(F^{-1}\right)^T \cdot F^{-1} \cdot n}} \quad . \tag{3a}$$

All particles with centres currently within a distance $2R$ of the reference particle centre will now have made contact with it, and overlap locally by an amount $2h$ where

$$2h = 2R - r \quad . \tag{3b}$$

Therefore, averaged over all particles, the number of contacts z per unit particle surface area at orientation n is given by

$$z(n) = \frac{Z}{4\pi R^2} + \frac{\Gamma}{4\pi R^2}\left(\frac{1}{\lambda(n)} - 1\right) \quad . \tag{4}$$

3. Viscoplastic Indentation

Consider two identical frictionless particles of radii R experiencing indentation by an overlap displacement of $2h$. It is assumed that the particles are made of isotropic material with a uniaxial viscoplastic stress response given by

$$\sigma = \sigma_0 \varepsilon^M \dot{\varepsilon}^N \tag{5}$$

where σ_0 is a strength parameter, ε is the axial strain, $\dot{\varepsilon}$ is its rate, M is a hardening exponent and N is a power law creep exponent. Indentation theory based on von Mises flow theory (see for example, Storåkers (1996) and Storåkers et al., 1996) shows that the contact area A is related to the overlap $2h$ by

$$A = \pi c^2 2hR \tag{6}$$

From detailed finite element solutions it was found, to a good approximation, that

$$c^2 = 1.43 \exp(-0.97(M+N)) \quad . \tag{7}$$

with the contact load P given by

$$P = \eta h^{(2+M-N)/2} \dot{h}^N \tag{8}$$

where

$$\eta = 6^{1-M-N}(1+2N)\pi c^2 \sigma_0 R_0 \left(\frac{2c^2}{R_0}\right)^{M/2}\left(\frac{c^2}{2R_0}\right)^{N/2} \quad . \tag{9}$$

Note that the above indentation theory is a first order theory assuming small straining.

4. Macroscopic Stress

The macroscopic stress Σ corresponding to a state of uniform macroscopic deformation F and strain rate D may be calculated via a macroscopic *creep potential* Φ defined by

$$\Phi(D) = \int_0^D \Sigma_{ij}(D')dD'_{ij} \qquad \Leftrightarrow \qquad \Sigma_{ij} = \frac{\partial \Phi}{\partial D_{ij}} \quad . \tag{10}$$

The macroscopic creep potential is the volumetric average of the local creep potential ϕ within the body of volume V,

$$\Phi(D) = \frac{1}{V}\int_V \phi(\dot{\varepsilon})dV \tag{11}$$

where

$$\phi(\dot{\varepsilon}) = \int_0^{\dot{\varepsilon}} \sigma_{ij}(\dot{\varepsilon}'_{ij})d\dot{\varepsilon}'_{ij} \quad . \tag{12}$$

On defining the creep potential ϕ_c for a single contact, via eqn. (8), as

$$\phi_c(\dot{h}) = \int_0^{\dot{h}} P(\dot{h}')dh' = \frac{P\dot{h}}{N+1} = \frac{\eta}{N+1} h^{(2+M-N)/2}\dot{h}^{N+1} \tag{13}$$

the macroscopic creep potential may be written

$$\Phi(D) = \frac{3D}{4\pi R^3}\int_S \frac{dS}{S}\int_{\hat{r}=2R}^{\hat{r}=2R/\lambda} d\hat{r}\left[\phi_c(\dot{h})\frac{dG}{d\hat{r}}\right] \quad . \tag{14}$$

where $S = 4\pi R^2$ is the surface area of the representative particle. Note that the radial integration in (14) is in the undeformed configuration and includes the contribution from both initial contacts at $F = 0$ and from new contacts formed over the compaction history. The macroscopic stress Σ is obtained by substituting (1), (3) and (13) into (14), to give via (10)

$$\Sigma_{ij} = \frac{\partial \Phi}{\partial D_{ij}} = \frac{-3D}{8\pi R^3}\int_S \frac{dS}{S}\int_{\hat{r}=2R}^{\hat{r}=2R/\lambda} d\hat{r}\left[n_i n_j \lambda \hat{r} P \frac{dG}{d\hat{r}}\right] \quad . \tag{15}$$

It is convenient to introduce the change of variables from $\hat{r}$ to $x \equiv 2R - \lambda\hat{r}$ and to expand (15) to leading order in small x/R. The radial integration in $\hat{r}$ can then be done analytically, giving to high accuracy

$$\Sigma_{ij} = \frac{-3D}{4\pi R^2}\int_S \frac{dS}{S}\left[n_i n_j \lambda P_0\left\{Z + 2\Gamma(1-\lambda)\lambda^{-(N+2)}\left(\frac{1}{4+M-N} - \frac{(N+1)(1-\lambda)}{6+M-N}\right)\right\}\right] \tag{16}$$

in terms of the contact load P_0 for a pair of particles initially in contact. For such a pair, P_0 is given by (8), with $h = (1 - \lambda)R$.

The expression (16) for the macroscopic stress can be simplified further upon assuming small straining, and neglecting the effects of new contacts, to give

$$\Sigma_{ij} = \frac{-3\eta D_0 Z}{4\pi} R^{\frac{M+N-2}{2}} \int_S \frac{dS}{S} \left[n_i n_j \left(-E_{kl} n_k n_l \right)^{\frac{2+M-N}{2}} \left(-\dot{E}_{pq} n_p n_q \right)^N \right] \tag{17}$$

where E is the infinitesimal strain tensor and D_0 is the initial relative density of the compact.

In order to assess the accuracy of the simplifying approximations of infinitesimal straining and no new contacts, it is useful to explore the predictions for closed die compaction.

5. Closed Die Compaction

Consider the strain state of axisymmetric closed die compaction, with the compression direction parallel to the x_3 axis. The radial strain vanishes, and the deformation gradient is given by

$$F_{ij} = \delta_{ij} \tag{18}$$

except for the axial term

$$F_{33} = \Lambda \tag{19}$$

where Λ is the compaction ratio. With ϕ being the angle between a given orientation n and the x_3-axis as shown in Fig. 1, eqn. (3) provides

$$\lambda^{-2} = 1 + \left(\Lambda^{-2} - 1 \right) \cos^2 \phi \ . \tag{20}$$

In addition,

$$D/D_o = 1/\Lambda \tag{21}$$

where, as before, it is assumed that compaction commences at random dense packing. The compaction ratio Λ will be used below interchangeably with the relative density.

In the case of $M = N = 0$ (i.e. perfectly plastic rate independent yielding) exact results can be obtained from eqn. (16). Axial symmetry allows eqn. (16) to be rewritten as

$$S = -\Sigma_{33} = \frac{3}{4\pi} \frac{\eta D}{R} \int_0^{\frac{\pi}{2}} d\phi \cos^2 \phi \ \sin\phi \ (1-\lambda)\lambda \left\{ Z + \frac{\Gamma(1-\lambda)}{2\lambda^2} \left[1 - \frac{2}{3}(1-\lambda) \right] \right\} \tag{22}$$

and

$$T = -\Sigma_{11} = \frac{3}{8\pi} \frac{\eta D}{R} \int_0^{\frac{\pi}{2}} d\phi \ \sin^3 \phi \ (1-\lambda)\lambda \left\{ Z + \frac{\Gamma(1-\lambda)}{2\lambda^2} \left[1 - \frac{2}{3}(1-\lambda) \right] \right\} \ . \tag{23}$$

These integrate to give

$$S = \frac{3}{20\pi} \frac{\eta Z(D - D_o)}{R} \left\{ G_1 + \frac{\Gamma}{Z} \left[(1 - \Lambda)G_2 + (1 - \Lambda)^2 G_3 \right] \right\} \tag{24}$$

$$T = \frac{1}{20\pi} \frac{\eta Z(D - D_o)}{R} \left\{ G_4 + \frac{\Gamma}{Z} \left[(1 - \Lambda)G_5 + (1 - \Lambda)^2 G_6 \right] \right\} \tag{25}$$

where

$$G_1 = \frac{5\Lambda}{2(1 - \Lambda^2)(1 - \Lambda)} \left[1 - \frac{\Lambda^2 \sinh^{-1}\left(\sqrt{\frac{1}{\Lambda^2} - 1}\right)}{\sqrt{1 - \Lambda^2}} - 2\Lambda + \frac{2\Lambda^2 \tan^{-1}\left(\sqrt{\frac{1}{\Lambda^2} - 1}\right)}{\sqrt{1 - \Lambda^2}} \right] \tag{26}$$

$$G_2 = \frac{5}{2(1 - \Lambda)^2} \left\{ -\frac{2}{3} + \frac{1}{8\Lambda(1 - \Lambda^2)} \left[2 + 3\Lambda^2 - \frac{5\Lambda^4 \sinh^{-1}\left(\sqrt{\frac{1}{\Lambda^2} - 1}\right)}{\sqrt{1 - \Lambda^2}} \right] \right\} \tag{27}$$

$$G_3 = \frac{5}{3(1 - \Lambda)^3} \left\{ 1 - \frac{1}{8\Lambda(1 - \Lambda^2)} \left[2 + 11\Lambda^2 - 8\Lambda^3 \right. \right.$$
$$\left. \left. - \frac{13\Lambda^4 \sinh^{-1}\left(\sqrt{\frac{1}{\Lambda^2} - 1}\right)}{\sqrt{1 - \Lambda^2}} + \frac{8\Lambda^4 \tan^{-1}\left(\sqrt{\frac{1}{\Lambda^2} - 1}\right)}{\sqrt{1 - \Lambda^2}} \right] \right\} \tag{28}$$

$$G_4 = \frac{15\Lambda}{2(1 - \Lambda^2)(1 - \Lambda)} \left[\frac{(2 - \Lambda^2)\sinh^{-1}\left(\sqrt{\frac{1}{\Lambda^2} - 1}\right)}{2\sqrt{1 - \Lambda^2}} - \frac{1}{2}(1 - 2\Lambda) - \frac{\tan^{-1}\left(\sqrt{\frac{1}{\Lambda^2} - 1}\right)}{\sqrt{1 - \Lambda^2}} \right] \tag{29}$$

$$G_5 = \frac{15}{4(1 - \Lambda)^2} \left\{ -\frac{4}{3} + \frac{1}{8\Lambda(1 - \Lambda^2)} \left[2 - 7\Lambda^2 + \frac{\Lambda^2(12 - 7\Lambda^2)\sinh^{-1}\left(\sqrt{\frac{1}{\Lambda^2} - 1}\right)}{\sqrt{1 - \Lambda^2}} \right] \right\} \tag{30}$$

$$G_6 = \frac{5}{2(1-\Lambda)^3}\left\{2 - \frac{1}{2\Lambda} - \frac{1}{8\Lambda(1-\Lambda^2)}\left[8\Lambda^3 - 11\Lambda^2 - 2\right.\right.$$

$$\left.\left. + \frac{(28-15\Lambda^2)\Lambda^2 \sinh^{-1}\left(\sqrt{\frac{1}{\Lambda^2}-1}\right)}{\sqrt{1-\Lambda^2}} - \frac{8\Lambda^2 \tan^{-1}\left(\sqrt{\frac{1}{\Lambda^2}-1}\right)}{\sqrt{1-\Lambda^2}}\right]\right\} . \tag{31}$$

When Λ is close to 1, and therefore the axial compressive strain

$$E = 1 - \Lambda \tag{32}$$

is much smaller than unity, asymptotic expansions give

$$G_1 = 1 - \frac{2}{7}E - \frac{19}{63}E^2 + O(E^3) \tag{33}$$

$$G_2 = \frac{5}{14}\left(1 + \frac{13}{9}E\right) + O(E^2) \tag{34}$$

$$G_3 = -\frac{5}{27} + O(E) \tag{35}$$

$$G_4 = 1 + \frac{3}{7}E + \frac{2}{21}E^2 + O(E^3) \tag{36}$$

$$G_5 = \frac{3}{14}\left(1 + \frac{17}{9}E\right) + O(E^2) \tag{37}$$

$$G_6 = -\frac{5}{63} + O(E) . \tag{38}$$

Substitution of (33)-(38) into (24) and (25) gives

$$S = \frac{3}{20\pi}\frac{\eta D_o Z}{R}\left\{E + \frac{5}{14}\left[2 + \frac{\Gamma}{Z}\right]E^2\right\} + O(E^3) \tag{39}$$

and

$$T = \frac{1}{20\pi}\frac{\eta D_o Z}{R}\left\{E + \frac{1}{14}\left[20 + 3\frac{\Gamma}{Z}\right]E^2\right\} + O(E^3) . \tag{40}$$

respectively. Note from (39) and (40) that the lead term is linear in E and represents the solution for small straining with no new contacts. The contribution from new contacts and from finite straining enter as higher order terms and are of approximately equal weight. Since the stage I analysis is only well-grounded for small strains $E < 0.1$, say, we conclude that the small-strain solution is adequate for practical purposes and there is little need to consider either the effects of finite strains or of new contacts. This is equivalent to taking G_1 and G_4 equal to unity and setting $G_2 = G_3 = G_5 = G_6 = 0$ in expressions (26-31) and (33-38).

6. Comparison of closed die compaction and isostatic compaction

Now consider the case of isostatic compaction, wherein the particles deform according to the viscoplastic description (5). It is assumed that the powder aggregate has an initial relative density D_0 and is subjected to a constant macroscopic pressure p from time $t=0$. After a time t the compact has densified to a relative density D and a current densification rate $\dot{D}$. Straightforward expressions can be given for the compaction pressure in terms of D and $\dot{D}$ by specialising eqn. (16) to

$$p = \frac{D}{4\pi R^2}\left[\lambda P_0\left\{Z + 2\Gamma(1-\lambda)\lambda^{-(N+2)}\left(\frac{1}{4+M-N} - \frac{(N+1)(1-\lambda)}{6+M-N}\right)\right\}\right] \qquad (41)$$

where the stretch ratio $\lambda^3 = D_0 / D$ is taken to be independent of orientation. As before, the contact load P_0 is given by (8), with $h = (1-\lambda)R$.

For the case of small straining, and neglecting the effect of new contacts, (41) reduces to

$$p = \frac{\eta D_0 Z}{4\pi} R^{\frac{M+N-2}{2}}\left[\left(\frac{D-D_0}{3D_0}\right)^{\frac{2+M-N}{2}}\left(\frac{\dot{D}}{3D_0}\right)^N\right] . \qquad (42)$$

In the limit of an ideally plastic solid, p may be written as

$$p = \frac{\eta D_0 Z}{4\pi R}\left(\frac{D-D_0}{3D_0}\right) \qquad (43)$$

Upon explicit evaluation of η by eqns. (7) and (9) and on taking $D_0 = 0.64$, p is given simply by

$$p = 5.22\,\sigma_0(D-D_0) . \qquad (44)$$

In like manner, the small-strain closed die solution follows from (39) and (40) as

$$S = 9.39\,\sigma_0(D-D_0) \qquad (45)$$

and

$$T = 3.13\,\sigma_0(D-D_0) \qquad (46)$$

upon setting $D_0 = 0.64$. The full non-linear solutions (24), (25) and (41), and the linearised small-strain solutions (44-46) are shown in Fig. 2, for both isostatic and closed die compaction of a rigid-perfectly plastic solid ($N = M = 0$). For both loading paths, the stress versus density response given by the full solution is almost linear, in support of the simplifying relations (44-46): it is concluded that the assumptions of small straining with no new contacts are accurate over the practical range of stage I compaction $D/D_0 < 1.2$. Further, the radial stress T in closed die compaction equals about one third of the axial stress S, as demanded by the leading order term in (39) and (40). The pressure p in isostatic compaction is equal to approximately 5/9 times the axial stress S in closed die compaction at the same value of relative density D, as seen by comparing the leading order terms in (44) and (45).

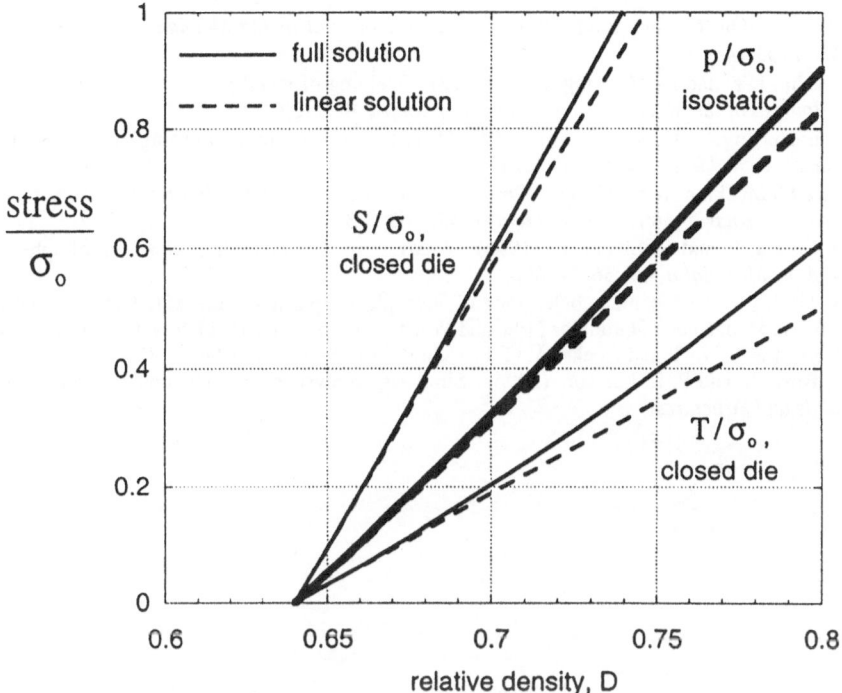

Figure 2. Predicted compaction response for closed die and isostatic compaction, for a rigid- perfectly plastic solid ($N = M = 0$). Initial relative density $D_0 = 0.64$.

Acknowledgement

R. M. McMeeking was supported during this research by a Visiting Professorship at Cambridge University and a sabbatical leave from the University of California, Santa Barbara. N. A. Fleck wishes to thank Dr. R. J. Fields of NIST for financial support in the form of a collaborative program on the Mechanics of Compaction of Powder Composites. B. Storåkers was supported by a sabbatical leave from the Royal Institute of Technology, Stockholm.

References

Arzt, E. (1982). The influence of an increasing particle co-ordination on the densification of spherical powder, *Acta Metallurgica*, **30**, 1883-1890.

Fischmeister, H.E., Arzt,. E. and Olsson, L.R. (1978). Particle deformation and sliding during compaction of spherical powders: a study of quantitative metallography. *Powder Metallurgy*, **21**, 179-187.

Fleck, N.A. (1995). On the cold compaction of powders, *Journal of the Mechanics and Physics of Solids*, **43**, 1409-1431.

Fleck, N.A., Kuhn, L.T. and McMeeking, R.M. (1992). Yielding of metal powder bonded by isolated contacts, *Journal of the Mechanics and Physics of Solids*, **40**, 1139-1162.

Helle, A.S., Easterling, K.E. and Ashby, M.F. (1985). Hot-isostatic pressing diagrams: new developments, *Acta Mettalurgica* , **33**, 2163-2174.

Kuhn, L.T. and McMeeking, R.M. (1992). Power law creep of powder bonded by isolated contacts, *International Journal of Mechanical Sciences*, **34**, 563-573.

Larsson, P.-L., Biwa, S. and Storåkers, B. (1996). Analysis of cold and hot compaction of spherical particles, *Acta Materialia*, **44**, 3655-3666.

Storåkers, B. (1996). Local contact behaviour of viscoplastic particles, presented at the IUTAM Symposium on Mechanics of Granular Flow and Powder Compaction, 15-17 July 1996, Cambridge, England, eds. Fleck, N. A. and Cocks, A. C. F. , Kluwer Academic Publishers, 173-184.

Storåkers, B., Biwa, S. and P.-L. Larsson, (1996). Similarity analysis of inelastic contact, to appear in *Int. J. Solids and Structures*.

INELASTIC BEHAVIOR OF RANDOM ARRAYS OF IDENTICAL SPHERES

JAMES T. JENKINS
Department of Theoretical and Applied Mechanics
Cornell University, Ithaca, NY 14853

1. Introduction

Our overall goal is to derive an incremental, rate-independent continuum theory for an idealized granular material consisting of spherical particles from a consideration of force and moment equilibrium of the individual grains and simple but appropriate assumptions regarding the force of contact between grains and the statistical geometry of their packing.

For frictional contacts, the incremental relation between the contact force and the displacement of a contact is independent of rate, so the macroscopic relation between the average stress and the average strain is also incremental and rate-independent.

Particles slide and rotate with respect to each other as the aggregate is deformed and the packing of the aggregate may change as contacts are lost and created. In a frictional material there is a distribution of contact forces even in an isotropically compressed aggregate; since the stiffness of a contact depends upon the contact force, the stiffness of similarly oriented contacts will be very different. As a consequence, intensities and anisotropies in both the geometry of the packing and in the contact forces can evolve. An example of what is meant by an intensity is the average number of contacts per particle, the coordination number k. An example of an anisotropy is the non-uniformity in the orientational distribution of nearest neighbors, the geometric "fabric" (Oda, 1976).

At any stage of the deformation, these intensities and anisotropies will influence the relation between the increments of the average strain and the average stress. In this regard, they characterize the present state of the material and, by virtue of their dependence on the path of the deformation, they endow the material with a memory of its prior history. The challenge is first, to identify the essential measures of these; second, to determine how

11

N. A. Fleck and A. C. F. Cocks (eds.),
IUTAM Symposium on Mechanics of Granular and Porous Materials, 11–22.
© 1997 *Kluwer Academic Publishers.*

these measures enter into the incremental stress-strain relation; and third, to establish how they change with the deformation.

Here, we restrict our attention to the behavior of an idealized granular material consisting of identical spheres that are first isotropically compressed and the sheared, perhaps with simultaneous changes in confining pressure, until the onset of strain localization. Because our primary interest is in problems related to soil mechanics, we restrict our attention to a relatively limited range of the ratio of shear stress to confining pressure. In this limited range, we anticipate that both the elasticity and the friction of the contacts will be important, but that changes in the geometry of the packing may be small and are likely to be dominated by the effective deletion of contacts as contact forces relax to zero.

The continuum theory that relates increments in the average stress $\mathbf{T}$, the state variables $\mathbf{\Sigma}$, and an average distortion $\mathbf{L}$ is expected to have the structure:

$$\dot{T}_{ij} = F_{ijkl}(\mathbf{\Sigma}, \mathbf{L})\dot{L}_{kl}$$

and

$$\dot{\Sigma}_{ij} = G_{ijkl}(\mathbf{\Sigma}, \mathbf{L})\dot{L}_{kl}$$

where the overdot denotes an objective increment. Such a theoretical structure is more general than the plasticity theories for compressible, frictional materials that have been employed to interpret physical experiments (e.g., Tatsuoka & Ishihara, 1974; Voyiadjis, Thiagarajan & Petrakis, 1995). The idea is that systems of stiff evolution equations provide a general structure that can include behavior that might be interpreted as yielding.

2. Theory

We focus our attention on a pair of contacting spheres, label them A and B, and denote the vector from the center of A to the center of B by $\mathbf{d}^{(BA)}$. We write the increment $\dot{\mathbf{F}}^{(BA)}$ in the contact force exerted by particle B on particle A in terms of the increment $\dot{\mathbf{u}}^{(BA)}$ in the relative displacement of the points of contact:

$$\dot{F}_i^{(BA)} = K_{ij}^{(BA)}\dot{u}_j^{(BA)},$$

where $\mathbf{K}^{(BA)}$ is the contact stiffness.

The contact stiffness is given in terms of the unit vector $\hat{\mathbf{d}}^{(BA)}$ in the direction of $\mathbf{d}^{(BA)}$ by

$$K_{ij}^{(BA)} = K_N^{(BA)}\hat{d}_i^{(BA)}\hat{d}_j^{(BA)} + K_T^{(BA)}(\delta_{ij} - \hat{d}_i^{(BA)}\hat{d}_j^{(BA)}).$$

Here $K_N^{(BA)}$ and $K_T^{(BA)}$ are the normal and tangential contact stiffness, given, for example, in terms of the normal and tangential components of the contact force, the diameter d of the spheres, and their material properties (e.g., Thornton & Randall, 1988).

The increment $\dot{\mathbf{u}}^{(BA)}$ in contact displacement may be written in terms of the increments $\dot{\mathbf{c}}^{(B)}$ and $\dot{\mathbf{c}}^{(A)}$ in the translations of the centers of the two spheres and the increments $\dot{\omega}^{(B)}$ and $\dot{\omega}^{(A)}$ in their rotations about their centers by

$$\dot{u}_i^{(BA)} = \dot{c}_i^{(B)} - \dot{c}_i^{(A)} - \frac{1}{2}\varepsilon_{ijk}\left(\dot{\omega}_j^{(B)} + \dot{\omega}_j^{(A)}\right) d_k^{(BA)}.$$

Alternatively, the relative displacement of the two contacting points may be written in terms of the increments in the averages of quantities and their fluctuations as

$$\dot{u}_i^{(BA)} = \left(\dot{E}_{ij} + \dot{W}_{ij}\right) d_j + \dot{\Delta}_i^{(BA)} - \varepsilon_{ijk}\dot{\Omega}_j d_k^{(BA)} - \frac{1}{2}\varepsilon_{ijk}\dot{S}_j^{(BA)} d_k^{(BA)},$$

where $\mathbf{E}$ and $\mathbf{W}$ are, respectively, the average strain and the average rotation based on the positions of the particle centers, $\mathbf{\Omega}$ is the average rotation of the particles about their centers, $\mathbf{\Delta}^{(BA)}$ is the fluctuation in the difference between the translations of the two centers, and $\mathbf{S}^{(BA)}$ is the sum of the fluctuations in the rotations about their centers.

It is necessary to distinguish between the average rotation based upon the inhomogeneous displacements of the particle centers and the average spin about the centers because if the initial state is anisotropic or as anisotropies develop in the state of the material, these need not be equal (Koenders, 1990; Jenkins, 1991). Their difference is then determined by the requirement that the stress be symmetric.

Given $\dot{\mathbf{F}}$, the incremental stress $\dot{\mathbf{T}}$ may be written as the average over all N particles in a region of homogeneous distortion that is identified with the continuum point as

$$\dot{T}_{ij} = \langle \frac{1}{V^{(A)}} \sum_{n=1}^{N^{(A)}} \dot{F}_i^{(nA)} d_j^{(nA)} \rangle \equiv \frac{1}{N}\sum_{A=1}^{N} \frac{1}{V^{(A)}} \sum_{n=1}^{N^{(A)}} \dot{F}_i^{(nA)} d_j^{(nA)} \qquad (1)$$

where $N^{(A)}$ is the number of particles in contact with particle A and $V^{(A)}$ is the volume occupied by particle A and its nearest neighbors.

Given the increments $\dot{\mathbf{E}}$ and $\dot{\mathbf{W}}$ in average strain and rotation, the calculation of the increment in stress requires the determination of the fluctuations $\dot{\mathbf{\Delta}}$ and $\dot{\mathbf{S}}$ for all pairs of particles in the region. As emphasized by Koenders (1987, 1993, 1994), these should be obtained as parts of the solutions of the equations of balance of force and moment for each of the N

particles. Then $\dot{\mathbf{\Omega}}$ follows from the condition that the increment in stress be symmetric.

However some important qualitative features of the behavior of the aggregate and, sometimes, some quantitative predictions can be obtained by taking a far simpler approach and ignoring the fluctuations entirely. Such a procedure seems to have a physical justification only when the ratio of the shear stress to the confining pressure is small.

2.1. MEAN FIELD THEORY

In this case, the increment in stress is given by

$$\dot{T}_{ij} = C_{ijkl}\dot{L}_{kl},$$

where

$$C_{ijkl} \equiv \langle \frac{1}{V(A)} \sum_{n=1}^{N(A)} K_{ik}^{(nA)} d_j^{(nA)} d_l^{(nA)} \rangle$$

and

$$\dot{L}_{kl} \equiv \dot{E}_{kl} + \dot{W}_{kl} - \varepsilon_{kml}\dot{\Omega}_m.$$

An analytical expression for C may be obtained by employing the continuous analog of equation 1. This is phrased in terms of the number of particles per unit volume n and a contact distribution function $f(\hat{\mathbf{d}})$ defined so that $f(\hat{\mathbf{d}})d\hat{\mathbf{d}}$ is the number of contacts in the element of solid angle $d\hat{\mathbf{d}}$ centered at $\hat{\mathbf{d}}$. For an isotropic distribution of contacts, $f = k/4\pi$, where k is the coordination number. Then

$$C_{ijkl} \equiv n \int f(\hat{\mathbf{d}}) K_{ik} \hat{d}_j d_l d\hat{\mathbf{d}}.$$

Using such a formula, Digby (1981) and Walton (1987) consider isotropic aggregates and calculate the effective shear and bulk moduli. Their expression for the shear modulus $\bar{G}$ is, for example,

$$\bar{G} = \frac{1}{15}nkd^3 \frac{G}{(1-\nu)} \frac{(5-4\nu)}{(2-\nu)} \left[\frac{9}{8}\frac{(1-\nu)}{G}\frac{p}{nkd^3}\right]^{1/3},$$

where p is the confining pressure and G and ν are, respectively, the shear modulus and Poissons ratio of the material of the spheres.

Using $k = 5.36$ as determined in their numerical simulations, Jenkins, Ishibashi, and Cundall (1989) find the predicted shear modulus to be three times that measured in their experiments and numerical simulations. On the other hand, Johnson and Norris (1996), adopt a value of $k = 9$ that may be appropriate to their much higher confining pressures, and find the predicted values of wave speeds based on these to be in reasonable agreement with those measured in experiments by Domenico (1977). These differing results highlight the importance of the coordination number in the mechanics of these materials and may provide an indication of its possible range of variation when friction is present.

The corresponding incremental relation between the pressure and the volumetric strain can be integrated to obtain the elastic volume strain e corresponding to the confining pressure (e.g. Jenkins, 1988):

$$e = \left[27\sqrt{3} \frac{1}{nkd^3} (1 - \nu) \frac{p}{G} \right]^{2/3}.$$

This particular value of the volume strain can be used to scale the shear strains to provide an appropriate measure of their strength. The shear stress should be scaled by the pressure in a similar way. It is only by employing such scaled variables that the results of physical experiments carried out on different materials at different confining pressures can be compared. In typical experiments relevant to the behavior of soils, both the scaled shear stress and the scaled shear strain range from zero to around one.

Jenkins and Strack (1993) take the mean field model somewhat further. Using Hertz elasticity for the normal component of the contact force and a crude model of linear elasticity followed by frictional sliding for the tangential component, they focus on triaxial compression. They predict the behavior of contacts as a function of their orientation with respect to the axis of compression and the compressive strain. When, after an initial isotropic compression, deviatoric straining commences, all contacts first deform elastically. As the compression proceeds, contacts oriented at a definite angle first begin to slide; the sliding then spreads to other contacts oriented further away from the axis. After sliding has reached the contacts perpendicular to the axis, continued deformation results in the reduction of their normal force to zero. Such contacts are considered to be deleted. Further compression results in an increasing region of deleted contacts near the perpendicular. The sum of such contact behavior is reflected in the relation between shear stress and shear strain: there is an initial region of stiff elastic response, a region of diminishing stiffness as contacts begin to slide, followed by a region of dramatically reduced stiffness as contacts are deleted.

This simple picture of the mean field behavior of contacts goes far in describing the observed features of the stress strain curve for loading. However, over a range of scaled shear stress between zero and one, the shear stresses predicted by such a theory are about three times larger than those measured in numerical simulations and experiments. The stresses are over-predicted because the relative displacement of the contacts are not so con-strained; the particles are free to displace and rotate in accord with stress and moment equilibrium.

Numerical simulations indicate that the rotational degrees of freedom are more important that the translational (Cundall, 1988; Kuhn, 1991; Bagi, 1992). For example, if the rotations are constrained to be zero in a simulation of triaxial compression, the measured relation between the shear stress and the shear strain is close to that predicted using the mean field assumption. A similar increase in stiffness is not observed when the translations are constrained to follow the mean fields and the rotations are left free.

2.2. PAIR FLUCTUATIONS

Motivated by the over-prediction of the shear stress by the mean-field the-ory and by the apparent need to incorporate additional internal degrees of freedom, we next analyze the simplest possible situation in which two con-tacting particles, A and B, have sufficient translational and rotational free-dom to satisfy force and moment equilibrium. In order that the equilibrium equations for the two particles determine these translations and rotations, we assume that the other particles in contact with the pair translate and rotate with the average deformation.

We denote the increment in the translation of center of the nth neighbor of particle A by $\dot{\mathbf{c}}^{(n)}$ and the increment in its rotation about its center by $\dot{\omega}^{(n)}$. Then

$$\dot{u}_i^{(nA)} = \dot{c}_i^{(n)} - \dot{c}_i^{(A)} - \frac{1}{2}\varepsilon_{ijk}(\dot{\omega}_j^{(n)} + \dot{\omega}_j^{(A)})d_k^{(nA)}.$$

We assume that for $n \neq B$,

$$\dot{c}_i^{(n)} - \dot{c}_i^{(A)} = \left(\dot{E}_{ij} + \dot{W}_{ij}\right)d_j^{(nA)} \quad \text{and} \quad \dot{\omega}_k^{(n)} = \dot{\Omega}_k.$$

The equations of force equilibrium for particle A are

$$0 = K_{ij}^{(BA)}\left(\dot{c}_j^{(B)} - \dot{c}_j^{(A)}\right) + \sum_{n \neq B}^{N^{(A)}} K_{ij}^{(nA)}\left(\dot{E}_{jk} + \dot{W}_{jk}\right)d_k^{(nA)}$$

$$-\frac{1}{2}\varepsilon_{jlk}K_{ij}^{(BA)}(\dot{\omega}_l^{(B)} + \dot{\omega}_l^{(A)})d_k^{(BA)} - \varepsilon_{jlk}\sum_{n \neq B}^{N^{(A)}} K_{ij}^{(nA)}\dot{\Omega}_l d_k^{(nA)};$$

while those of moment equilibrium are

$$
\begin{aligned}
0 = & \; \varepsilon_{pqr} d_q^{(BA)} K_{ri}^{(BA)} \left(\dot{c}_i^{(B)} - \dot{c}_i^{(A)} \right) \\
& + \varepsilon_{pqr} \sum_{n \neq B}^{N^{(A)}} d_q^{(nA)} K_{ri}^{(nA)} \left(\dot{E}_{ij} + \dot{W}_{ij} \right) d_j^{(nA)} \\
& - \frac{1}{2} \varepsilon_{pqr} \varepsilon_{ijk} d_q^{(BA)} K_{ri}^{(BA)} (\dot{\omega}_j^{(B)} + \dot{\omega}_j^{(A)}) d_k^{(BA)} \\
& - \varepsilon_{pqr} \varepsilon_{ijk} \sum_{n \neq B}^{N^{(A)}} d_q^{(nA)} K_{ri}^{(nA)} \dot{\Omega}_j d_k^{(nA)}.
\end{aligned}
$$

In order to solve these and the corresponding equilibrium equations for particle B, we neglect the variation from particle to particle of the number and relative location of neighbors and the differences in the incremental contact forces associated with them and assume that the neighborhood of each of the particles A and B is the average neighborhood. In this event, because in an average over the aggregate, each contact between a pair of contacts is counted twice, once with an orientation $\mathbf{d}^{(BA)}$ and once with and an orientation $\mathbf{d}^{(AB)} = -\mathbf{d}^{(BA)}$,

$$
\langle \sum_{n=1}^{N^{(A)}} K_{ij}^{(nA)} d_l^{(nA)} \rangle = 0; \tag{2}
$$

and the equations for $\dot{\mathbf{\Delta}}^{(BA)}$ and $\dot{\mathbf{S}}^{(BA)}$ may be uncoupled from those involving the increments in the sum of the fluctuations in translations and the difference in the fluctuations in rotations. They may be written in terms of the averages

$$
A_{ij} \equiv \langle \sum_{n=1}^{N^{(A)}} K_{ij}^{(nA)} \rangle,
$$

$$
C_{pij} \equiv \varepsilon_{pqr} \langle \sum_{n=1}^{N^{(A)}} K_{ri}^{(nA)} d_q^{(nA)} d_j^{(nA)} \rangle,
$$

and

$$
B_{pk} \equiv \varepsilon_{ijk} C_{pij}. \tag{3}
$$

In the context of this simple model, these average fields characterize the state of a pair of contacting particles with their line of centers parallel to $\hat{\mathbf{d}}^{(BA)}$.

The equilibrium equations for this pair that are relevant to the calculation of the contact force are the difference of their force equilibrium:

$$K_{ij}^{(BA)}\left(\dot{\Delta}_j^{(BA)} + \frac{1}{2}\varepsilon_{jlk}d_l^{(BA)}\dot{S}_k^{(BA)}\right) + A_{ij}\dot{\Delta}_j^{(BA)} = 0,$$

and the sum of their moment equilibrium:

$$\varepsilon_{pqr}d_q^{(BA)}K_{rj}^{(BA)}\left(\dot{\Delta}_j^{(BA)} + \frac{1}{2}\varepsilon_{jlk}d_l^{(BA)}\dot{S}_k^{(BA)}\right) + \frac{1}{2}B_{pk}\dot{S}_k^{(BA)} = -2C_{pqs}\dot{L}_{qs}.$$

On the way to obtaining an approximate solution to these equations, we first write them in the form

$$\left\{\begin{bmatrix} A_{ij} & 0 \\ 0 & B_{pk} \end{bmatrix} + \begin{bmatrix} K_{ij}^{(BA)} & \frac{1}{2}K_{ij}^{(BA)}\varepsilon_{jlk}d_l^{(BA)} \\ 2\varepsilon_{pqr}d_q^{(BA)}K_{rj}^{(BA)} & \varepsilon_{pqr}d_q^{(BA)}K_{rj}^{(BA)}\varepsilon_{jlk}d_l^{(BA)} \end{bmatrix}\right\}\begin{pmatrix} \dot{\Delta}_j \\ \dot{S}_k \end{pmatrix}$$

$$= \begin{pmatrix} 0 \\ -4C_{pqs}\dot{L}_{qs} \end{pmatrix}.$$

We note that the terms in the first matrix each involve the average number of contacts k per particle while those in the second matrix each involve a single contact. We exploit this and obtain an approximate inverse by ignoring terms of order $1/k^2$. The approximate solutions for $\dot{\mathbf{\Delta}}^{(BA)}$ and $\dot{\mathbf{S}}^{(BA)}$ are, then, given by

$$\dot{\Delta}_i^{(BA)} = 2A_{ij}^{-1}K_{jl}^{(BA)}\varepsilon_{lkm}d_k^{(BA)}B_{mp}^{-1}C_{pqs}\dot{L}_{qs} \tag{4}$$

and

$$\dot{S}_i^{(BA)} = -4B_{ij}^{-1}\left(\delta_{jp} - \varepsilon_{jkl}d_k^{(BA)}K_{lm}^{(BA)}\varepsilon_{mno}d_n^{(BA)}B_{op}^{-1}\right)C_{pqs}\dot{L}_{qs}. \tag{5}$$

As fluctuations, $\dot{\mathbf{\Delta}}^{(BA)}$ and $\dot{\mathbf{S}}^{(BA)}$ must each average to zero. The average of the right-hand side of equation 4 over all orientations is zero by virtue of equation 2; the average of the right-hand side of equation 5 over all orientations is zero because of the definition 3 of $\mathbf{B}$.

Using these solutions in the expression for the contact force and employing the result in the definition 1 of the stress yields an incremental stress given by

$$\dot{T}_{ij} = \left[\mathcal{C}_{ijmn} - 2\left(\varepsilon_{plu}\mathcal{C}_{ilpj} - A_{sp}^{-1}\mathcal{D}_{pisju} - B_{ks}^{-1}\mathcal{E}_{kisju}\right)B_{ut}^{-1}C_{tmn}\right]\dot{L}_{mn}, \tag{6}$$

where, as before,

$$C_{ilpj} \equiv \langle \frac{1}{V(A)} \sum_{n=1}^{N(A)} K_{ip}^{(nA)} d_l^{(nA)} d_j^{(nA)} \rangle;$$

while

$$\mathcal{D}_{pisju} \equiv \varepsilon_{klu} \langle \frac{1}{V(A)} \sum_{n=1}^{N(A)} K_{is}^{(nA)} K_{pk}^{(nA)} d_l^{(nA)} d_j^{(nA)} \rangle$$

and

$$\mathcal{E}_{kisju} \equiv \varepsilon_{plk} \varepsilon_{sqw} \varepsilon_{ovu} \langle \frac{1}{V(A)} \sum_{n=1}^{N(A)} K_{ip}^{(nA)} K_{wo}^{(nA)} d_l^{(nA)} d_q^{(nA)} d_v^{(nA)} d_j^{(nA)} \rangle.$$

Given the statistical properties of an initial state, and supposing that the geometry of the contacts does not change, the evolution of the stiffness can be determined by integrating the incremental expression for the contact force using the solution for the fluctuations. In the context of the present calculation, an appropriate characterization of the initial state might include the confining pressure, the number of particles per unit volume, the average number of contact per particle, and, perhaps, the distributions of the strengths of the normal and tangential components of the contact force (e.g., Cundall, Jenkins & Ishibashi, 1989). With the stiffness known for a contact of a given orientation, the continuous version of the average may be employed to calculate the relationship between the increment in stress and the increment in distortion. When the geometry of the contacts does not change, it may be appropriate to assume that

$$\langle \frac{1}{V(A)} \sum_{n=1}^{N(A)} K_{ip}^{(nA)} d_l^{(nA)} d_j^{(nA)} \rangle = \langle \frac{1}{V(A)} \rangle \langle \sum_{n=1}^{N(A)} K_{ip}^{(nA)} \rangle \langle \sum_{n=1}^{N(A)} d_l^{(nA)} d_j^{(nA)} \rangle.$$

This reduces the number of state variables and introduces the geometric fabric in a natural way. Here, we don't carry out these calculations, but are content with making a few observations about the form of the incremental relationship 6.

As can easily be verified from their definitions, **B** and **C** depend only upon K_T, while **A** depends upon both K_N and K_T. The terms contributed by the fluctuations involve the distortion only through

$$B_{kp}^{-1} C_{pij} \dot{L}_{ij} = B_{kp}^{-1} \varepsilon_{pqr} \langle \sum_{n=1}^{N(A)} K_{ri}^{(nA)} d_q^{(nA)} d_j^{(nA)} \rangle \dot{L}_{ij}.$$

For an axisymmetric compression with an axis of symmetry parallel to the unit vector $\hat{\mathbf{k}}$, $\mathbf{B}^{-1}\mathbf{C}$ has the structure

$$B_{kp}^{-1}C_{pij} = a_1\varepsilon_{kij} + a_2\hat{k}_k\hat{k}_p\varepsilon_{pij} + a_3\varepsilon_{kip}\hat{k}_p\hat{k}_j + a_4\varepsilon_{kjp}\hat{k}_p\hat{k}_i$$
$$+a_5\hat{k}_k\hat{k}_p\varepsilon_{piq}\hat{k}_q\hat{k}_j + a_6\hat{k}_k\hat{k}_p\varepsilon_{pjq}\hat{k}_q\hat{k}_i,$$

while

$$\dot{L}_{ij} = \dot{l}_1\delta_{ij} + \dot{l}_2\hat{k}_i\hat{k}_j.$$

Consequently,

$$B_{kp}^{-1}C_{pij}\dot{L}_{ij} = 0.$$

So, although the fluctuations that we have considered may make important contributions to the stiffness in distortions that involve significant rotations, in triaxial compression they contribute nothing. In order to obtain the appropriate response in triaxial compression, fluctuations driven by the variability of the pair's interactions with their neighbors must be incorporated rather than those driven by the anisotropy of their average neighborhoods. A framework for this has been established by Koenders (1993,1994).

3. Conclusion

We have outlined a continuum mechanical description for granular materials that involves incremental constitutive relations for the stress and internal state of the material that are based upon the mechanics of grain contact and a statistical characterization of the grain interactions.

The simplest such theory, based upon the average strain and rotation captures some important features of the behavior of the aggregate. Formulas for effective moduli based upon the mean field kinematics highlight the importance of the coordination number, the average number of contacts per grain, as an independent internal variable. Calculation of the incremental stress also leads to an expression for the elastic volume strain associated with the confining pressure that provides a natural scale for shear strains. In the context of this theory, the succession of contact mechanisms can be identified that determine the shape of the stress-strain curve.

The hope is that by incorporating the translational and rotational degrees of freedom of only pairs of particles at various orientations, the predicted magnitude of the shear stress may be reduced to values more in accord with experiments and numerical simulations. Here, we considered the fluctuations of a pair of particles, each in an average anisotropic neighborhood. In this case, a near exact analysis is possible; however, the fluctuations do not contribute to the stiffness in axisymmetric deformations.

4. Acknowledgment

The author benefited from useful discussions with M. A. Koenders during the course of this work. This research was supported under a contract from Sandia National Laboratories, Albuquerque, New Mexico.

References

Bagi, K. (1992) Personal communication.

Cundall, P. A. (1988) Personal communication.

Cundall, P. A., Jenkins, J. T., and Ishibashi, I. (1989) Micromechanical modeling of granular materials with the assistance of experiments and numerical simulations, in J. Biarez and R. Gourvès (eds.), *Powders and Grains*, Balkema, Rotterdam, pp. 319–322.

Digby, P. J. (1981) The effective elastic moduli of porous granular rocks, *Journal of Applied Mechanics* **48**, 803–808.

Domenico, S. N. (1977) Elastic properties of unconsolidated sand reservoirs, *Geophysics* **42**, 1339–1368.

Jenkins, J. T. (1988) Volume change in small strain axisymmetric deformations of a granular material, in M. Satake and J. T. Jenkins (eds.), *Micromechanics of Granular Materials*, Elsevier Science Publishers, Amsterdam, pp. 245–252.

Jenkins, J. T. (1991) Anisotropic elasticity for random arrays of identical spheres, in J. Wu, T. C. T. Ting, and D. M. Barnett (eds.) *Modern Theory of Anisotropic Elasticity and Its Applications*, Society for Industrial and Applied Mathematics, Philadelphia, pp. 368–377.

Jenkins, J. T., Cundall, P. A. and Ishibashi, I. (1989) Micromechanical modeling of granular materials with the assistance of experiments and numerical simulations, in J. Biarez and R. Gourvès (eds.), *Powders and Grains*, Balkema, Rotterdam, pp. 257–264.

Jenkins, J. T. and Strack, O. D. L. (1993) Mean-field inelastic beavior of random arrays of identical spheres, *Mechanics of Materials* **16**, 25–33.

Koenders, M. A. (1987) The incremental stiffness of an assembly of particles, *Acta Mechanica* **70**, 31–49

Koenders, M. A. (1990) Localized deformation using higher order stress/strain theory, *Journal of Energy Resources Technology* **112**, 51–53.

Koenders, M. A. (1993) Analytical estimates for constitutive relations of assemblies of particles in elasto-frictional contact, in C. Thornton (ed.), *Powders and Grains 93*, Balkema, Rotterdam, pp. 111–116.

Koenders, M. A. (1994) Least squares method for the mechanics of nonhomogeneous granular assemblies, *Acta Mechanica* **106**, 23–40.

Kuhn, M. W. (1991) Factors affecting the incremental stiffness of particles assemblies, in H. Adeli and R. L. Sierakowski (eds.) *Mechanics Computing in the 1990's and Beyond*, American Society of Civil Engineers, New York, pp. 1229-1233.

Norris, A. N. and Johnson, D. L. (1996) Nonlinear elasticity of granular media. *Journal of Applied Mechanics* (In press).

Oda, M. (1976) *Fabrics and their effects on the deformation behaviours of sand*, Department of Foundation Engineering, Saitama University, 59 pp.

Tatsuoka, F. and Ishihara, K. (1974) Yielding of sand in triaxial compression, *Soils and Foundations* **14**, 63–76.

Thornton, C. and Randall, C. W. (1988) Application of theoretical contact mechanics to solid particle system simulation, in M. Satake and J. T. Jenkins (eds.), *Micromechanics of Granular Materials*, Elsevier Science Publishers, Amsterdam, pp 133–142.

Voyiadjis, G. Z., Thiagarajan, G., and Petrakis, E. (1995) Constitutive modeling for granular media using an anisotropic distortional yield model, *Acta Mechanica* **110**, 151–171.

Walton, K. (1987) The effective elastic moduli of a random packing of spheres, *Journal of the Mechanics and Physics of Solids* **35**, 213–226.

FRACTAL FRAGMENTATION AND FRICTIONAL STABILITY IN GRANULAR MATERIALS

CHARLES G. SAMMIS

Department of Earth Sciences
University of Southern California
Los Angeles, CA 90089-0740

Abstract

The wall rocks of most natural faults are separated by a layer of finely crushed rock called "fault gouge". Particle-size studies of such unaltered gouge have found a fractal (power-law) distribution with a fractal dimension of 1.6 ± 0.1 in 2D planar section (or 2.6 ± 0.1 for the isotropic distribution in 3D). A simple fragmentation mechanism which leads to a fractal distribution with this dimension can be formulated. It assumes that a particle's fracture probability is based solely on the relative size of its nearest neighbors - a particle loaded at opposing poles by neighbors of the same size develops the largest internal tension, and is most likely to fragment. The ultimate result is a particle distribution in which no particle has a same-sized neighbor at any scale. Such a distribution is fractal; a perfect, geometrical, discrete fractal having this property is the Sierpinski gasket which has a dimension of 1.58. We have coined the term "constrained comminution" for the neighbor-dominated process which leads to a fractal distribution. Constrained comminution has been simulated in a double-direct shear friction apparatus and has been shown to offer a physical explanation for the observed evolution of the frictional behavior of a granular layer from stable velocity strengthening behavior to potentially unstable velocity weakening as shear deformation proceeds. The initial velocity strengthening occurs when strain is mostly accommodated by the crushing of grains which has no velocity dependent weakening mechanism. As the fractal distribution emerges, an increasing proportion of the strain is accommodated by slip between the grains. Such slip is enhanced by the geometrical fact that a fractal distribution minimizes the dilatancy associated with slip. Slip between grains produces the observed velocity weakening and consequent stick-slip instability.

N. A. Fleck and A. C. F. Cocks (eds.), IUTAM Symposium on Mechanics of Granular and Porous Materials, 23–34.
© *1997 Kluwer Academic Publishers.*

1. Introduction

It has been observed that many processes which fragment large rock-masses *in situ* tend to produced a power-law distribution of particles of the form $P_v(L) = AL^{-m}$ where $P_v(L)$ is the cumulative fraction of the initial volume corresponding to particles with linear dimensions up to L (Turcotte, 1986). Sammis et al. (1987) proposed that the power law distribution results from a process which they termed "constrained comminution". In constrained comminution, the fragments retain their relative positions throughout the fragmentation process. Under these conditions, a particle's fragmentation probability is mainly determined by the relative size of its nearest neighboring particles, and only secondarily by its inherent strength or the distribution of its internal flaws. The result is a fractal distribution of fragments which, therefore, has a power law distribution of sizes.

The particle size distribution and spatial structure in fault zone materials is of interest because it influences the frictional stability of the fault. Dieterich (1981) studied the frictional behavior of a layer of granular material deformed in simple shear between sliding blocks of solid rock. He found that, phenomenalogically, the material could be described by the same rate- and state-dependent friction laws which described the frictional behavior of two barren rock surfaces in contract (Dieterich, 1979a,b, 1981; Ruina, 1983; Okubo and Dieterich, 1984). However, the physical interpretation of the model parameters in terms of the Bowden-Tabor (1950, 1964) asperity model which proved so satisfying for two surfaces in contact, is no longer appropriate. What, for example, are the asperities when the two sliding surfaces are separated by a layer of granular rock? Biegel et al. (1989) repeated Dieterich's experiments for a range of particle sizes and layer thicknesses. They documented the evolution of an initially narrow distribution of fragment sizes into a broad fractal distribution with progressive strain exactly as hypothesized by Sammis et al. (1987). Associated with this evolution of a fractal distribution, they also observed the evolution of frictional behavior from "velocity strengthening" at small strain to "velocity weakening" at large strain. As will be discussed below, velocity strengthening behavior always produces stable-sliding whereas velocity weakening can (depending on the stiffness of the loading apparatus) lead to a stick-slip instability.

In this paper, we review the constrained comminution model for fractal fragmentation and its implications for the frictional behavior of granular layers in general and the nucleation of stick-slip instabilities which lead to earthquakes in specific.

2. Constrained Comminution

The wall rocks of natural faults are usually separated by a layer of finely crushed rock called fault gouge. In most cases, ground water alters the mineralogy of gouge to clay, but in some instances the gouge retains the mineralogy of the wall rock from which it was derived (Anderson et al., 1983). Sammis et al. (1987) measured the fragment size distribution in unaltered granitic gouge from the Lopez canyon fault in southern California. They prepared a series of photomosaics of epoxy impregnated polished sections at magnifications ranging (in steps 2x) from 12x to 1200x. An example of one of these sections is shown in Fig.1. They demonstrated that, over this range, the distribution of particles in each mosaic were statistically indistinguishable from the others. Since the particle distribution looked the same at each scale, it was statistically self-similar. The size distribution measured on each photomosaic was a power law of the form $N(L) = AL^{-D}$ where D is the fractal capacity dimension (see, e.g., Feder, 1988; Mandelbrot, 1989). Taken together, the entire suite of 12 mosaics yielded $D=1.6\pm0.1$ for fragment dimensions in the range $10\mu m < L < 1mm$. Self similarity was observed to break down at $10\mu m$ where individual grains were observed to be breaking into long splinters. The 1mm upper limit was imposed by the size of the polished sections.

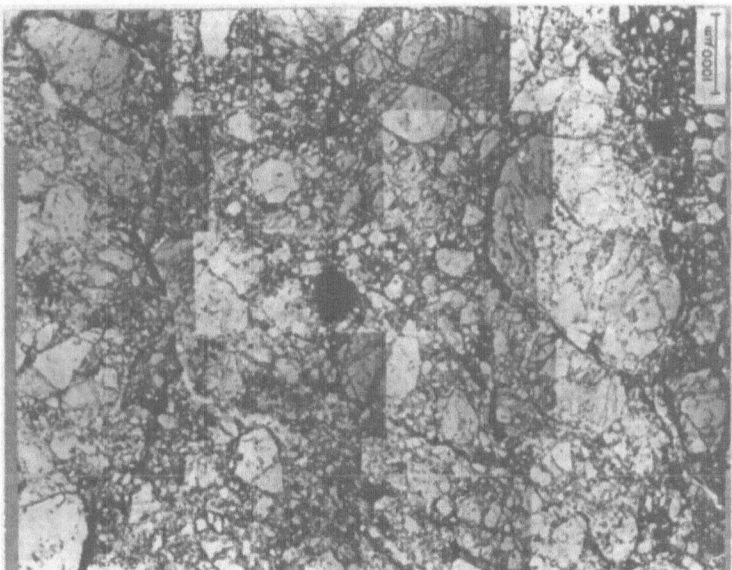

Figure 1. Optical photomosaic of a section of fault gouge from the Lopez Canyon fault zone in southern California (Sammis et al., 1987).

Sammis and Biegel (1989) extended these measurements using oversized polished sections and direct field measurements on the gouge exposure. They found that self-similarity extended only to fragments having a dimension less than 1 cm. Above 1 cm there were fewer fragments than required by an extrapolation from the smaller ones using the power law.

An and Sammis (1994) measured the particle distributions of a suite of natural fault gouges by direct counting. They used sieves for particles in the range $62.5\mu m<L<16mm$ and a Coulter-Counter for those in the range $1\mu m<L<62.5\mu m$. The fractal dimension of the Lopez Canyon gouge was found to be 2.7 ± 0.2 in agreement with $D=1.6\pm0.1$ measured in 2D section by Sammis et al. (1987). They also measured gouges from the San Andreas and San Gabriel fault zones in southern California for which they found a correlation between the peak particle size (by weight) and the fractal dimension. They observed that finer gouges tended to have a higher fractal dimension which they interpreted as being the consequence of the existence of a "grinding limit" at about $1\mu m$ (see, e.g., Prasher, 1987). They used a computer simulation to show that their observed distributions could be explained by a progressive fractal fragmentation in which finer particles piled up at the grinding limit.

Constrained comminution of the type produced in natural fault zones has been simulated in the laboratory using a double-direct shear apparatus described in Dieterich (1978). This apparatus measures the coefficient of friction between the surfaces of a central sliding slab and the inner surfaces of two symmetrical, fixed outer slabs which apply a controlled normal stress. Biegel et al. (1989) introduced a 3mm thick layer of $750\mu m$ rock fragments between the sliding surfaces and observed the evolution of the particle size distribution with progressive shear deformation. After a strain of about 3 the distribution had evolved into one which looked very similar to the cross-sections from natural fault zones (Fig. 1). Measurements of the fragment size distribution yielded D=1.6 in 2-D section, consistent with measurements in natural gouges. Marone and Scholz (1989) also produced fractal gouge in the laboratory using a triaxial saw-cut testing configuration.

Why does shear deformation produce a fractal particle size distribution, what is the significance of the observed fractal dimension $D=2.6$ in 3D ($D=1.6$ in 2D section), and what controls the upper and lower fractal limits of this distribution? A rational first guess might be that the fragmentation process itself is scale-independent. However, it is well known that a fragmentation process in which all particles are equally likely to fracture, and for which each fracture event produces the same distribution of fragment sizes (scaled to the size of the original particle) leads to a log-normal distribution (see, e.g., Epstein, 1947). In the constrained comminution model, all particles are not equally likely to fracture at any given time. Rather, a particle's fracture probability is completely determined by the relative size of its nearest neighbors. The key to understanding constrained comminution is the observation that the stress field in granular layers loaded by boundary shear displacements is extremely heterogeneous. Strain is accommodated by the continuous failure and reformation of grain bridges which span the layer (Mandl et al., 1977). In a grain bridge, the particles load each other in compression. A particle loaded in compression at opposite poles develops an internal tensile stress which is proportional to the applied point force and inversely proportional to the area of the particle. When this tensile stress reaches the strength of the particle it fragments and its grain bridge fails. Now, if a particle carries its share of the applied stress, then the force at its poles should be proportional to its area, so the tensile stress developed in a particle should be independent of its size. If all else is equal, the largest particles should be most likely to fail since they are intrinsically the weakest (they contain the largest flaws). However, there is another consideration which turns out to be more important than the intrinsic strength of a particle; the relative size of its nearest neighbors which transmit the load. A particle is most fragile when it is loaded by nearest neighbors which are the same size. Only then is a single point force applied to opposite poles as illustrated in Fig. 2a. Smaller neighbors tend to distribute the load over its surface thus reducing the internal tension and cushioning the particle as in Fig. 2b. Larger neighbors tend not to load the particle at opposing poles, particularly in dense packing configurations as in Fig.2c.

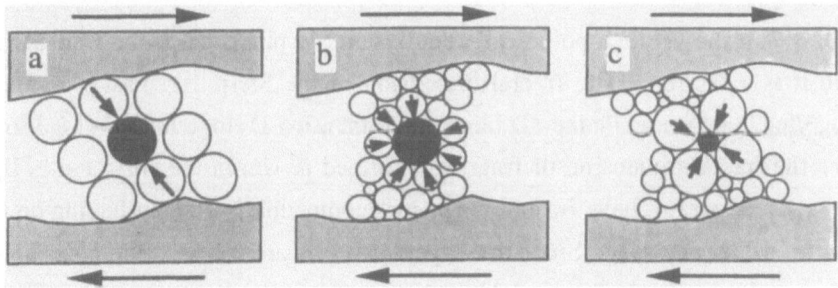

Figure 2. Loading geometry of a particle (a) when it is loaded by neighbors which are its same size, (b) when it is loaded by smaller neighbors, and (c) when it is loaded by larger neighbors (after King and Sammis, 1992).

The fragmentation process thus tends to eliminate particles which have same-sized neighbors. The end result is a distribution in which no two particles the same size are neighbors <u>at any scale</u>. Such a distribution is fractal. The well known Sierpinski gasket (Fig. 3) is a perfect geometrical fractal which has exactly this property: no two neighbors are the same size at any scale.

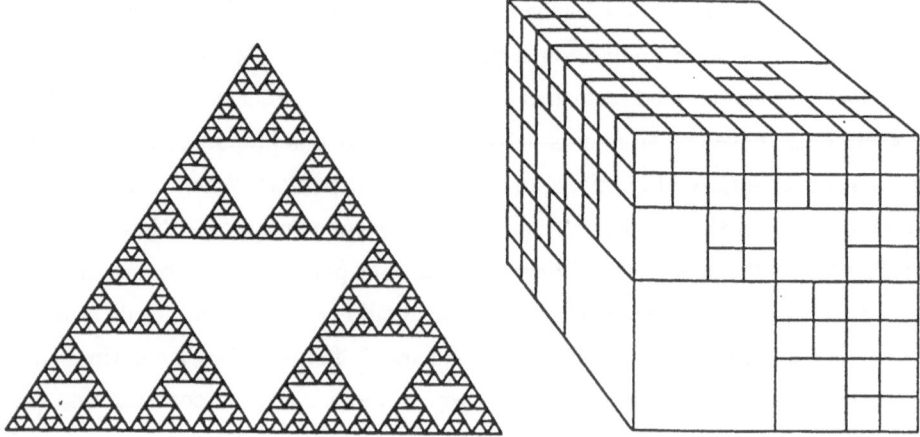

Figure 3. The left figure shows the 2D Sierpinski gasket which has fractal dimension of $D=1.58$. The right figure shows a 3D fractal arrangement of cubes for which has a fractal dimension of $D=2.58$. Each face of the cube is also a 2D fractal which is topologically equivalent to the Sierpinski carpet and also has $D=1.58$. Note that the two 2D fractals and the 3D fractal cube all have the property that no two same-sized pieces are nearest neighbors at any scale length longer than the lower fractal limit.

Figure 3 also shows the topologically equivalent Sierpinski carpet and the 3D fractal of which it is a section. The fractal dimension of the Sierpinski gasket and carpet are $D=\log3/\log2=1.58...$ while the 3D block has dimension $D=\log6/\log2=1+\log3/\log2=2.58..$ Hence the fractal dimension of fragments formed *in situ* in the crust and in the double direct-shear apparatus have fractal dimensions compatible with their having no neighbors the same size at any scale. Since the largest particles are the weakest, comminution first eliminates all large same sized neighbors before proceeding to the next smaller size and so on. The lower fractal limit is therefore expected to depend on total strain. Eventually the process reaches the grinding limit (An and Sammis, 1994) for particle sizes near 1μm. At this point two things appear to happen: 1) small particles pile up leading to an increase in the measured fractal dimension and 2) fragmentation is replaced by shear localization and slip as the dominant strain mechanism, as will be discussed below.

3. Frictional Stability in Granular Layers

Since most earthquakes nucleate on a preexisting fault-plane, they have commonly been modeled as a stick-slip frictional instability. A stick-slip instability results when the coefficient of friction on a surface decreases more rapidly with slip than does the applied load. Extensive laboratory studies of the stick-slip instability in rock on rock sliding experiments have shown that it can be produced by a velocity weakening phenomenon in which the coefficient of friction decreases with increased sliding velocity. Velocity weakening is usually studied in an experiment where the sliding velocity is controlled and stepped up and down while the shear stress is monitored. Since the normal stress is held constant, this is equivalent to monitoring the coefficient of friction.

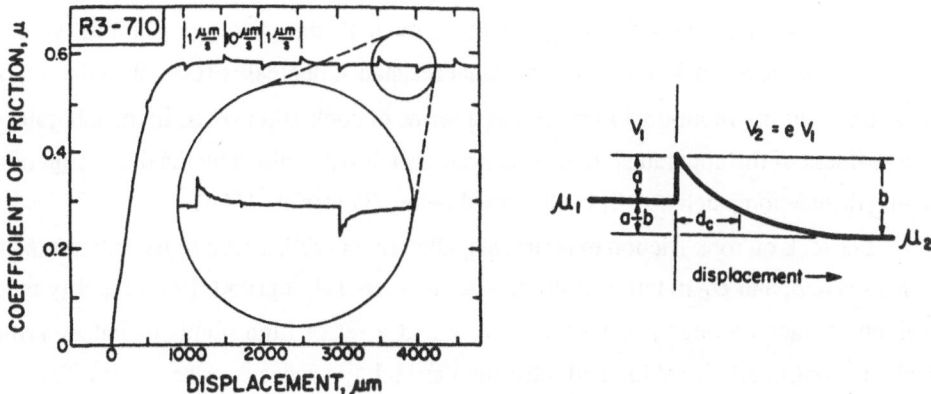

Figure 4. The left diagram shows data from a typical velocity stepping experiment. The coefficient of friction is plotted as a function of the total sliding displacement as the sliding velocity is cycled between 1μm/s and 10μm/s. The transient response to a velocity increase and decrease is enlarged in the inset. The right diagram shows the idealized response to a step increase in sliding velocity by a factor of e, and illustrates the definition of the parameters a, b, and D_c (after Biegle et al., 1989).

Figure 4 shows the coefficient of friction as a function of time in a velocity stepping experiment. When the sliding velocity is increased by a step factor e, the shear stress instantly increases by an amount which is parameterized as "a" and called the "direct effect". As sliding proceeds at the higher rate, the coefficient gradually decreases by an amount "b", which is called the "evolutionary effect" to a new steady-state value. Roughly speaking, the distance required to reach equilibrium is called the "characteristic displacement", "D_C".

If $a>b$, then the surface is said to "velocity strengthen" and sliding is always stable. If $b>a$ then the surface is said to "velocity weaken" and a stick slip instability is possible. Whether or not a stick-slip instability occurs depends on the stiffness k of the system which is driving the motion. Ruina (1983) showed that the condition for stick-slip can be written in terms of a critical stiffness, k_{crit}, defined as

$$k_{crit} = \frac{(b-a)\sigma_n}{D_c},\qquad(1)$$

where a, b, and D_C are the friction parameters and σ_n is the normal stress across the surface. If $k>k_{crit}$, sliding will be stable, even if the rheology exhibits velocity weakening. However, if $k<k_{crit}$, the frictional resistance decreases faster than the applied load and the sliding motion will proceed as a series of stick-slip events. In the laboratory k is the stiffness of the apparatus. In natural systems k is determined by the dimension of the slipping dislocation patch as will be discussed below (Dieterich, 1986).

For rock on rock friction experiments, Dieterich (1978, 1979a,b) has interpreted the parameters a, b, and D_C in terms of the classic Bowden-Tabor (1950, 1964) asperity model in which contact between surfaces only occurs at a set of high points called asperities. which are continually breaking and reforming as sliding proceeds. The direct affect, a, is ascribed to the effect of loading rate on strength. The compressive strength of rock increases by about 3% for every order-of-magnitude increase in loading rate (Paterson, 1978; Brace and Martin, 1968) which is about the right size to explain the observed a (Tullis and Weeks, 1987). The evolutionary effect, b, is ascribed to the decrease in average asperity strength caused by their shorter average lifetime at the higher sliding velocity. Dieterich (1978) demonstrated that the size of b is consistent with the observed logarithmic increase in the static coefficient of friction with time of contact. The characteristic displacement, D_C, is a measure of the sliding distance required to change the population of contacting asperities from those having a strength corresponding to the old sliding velocity to those corresponding to the new sliding velocity. The average age of the asperities thus defines the "state" of the surface.

The frictional properties of two rock surfaces separated by a layer of crushed rock have also been studied (Dieterich, 1981; Biegel et al., 1989; Marone and Scholz, 1989; Marone et al., 1990; Marone and Kilgore, 1993). Velocity stepping experiments yield a transient response which can be parameterized using a, b, and D_C as above, but there are significant differences which indicate that the micromechanics might be quite different from the asperity model. Marone and Kilgore (1993) show that an increase in sliding velocity produces a slight dilation of the layer while a decrease in velocity produces a compaction. These changes in gouge volume do not occur instantaneously, but develop over a sliding distance comparable to the characteristic displacement. Because both D_C and the magnitude of the dilatation and compaction decrease with progressive sliding displacement, they argue that this decrease in D_C is due to shear localization in which strain is concentrated in a progressively narrowing band or set of bands of width h. In this case, $D_c = \gamma_c h$ where the more fundamental parameter is the characteristic strain γ_C, and the observed D_C decreases with h.

Sammis and Steacy (1994) point out other significant differences brought out by the experiments of Biegel et al. (1989). First, although the transient stress changes produced by a step change in velocity were of the same form as those on barren surfaces shown in Fig. 3, the magnitudes of both a and b were almost twice as large. Second, both a and b changed with progressive displacement. A slight decrease in a and a large increase in b produces a transition from velocity strengthening ($a>b$) at small displacements to velocity weakening ($b>a$) at large displacements.

Sammis and Steacy (1994) showed how these observations can be interpreted in the context of the grain bridge model discussed above. The grain bridge model is more complex than the asperity model since grain bridges can fail in at least four ways: by the crushing of a grain, by slip at the contact between two grains, by fracture of the bounding wall rock, or by slip at the contact between a grain and the wall rock. The observed a and b values are largely determined by the way in which the bridges fail. Failure of the contact between two grains or of the contact between a grain and the wall rock is expected to show the same basic rate dependence as the failure of a contact between asperities in the classical friction model. However, the failure of a grain (or the wall rock) in compression has a different state dependence. While the strength still depends on the loading rate, the intrinsic strength of a grain does not increase with time under load. Unlike a contact, there is no evolutionary effect.

The dilatation and compaction of granular layers associated with step changes in velocity also produce small changes in the coefficient of friction (Marone et al., 1990). A simple energy argument equating the extra shear work $\Delta W_s = \Delta\sigma_s A dx$ to the dilatational work $W_\theta = \sigma_n A dh$ gives the change in the coefficient of friction $\Delta\mu = \dfrac{\Delta\sigma_s}{\sigma_n} = \dfrac{dh}{dx}$ Hence the perturbation in the coefficient of friction is proportional to the dilation rate. The dilation and compaction produce a transient response to a step change in sliding rate having exactly the same form as that produced by the strength and lifetime of asperities (Fig. 4). The a parameter for dilatation is given by the initial dilatation rate $a_\theta = \left[\dfrac{dh}{dx}\right]_0$. Note that the perturbation in μ goes to zero once the layer attains an equilibrium volume appropriate to the new sliding rate. Since dh/dx falls to zero over the characteristic distance D_c, this looks like the evolutionary effect, but note that the apparent evolutionary parameter b_θ associated with dilatation is equal to a_θ exactly.

Sammis and Steacy (1994) used these ideas to fit the data for granular layers collected by Biegel et al. (1989). By assuming a gradual transition from grain-bridges which fail by the crushing of a grain to ones which fail by slip between grains, they were able to quantitatively model the observed evolution of the individual parameters a and b which produced the observed evolution from velocity strengthening to velocity weakening

4. Fractal Fragmentation and the Frictional Stability of Faults

The transition from stable sliding to stick-slip in granular materials is thus directly related to the emergence of a fractal grain size distribution. As particles become increasingly isolated from same-size neighbors at all scales, particle fracture becomes increasingly difficult. At the same time, slip between particles becomes easier since the condition of different sized neighbors minimizes the dilatation required to slip. The result is an increase in (b-a) in the stability condition (1)

What are the implications for the stability of natural faults? The stability condition (1) requires that we know the stiffness k of the fault. Dieterich (1986) has estimated fault stiffness $k_{crack} = \dfrac{7\pi G}{24r}$ from the displacement and stress-drop at the center of a circular crack of radius r in a medium with shear modulus G. Combining this with (1) gives the minimum sized crack required to nucleate a stick-slip instability,

$$r_{min} = \frac{7\pi GD_c}{24\sigma_n(b-a)}.$$ (2)

Since $D_c = \gamma_c h$, faults which have had more total displacement and hence thicker gouge layers should have a larger D_C. According to (2) such faults should nucleate only larger events in the sense that a larger slipping patch is required to nucleate an instability. On the other hand, gouge layers should become less stable with increasing strain due to the increase of (b-a) associated with the emergence of a fractal structure.

References

An, L-J, and Sammis C.G. (1994) Particle size distribution of cataclastic fault materials from southern California, *Pure and Appl. Geophys*, **143**, 203-227.

Anderson, J.L., Osborne, R.H., and Palmer D. (1983) Cataclastic rocks of the San Gabriel fault zone--an expression of deformation at deeper crustal levels in the San Andreas fault zone, *Tectonophysics*, **98**, 209-251.

Biegel, R.L., Sammis C.G., and Dieterich J.H. (1989) The frictional properties of a simulated gouge having a fractal particle distribution, *J. Structural Geol.*, **11**, 827-846.

Bowden, F.P., and Tabor D. (1950) *The Friction and Lubrication of Solids. Part I*, Clarenden Press, Oxford.

Bowden, F.P., and Tabor D. (1964) *The Friction and Lubrication of Solids. Part II*, Clarenden Press, Oxford.

Brace, W.F., and Martin, R.J. (1968) A test of the law of effective stress for crystalline rocks of low porosity, *Int. J. Rock Mech. Min Sci.*, **5**, 415-426.

Dieterich, J. H. (1978) Time-dependent friction and the mechanics of stick-slip, *Pure and Appl. Geophys.*, **116**, 790-806.

Dieterich, J.H. (1979a) Modeling of rock friction: 1. Experimental results and constitutive equations, *J. Geophys. Res.*, **84**, 2161-2168.

Dieterich, J.H. (1979b) Modeling of rock friction: 2. Simulation of preseismic slip, *J. Geophys. Res.*, **84**, 2169-2175.

Dieterich, J. H. (1981) Constitutive properties of faults with simulated gouge, in *Mechanical Behavior of Crustal Rocks., Geophysical Monograph 24*. American Geophysical Union, 108-120.

Dieterich, J.H. (1986) A model for the nucleation of earthquake slip, in *Earthquake Source Mechanics. AGU Geophys. Mono. 37.* Washington, D.C., American Geophysical Union, 37-49.

Epstein, B. (1947) The mathematical description of certain breakage mechanisms leading to the logarithmic- normal distribution, *J. Franklin Inst.*, **244**, 471-477.

Feder, J. (1988), *Fractals*, Plenum, New York.

King, G.C.P. and Sammis, C.G. (1992) The mechanisms of finite brittle strain, *Pure and Appl. Geophys.*, **138**, 611-640.

Mandelbrot, B.B. (1989) Multifractal measures, especially for the geophysicist, *Pure and Appl. Geophys.*, **131**, 5-42.

Mandl, G., deJong, L.N.J., and Maltha, A. (1977) Shear zones in granular material: An experimental study of their structure and mechanical genesis, *Rock Mechanics*, **9**, 95-144.

Marone, C., and Scholz C.H. (1989) Particle-size distribution and microstructures within simulated fault gouge, *J. Struct. Geology*, **11**, 799-814.

Marone, C., Raleigh, C.B., and Scholz, C.H. (1990) Frictional behavior and constitutive modeling of simulated fault gouge, *J. Geophys. Res.*, **95**, 7007-7026.

Marone, C. and Kilgore, B. (1993) Scaling of the critical slip distance for seismic faulting with shear strain in fault zones, *Nature*, **362**, 618-621.

Okubo, P.G., and Dieterich, J.H. (1984) Effects of physical fault properties on frictional instabilities produced on simulated faults, *J. Geophys. Res.*, **89**, 5817-5827.

Paterson, M.S. (1978) *Experimental Rock Deformation: The Brittle Field*, Springer-Verlag, New York.

Prasher, C. (1987). *Crushing and Grinding Process Handbook*, John Wiley and Sons Ltd., New York.

Ruina, A.L. (1983) Slip instability and state variable friction laws, *J. Geophys. Res.*, **88**, 10359-10370.

Sammis, King, C.G., and Biegel, R. (1987) The kinematics of gouge deformation, *Pure Appl. Geophys.*, **125**, 777-812.

Sammis, C.G., and Biegel, R. (1989) Fractals, fault-gouge, and friction, *Pure Appl. Geophys.*, **131**, 255-271.

Sammis, C.G., and Steacy, S.J. (1994), The micromechanics of friction in a granular layer, *Pure and Appl. Geophys*, **142**, 777-794.

Tullis, T.E., and Weeks, J.D. (1986) Constitutive behavior and stability of frictional sliding of granite, *Pure Appl. Geophys.*, **124**, 383-414.

Turcotte, D. L. (1986). Fractals and Fragmentation. *J. Geophys. Res.*, **91**, 1921-1926.

CLASTIC MECHANICS

M. D. BOLTON and G. R. McDOWELL
Cambridge University Engineering Department
Trumpington Street, Cambridge, CB2 1PZ, U.K.

Abstract

A study has been made of the micro mechanical origins of hardening in aggregates which comprise elastic-brittle grains. We consider the compression of an aggregate of uniform grains and explain hardening at very small strains in terms of elastic contacts between particles. In the discipline of soil mechanics the terms "yielding" and "plastic hardening" are used to describe the post-elastic behaviour of granular media. Here we propose mechanisms of "clastic yielding" which is the onset of fracture of the weakest particles in the aggregate, followed by "clastic hardening", whereby particles split probabilistically depending on the applied macroscopic stress and the co-ordination number and size of each particle. This results in the development of a fractal geometry, and the subsequent unload-reload behaviour of such a material will be strongly affected by the disparity in sizes of neighbouring particles.

1. Introduction

Figure 1(a) shows a plot of voids ratio e (volume of voids per unit volume of solids) against mean effective stress p' for a typical compression test on soil. Fig. 1(b) shows the same data with pressure plotted on a logarithmic scale. Engineers interpret regions 1, 2 and 3 as "elastic stiffening", "yielding" and "plastic hardening" respectively. The plastic hardening curve in Fig. 1(b) is remarkably linear: from the earliest publications in soil mechanics, it has been accepted that the isotropic plastic compression of granular media satisfies an approximately linear relationship between voids ratio e and the logarithm of effective macroscopic pressure p'. This linearity applies to a wide range of granular materials (Novello and Johnston, 1989). Engineers describe the behaviour in Fig. 1 by writing:

$$e = f(p') \tag{1}$$

or for the case of plastic compression (region 3)

35

N. A. Fleck and A. C. F. Cocks (eds.),
IUTAM Symposium on Mechanics of Granular and Porous Materials, 35–46.
© 1997 *Kluwer Academic Publishers.*

$$e = e_o - \lambda \ln p' \qquad (2)$$

Equations (1) and (2) are dimensionally incorrect: clearly p' should be normalized by a parameter X,

$$e = f(p'/X) \qquad (3)$$

where X has dimensions of stress, and must, for objectivity relate to the soil particles themselves.

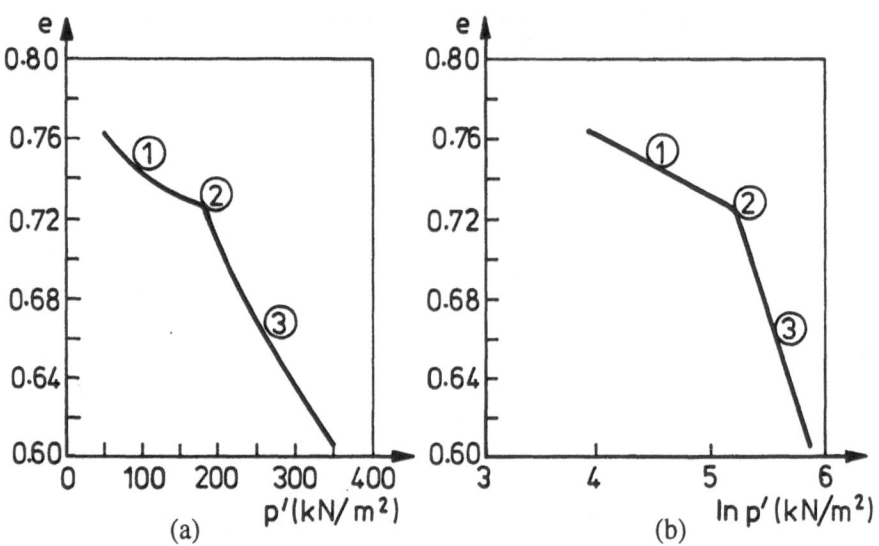

Figure 1. Typical compression curves.

2. Elastic stiffening

At very small strains (less than about 10^{-5}), an aggregate of soil particles behaves entirely elastically. This behaviour must originate solely from the elastic behaviour of individual particle contacts. In this case it is clear that the parameter X in (3) should be an elastic modulus of the particle material. It is found that at very small strains, neglecting changes of voids ratio, the elastic shear modulus of a soil aggregate increases with confining pressure p as $p^{1/2}$ (Hardin and Black,1966,1968; Viggiani and Atkinson, 1995). We therefore write:

$$G_o \propto p^{1/2} G_p^{1/2} \qquad (4)$$

where G_o is the very small strain shear modulus of soil, and G_p is the elastic shear modulus of the particle material. For isotropic soil with constant Poisson's ratio, the bulk and shear moduli are proportional to one another, so that the bulk modulus has the same form of pressure dependence. We now examine the micro mechanical origins of the $p^{1/2}$ dependence.

We first consider a mean-field estimate for the stress-strain response of a random array of identical spheres under isotropic compression, following Jenkins and Strack (1993). For an array of particles with an initially isotropic distribution of Hertzian contacts, the bulk modulus of the aggregate, K_v is given by:

$$K_v \propto p^{1/3} G_p^{2/3} C^{2/3} \tag{5}$$

where C is co-ordination number (number of contacts with neighbours). An apparent discrepancy exists between the power of p found to influence stiffness empirically (4) and that developed in mean field theory (5). Goddard (1990) develops two possible explanations. First, the co-ordination number may be a function of mean stress. He analysed a mechanism in which sample-spanning particle chains buckle under compression, until sufficient lateral force is provided by the formation of new contacts. He showed that small rotations at particle contacts increase the average co-ordination number C such that

$$C \propto \varepsilon_v^{1/2} \tag{6}$$

Substituting (6) in (5), it is readily seen that the aggregate elastic modulus then increases with the square root of the confining pressure.

Secondly, he analysed the contact between a plane and a sphere with an obtuse conical asperity, and deduced that for an aggregate of particles with very shallow conical contacts and a fixed co-ordination number, the bulk modulus of the aggregate would be related to the confining pressure by:

$$K_v \propto p^{1/2} G_p^{1/2} \tag{7}$$

Either, or both, of these mechanisms may be involved in raising the power of p which is found empirically to influence the small-strain stiffness of soils.

3. Clastic yielding

For particles of a specified material and of a given size, there is a statistical variation in strength (Moroto and Ishii, 1990). This statistical variation is inherent in the strength of ceramics (Ashby and Jones, 1986). For a sample of uniform grains under compression, the sudden decrease in the rate of hardening which is evident in region 2 of Fig. 1, must be due to the fracture of the weakest particles in the aggregate. We call this "clastic yielding". In this case the parameter X in (3) should relate to the tensile strength of a particle. Lee (1992) measured the tensile strengths of particles by loading grains diametrically between flat platens. When a spherical grain is loaded diametrically under a pair of forces F, the characteristic tensile stress induced within it can be defined as

$$\sigma = \frac{F}{d^2} \qquad (8)$$

Lee used this to calculate the tensile stress at fracture as

$$\sigma_t = \frac{F_f}{d^2} \qquad (9)$$

where fracture is interpreted as "particle splitting". We therefore write $X = \sigma_t$ for irrecoverable strains in region 2 of Fig. 1.

4. Evolution of an aggregate

The crushing strength of a particle is a function of its size as a consequence of the statistical variation in the strength of ceramics. Furthermore, if a particle has a high co-ordination number, the load on it is well distributed and the probability of fracture is much lower than that at low co-ordination numbers. In addition, particles are more likely to crush as the stress on a sample of granular material increases. We now use these three criteria to model the evolution of an aggregate of elastic-clastic grains for the simple case of one-dimensional compression.

4.1 A SIMPLE NUMERICAL MODEL

In order to establish a demonstration of the principles involved, a highly simplified two-dimensional numerical model was developed (McDowell, Bolton and Robertson, 1996). The initial sample of material in the model comprises uniformly sized grains, which appear as an array of 50 identical isosceles triangles (Fig. 2(a)). Each particle (triangle) can then split into two identical self-similar triangles, and so on. The triangular laminae used in the model are intended to represent real soil particles, in rather the same way that Palmer and Sanderson (1991) used a hierarchy of splitting cubes to model the crushing of ice. The grains are allowed to split with a probability which increases with the applied macroscopic compressive stress σ and reduces with either an increase in the co-ordination number C or a reduction in particle size d (Weibull, 1951). The model does not deal with local equilibrium or kinematics, but simply with the probability of splitting of grains.

The survival probability $P_s(d)$ for a particle of size d was calculated as:

$$P_s(d) = \exp\left\{-\left(\frac{d}{d_o}\right)^2 \frac{(\sigma/\sigma_{t,o})^m}{(C-2)^a}\right\} \qquad (10)$$

The Weibull modulus m is a measure of the uniformity of the material, and $\sigma_{t,o}$ is a characteristic tensile strength of a particle of size d_o. The factor a can be used to vary the degree to which the co-ordination number C affects the probability of fracture, and may be related to particle angularity.

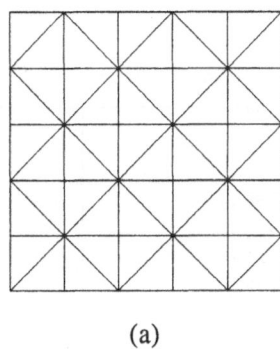   

(a)   (b)

Figure 2. Initial and fractured array of triangles.

For normalisation of the behaviour, the tensile strength $\sigma_{t,o}$ of the initial triangles was set at unity. The initial value of σ was chosen so that only one triangle is likely to break. The stress was then successively incremented and particles were selected for fracture according to (10).

Fig. 3(a) shows the particle size distribution which evolves from taking $m=5$ which fits statistical data, and using $a=5$ which guarantees that co-ordination number is a dominant factor. Fig. 3(b) shows the variation of uniformity coefficient U with increasing macroscopic stress ($U=d_{60}/d_{10}$, where d_{60} is the particle size which 60% by mass of particles are finer than) and Fig. 2(b) shows the resulting array of broken triangles.

4.2 EMERGENCE OF FRACTALS

Fig. 3(a) shows that the curve which evolves of percentage by "mass" (i.e. "area" in the 2-D simulation) of particles smaller than d, versus d on a logarithmic scale is an exponential, which implies a fractal geometry. For a fractal distribution in two dimensions, the percentage by mass of particles smaller than size d, $M(L<d)$ is given by:

$$M(L < d) \propto d^{2-D} \qquad (11)$$

D is the fractal dimension, and usually has a value between 2 and 3 for granular materials (Turcotte, 1986). For a corresponding 2-D simulation, the fractal dimension should lie between 1 and 2. For $m=5$ and $a=5$, D was calculated to be 1.36. Only the smaller particles are fracturing each time the applied stress σ is incremented, because the high co-ordination numbers of the larger particles give them low probabilities of fracture.   As the final particle size distribution is approached, the uniformity coefficient approaches a constant value as shown in Fig. 3(b). The form of the curves in Fig. 3(a) is consistent with data for one-dimensionally compressed Ottawa sand (Fig. 4).

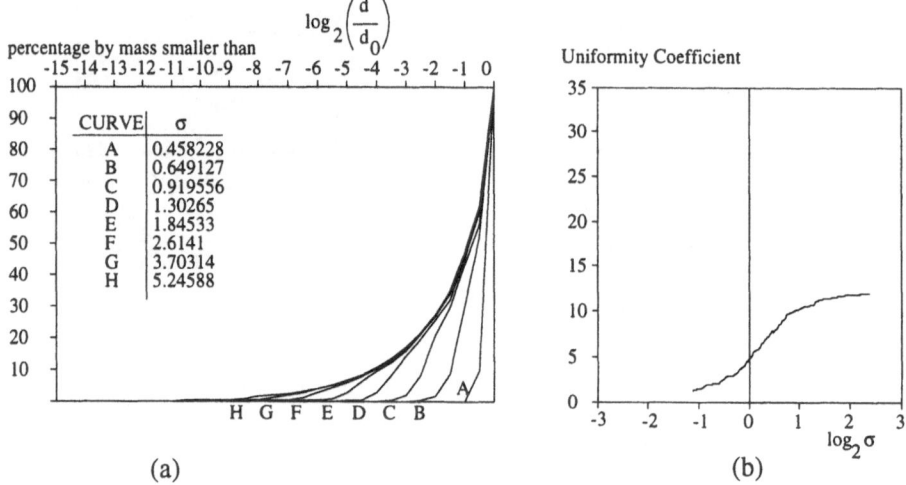

(a)   (b)

Figure 3. Evolving particle size distributions and uniformity coefficient.

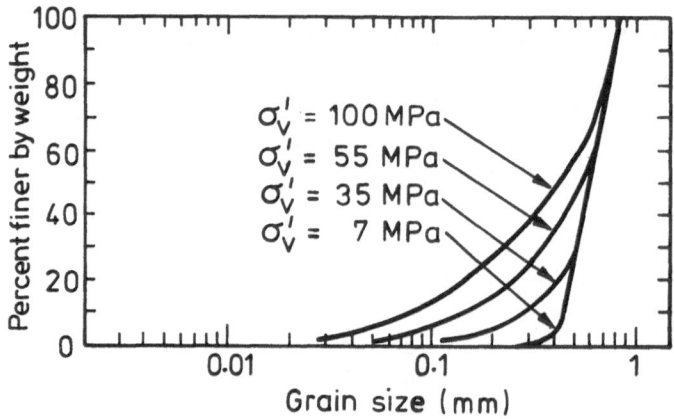

Figure 4. Evolving particle size distributions for Ottawa sand (Fukumoto, 1992).

As with Palmer and Sanderson's fractal model, the range of particle sizes increases with applied stress. The probability of fracture f for the smallest particles of size d_s must be the same each time they fracture. Consequently, Palmer's fractal probability of fracture f can be related to Weibull's survival probability in (10) for the 2-D crushing of triangles:

$$1 - f = P_s = \exp\left\{-\left(\frac{d_s}{d_o}\right)^2 \frac{(\sigma/\sigma_{t,o})^m}{(C-2)^a}\right\} \qquad (12)$$

so that

$$\frac{d_s^2 \sigma^m}{d_o^2 \sigma_{t,o}^m} = \text{constant} \qquad (13)$$

A repetition of numerical simulations using different Weibull moduli m, confirmed that equation (13) holds (McDowell, Bolton and Robertson, 1996).

5. Clastic hardening

It is now shown that the evolution of a distribution of particle sizes with constant uniformity coefficient is consistent with the form of the plastic compression curve in region 3 of Fig. 1(b).

5.1 A NEW WORK EQUATION

We modify a work equation proposed by Roscoe, Schofield and Thurairajah (1963) and Schofield and Wroth (1968) to include energy dissipated in fracture and get:

$$q\delta\varepsilon_q^p + p'\delta\varepsilon_v^p = Mp'\delta\varepsilon_q^p + \frac{\Gamma \, dS}{V_s(1+e)} \qquad (14)$$

The left hand side represents the work done per unit volume by deviatoric stress q and mean effective stress p' (with corresponding irrecoverable strain increments $\delta\varepsilon_q^p$, $\delta\varepsilon_v^p$). The first term on the right hand side was identified as internal frictional dissipation, and the new term is the energy dissipated in the creation of new surface area dS for a volume V_s of solids distributed in a gross volume of $V_s(1+e)$ and Γ is the "surface energy". For the case of one-dimensional compression, we decompose the compression of voids ratio δe into elastic and plastic components δe^e and δe^p, following Schofield and Wroth (1968) who showed that if stress is monotonically increased from σ to $\sigma + d\sigma$, $\log \sigma$ increases by $d\sigma/\sigma$ and the plastic component of reduction in voids ratio is given by:

$$de^p = \Lambda d\sigma/\sigma \qquad (15)$$

where Λ is expressed in terms of Schofield and Wroth's parameters λ and κ by $\Lambda = \lambda - \kappa$. We will shortly derive the conditions necessary for Λ to be a constant. Equation (14) gives for the case of uniaxial compression,

$$de^p = -\frac{\Gamma \, dS}{(1-\mu)\sigma V_s} \qquad (16)$$

where μ is solely a function of the angle of friction ϕ. The form of (16) is consistent with available triaxial data for decomposed granite soil (Miura and O-Hara, 1979).

5.2 POWER LAW COMPRESSION

Equation (16) was used in the numerical model to calculate the hypothetical reduction in voids ratio as the smallest particles fracture with increasing macroscopic stress. Fig. 5 shows the resulting plot for $m=5$, $a=5$. It is evident that an approximately linear-log compression curve develops.   This is due to the formation of a fractal geometry. Equation (11) can be used to calculate the total sectional surface area for all the particles in the 2-D sample:

$$S(L > d_s) \propto d_s^{1-D} \qquad (17)$$

Substituting (13) and (17) into (16) gives the plastic reduction in voids ratio with applied stress increment $d\sigma$:

$$de^p = -\Lambda \sigma^{\frac{m}{2}(D-1)-1} \frac{d\sigma}{\sigma} \qquad (18)$$

In the particular case

$$D = 1 + 2/m \qquad (19)$$

equation (18) reduces to equation (15) and can be integrated to give an equation similar in form to (2). In three dimensions (19) becomes

$$D = 2 + 3/m \qquad (20)$$

For most soils, m will be between 5 and 10; D is often around 2.5 (Turcotte, 1986; Palmer and Sanderson, 1991), and these values fit (20) rather well.   It is therefore proposed that the difference in linearity between normal compression curves for various soils of equal fractal dimension is due to the difference in the variability of the tensile strengths of the particles themselves.

It is now evident that the development of linear-log plastic compression lines in soils has a sound basis in the evolution of a fractal distribution of particle sizes and a constant uniformity coefficient.   This is consistent with data for petroleum coke (Biarez and Hicher, 1994, Fig. 6).

5.3 PLASTIC COMPRESSION INDEX Λ AND COMPRESSION INDEX λ

McDowell, Bolton and Robertson (1996) showed that the value of Λ in (18) has the right order of magnitude, comparing with available data for sands and clays. Schofield and Wroth's elastic parameter κ can be added to the plastic compression index Λ to give the compression index λ. Clearly, κ will be a function of the nature of particle-particle contacts (Hertzian, conical), but the precise micro mechanical origins of κ are not explored here.

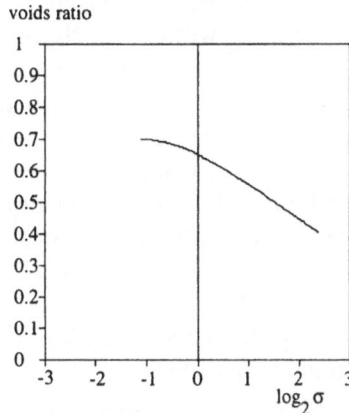

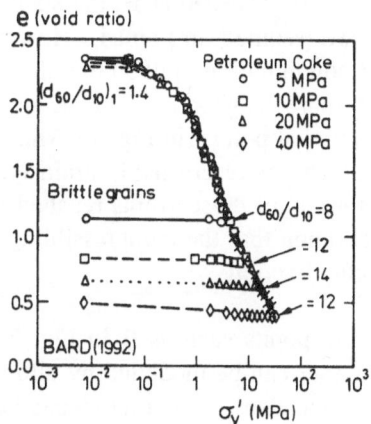

Figure 5. Compression curve for $m=5$, $a=5$.

Figure 6. Typical e-log σ plot for petroleum coke.

6. The evolution of soils

If the successive fracture of grains is accepted as the mechanism for clastic (plastic) hardening of soils, particle size disparity must be a hidden feature of all constitutive behaviour. Fig. 7 depicts the qualitative evolution of a soil aggregate in terms of changes of voids ratio with mean effective stress.

A suspended sediment O may be deposited under water as an aggregate of soil particles A. Sands may have been sorted by river and ocean currents and deposited as a uniform aggregate under still water at a relatively low voids ratio. Clay platelets are sub-micron in size and electrically active, settling so slowly that they can agglomerate to form porous macro-particles, here simply called grains, before they finally aggregate as a sediment with a very high voids ratio. Point A in Fig. 7 can represent either grains of sand with internal flaws, or grains of clay composed of electrically bonded platelets. The soil sediment at A may be described as a "genus", from which other soils evolve by cycles of burial and erosion.

As effective stress increases to point B, some grains fracture so that the soil at B will be seen to be a different "species". The broken pieces pack more efficiently, and the voids ratio at B is irrecoverably reduced. Between B and C many grains fracture, and many of the broken fragments fracture again. After the successive fracture of some grains, the size distribution approaches a fractal for the first time at point C. Although soil species D evolves to have more fines than species C, each now approximates to a given fractal geometry which persists as the stress increases.

At D the effective stress may be taken through an unloading-reloading cycle E, F, G, H. This cycle will be conducted on the same soil species, since little extra damage will occur until stresses exceed their previous maximum. As the pre-consolidation pressure is exceeded at H, the states of fractal evolution on the normal consolidation line are resumed, to point I. However, the smallest fragments will eventually reach their comminution limit so that the larger grains split for preference even though they have more neighbours. The accumulation of unbreakable fines would lead to the curvature at point J in Fig. 7. Mixtures of grain types can be considered in the same way. The more crushable grain type would tend to control the compression of the mixture, until those grains reached their comminution limit. This would explain the observation that the compressibility of a clay-sand mixture is simply proportional to the clay fraction.

Soils at points such as B in Fig. 7 are observed to creep, to G for example. This behaviour can be modelled by allowing the tensile strength of particles in (10) to be time-dependent. This time-dependence of strength in oxide ceramics arises from the slow propagation of micro-cracks due to the chemical interaction with water in the environment. This one-parameter addition creates realistic rate effects linked to the popular concept of "apparent pre-consolidation pressure". The species after creep to G is the same as after load cycling to G, and the pre-consolidation pressure is D in either case.

7. Behaviour of a soil species

The transition from recoverable to irrecoverable behaviour (point H in Fig. 7) would be characterised in stress space by a "yield surface". Constitutive models derived from conventional plasticity theory usually invoke the concept of "isotropic hardening" which expands the yield surface homologously as the voids are irrecoverably compressed (e.g. Cam Clay, Schofield and Wroth, 1968). Our view is that the process of isotropic (or anisotropic) hardening occurs by the successive fracture of grains and is better termed "clastic hardening" which follows "clastic yielding".

The behaviour of "over-consolidated soil" at points D, E, F, G, H which would be "inside the yield surface" can be shown not simply to be non-linear elastic, but to involve "kinematic yielding" of a soil species, which seems more appropriate than the previous term "kinematic hardening". Fig. 8 outlines the consequences of a disparity of particle sizes for the deformation of a species such as D. A hypothetical unit cell is considered to contain a rigid kernel, representing a large grain surrounded by a fractal matrix of finer particles. Initial small-strain behaviour will be elastic, albeit inhomogeneous. The lack of strain within the kernel eventually induces a zone of kinematic yielding (particle sliding against particle) in the surrounding matrix, leading to a partial loss of shear stiffness while some normal contact stiffness remains.

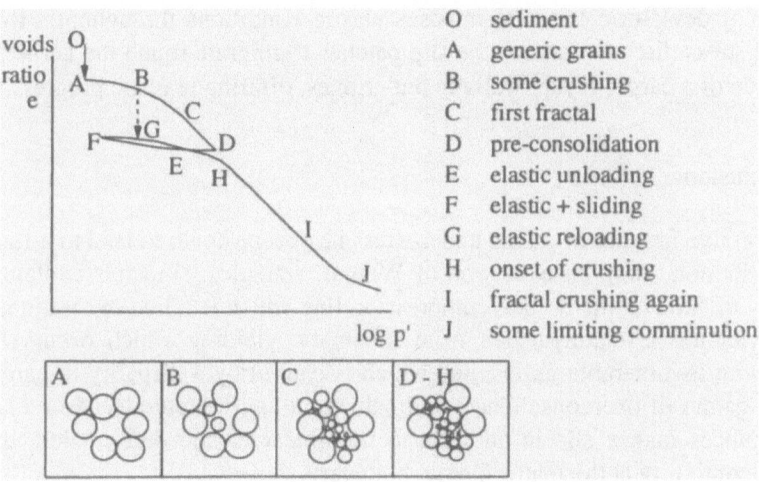

Figure 7. Evolution of soil species.

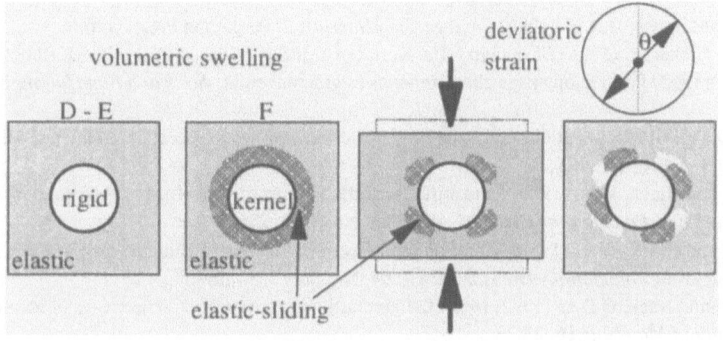

Figure 8. Kinematic yielding within a kernel cell.

From D to E the volumetric expansion will cause the kernel to lose radial stress. At E slip will commence in the matrix at the kernel boundary, due to the local mobilisation of the internal angle of friction between the fine grains. This will successively affect a larger annulus as the boundary stress reduces further, to point F. On re-imposition of spherical compression at F the regime of inter-particle slipping is immediately suspended, since the motion is reversed. The behaviour is elastic on re-loading in the matrix through G but the inhomogeneous stresses may induce clastic yielding "early" so that H falls below D in Fig. 7. This system can provide realistic hysteresis loops such as DEFGH featuring shake-down on cyclic loading, offering compaction if the pores are drained or partial liquefaction if volume is constrained to be constant.

Deviatoric strain starts elastic but eventually induces slip between the matrix and the kernel, on patches which are parallel to planes of maximum shear in the matrix.

Reversal of deviatoric strain re-imposes elastic conditions throughout. Rotation of principal stress direction causes the slip patches to migrate round the kernel, rather in the fashion of a carpet-layer "walking out" ridges, offering realistic progressive strain.

8. Conclusions

The successive fracture of grains under stress has been shown to lead to a fractal grain size distribution using an extension of Weibull statistics. Fractals explain the self-similarity of soils on the normal compression line which is a line of clastic hardening. Clastic yielding is distinguished from kinematic yielding which occurs for a soil species with invulnerable grains, and which is caused by a disparity in particle sizes. The rich gamut of overconsolidated soil behaviour can be reproduced in a kernel cell which induces matrix slip in addition to non-linear elastic deformation, due to the strain discontinuity at the matrix/kernel boundary.

9. References

Ashby, M.F. and Jones, D.R.H. (1986) *Engineering Materials 2*, Pergamon Press, Oxford.

Biarez, J. and Hicher, P. (1994) *Elementary Mechanics of Soil Behaviour*, A.A. Balkema, Rotterdam.

Fukumoto, T. (1992) Particle breakage characteristics in granular soils, *Soils and Foundations* **32**, No. 1, 26-40.

Goddard, J.D. (1990) Nonlinear elasticity and pressure-dependent wave speeds in granular media, *Proc. R. Soc. Lond. A* **430**, 105-131.

Hardin, B.O. and Black, W.L. (1966) Sand stiffness under various triaxial stresses, *Journal of the Soil Mechanics and Foundations Division*, ASCE **92**, No. SM2, 27-42.

Hardin, B.O. and Black, W.L. (1968) Vibration modulus of normally consolidated clay, *Journal of the Soil Mechanics and Foundations Division*, ASCE **94**, No. SM2, 353-369.

Jenkins, J.T. and Strack, O.D.L. (1992) Mean field inelastic behaviour of random arrays of identical spheres, *Mechanics of Materials* **16**, 25-33.

Lee, D.M. (1992) *The angles of friction of granular fills*, Ph. D. dissertation, Cambridge University.

McDowell, G.R., Bolton, M.D. and Robertson, D. (1996) The fractal crushing of granular materials, *J. Mech. Phys. Solids* **44**, No. 12, 2079-2102.

Miura, N. and O-Hara, S. (1979) Particle-crushing of a decomposed granite soil under shear stresses, *Soils and Foundations* **19**, No. 3, 1-14.

Moroto, N. and Ishii, T.(1990) Shear strength of uni-sized gravels under triaxial compression, *Soils and Foundations* **30**, No. 2, 23-32.

Novello, E.A. and Johnston, I.W. (1989) Normally consolidated behaviour of geotechnical materials, *Proc. 12th Int. Conf. Soil Mech., Rio de Janeiro* **3**, 2095-2100.

Palmer, A.C. and Sanderson, T.J.O. (1991) Fractal crushing of ice and brittle solids, *Proc. R. Soc. Lond. A* **433**, 469-477.

Roscoe, K.H., Schofield, A.N. and Thurairajah, A. (1963) Yield of clays in states wetter than critical, *Geotechnique* **13**, 211-240.

Schofield, A.N. and Wroth, C.P. (1968) *Critical State Soil Mechanics*, McGraw-Hill, London.

Turcotte, D.L. (1986) Fractals and Fragmentation, *Journal of Geophysical Research* **91**, 1921-1926.

Viggiani, G. and Atkinson, J.H. (1995) Stiffness of fine-grained soils at very small strains, *Geotechnique* **45**, No. 2, 249-265.

Weibull, W. (1951) A statistical distribution of wide applicability, *J. Appl. Mech.* **18**, 293-297.

COMPRESSIBLE POROUS MEDIA: TOWARD A GENERAL THEORY

R. DE BOER
Institute of Mechanics, FB 10,
University of Essen,
D - 45117 Essen, Germany

1. Introduction

The porous media theory with incompressible constituents had already come to well founded conclusions in the early stage of its development (see Fillunger, 1936). This concerns, in particular, the discovery of the effective stress principle that states, that the stress state in a saturated porous solid can be additively decomposed into a weighted pore pressure and a stress which is governed by the motion of the skeleton. This principle was formulated by von Terzaghi (1936), although the first theoretical and experimental proofs were carried out by Fillunger (1913, 1914, 1915) (see also de Boer and Ehlers, 1990). With the revival of the porous media theory in the 1970's and 1980's Bowen (1980) constructed a general porous media theory including thermal effects, with incompressible constituents, using mixture theory restricted by the volume fraction concept. Its main results are consistent with the current point of view.

The transition of the effective stress principle, developed for incompressible constituents, to compressible porous solids, in particular in rock mechanics, has been questioned for a long time (see Lade and de Boer, 1994). Observations have shown that von Terzaghi's principle should be extended. The first attempt to correct the effective stress principle was by Biot (see, e.g., Biot and Willis, 1957). Later Nur and Byerlee (1971) developed an ad hoc theory for saturated rocks and proved their results with experimental observations. However, their theoretical investigations are not based on fundamental thermodynamical principles. This was done by Bowen (1982). He had recognized (among others) that by introducing the volume fractions the closure problem is violated and concluded that ad-

47

N. A. Fleck and A. C. F. Cocks (eds.),
IUTAM Symposium on Mechanics of Granular and Porous Materials, 47–56.
© 1997 *Kluwer Academic Publishers.*

ditional relations for the volume fractions are required. Bowen (1982) chose
evolution equations for the volume fractions in order to close the problem.
However, this attempt does not contain the effective stress principle and
contradicts the aforementioned experimental observations.

Another evaluation of the closure problem was undertaken by de Boer
(1996) and it was recognized that the saturation condition, i.e. the sum of
all volume fractions is equal to unity, is a constraint independent of the
real material behavior (see also Passman, Nunziato and Walsh, 1984) and
must always be considered in the evaluation of the dissipation principle.
Since the volume fractions relate to both the macro- and the microlevel
and unknown quantities at the microlevel cannot be determined by balance
equations of the macroscopic mixture theory, constitutive relations for the
deformation of a real compressible material must be introduced in order to
achieve closure (see de Boer, 1996).

The goal of this paper is to develop a general theory for saturated and
empty porous solids with compressible constituents where in particular, the
effects of the effective stress principle will be discussed.

2. Material independent relations

In this section a brief survey of the basic relations for the mechanical behavior of saturated and empty porous solids in terms of the volume fraction
concept, kinematics, balance equations, and the entropy inequality will be
given.

2.1. VOLUME FRACTION CONCEPT

The volume fraction concept connects properties of the macroscale with
those of the microscale. Thus, the appropriate way to formulate this concept
would be to start from micromechanical considerations (see de Boer and
Didwania, 1995). However, this procedure leads to a lengthy formalism
which exceeds the scope of this paper. Therefore, only the main results are
reported. We assume the existence of volume fractions

$$n^\alpha = n^\alpha(\mathbf{x}, t) , \tag{2.1}$$

where $\mathbf{x}$ is the position vector of the spatial point x, t is the time and
the index α denotes the individual constituents of the porous media. The
realistic volume v^α and volume element dv^α in the control space $\mathcal{B}_S$, which
is shaped by the solid skeleton are defined as

$$v^\alpha = \int_{\mathcal{B}_S} n^\alpha dv , \quad dv^\alpha = n^\alpha dv , \quad dv = dv(\mathbf{x}, t) , \tag{2.2}$$

where dv denotes the actual volume element of the control space. With the aid of (2.1) and (2.2)

$$v = \int_{B_S} dv = \sum_{\alpha=1}^{\kappa} v^\alpha = \int_{B_S} \sum_{\alpha=1}^{\kappa} dv^\alpha = \int_{B_S} \sum_{\alpha=1}^{\kappa} n^\alpha dv \tag{2.3}$$

holds. From (2.3) it is easily recognized that the volume fractions satisfy the volume fraction condition

$$\sum_{\alpha=1}^{\kappa} n^\alpha = 1 , \tag{2.4}$$

which becomes the so-called saturation condition if the porous solid is completely saturated. In this case (2.4) represents an important constraint and plays a decisive role in the constitutive theory.

With the volume fraction concept "smeared" continua, with reduced densities are created, namely

$$\rho^\alpha(\mathbf{x}, t) = n^\alpha(\mathbf{x}, t)\rho^{\alpha R}(\mathbf{x}, t) , \tag{2.5}$$

where the realistic density $\rho^{\alpha R}(\mathbf{x}, t)$ is the average density of the porous media constituents.

2.2. KINEMATICS

Concerning the kinematics of the "smeared" continua it is assumed that the particles of all constituents φ^α simultaneously occupy the spatial point x at any time t and that the particles X_α proceed from different reference positions $\mathbf{X}_\alpha$ at time $t = t_0$. Thus, each constituent is assigned its own independent motion function χ_α from which the velocity $\mathbf{x}'_\alpha$, the acceleration $\mathbf{x}''_\alpha$ and the deformation gradient $\mathbf{F}_\alpha$ can be derived:

$$\begin{aligned} \mathbf{x} &= \chi_\alpha(\mathbf{X}_\alpha, t) , & \mathbf{x}'_\alpha &= \frac{\partial \chi_\alpha}{\partial t}(\mathbf{X}_\alpha, t) , \\ \mathbf{x}''_\alpha &= \frac{\partial^2 \chi_\alpha}{\partial t^2}(\mathbf{X}_\alpha, t) , & \mathbf{F}_\alpha &= \mathrm{Grad}_\alpha \, \chi_\alpha(\mathbf{X}_\alpha, t) , \end{aligned} \tag{2.6}$$

where the operator Grad_α denotes the derivative with respect to $\mathbf{X}_\alpha$.

It is well-known in porous media theory that the deformation of the saturated porous solid is composed of two parts, namely the deformation of the real material and the change in size and shape of the pores. In order to be able to investigate some properties of the deformation of real materials it is convenient to decompose the deformation gradient $\mathbf{F}_\alpha$ into a tensor $\mathbf{F}_{\alpha R}$ which reflects the microscopic deformation of the real material at the macroscale and a tensor $\mathbf{F}_{\alpha N}$ which describes the deformation of the pores

$$\mathbf{F}_\alpha = \mathbf{F}_{\alpha N} \mathbf{F}_{\alpha R} \tag{2.7}$$

(for more information see Bluhm and de Boer, 1994, de Boer, 1996, Bluhm, 1996).

In the following, some additional deformation measurements of the partial bodies as well as deformation measurements of realistic materials and the deformation of the pores will be introduced. At first, the right Cauchy-Green deformation tensors and the determinants of the deformation gradients

$$\begin{aligned}
\mathbf{C}_\alpha &= \mathbf{F}_\alpha^T \mathbf{F}_\alpha \,, \quad \mathbf{C}_{\alpha R} = \mathbf{F}_{\alpha R}^T \mathbf{F}_{\alpha R} \,, \\
J_{\alpha R} &= \det \mathbf{F}_{\alpha R} \,, \quad J_{\alpha N} = \det \mathbf{F}_{\alpha N}
\end{aligned} \tag{2.8}$$

as well as the symmetric parts of the velocity gradients are introduced.

$$\mathbf{D}_\alpha = \frac{1}{2}(\mathbf{L}_\alpha + \mathbf{L}_\alpha^T) \,, \quad \mathbf{D}_{\alpha R} = \frac{1}{2}(\mathbf{L}_{\alpha R} + \mathbf{L}_{\alpha R}^T) \,, \tag{2.9}$$

where

$$\mathbf{L}_\alpha = (\mathbf{F}_\alpha)_\alpha' \mathbf{F}_\alpha^{-1} \,, \quad \mathbf{L}_{\alpha R} = (\mathbf{F}_{\alpha R})_\alpha' \mathbf{F}_{\alpha R}^{-1} \tag{2.10}$$

and where $(...)_\alpha'$ denotes the material time derivative.

2.3. BALANCE EQUATIONS AND ENTROPY INEQUALITY

In porous media theory the balance equations – balance of mass, balance of momentum, balance of moment of momentum, and balance of energy – must be established for each constituent φ^α, taking into account all interactions and external agencies.

For simplicity, it is assumed that no mass and moment of momentum exchange occur.

2.3.1. *Balance of mass*
With respect to the volume element dv, the balance equation of mass for each constituent φ^α with the partial density ρ^α is the local statement

$$(\rho^\alpha)_\alpha' + \rho^\alpha \operatorname{div} \mathbf{x}_\alpha' = 0 \,. \tag{2.11}$$

2.3.2. *Balance of momentum and moment of momentum*
The axiom of the balance of momentum yields the local statement

$$\operatorname{div} \mathbf{T}^\alpha + \rho^\alpha (\mathbf{b}^\alpha - \mathbf{x}_\alpha'') + \hat{\mathbf{p}}^\alpha = \mathbf{0} \,, \tag{2.12}$$

where $\mathbf{T}^\alpha$ represents the partial Cauchy stress tensor of the constituent φ^α and $\rho^\alpha \mathbf{b}^\alpha$ is the external body force. The quantity $\hat{\mathbf{p}}^\alpha$ is a local interaction force between φ^α and the other $\kappa - 1$ constituents. These interaction forces obey the restriction

$$\sum_{\alpha=1}^{\kappa} \hat{\mathbf{p}}^\alpha = \mathbf{0} \, . \tag{2.13}$$

The axiom of the balance of moment of momentum in the absence of any moment of momentum supply leads to the symmetry of Cauchy's stress tensor $\mathbf{T}^\alpha$

$$\mathbf{T}^\alpha = (\mathbf{T}^\alpha)^{\mathbf{T}} \, . \tag{2.14}$$

2.3.3. *Balance of energy*

The axiom of the energy balance of the constituent φ^α is introduced in the following local form

$$\varphi^\alpha (\varepsilon^\alpha)'_\alpha = -\hat{\mathbf{p}}^\alpha \cdot \mathbf{x}'_\alpha + \mathbf{T}^\alpha \cdot \mathbf{D}_\alpha + \rho^\alpha r^\alpha - \operatorname{div} \mathbf{q}^\alpha + \hat{e}^\alpha \, . \tag{2.15}$$

Herein ε^α stands for the partial specific internal energy of φ^α per mass element, r^α is the partial external heat supply, and $\mathbf{q}^\alpha$ is the external heat flux vector. The quantity $\hat{e}^\alpha$ represents the local energy supply to φ^α from the other $\kappa - 1$ constituents. The energy supplies must meet the restriction

$$\sum_{\alpha=1}^{\kappa} \hat{e}_\alpha = 0 \, . \tag{2.16}$$

See, for example, de Boer *et al.* (1991), for derivation of the balance equations.

2.3.4. *Entropy inequality*

In contrast to the balance equations, the second axiom of thermodynamics is an inequality for general irreversible processes. In the mixture theory only one entropy inequality for the mixture as a whole is used. If all constituents φ^α have the same absolute Kelvin temperature Θ, the derivation of the entropy inequality results in (see, e.g., de Boer *et al.*, 1991)

$$\sum_{\alpha=1}^{\kappa} [-\rho^\alpha (\psi^\alpha)'_\alpha - \Theta'_\alpha \rho^\alpha \eta^\alpha - \hat{\mathbf{p}}^\alpha \cdot \mathbf{x}'_\alpha + \mathbf{T}^\alpha \cdot \mathbf{D}_\alpha -$$
$$-\frac{1}{\Theta} \mathbf{q}^\alpha \cdot \operatorname{grad} \Theta] \geq 0, \tag{2.17}$$

where ψ^α is the free Helmholtz energy function and η^α the specific entropy. The free Helmholtz energy function is related to the specific energy and the specific entropy by

$$\psi^\alpha = \varepsilon^\alpha - \Theta \eta^\alpha \, . \tag{2.18}$$

The form (2.17) of the entropy inequality is the most simple one, in particular, with respect to the development of thermodynamic restrictions for constitutive equations.

3. Constitutive theory

3.1. THE CLOSURE PROBLEM AND THE MODIFIED ENTROPY INEQUALITY

The basis of porous media theory, namely that mixture theory is closed, i.e. the number of unknown fields is equal to the sum of the balance equations and the constitutive relations, as can easily be proved. However, by the introduction of α volume fractions n^α for the constituents φ^α the problem arises that $\alpha - 1$ field equations are missing if one takes into account the volume fraction condition (2.4). It is extremely difficult to gain additional fields since the volume fractions relate to quantities at the microscale, see (2.5), for which balance equations or constitutive equations are not contained in the mixture theory. Therefore, there has been much effort to overcome this crucial point, which starts from the introduction of additional balance equations and proceeds to the formulation of evolution equations for the volume fractions. However, this procedure seems to be insufficient. For one has to be aware of an important constraint. In order to recognize this constraint, one must consider the partial bodies (macroscale) and also the real materials (microscale). Since both are involved in porous media theory, there is no reason for excluding the microscale from general considerations. Now, it is obvious that mixture theory (macroscale) and also the continuum theory of real materials (microscale) are closed. If one could solve the field equations of the macroscale and the microscale, the partial and the realistic densities ρ^α and $\rho^{\alpha R}$ would be known. Then, according to (2.5), also the volume fractions would be determined. In this case, the volume fraction condition (2.4) provides an equation in excess. Thus, in the constitutive theory, this equation (in the rate formulation) together with a Lagrange multiplier must be added to the entropy inequality and considered in the evaluation of this inequality.

From the saturation condition (2.4) considering the kinematics in Section 2.2 we obtain (see de Boer 1996):

$$-n^S(\mathbf{D}_S \cdot \mathbf{I}) + n^S(\mathbf{D}_{SR} \cdot \mathbf{I}) - n^F(\mathbf{D}_F \cdot \mathbf{I}) + \\ +n^F(\mathbf{D}_{FR} \cdot \mathbf{I}) - \operatorname{grad} n^F \cdot (\mathbf{x}_F' - \mathbf{x}_S') = 0 \ . \tag{3.1}$$

A consequence of the fact that the saturation condition is a constraint is that additional field equations must be formulated in order to achieve closure. This will be done in the next section.

Together with the Lagrange multiplier λ Eqn. (3.1) can be added to the entropy inequality (2.17). In the remainder of this paper we consider only isothermal deformations for a binary model (*solid* $= F$, *fluid* $= F$), then

$$
\begin{aligned}
&-\rho^S(\psi^S)'_S - \rho^F(\psi^F)'_F + \mathbf{D}_S \cdot (\mathbf{T}^S + n^S \lambda \mathbf{I}) + \\
&+ \mathbf{D}_F \cdot (\mathbf{T}^F + n^F \lambda \mathbf{I}) - \mathbf{D}_{SR} \cdot n^S \lambda \mathbf{I} - \\
&- \mathbf{D}_{FR} \cdot n^F \lambda \mathbf{I} - (\hat{\mathbf{p}}^F - \lambda \operatorname{grad} n^F) \cdot (\mathbf{x}'_F - \mathbf{x}'_S) \geq 0 .
\end{aligned}
\tag{3.2}
$$

The evaluation of inequality (3.2) yields important restrictions for the constitutive relations.

3.2. CONSTITUTIVE RELATIONS

As has already been mentioned constitutive equations (response functions) need to be introduced in order to achieve closure. At first sight, response functions must be formulated for

$$
\mathcal{R} := \left\{ \psi^\alpha, \mathbf{T}^\alpha + n^\alpha \lambda \mathbf{I}, \quad \hat{\mathbf{p}}^F - \lambda \operatorname{grad} n^F \right\} .
\tag{3.3}
$$

The principles of determinism and local action state that the response functions in (3.3) for a material point X_α, at any time t, and any place $\mathbf{x}$, must be determined by the history of an arbitrary small neighbourhood of this material point. The history of this small neighbourhood is determined by the process variables $s(\mathbf{x}, t)$ which reflect the history of the motions of the two partial constituents solid and fluid. In the following we study a purely elastic model for which the process variables

$$
s = \left\{ \mathbf{C}_S, \quad \rho^{FR}, \quad \mathbf{x}'_F - \mathbf{x}'_S \right\}
\tag{3.4}
$$

are postulated. The choice of the process variables is determined as follows: The elastic deformation of the partial solid is described by $\mathbf{C}_S$ and the deformation of the real fluid is reflected by ρ^{FR}. Finally, the velocity difference $\mathbf{x}'_F - \mathbf{x}'_S$ governs dissipative effects.

In order to meet the requirements discussed in Section 3.1 concerning the closure problem, constitutive equations must also be formulated for process variables reflecting the elastic behavior of real materials (microscale). That means constitutive equations must be introduced for the right Cauchy-Green deformation tensor $\mathbf{C}_{SR}$ of the solid phase and the Lagrange multiplier λ. For our purpose it is sufficient to choose rather simple response functions:

$$
\mathbf{C}_{SR} = \mathbf{C}_{SR}(\mathbf{C}_S), \quad \lambda = \lambda(\rho^{FR}) .
\tag{3.5}
$$

These constitutive assumptions must be formulated in such a way that all restrictions mentioned previously are fulfilled.

For the present model, thermodynamical restrictions result from the entropy inequality, relation (3.2), together with the constitutive assumptions (3.3), (3.4), and (3.5). Using standard arguments, we obtain a model which is goverened by constitutive equations which are relatively simple (the complete derivations of the thermodynamical restrictions can be found in Bluhm, 1996). In the following only the main results are listed. Evaluation of the entropy inequality (3.2), considering (3.4) and (3.5), yields:

$$
\mathbf{T}^S = -n^S \lambda (\mathbf{I} - \boldsymbol{\mathcal{F}}_S) + \rho^S \mathbf{F}_S \frac{\partial \psi^S}{\partial \mathbf{C}_S} \mathbf{F}_S^T +
$$
$$
+ \rho^S \frac{\partial \psi^S}{\partial (\mathbf{x}_F' - \mathbf{x}_S')} \otimes (\mathbf{x}_F' - \mathbf{x}_S') \,,
\tag{3.6}
$$

where

$$
\boldsymbol{\mathcal{F}}_S = \mathbf{F}_S [(\frac{\partial \mathbf{C}_{SR}}{\partial \mathbf{C}_S})^T \mathbf{C}_{SR}^{-1}] \mathbf{F}_S^T \,,
\tag{3.7}
$$

$$
\mathbf{T}^F = -n^F \lambda + \rho^F \frac{\partial \psi^F}{\partial (\mathbf{x}_F' - \mathbf{x}_D')} \otimes (\mathbf{x}_F' - \mathbf{x}_S') \,,
$$
$$
\lambda = (\rho^{FR})^2 \frac{\partial \psi^F}{\partial \rho^{FR}} \quad \text{(see de Boer and Kowalski, 1995)}.
\tag{3.8}
$$

We identify λ as the pore-fluid pressure p.

$$
\lambda = p = (\rho^{FR})^2 \frac{\partial \psi^F}{\partial \rho^{FR}} \,.
\tag{3.9}
$$

Moreover, a dissipation inequality remains

$$
-(\hat{\mathbf{p}}^F - p \operatorname{grad} n^F) \cdot (\mathbf{x}_F' - \mathbf{x}_S') \geq 0 \,.
\tag{3.10}
$$

The above stated constitutive relations simplifiy close to the so-called mixture equilibrium state, which is characterized by a vanishing relative velocity, i.e. $\mathbf{x}_F' - \mathbf{x}_S' = \mathbf{0}$. In this case we have

$$
\mathbf{T}^S = -n^S p (\mathbf{I} - \boldsymbol{\mathcal{F}}_S) + 2\rho^S \mathbf{F}_S \frac{\partial \psi^S}{\partial \mathbf{C}_S} \mathbf{F}_S^T \,,
\tag{3.11}
$$

$$
\mathbf{T}^F = -n^F p \mathbf{I} \,,
\tag{3.12}
$$

and

$$
\hat{\mathbf{p}}^F = p \operatorname{grad} n^F \,.
\tag{3.13}
$$

In the case of incompressible fluids, p remains undetermined (see de Boer, 1996). The undetermined Lagrange multiplier $\lambda = p$ can be determined with the help of the balance equation of momentum (2.12) for the fluid phase considering (3.12), (3.13), and the boundary condition for the realistic fluid phase. If the pores are empty the Lagrange multiplier $\lambda = p$ in (3.6) – (3.13) becomes zero.

With the derivation of the above constitutive relations (3.6) to (3.12), considering in addition the kinematics and the balance equations, the mechanical behavior in the elastic range of a binary model can be described. It is easy to prove that this general model reduces to models with incompressible phases (see de Boer, 1996). In this case the validity of known concepts for the effective stress principle can easily be proved. We consider an elastic compressible porous solid and an incompressible liquid in a purely hydrostatic stress state. The governing constitutive relations for this state are derived from (3.11) considering (3.7) and (3.12) (see also de Boer, 1996):

$$p^S = - n^S p (1 - J_{SN} \frac{\partial J_{SR}}{\partial J_S}) + p_E^S , \tag{3.14}$$

$$p^F = - n^F p . \tag{3.15}$$

In (3.14) and (3.15) p^S and p^F are the partial hydrostatic pressures and p_E^S is the effective hydrostatic stress of the partial solid. The total hydrostatic stress $\bar{p}$ determined by adding the two response functions (3.14) and (3.15)

$$\bar{p} = p^S + p^F = -p(1 - n^S J_{SN} \frac{\partial J_{SR}}{\partial J_S}) + p_E^S , \tag{3.16}$$

With the appropiate relation for J_{SR}

$$J_{SR} = J_S^{K_S/K_{SR}} , \tag{3.17}$$

where K_S amd K_{SR} are the compressibility moduli of the partial solid and of the real solid material. Using relation (3.17), (3.16), Suklje's formula (see de Boer, 1996), is obtained:

$$p_E^S = \bar{p} + p(1 - n^S \frac{K_S}{K_{SR}}) . \tag{3.18}$$

Suklje's formula (3.18) reduces to von Terzaghi's (1936) statement for a porous solid with incompressible constituents. In this case the modulus K_{SR} of the real material is much larger than the modulus K_S of the partial porous solid and the second term in the brackets of (3.20) can be neglected.

Acknowledgement: The financial support for the investigations of compressible porous media by the Deutsche Forschungsgemeinschaft (DFG) is gratefully acknowledged.

References

Biot, M.A. and Willis, D.G. (1957) The elastic coefficients of the theory of consolidation, *J. of Appl. Mechanics* **24**, pp 594–601.

Bluhm, J. (1996) A consistent model for saturated and empty porous media, *Habilitation thesis*, Universität Essen.

Bluhm, J. and de Boer, R. (1994) The volume fraction concept in the porous media theory, Report Mech 94/5, FB 10/Mechanik, UniversitätGH Essen, to appear in *Zeitschr. f. Ang. Math. Mech. (ZAMM)*.

de Boer, R. (1996) Highlights in the historical development of the porous media theory: Toward a consistent macroscopic theory, *Appl. Mech. Rev.*,**49**, (4), pp 201–262.

de Boer, R. and Ehlers, W. (1990) The development of the concept of effective stresses, *Acta Mechanica* **83**, pp 77–92.

de Boer, R., Ehlers, W., Kowalski S., and Plischka, J. (1991) Porous Media – a survey of different approaches, *Forschungsberichte aus dem Fachbereich Bauwesen*, **54**, Universität-GH Essen.

de Boer, R. and Kowalski, S.J. (1995) Thermodynamics of fluid-saturated porous media with a phase change, *Acta Mechanica* **109**, pp 167–189.

Bowen, R.M. (1980) Incompressible porous media models by use of the theory of mixtures, *Int. J. Eng. Sci.*, **18**, pp 1129-1148.

Bowen, R.M. (1982) Compressible porous media models by use of the theory of mixtures, *Int. J. Eng. Sci.*, **20**, pp 697–735.

Fillunger, P. (1913) Der Auftrieb in Talsperren, *Österr. Wochenschrift für den öffentl. Baudienst*, **19**, pp 532–556, 567–570.

Fillunger, P. (1914) Neuere Grundlagen für die statische Berechnung von Talsperren, *Zeitschrift des Österr. Ing.- und Arch.-Vereines*, **23**, pp 441-447.

Fillunger, P. (1915) Versuche über die Zugfestigkeit bei allseitigem Wasserdruck, *Österr. Wochenschrift für den öffentl. Baudienst*, H. 29, pp 443–448.

Fillunger, P. (1936) Erdbaumechanik?, Selbstverlag des Verfassers, Wien.

Lade, P.V. and de Boer, R. (1994) The concept of effective stress for soil, concrete and rock. Report Mech 94/4, FB 10/Mechanik, Universität-GH Essen, to appear in *Geotechnique*.

Nur, A. and Byerlee, J.D. (1971) An exact effective stress law for elastic deformation of rock with fluids, *J. Geophys. Res.*, **76**, pp 6414–6419.

Passman, S.L., Nunziato, J.W. and Walsh, E.K. A theory of multiphase mixtures, in *Rational Thermodynamics*, ed. by C. Truesdell, Mc. Graw-Hill, New York, pp 286–325.

von Terzaghi, K. (1936) The shearing resistance of saturated soils and the angle between the planes of shear, *Proc. of Int. Conf. on Soil Mech. and Foundation*, Vol. 1, Harvard Univ. Cambridge, Mass. (ed. Casagrande *et al.*), pp 54–56.

MECHANICAL BEHAVIOUR OF MIXTURES OF KAOLIN AND COARSE SAND

G.V. KUMAR
Department of Civil Engineering, Andhra University
Viskhapatnam, India
D. MUIR WOOD
Department of Civil Engineering, University of Bristol
Bristol, United Kingdom

Summary

Experimental studies have been performed in order to discover the ways in which the presence of different proportions of coarse sand (particle size ~2mm) affects the mechanical properties of a clay (kaolin). The behaviour of the mixtures can be understood through the use of appropriate volumetric variables. At one extreme, if the coarse particles do not interact then the response of the mixture may be dependent on volumetric packing of the clay matrix which can be described through the use of clay specific volume and clay volumetric strain which treat the sand as inert space filler. At the other extreme, when the sand particles become sufficiently close to interact, a granular specific volume characterises the potential for interaction between individual sand grains.

Clayey soils are classified according to the values of index properties measured using standard procedures. The liquid limit, determined using a cone penetrometer, is found to be proportional to clay content for clay contents of the mixture between 40 and 100%. It can be inferred that the strength of the clay matrix controls the strength of the mixtures for this range of clay contents.

Direct measurements of the flow characteristics of the mixtures show that a simple relationship can be obtained between permeability of the mixture and the volumetric packing of the clay matrix, for clay contents down to 40%. The relationship in one-dimensional compression between vertical effective stress and volumetric packing of the clay matrix is also unique for vertical stresses up to 1600kPa and for clay contents down to about 40%.

The stiffness and strength characteristics of the mixtures have been studied in conventional triaxial tests. The shear stiffness, shear strength and pore pressure response are unaffected by the presence of coarse material for clay contents down to 40%. The volume changes of the different mixtures are identical when expressed in terms of the clay volumetric strains.

N. A. Fleck and A. C. F. Cocks (eds.),
IUTAM Symposium on Mechanics of Granular and Porous Materials, 57–68.
© 1997 *Kluwer Academic Publishers.*

1. Introduction

Experimental and theoretical studies of soil behaviour have often concentrated on ideal soils: pure clays or uniform sands. However, many natural soils - for example, glacial soils and residual soils - and construction fills consist of coarse granular particles in a matrix of clay paste; and granular material may be added as a filler to weak clays in order to improve their mechanical behaviour. The properties of such natural or man-made mixtures are expected to be intermediate between the properties of the constituent materials. The mineral structures of clays and sands are quite different and it is expected that the behaviour of the composite material will be influenced both by the particle sizes and by the physico-chemical characteristics of the materials present.

As an initial step towards the understanding of the origins of the mechanical response of these composite soils the experimental study presented here describes various mechanical tests on mixtures of clay with increasing proportions of coarse sand: the two constituents being deliberately chosen to be hugely different in both particle size and mineral character. This contrasts with previous work on clayey sands described by Georgiannou, Burland and Hight (1990) in which kaolin clay has been added to Ham river sand with the majority of tests being performed with clay contents below 10%. They found that the familiar dilatant character of the mechanical behaviour of the sand began to be affected when the clay content of the mixture reached about 30%.

The tests performed on the mixtures have included the liquid limit, permeability, one dimensional compression, and drained and undrained triaxial compression. Full details of all the tests and procedures can be found in Kumar (1996).

2. Volumetric variables

For a soil which is made up of quite distinct particle sizes as shown in Fig 1, conventional definitions of volumetric packings may no longer be appropriate. It is helpful to consider the three separate constituents : water (filling the voids), clay solids, and granular solids (coarse sand) with volumes V_w, V_c, V_g and masses M_w, M_c, M_g respectively. It is convenient (and not greatly in error) to assume that the specific gravity of clay and coarse sand particles is the same so that the clay content C is given exactly by

$$C = \frac{M_c}{M_c + M_g} \tag{1}$$

and approximately by

$$C \approx \frac{V_c}{V_c + V_g} \tag{2}$$

The void ratio e can be defined conventionally as

$$e = \frac{V_w}{V_c + V_g} \tag{3}$$

but it is also useful to define a clay void ratio e_c for the clay paste, assuming that the coarse sand plays no part:

$$e_c = \frac{V_w}{V_c} = \frac{e}{C} \qquad (4)$$

and a clay specific volume (because natural volumetric strain is directly related to logarithm of specific volume)

$$v_c = \frac{V_w + V_c}{V_c} = \frac{e + C}{C} \qquad (5)$$

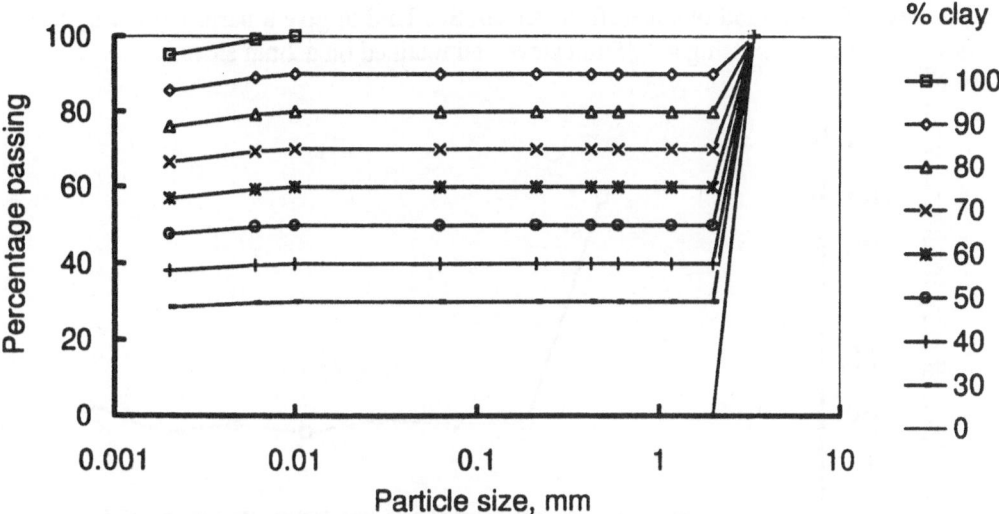

Figure 1. Particle size distributions for mixtures of clay and coarse sand

Equally, a granular void ratio e_g can be defined (proposed by Lupini et al (1981) as a useful variable with which to assess residual strength of cohesive soils), assuming that the voids in the coarse sand are filled with a mixture of clay and water:

$$e_g = \frac{V_w + V_c}{V_g} = \frac{e + C}{1 - C} \qquad (6)$$

but a granular specific volume is more appropriate in the context of the present work:

$$v_g = \frac{V_w + V_c + V_g}{V_g} = \frac{1 + e}{1 - C} \qquad (7)$$

The water content w is defined as*

$$w = \frac{M_w}{M_c + M_g} = \frac{e}{G_s} \qquad (8)$$

and a clay water content w_c is defined as

$$w_c = \frac{M_w}{M_c} = \frac{w}{C} \qquad (9)$$

for the clay paste alone without regard for the presence of the coarse particles.

3. Preparation of mixtures

Kaolin was chosen to be the clay constituent because of its wide use for laboratory testing. The kaolin was obtained in powder form from English China Clays, Lovering and Pochin Limited, St Austell, Cornwall. Plastic, clayey soils are classified according to the water contents at which the nature of their mechanical response changes using techniques devised by Atterberg (1911) and described in BS1377 (British Standards Institution, 1991). The kaolin used has a liquid limit 80.5%, plastic limit 34.5%, and 95% of the solid material is finer than 2 μm. The coarse sand used for the mixtures was sieved from a sand obtained from Arran, Scotland to give a narrow grading curve as shown in Fig 1, passing a 3.35mm sieve and retained on a 2mm sieve.

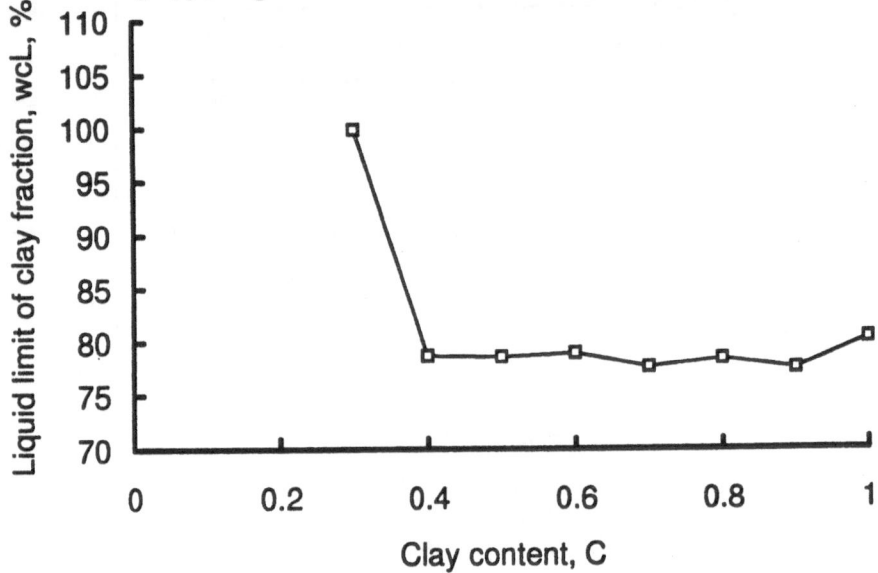

Figure 2. Liquid limit of clay fraction and clay content

Kaolin powder was mixed with distilled water as a slurry at 120% water content under vacuum. The required weight of coarse sand was added to the slurry and allowed to mix. The resulting mixture of slurry and coarse sand was then stored until required for testing. Typical particle size distributions for the mixtures are shown in Fig 1. No evidence of segregation was found in any samples - the distribution of coarse particles was uniform throughout the height of the samples.

4. Liquid limit

The liquid limit w_L of the clay/sand mixtures was determined by the standard procedure specified in BS1377 (1991) using a fall-cone penetrometer as the water content at which a cone of mass 80g with tip angle 30° penetrates 20mm when allowed to fall under its own weight from contact with the soil surface. The British Standard specifies that all particles with diameter greater than 475μm (which would include all the coarse sand) should be removed before the fall-cone test is performed but this was not done in the present tests: there is nothing inherent in the fall-cone test which in principle prevents it from being used with soil containing coarser particles.

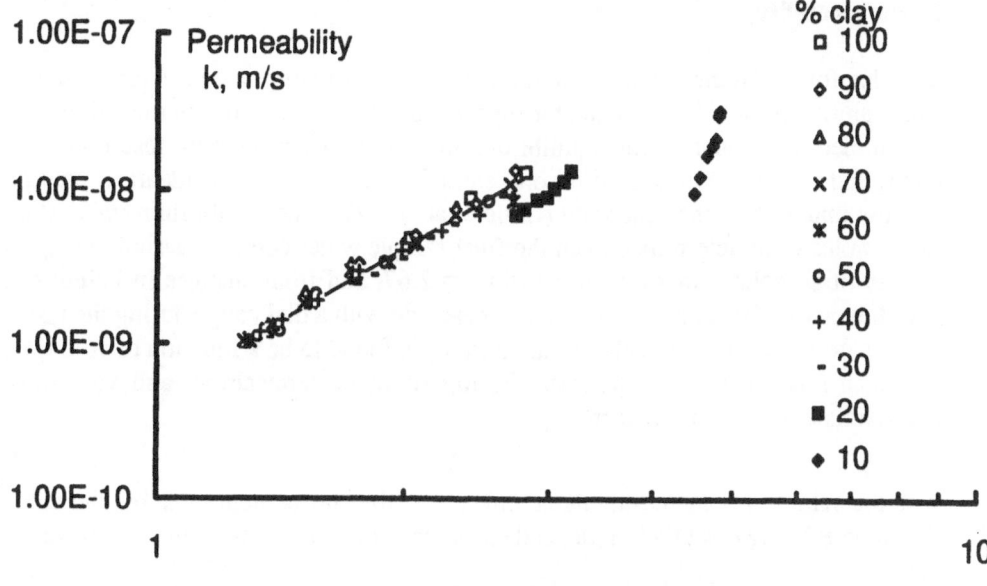

Figure 3. Permeability of mixtures as function of clay void ratio

It is found that the liquid limit w_L varies approximately linearly with clay content. If the clay water content w_{cL} (see (9)) is calculated for the measured liquid limit of each mixture then it is found that this is more or less independent of clay content until the clay content falls below about 40% (Fig 2). This implies that the presence of the coarse sand is not affecting the penetration resistance of the clay at the low strengths that it has for water contents around the liquid limit (where the undrained strength of the clay is of the order of 2kPa (Muir Wood, 1990)) until 60-70% of the volume of solid material is made up of hard sand particles.

If it is assumed that the sand particles are spheres and arranged according to a regular close-packed hexagonal array then it is possible to estimate the centre to centre spacing s of these particles as a function of the particle diameter d and granular specific volume

$$v_g = \frac{3\sqrt{2}}{\pi} \left(\frac{s}{d} \right)^3 \tag{10}$$

For the mixture with clay content C = 40%, and with clay water content equal to the liquid limit of the pure clay, the implied granular specific volume is 3.1 and corresponding particle spacing 1.3d. The volume of soil displaced by 20mm penetration of a 30° cone is 600mm³ and with granular specific volume 3.1 and typical particle diameter 2mm it can be deduced that roughly 50 sand particles are displaced together with the clay paste as the cone penetrates. It is perhaps surprising that even at this density of packing the presence of the sand particles appears to have no mechanical effect on the penetration of the cone.

5. Permeability

Samples for one-dimensional compression and permeability testing were compressed from slurry in a 75mm diameter hydraulic oedometer with permeability being determined by measuring the equilibrium differential water pressure resulting when a controlled inflow was imposed on one end of the sample and an identical controlled outflow imposed on the other end (Little et al, 1992). The equilibrium void ratios of the samples were determined from the final sample water content, assuming a specific gravity of all solid mineral material of $G_s = 2.67$, and from changes in height of the one-dimensionally compressed samples measured with a dial gauge during the test.

The permeability of the clay/sand mixtures is found to be a function of void ratio e. For each mixture the link between the logarithm of permeability and void ratio is approximately linear of the form

$$k = A e^B \tag{11}$$

with the values of A increasing monotonically as the clay content falls, but the value of the slope B being essentially independent of clay content for clay contents above about 30%.

This becomes clear when the data are presented in terms of clay void ratio e_c (Fig 3) and it is found that the relationship is essentially unique for clay contents down, in this case, to about 30% (sand contents above 70%). It is only for higher sand contents that the unique relationship breaks down - and for high sand contents, of 80% and 90%, the permeability will tend to be sufficiently high that techniques of permeability determination devised for clays will not be appropriate.

Presented in terms of clay void ratio the data in Fig 3 imply that

$$k = A^* e_c^B \tag{12}$$

with B = 3.16 and $A^* = 4.6 \times 10^{-10}$ m/s. Using (4), and comparing (12) with (11), it is seen that

$$AC^B = A^* \tag{13}$$

or

$$A = \frac{A^*}{C^B} \tag{14}$$

for clay contents greater than 40%. .

When the content of impermeable coarse sand particles is high it might be anticipated that the space that they occupy would itself impede some of the flow of water through the soil. The permeability k links flow velocity with hydraulic gradient, but velocity is calculated from the measured volumetric flow rate and the cross sectional area through which the water flows. If it were assumed that the flow area is only a proportion equal to the clay content C of the total cross sectional area then values of 'clay permeability' could be calculated which would be higher than the values

shown in Fig 3.   However, use of this 'clay permeability' destroys the unique
relationship that has been shown in Fig 3 between permeability and clay void ratio.
The sand particles seem to act as filler in some ways - occupying volume within the
soil mixture - but not in others - not apparently restricting the area through which flow
occurs.   The process of permeation through an impermeable clay is concerned with
flow of water through microscopically tortuous passages on which the presence of
occasional sand particles has little influence.

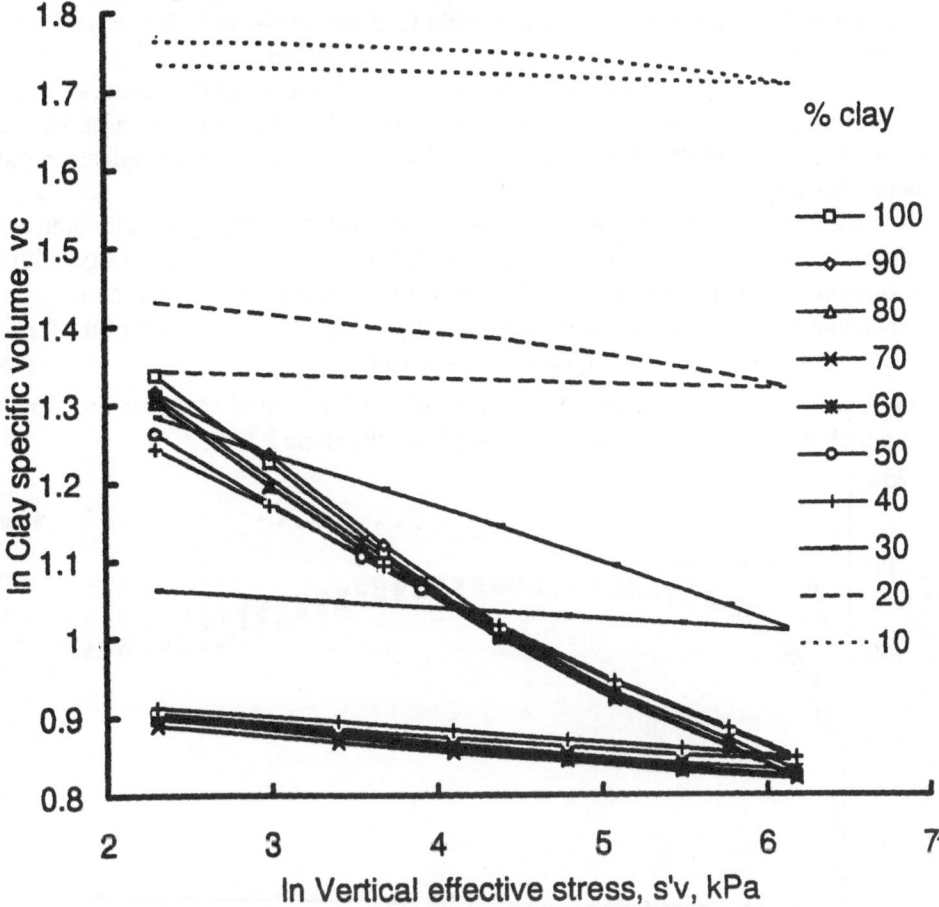

Figure 4. One-dimensional compression of mixtures: vertical effective stress and clay
specific volume

6. One-dimensional compression

One-dimensional compression of the clay/sand mixtures interpreted in the traditional
way produces a series of relationships between void ratio and vertical effective stress.
The pattern within these relationships is revealed when the data are plotted in terms of
clay void ratio instead of void ratio (Fig 4), and it is again found that for clay contents

down to about 40% the relationship is more or less unique (within the range of experimental scatter) and it is only for lower clay contents - sand contents above 60% - that the sand particles interact sufficiently to affect the one-dimensional compression relationship.

It would be expected that the effect of the coarse particles would become progressively more significant as the water is squeezed out of the clay paste, as the clay void ratio decreases with increasing vertical effective stress. The values of clay void ratio e_c for each vertical stress σ'_v show a slight tendency to rise as C falls especially at the higher values of vertical stress, but this tendency is almost negligible. A small number of one-dimensional compression tests taken to much higher stresses - using dead loading on a piston in a 38mm diameter tube - confirm that even at vertical stresses of 1580kPa the effect of the presence of granular particles is negligible for clay contents above 40%.

It is natural to propose that it is granular specific volume which primarily indicates the potential for interaction of the coarse sand particles. The data from the liquid limit tests suggest a transitional value $v_g \approx 3$. It might be assumed that the one-dimensional normal compression of the clay paste is always described by a relationship of the form

$$\ell n\ v_c = \ell n\ A - \lambda\ \ell n\ \sigma'_v \qquad (15)$$

where A is a reference value of clay specific volume for vertical effective stress $\sigma'_v = 1kPa$ which, from the data in Fig 4 is $A \approx 4.67$ and the slope $\lambda \approx 0.1$.

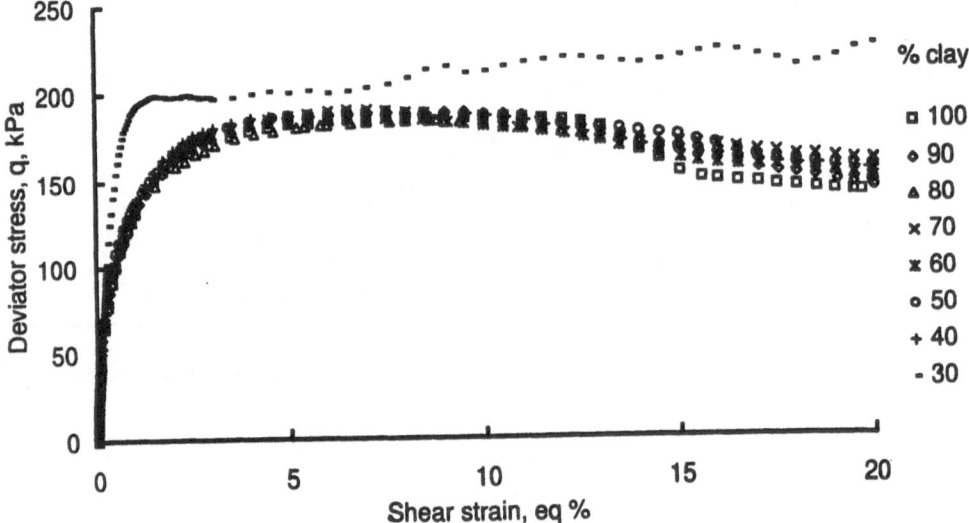

Figure 5. Undrained triaxial compression of mixtures (a) stress:strain response, normally consolidated samples.

Using expressions (5), (7) and (15) a link can be established between clay content and the vertical effective stress required to reach a specified granular specific volume:

$$\sigma_v' = \left[\left(\frac{1-C}{C} \right) \left(\frac{v_g - 1}{A} \right) \right]^{-\frac{1}{\lambda}} \tag{16}$$

The vertical stress required to give any particular value of granular specific volume
(that is to say packing of coarse sand particles) falls off extremely rapidly with clay
content. It is clear that, on the basis of the simple approach proposed here, if the
threshold value of granular specific volume is set around 3 then the threshold will only
be crossed at stresses of common engineering concern for clay contents around 0.4.
The implied vertical consolidation stress in the use of the fall cone to determine the
liquid limit is of the order of 8kPa (Muir Wood, 1990) so it would be expected that the
effect of the presence of the coarse sand on the liquid limit determination would only
manifest itself at lower values of clay content.

7. Triaxial compression

Specimens to be tested in the triaxial apparatus were prepared from slurry consolidated
one dimensionally to a vertical stress of 400kPa, then extruded and set up in the
triaxial cell. All samples were then isotropically consolidated to 400kPa. Undrained
and drained compression tests were performed on isotropically normally compressed
samples, and on samples with isotropic overconsolidation ratios of 1.33 and 4.

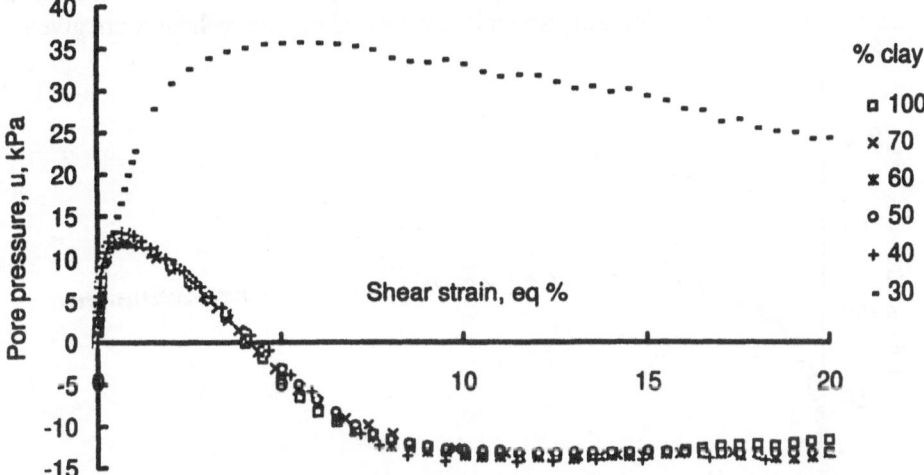

Figure 5. Undrained triaxial compression of mixtures: (b) pore pressure response,
overconsolidation ratio=4.

Typical results of undrained compression tests are shown in Fig 5. It is very clear
that the relationships between both the shear stress and the shear strain (Fig 5a) and
the pore pressure and the shear strain (Fig 5b) are almost independent of clay content
for clay contents down to 40% but that thereafter, with 30% clay, there is a marked
change. In particular, all the overconsolidated samples with clay contents $C \geq 40\%$
show a pore pressure response typical of overconsolidated clays (Fig 5b) with an initial

pore pressure increase followed by a marked pore pressure decrease. However, the 30% clay, 70% sand overconsolidated sample shows a pore pressure response more typical of a loose sand, with much greater increase of pore pressure and much less dramatic subsequent reduction.

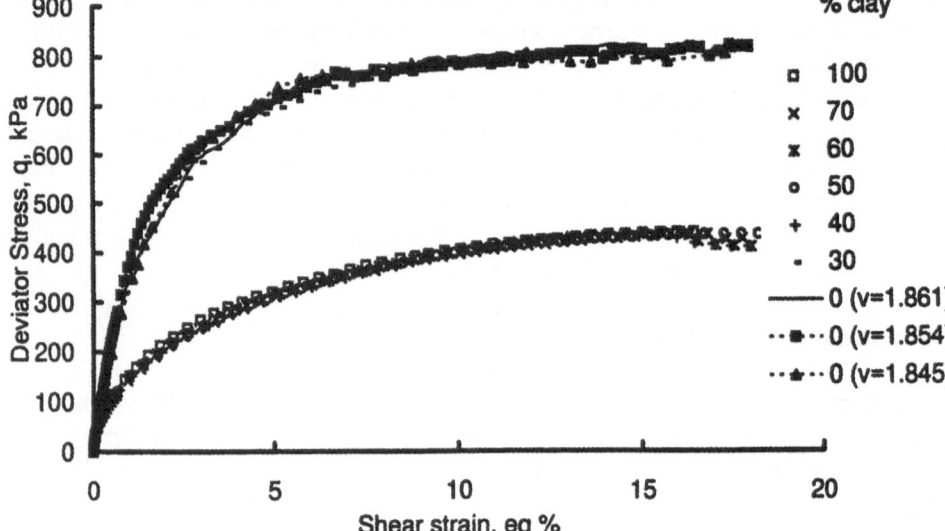

Figure 6a. Drained triaxial compression of mixtures: stress:strain response, normally consolidated samples; for pure sand initial values of specific volume v are given.

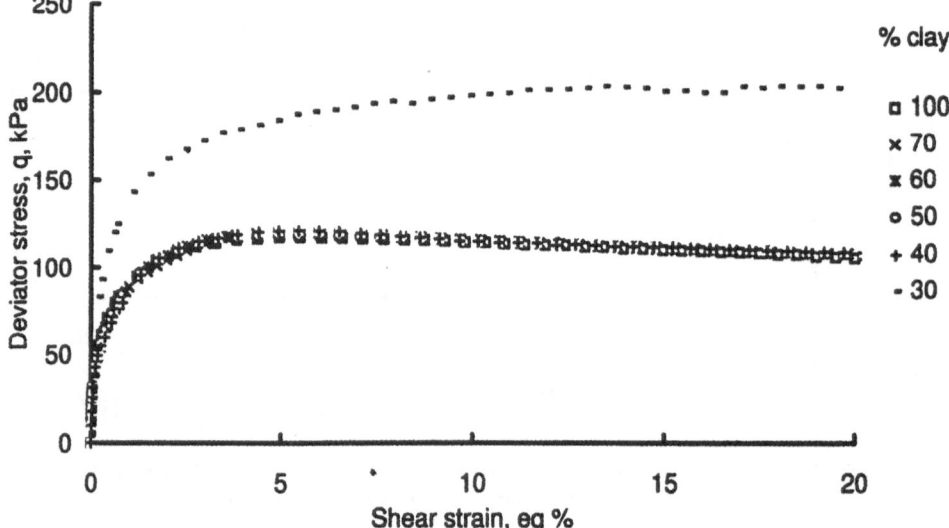

Figure 6b. Drained triaxial compression of mixtures: stress:strain response, overconsolidation ratio=4.

The conclusions concerning the relationships between shear stress and shear strain response under drained conditions are similar (Fig 6a, b) with the behaviour being independent of clay content down to 40% clay. However, the volumetric strains are strongly affected by clay content and it is only when clay volumetric strain rather than volumetric strain is plotted against shear strain that the volumetric response once again becomes remarkably unique (Fig 6c, d).

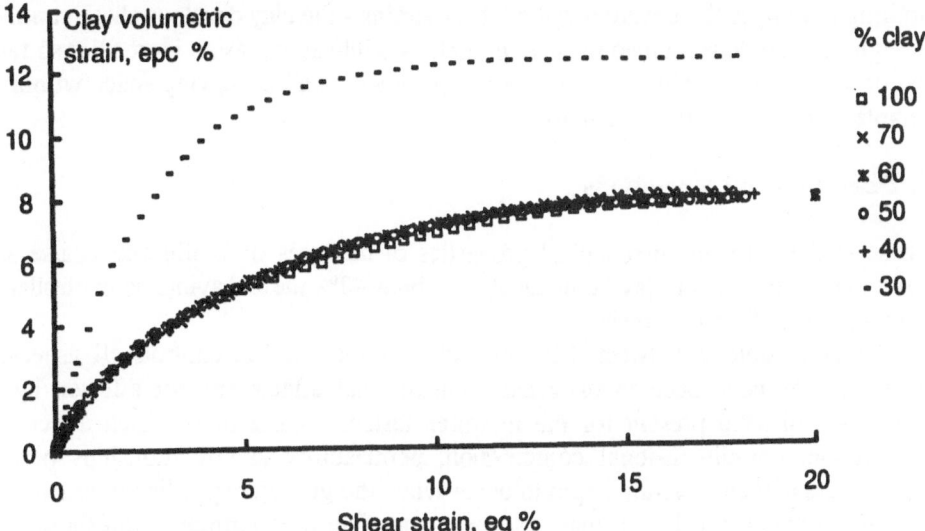

Figure 6c. Drained triaxial compression of mixtures: clay volumetric strains, normally consolidated samples.

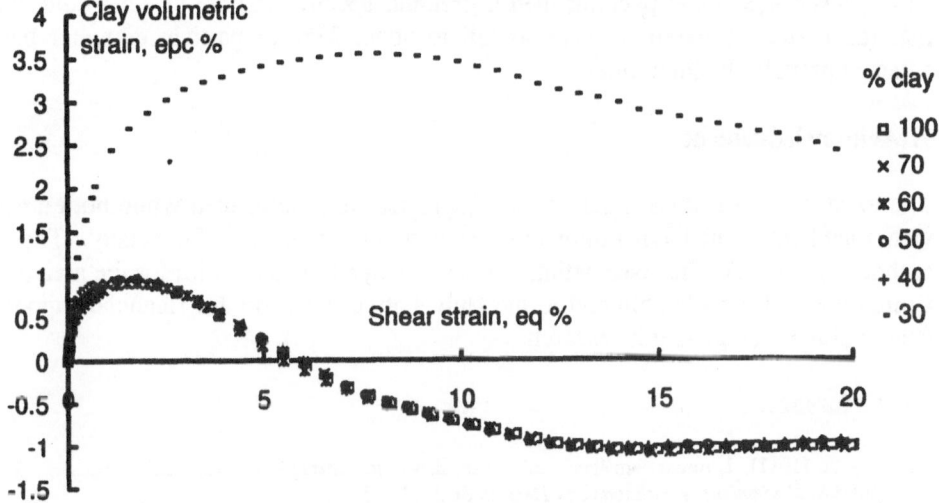

Figure 6d. Drained triaxial compression of mixtures: clay volumetric strains, overconsolidation ratio=4.

It might seem reasonable to propose that the presence of the sand particles should influence the development of shear strains in the mixture as well as volumetric strains. A 'clay axial strain' can be defined in which part of the length of the sample is imagined as consisting of rigid sand material so that the axial strains should be

68 G. V. KUMAR AND D. MUIR WOOD

calculated with reference to a reduced sample height. However, use of clay axial strain destroys the uniqueness of the relationships between shear stress and shear strain and between volumetric strain and shear strain. The presence of the sand apparently does nothing to impede the development of shear strains - the clay develops shear strains as though the sample contained no sand particles - while acting as an inert filler so far as the development of volumetric strains is concerned - and occupying space within the sample which affects the volumetric response.

8. Discussion and conclusions

These studies of some mechanical properties of mixtures of kaolin and coarse sand have indicated that for clay contents above about 40% the behaviour is controlled by the properties of the clay phase alone.

For clay contents between 100 and 40% the clay matrix controls all aspects of response that have been investigated. Distortional effects are not affected by the proportion of sand present for the mixtures tested. Interactions which affect cone penetration, one-dimensional compression, permeability and relationships between shear stress and shear strain begin to occur when the granular specific volume reaches a value of about 2.5, but it does not appear possible to determine a threshold value which is valid for all the mechanical configurations that have been studied. If it were supposed that the coarse sand consisted of spherical particles arranged according to close packed hexagonal packing then a granular specific volume of 2.5 would imply that the particle separation needs to fall to about 23% of particle diameter before adjacent particles begin to interact.

9. Acknowledgement

The experimental studies described in this paper were performed while both authors were members of the Civil Engineering Department at Glasgow University. The first author is grateful to the Association of Commonwealth Universities for the award of a Commonwealth Scholarship and to the University of Glasgow for financial support to enable him to complete this research.

10. References

Atterberg, A. (1911) Lerornas forhållande till vatten, deras plasticitetsgränser och plasticitetsgrader. *Kungl. Lanbruks akademiens Handlingar och Tidskrift* 50 2, 132-158.

British Standards Institution (1991) *Methods of test for soils for civil engineering purposes.* BS1377:1991.

Georgiannou, Burland and Hight (1990) The undrained behaviour of clayey sands in triaxial compression and extension. *Géotechnique* 40 3, 431-449.

Kumar, G.V. (1996) *Some aspects of the mechanical behaviour of mixtures of kaolin and coarse sand.* PhD thesis, University of Glasgow.

Little, J.A., Muir Wood, D., Paul, M.A. and Bouazza, A. (1992) Some laboratory measurements of permeability of Bothkennar clay in relation to soil fabric. *Géotechnique* 42 2, 355-361.

Lupini, J.F., Skinner, A.E. and Vaughan, P.R. (1981) The drained residual strength of cohesive soils. *Géotechnique* 31 2, 181-213.

Muir Wood, D. (1990) *Soil behaviour and critical state soil mechanics.* Cambridge University Press.

CONSTITUTIVE MODELLING OF RESPONSE OF GRANULAR INTERFACE LAYERS

Z. MRÓZ

Institute of Fundamental Technological Research,
Świętokrzyska 21, 00-049 Warsaw, Poland.

Abstract

The constitutive model of the interface layer is first discussed using the concept of the critical state model. Next, the dynamic dilatancy effect occuring in rapid flows of granular materials is described by applying the viscoplastic model. It is assumed that the critical state line is a unique function of actual material density. Static and kinematic conditions at the interface are discussed and the effect of strain rate on dilatancy or transverse pressure is analysed.

1. Introduction

Granular material flow in quasistatic conditions is usually described by the elasto-plastic or rigid-plastic material models. The hardening or softening response is associated with density variation and the critical state is specified by a unique relation between the effective pressure and density or void ratio. During the progresive flow, localised deformation zones develop with the associated dilatancy and softening phenomena. The treatment of boundary-value problems assuming existence of localised shear zones was discussed by Michalowski (1990) and Mróz and Maciejewski (1994) who treated these zones as material interfaces with strain softening. The phenomenon of flow mode switching in problems of punch or wall penetration into a soil was exhibited and confirmed experimentally. The constitutive models for interface layer was proposed by Mróz and Jarzębowski [1994],assuming contact state evolution toward a critical steady state and following the previous formulations of anisotropic hardening rules for sands,cf. Mróz and Norris [1982], Mróz and Zienkiewicz [1984] and Jarzębowski and Mróz [1988].

69

N. A. Fleck and A. C. F. Cocks (eds.),
IUTAM Symposium on Mechanics of Granular and Porous Materials, 69–82.
© 1997 *Kluwer Academic Publishers.*

Besides the plasticity based models, an alternative approach was developed in the description of the rapid granular material flow. The kinetic theory of gases was modified and adapted in deriving the governing equations. Hard sphere models previously developed for dense fluids were used and the dissipative effect due to contact friction was accounted for, Savage and Jeffrey (1981), Jenkins and Savage (1983), Lun et al. (1984), Jenkins and Richman (1985), Goldsthein and Shapiro (1995), et al. The general representation of constitutive models for granular materials was presented by Goodman and Cowin [1971, 1972] and modified and improved by numerous authors, for example, Yalamanchili et al. [1994]. The usual assumption is that Cauchy stress σ depends on the volume fraction of particles v gradient ∇v, and the strain rate tensor $\dot{\varepsilon}$, thus

$$\sigma = \hat{\sigma}(v, \nabla v, \dot{\varepsilon})$$

The representation indicates strain rate dependence and results in specific forms containing numerous material parameters.

In this paper, an alternative approach will be presented. Neglecting density gradient dependence, we assume that the equilibrium states correspond to the critical state model equations, but the viscoplastic flow develops for stress states exceeding the instantaneous yield surface. The static and dynamic dilatancy effects will be predicted and identification of material parameters willl be obtained from tests of rapid shear of granular materials in annular shear apparatus. The model can be applied to study flows of granular materials or suspensions and also in the analysis of post-liquefaction response of soils when rapid shear deformation occurs in localized zones.

2. Critical state model of the interface layer.

Consider an interface surface element S specified by the unit normal vector $\mathbf{n}$. The traction vector $\mathbf{t} = \sigma\mathbf{n}$ can be decomposed into normal and tangential components

$$\mathbf{t} = \mathbf{t}_N + \mathbf{t}_T \tag{1}$$

where

$$\mathbf{t}_N = (\mathbf{n} \otimes \mathbf{n})\mathbf{t} = (\mathbf{t} \cdot \mathbf{n})\mathbf{n} = -\sigma_n \mathbf{n}$$

$$\tag{2}$$

$$\mathbf{t}_T = (1 - \mathbf{n} \otimes \mathbf{n})\mathbf{t} = \mathbf{t} - t_N \mathbf{n} = \tau_n \mathbf{m}$$

where $\mathbf{m}$ is the unit vector in the tangential direction. Here σ_n denotes the compressive normal stress to the interface and τ_n is the shear stress. Denote by $\overline{v}$

the relative velocity at the interface

$$\overline{\mathbf{v}} = \mathbf{v}^+ - \mathbf{v}^- \tag{3}$$

where $\mathbf{v}^+$ and $\mathbf{v}^-$ denote the velocities on both sides of the interface. Decompose $\overline{\mathbf{v}}$ into normal and tangential components, namely

$$\overline{\mathbf{v}}_N = (\mathbf{n} \otimes \mathbf{n})\overline{\mathbf{v}} = (\overline{\mathbf{v}} \cdot \mathbf{n})\mathbf{n} = \overline{v}_N \mathbf{n}$$

$$\overline{\mathbf{v}}_T = (1 - \mathbf{n} \otimes \mathbf{n})\overline{\mathbf{v}} = \overline{\mathbf{v}}_N \mathbf{n} = \overline{v}_T \mathbf{m} \tag{4}$$

Let us denote the compactive normal velocity component by $v_N = -\overline{v}_N$. We use the notation $v_N = -\overline{v}_N$, $v_T = \overline{v}_T$ and decompose the relative velocities into elastic and inelastic components, thus

$$v_N = v_N^e + v_N^p, \qquad v_T = v_T^e + v_T^p \tag{5}$$

The interface surface element can now be associated with the interface layer of thickness h several times greater than the average grain size, but small compared to a typical dimension of the problem.

Denote by $\sigma_n, \tau_n, \sigma,$ the stress components within the layer and by $\dot{\varepsilon}_n, \dot{\gamma}_n, \dot{\varepsilon}_t$ the respective strain rate components. The external components σ_n, τ_n are generated by the traction vector $\mathbf{t}$, the internal component σ_t acts within the interface layer on the plane normal to the interface. Similarly, the external strain rate components $\dot{\varepsilon}_n, \dot{\gamma}_n$ are specified in terms of relative velocity vector, namely

$$\dot{\gamma}_n = \frac{v_T}{h}, \qquad \dot{\varepsilon}_n = \frac{v_N}{h} \tag{6}$$

and the internal component $\dot{\varepsilon}_t$ specifies the stretching of the interface layer. Let us note that the external strain rate components do not induce any interface stretching. The specific rate of dissipation per unit area of the interface equals

$$D = (\sigma_n \dot{\varepsilon}_n + \tau_n \dot{\gamma}_n)h = \sigma_n v_N + \tau_n v_T \tag{7}$$

In formulating the constitutive equations of the interface layer, two approaches can be distinguished

i)   the response of the layer is governed only by the external stress components associated with tractions σ_n, τ_n and the conjugate strain rates $\dot{\varepsilon}_n, \dot{\gamma}_n$, or relative velocities v_N, v_T.

ii) the response of the layer is also affected by the internal stress and strain components $\sigma_t, \varepsilon_t,$ playing the role of internal state variables.

In this paper, we shall be concerned with interfaces of the first kind for which the surface tractions and relative velocities specify the interface layer response.

Let us now discuss the density hardening model implying existence of the critical state. In the plane σ_n, τ_n consider a set of elliptical yield loci depending on varying material density and generated by similarity mapping with respect to the origin O, Fig. 1. The yield locus has the form

$$F(\sigma_n, \tau_n, c, a) = \left[(\sigma_n - c)^2 + \frac{\tau_n^2}{m^2} \right]^{\frac{1}{2}} - a = 0 \qquad (8)$$

where m, a, and c are the geometrical parameters, namely a is the larger semiaxis, c is the position of ellipse centre and m is the ratio of semiaxes. The similarity coefficient α is specified as follows

$$\alpha = \frac{a-c}{a+c} \quad \text{or} \quad c = \frac{1-\alpha}{1+\alpha} a = ka, \quad 0 \le k \le 1 \qquad (9)$$

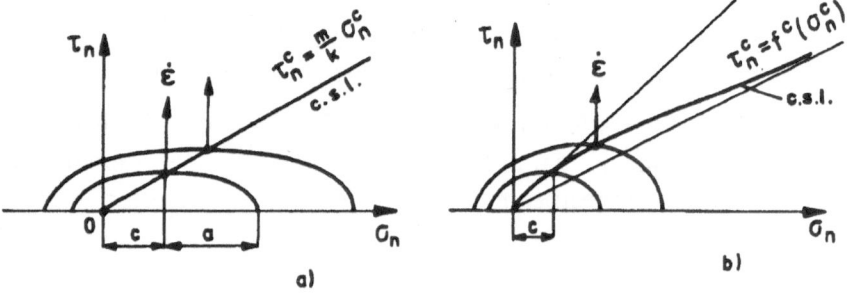

Figure 1 Yield surface and critical state lines for varying material density :
    a) c=ka, k=const. b) $c = k(\rho)a$. The notation c.s.l. means critical state line.

Consider first the rigid-plastic model and assume that m=const. and α=const. Then $c = ka$ and only one material function $a = a(\rho)$ specifies the yield condition . Here ρ denotes the bulk density of a granular material. Note that for $\alpha = 0$ there is $k = 1$ and $c = 0$ so the yield ellipses pass through the origin O. On the other hand, when $\alpha = 1$, then $k = 0$, $c = a$ and ellipse centres coincide with the origin O.

The associated flow rule now provides

$$\dot{\gamma}_n = \frac{v_T}{h} = \frac{1}{h}\dot{\lambda}\frac{\partial F}{\partial \tau_n} = \frac{\dot{\lambda}}{h}\frac{\tau_n}{am^2}, \qquad \dot{\lambda} > 0$$

(10)

$$\dot{\varepsilon}_n = \frac{v_N}{h} = \frac{1}{h}\dot{\lambda}\frac{\partial F}{\partial \sigma_n} = \frac{\dot{\lambda}}{h}\frac{(\sigma_n - ka)}{a}$$

where

$$\dot{\lambda} = \left[v_N^2 + m^2 v_T^2\right]^{\frac{1}{2}} = h\left[\dot{\varepsilon}_n^2 + m^2\gamma_n^2\right]^{\frac{1}{2}}$$

(11)

The dissipation function has the form

$$D = a(\rho)\left\{kv_N + \left[v_N^2 + m^2 v_T^2\right]^{\frac{1}{2}}\right\}$$

(12)

where $a(\rho)$ is a function of density that can be assumed

$$a = a_m\left(\frac{\rho - \rho_{min}}{\rho_{max} - \rho_{min}}\right)^p$$

(13)

where ρ_{min} and ρ_{max} denote minimum and maximum density values for a granular material and a_m corresponds to the maximal density ρ_{max}. The rate of variation of a is expressed as follows

$$\dot{a} = a'(\rho)\rho\dot{\varepsilon}_n = a'(\rho)\rho\dot{\lambda}\frac{\partial F}{\partial \sigma_n}$$

(14)

where $a' = da/d\rho > 0$. Using (14), the consistency condition imposed on (8) provides the expression of $\dot{\lambda}$ in terms of the stress rate, thus

$$\dot{\lambda} = \frac{\dfrac{\partial F}{\partial \tau_n}\dot{\tau}_n + \dfrac{\partial F}{\partial \sigma_n}\dot{\sigma}_n}{\left(\dfrac{\partial F}{\partial \sigma_n}k + 1\right)a'(\rho)\dfrac{\partial F}{\partial \sigma_n}} = \frac{\dfrac{\tau_n}{m^2}\dot{\tau}_n + (\sigma_n - ka)\dot{\sigma}_n}{H}$$

(15)

where

$$H = \frac{\left[k\sigma_n + a(1 - k^{2)}\right]\left[\sigma_n - ka\right]}{a} \rho a'(\rho) \tag{16}$$

is the hardening modulus. Let us note that $H > 0$ when $\sigma_n > ka$, $H < 0$ when $\sigma_n < ka$ and the critical state line $H=0$ corresponds to $\sigma_n^c = c = ka$, $\tau_n^c = ma = \frac{m}{k}\sigma_n^c$. In Fig. 1 it is represented by a straight line.

A more general model can be proposed by assuming that k depends on the varying material density, thus

$$c = k(\rho)a(\rho), \ \sigma_n^c = c(\rho), \ \tau_n^c = ma(\rho) \tag{17}$$

Assume that $k = \rho / \rho_{max}$. In view of (13), the critical state line is then described by the equation

$$m\sigma_n = \tau_n \left[\frac{\rho_{min}}{\rho_{max}} + \left(1 - \frac{\rho_{min}}{\rho_{max}}\right)\left(\frac{\tau_n}{ma_o}\right)^{\frac{1}{p}} \right] \tag{18}$$

and is represented by a curve in the (τ_n, σ_n)-plane, Fig. 1.b.

3. Dynamic dilatancy description

Consider now the constitutive relations for a granular material applicable in the range of fairly large rates of flow where the dynamic collision forces between grains play equal or even major role with respect to contact forces. The granular material then exhibits viscous effects with stress dependent on rate of straining and dilatancy increasing with the strain rate. The experiments carried out by Savage and Mc Keown [1983] in the rotating shear apparatus clearly exhibited this effect. The constant volume test demonstrated that the normal stress increases with the rate of straining. On the other hand, the tests carried at constant normal stress indicated growth of volume with the rate of shearing. To describe this dynamic dilatancy effect, let us assume the viscoplastic material model, regarding the critical state model of the previous section as the static model representing material response at vanishing strain rate.

Consider now the viscoplastic model for which the static yield condition is specified by (8). Assume the dynamic yield surface in the form

$$F_d = \left[\left(\sigma_n - c_d \right)^2 + \frac{\tau_n^2}{m^2} \right]^{\frac{1}{2}} - a_d = 0 \qquad (19)$$

This surface has a similar shape to the static surface (8) but its centre is now at $(c_d, 0)$ and its major semiaxis equals a_d. Assuming the similarity mapping (9) with constant k, we have $c_d = k a_d$, so a_d can be assumed as the size parameter of the dynamic yield surface. Assume the viscoplastic flow rule in the form, cf. Perzyna (1971)

$$\tilde{\mu} \dot{\varepsilon}_n = \frac{\partial F_d}{\partial \sigma_n} = \frac{\sigma_n - c_d}{a_d} \left(\frac{a_d}{a} - 1 \right)^n, \quad \tilde{\mu} \dot{\gamma}_n = \frac{\partial F_d}{\partial \tau_n} = \frac{\tau_n}{m^2 a_d} \left(\frac{a_d}{a} - 1 \right)^n, \qquad (20)$$

where $\tilde{\mu} = \mu h$ and μ, n are material parameters. Note that $\dot{\varepsilon}_n = \dot{\gamma}_n = 0$ when $a_d = a$, and we pass to the static case. From (20) it follows that

$$\frac{a_d}{a} = 1 + \tilde{\mu}^{\frac{1}{n}} \left(\dot{\varepsilon}_n^2 + m^2 \dot{\gamma}_n^2 \right)^{\frac{1}{2n}} = 1 + \tilde{\mu}^{\frac{1}{n}} \dot{\varepsilon}_e^{\frac{1}{n}} \qquad (21)$$

Thus $\dot{\varepsilon}_e = \left(\dot{\varepsilon}_n^2 + m^2 \dot{\gamma}_n^2 \right)^{\frac{1}{2}}$ vanishes when $a_d = a$.

3.1 SHEAR TEST AT CONSTANT h AND SPECIFIED $\dot{\gamma}_n$

Consider first the simple shear program for which $\dot{\varepsilon}_n = 0$. This test corresponds to a fixed thickness h of the shear layer when the shear strain is imposed, and the normal stress σ_n is measured. From (20) and (21), we have

$$\sigma_n^c = c_d = k a_d = k a(\rho) \left[1 + \tilde{\mu}^{\frac{1}{n}} \dot{\varepsilon}_e^{\frac{1}{n}} \right] = k a(\rho) \left[1 + \left(\tilde{\mu} m \dot{\gamma}_n \right)^{\frac{1}{n}} \right]$$

$$\tau_n^c = m a_d = m a(\rho) \left[1 + \left(\tilde{\mu} m \dot{\gamma}_n \right)^{\frac{1}{n}} \right]$$

$$(22)$$

and the stress state is represented by the static critical state line

$$\tau_n^c = \frac{m}{k}\sigma_n^c \qquad (23)$$

In other words, the constant volume shear test is represented by the critical state line but the values of both normal and shear stresses increase with the shearing strain rate. To specify the exponent n, the steady shear and normal stresses should be plotted against the shear rate, Fig.2

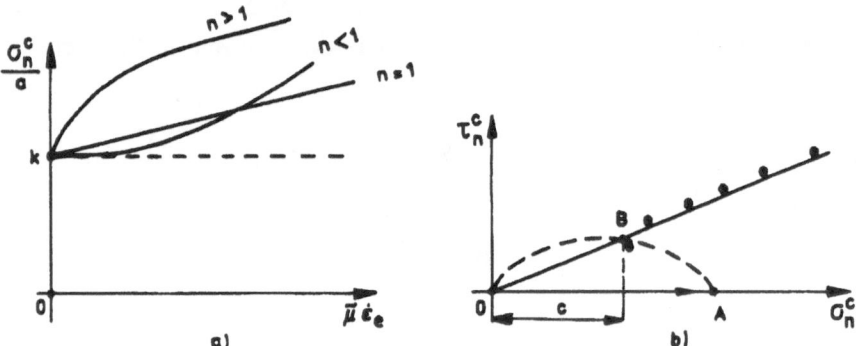

Figure. 2   a) Dependence of the critical normal stress on strain rate, b) experimental verification of the critical state line. Data taken from Savage and Mc Keown [1983].

The tests data carried out by Savage and Mc Keown [1983] in a rotating shear apparatus for constant height of shear layer provide us the values of n, namely, for glass $n=0.50$, beads for polystyrene beads $n=0.55$. The measured values of shear and normal stress at the steady state provide nearly linear relationship between τ_n^c and σ_n^c, Fig.2b.

3.2. SHEAR TEST AT CONSTANT σ_n AND SPECIFIED $\dot{\gamma}_n$

Consider now the shear test program carried out at a specified value of $\sigma_n = const.$ but with the varying height of the shear layer. From (20), we now obtain

$$\frac{\dot{\varepsilon}_n}{m^2\dot{\gamma}_n} = \frac{\sigma_n - c_d}{\tau_n}, \qquad \frac{\tau_n}{m^2} = \tilde{\mu}\dot{\gamma}^n a_d\left(\frac{a_d}{a} - 1\right)^{-n} \qquad (24)$$

The first equation (24) specifies the dilatancy relation. When $\sigma_n > c_d$, the shear deformation is accompanied by compaction, $\dot{\varepsilon}_n > 0$. However, for larger strain rates, there is $\sigma_n < c_d$ and $\dot{\varepsilon}_n < 0$. The dynamic dilatancy effect increases with the strain rate and induces the subsequent material softening. Consider, for instance, the material of density ρ_o for which the static value of a_0 is specified by (13) and the corresponding static normal stress equals $\sigma_{n0} = c + a = (k + 1)a_0$. From (13), we have

$$\rho_o = \rho_{min}\left[1 + (\beta - 1)\left(\frac{a_0}{a_m}\right)^{\frac{1}{p}}\right] = \rho_{min}\left[1 + (\beta - 1)\left(\frac{\sigma_{n0}}{\sigma_{nm}}\right)^{\frac{1}{p}}\right] \qquad (25)$$

where $\beta = \rho_{max} / \rho_{min}, \sigma_{nm} = (k + 1)a_m$.

Assume now that the stress is reduced to the critical value $\sigma_n^c = c_0 = ka_0$ and the granular layer is subjected to shear with the specified strain rate. The respective value of a_d is then specified by (21) and $c_d = ka_d$. The variation of τ_n and $\dot{\varepsilon}_n$ can be determined from (24). Fig. 3a presents typical shear curves at $\sigma_n = c$ obtained for different shear strain rates $\dot{\gamma}_n$.

Due to dynamic dilatancy effect, the shear stress curves exhibit softening and asymptotically tend to critical state. Fig. 3.b presents the respective curves for shear programs carried out at normal stress $\sigma_n > c_0$. For small strain rates the material consolidates, similarly as for quasistatic shear. However, for larger strain rate values the dynamic dilatancy and softening effect occurs.

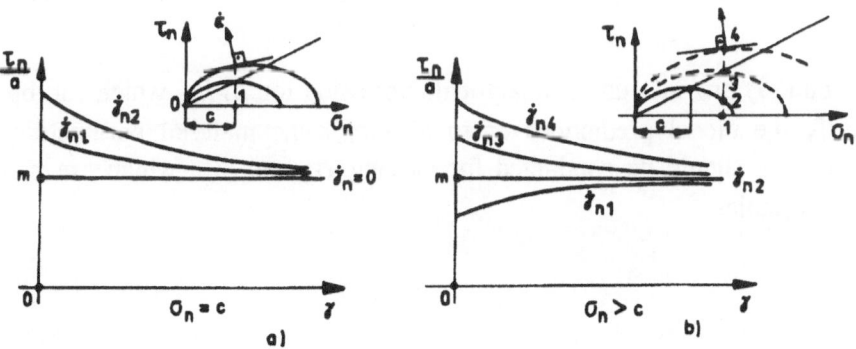

Figure 3. Shear curves for varying strain rates: a) for the case $\sigma_n = c$, b) for the case $\sigma_n > c$

Since the steady state is again represented by the critical state line, we can determine the variation of initial height of the shear layer due to dynamic dilatancy. Since at the critical state there is $\sigma_n = \sigma_n^c = ka_d$, $\dot{\varepsilon}_n = 0$, we can write

$$\frac{\sigma_n}{k} = a_d = a\left[1 + \left(\tilde{\mu}m\dot{\gamma}_n\right)^{\frac{1}{n}}\right], \quad ka = \frac{\sigma_n}{1 + \left(\tilde{\mu}m\dot{\gamma}_n\right)^{\frac{1}{n}}} \tag{26}$$

so the corresponding material density equals

$$\frac{\rho}{\rho_{min}} = \left[1 + (\beta - 1)\left(\frac{a}{a_m}\right)^{\frac{1}{p}}\right] = 1 + (\beta - 1)\frac{\left(\frac{\sigma_n}{ka_m}\right)^{\frac{1}{p}}}{\left[1 + \left(\tilde{\mu}m\dot{\gamma}_n\right)^{\frac{1}{n}}\right]^{\frac{1}{p}}} \tag{27}$$

For the initial density ρ_0 we have

$$\frac{\rho_0}{\rho_{min}}\left[1 + (\beta - 1)\left(\frac{a_0}{a_m}\right)^{\frac{1}{p}}\right] \tag{28}$$

In view of (27) and (28), the ratio of steady state and initial shear layer thickness can be presented in the form

$$\frac{h}{h_o} = \frac{\rho_o}{\rho} = \frac{1 + (\beta - 1)\left(\frac{a_0}{a_m}\right)^{\frac{1}{p}}}{1 + (\beta - 1)\left(\frac{a}{a_m}\right)^{\frac{1}{p}}} = \left[1 + \left(\tilde{\mu}m\dot{\gamma}_n\right)^{\frac{1}{n}}\right]^{\frac{1}{p}} \frac{1 + (\beta - 1)\left(\frac{a_0}{a_m}\right)^{\frac{1}{p}}}{\left[1 + \left(\tilde{\mu}m\dot{\gamma}_n\right)^{\frac{1}{n}}\right]^{\frac{1}{p}} + (\beta - 1)\left(\frac{\sigma_n}{ka_m}\right)^{\frac{1}{p}}}$$

$$\tag{29}$$

Equation (29) provides us the analytical expression for h/h_o which can be used to verify the model predictions or to identify some material parameters. The measurement should be conducted for varying σ_n and $\dot{\gamma}_n$ which are now the control variables.

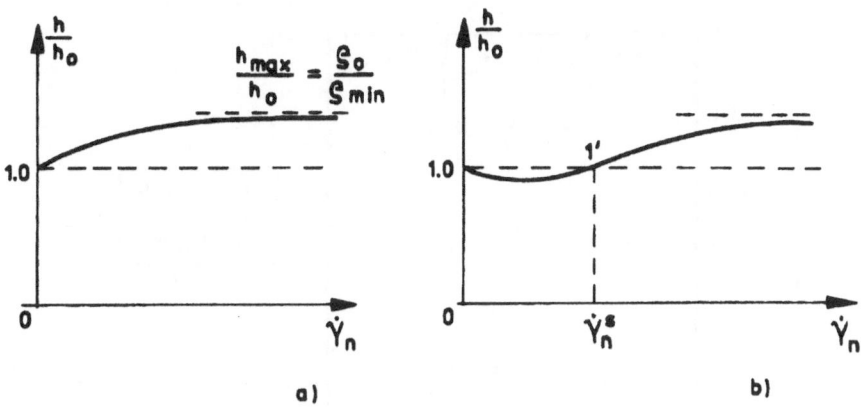

Figure 4. Layer thickness evolution with increasing strain rate $\dot{\lambda}_n$ for steady states: a) for $\sigma_n \langle ka_0$, b) for $\sigma_n \rangle ka_0$.

Fig.4 presents the typical dependence of h/h_o on the rate of shear $\dot{\gamma}_n$. When $\sigma_n \leq c_0 = ka_0$, the layer shearing is accompanied by the continuing dilatancy and h/h_o increases monotonically to its asymptotic value $h_{max}/h_o = \rho_{max}/\rho_{min}$, cf. Fig.4a. On the other hand, when $\sigma_n > ka_0$, then for small rates of shear the layer consolidation occurs and for large values of $\dot{\gamma}_n$, the material dilates, Fig.4b. The switching point 1' from consolidation to dilatancy corresponds to the strain rate

$$\tilde{\mu} m \dot{\gamma}_n^s = \left(\frac{\sigma_n}{ka_0} - 1 \right)^n \tag{30}$$

The constant normal stress experiments of shear of suspensions of 2 mm glass spheres in water/glycerin mixture for three values of normal stress were carried out by Kyotomaa and Prasad (1993). Fig.5 presents the solid fraction variation for increasing rates of shear. It is seen that the solid fraction first increases and for a sufficiently large strain rate value, it starts to decrease. The increasing normal pressure shifts the diagrams upwards, thus generating increasing solid fraction values. These experimental data correspond qualitatively to the present model predictions, though the full identification of model parameters is difficult in view of incomplete empirical data.

Z. MRÓZ

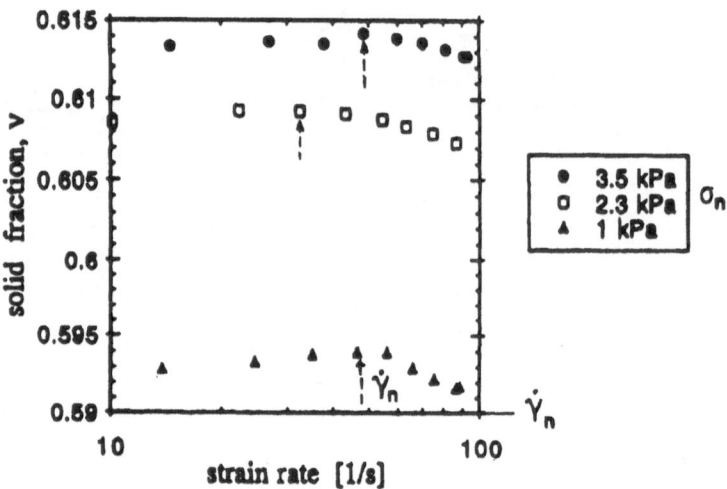

Figure 5. Variation of solid fraction ν with increasing rates of shear at constant σ_n (after Kyotomaa and Prasad (1993)).

3.3. COMPOSITE SHEAR LAYER, $h = h_0 + h_s = \text{const}$

Consider now a more complex case when the granular shear layer interacts with an elastic layer of specified normal compliance K_n. The thickness of the granular layer is denoted by h_m and that of elastic layer by h_s. The initial state is specified by the normal stress σ_{n0} and the thickness $h_0 = h_{m0} + h_{s0}$. The subsequent shearing process occurs at constant $h = h_0$, so we have

$$\Delta h_m + \Delta h_s = 0 \tag{31}$$

where Δ denote the thickness increment.
Denoting the normal stress during shearing process by σ_{nd}, we have

$$\frac{\Delta h_s}{h_{s0}} = \frac{\Delta h}{h_{m0}} \frac{h_{m0}}{h_{s0}} = \frac{\sigma_{nd} - \sigma_{n0}}{K_n} \tag{32}$$

or

$$\frac{\Delta h_s}{h_{m0}} = \frac{\sigma_{nd} - \sigma_{n0}}{K_n}, \qquad \overline{K}_n = K_n \frac{h_{m0}}{h_{s0}} \tag{33}$$

Using the relations

$$\frac{\rho}{\rho_{min}} = 1 + (\beta - 1)\left(\frac{a}{a_m}\right)^{\frac{1}{p}}, \qquad \frac{\rho_0}{\rho_{min}} = 1 + (\beta - 1)\left(\frac{a_0}{a_m}\right)^{\frac{1}{p}} \qquad (34)$$

and (31), we obtain

$$\frac{h_m}{h_{om}} = \frac{\rho_0}{\rho} = \frac{1 + (\beta - 1)\left(\dfrac{a_0}{a_m}\right)^{\frac{1}{p}}}{1 + (\beta - 1)\left(\dfrac{a}{a_m}\right)^{\frac{1}{p}}} = 1 + \frac{\Delta h_m}{h_{m0}} = 1 - \frac{\sigma_{nd} - \sigma_{ns}}{K} \qquad (35)$$

Equation (35) specifies the normal stress σ_{nd} during the shearing process. Let us note that now $\sigma_{nd} = \sigma_{nd}(\sigma_{n0}, \rho_0, \dot{\gamma}_n, \overline{K})$. Equation (35) can be written in a more explicit form

$$\left[1 + (\tilde{\mu}m\dot{\gamma}_n)^{\frac{1}{n}}\right]^{\frac{1}{p}} \frac{1 + (\beta - 1)\left(\dfrac{a_0}{a_m}\right)^{\frac{1}{p}}}{\left[1 + (\tilde{\mu}m\dot{\gamma}_n)^{\frac{1}{n}}\right]^{\frac{1}{p}} + (\beta - 1)\left(\dfrac{\sigma_{nd}}{ka_m}\right)^{\frac{1}{p}}} - 1 - \frac{\sigma_{nd} - \sigma_{ns}}{\overline{K}} \qquad (36)$$

4. Concluding remarks.

In this paper, we have introduced the concept of static and dynamic dilatancy and formulated the respective constitutive equations. The present formulation can be used in description of flow of a granular material with fairly high rates where dynamic interaction modes between grains become significant. The localised modes within interface layers can therefore be treated using the present formulation. A more extensive treatment will be presented separately with model calibration data and application examples.

5.   References

Drescher, A. and Michałowski, R.L.(1984). Density variation in pseudo-steady plastic flow of granular media. *Geotechnique,* **34**, 1-10.

Goodman, M.A. and Cowin, S.C. (1971). Two problems in the gravity flow of granular materials. *J.Fluid Materials,* **45**, 321-339.

Goodman, M.A. and Cowin, S.C. (1972). A continuum theory for granular materials. *Arch. Rat. Mech. Anal.,* **44**, 249-266.

Goldsthein, A. and Shapiro, M. (1995). Mechanics of collisional motion of granular materials, Part I. General hydrodynamic equations, *J. Fluid Mech.,* **282**, 75-114.

Jenkins, J.T. and Savage, S.B. (1983). A theory for the rapid flow of identical, smooth nearly elastic, circular disk. *J. Fluid Mech.,* **171**, 53-69.

Kytomaa, H.K. and Prasad, D. (1993). Transition from quasistatic to rate dependent shearing of concentrated suspensions. *Powders and Grains 93, Proc. 2-nd Int. Conf. Micromechanics of Granular Media,* (Ed. C.Thorton and A.A.Balkema0, 281-287.

Lun, C.K.K., Savage, S.B., Jeffrey, D.J. and Chepurny, N. (1984). Kinetic theories for granylar flow: inelastic particles in Couette flow and slightly inelastic particles in a general flow field.

Yalamanchili, R.C., Gudha, R. and Rajagopal, K.R. (1994). Flow of granular materials in a vertical channel under the action of gravity. *Powder Technology,* **81**, 65-73.

Michałowski, R. L. (1990). Strain localisation and periodic fluctuations in granular flow processes from hoppers. *Geotechnique,* **40**, 389-403.

Mróz, Z. and Norris, V.A. (1982). Elastoplastic and viscoplastic constitutive models for soils with application to cyclic loading. In *Soil Mechanics, Transient and Cyclic Load,* Chapter 8.(Ed. G.N. Pande and O.C. Zienkiewicz). J.Wiley.

Mróz, Z. and Zienkiewicz, O.C.(1984). Uniform formulation of constitutive models for clays and sands. In *Constitutive Equations for Engineering Materials,* (Ed. C.S. Desai, and R. Gallagher). J.Wiley.

Mróz, Z. and Jarzębowski, A. (1994). Phenomenological model of contact slip. *Acta Mech.* **102**, 59-72.

Mróz, Z. and Maciejewski, J. (1994). Post-critical response of soils and shear band evolution. In *Localisation Phenomena in Geomaterials* (Ed. R. Chambon et al.). Balkema.

Pietruszczak, S. and Mróz, Z. (1982). Finite element analysis of deformation of strain-softening materials. *Int. J. Num. Meth. Eng.,* **17**, 327-334.

Perzyna, P. (1971). Thermodynamic theory of viscoplasticity. *Advances in Applied Mech.,* **11**, 313-354.

Savage, S. B. and Mc Keown, S. (1983). Shear stresses developed during rapid shear of concentrated suspensions of large spherical particles between concentric cylinders. *J. Fluid Mech.,* **127**, 453-472.

Zienkiewicz, O.C. and Mróz, Z. (1984). Generalized plasticity formulation and application to geomechanics. In *Mechanics of Engineering Materials,* (Ed. C.S.Desai, and R. Gallagher). J.Wiley 655-680.

A CHARACTERISTIC STATE PLASTICITY
MODEL FOR GRANULAR MATERIALS

S. KRENK
Division of Mechanics
Lund Institute of Technology
Lund University
Box 118, S-221 00 Lund
Sweden

1. Introduction

The irreversible nonlinear deformation of granular materials is dominated by the effect of friction forces between grains, and therefore the ratio between shear stresses and mean stress plays a dominant role. Within the framework of plasticity theory this leads to a family of self-similar yield surfaces. Classical critical state theory was developed in terms of a two-dimensional stress space with mean stress p and maximum shear stress $\frac{1}{2}q$. The ultimate stress states are located on the critical line $q/p = M$ corresponding to a state of plastic shear without dilatation, (Schofield and Wroth, 1968). While representing some of the basic features of granular materials well, the critical state theory has shortcomings in its representation of the effect of triaxial stress states, seen e.g. in the triangular shape of the failure envelope, the representation of dilatation at failure, and the need for a tension cut-off of the yield surface.

In the present theory the concepts from classical critical state theory are generalized to provide a hardening plasticity theory in terms of stress invariants. The theory is formulated in terms of a family of smooth self-similar yield surfaces that satisfy the no-tension condition. The shape of the yield surface is determined by a single parameter, specifying a characteristic ratio of shear to mean stress, at which the material behaviour changes from compression to dilatation. While the classical critical state theory identifies this state with the ultimate state, the ultimate state is defined independently in the present theory and may correspond to dilatation.

N. A. Fleck and A. C. F. Cocks (eds.),
IUTAM Symposium on Mechanics of Granular and Porous Materials, 83–94.
© 1997 *Kluwer Academic Publishers.*

The development of plastic strain is controlled by a basically linear work-hardening rule, in which the shear contribution is given a small weight. The relative weight parameter in the hardening rule determines the plastic strain ratio and the shear to mean stress ratio of the ultimate stress states. Thus the plasticity effects of the present theory are specified by only three parameters: a plastic stiffness, the inclination M_c of the characteristic line, and the inclination M_u of the ultimate line.

2. The yield surface

Figure 1a shows an isotropic yield surface in principal stress space $\sigma_1, \sigma_2, \sigma_3$. The stresses are positive in compression, and when the material is assumed to be without cohesion, the yield surface is entirely within the first octant. The yield surface is assumed to be isotropic, and when the yield mechanism is friction, the yield surface will only depend on the ratio between the stress components. With these assumptions only a single yield surface is needed, as it will grow in a self-similar way.

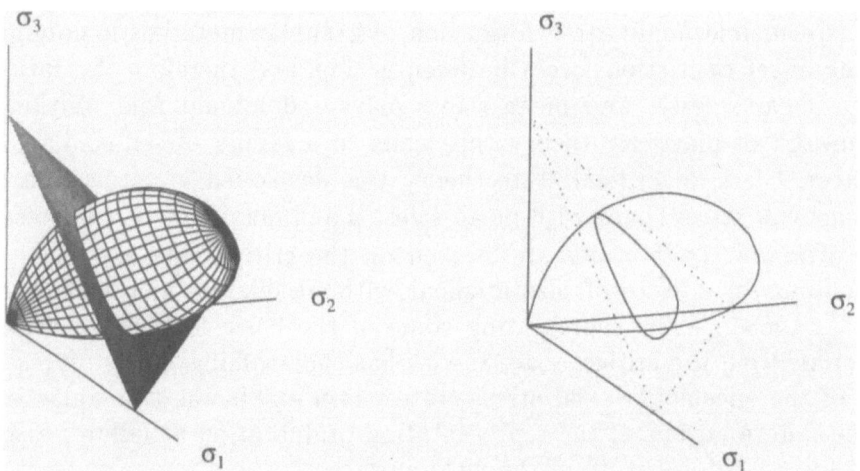

Figure 1. a) Yield surface in principal stress space, b) Generating curves in octahedral and meridian planes.

The yield surface is generated by a family of curves in the octahedral planes, of which one is shown in Fig. 1b. There is experimental evidence that the failure envelope is of rounded triangular shape, e.g. (Lade and Duncan, 1975). The minimum requirement of the octahedral curves therefore is a triangular format permitting the specification of the 'corner' and the midpoint of the 'side' of the triangle. Several such formats are available, e.g. (Lade and Duncan, 1975), (Lade, 1977), (Weidner, 1990) and (Krenk, 1996). These formats all have the common feature that the triangularity is generated via the third invariant of the total or deviatoric stresses. In the next section we describe the structure of the simple cubic format with two parameters, and then introduce a suitable format of the meridian curve to complete the definition of the yield surface.

2.1. THE OCTAHEDRAL CONTOURS

In the formulation of the model it is convenient to use the mean stress p and the deviatoric stresses s_{ij},

$$p = \tfrac{1}{3}\sigma_{ii} \quad , \quad s_{ij} = \sigma_{ij} - p\,\delta_{ij} \tag{1}$$

The simplest form of an octahedral contour that satisfies symmetry with respect to the three principal deviator stress components (s_1, s_2, s_3) is the cubic polynomial

$$(s_1 + d)(s_2 + d)(s_3 + d) = \eta\, d^3 \tag{2}$$

where the stress parameter d determines the size of the circumscribing triangle, and η is a non-dimensional shape parameter, (Krenk, 1996). The family of curves corresponding to parameter values $0 \leq \eta \leq 1$ are illustrated in Fig. 2. For $\eta = 0$ the curve is composed of the three lines $s_j = -d$, $j = 1, 2, 3$. The relevant part is the isosceles triangle corresponding to the corner points $(s_1, s_2, s_3) = (2d, -d, -d)$ etc. and midside points $(s_1, s_2, s_3) = (-d, \tfrac{1}{2}d, \tfrac{1}{2}d)$ etc.. The center $s_1 = s_2 = s_3 = 0$ corresponds to $\eta = 1$, and for any $0 < \eta < 1$ equation (2) defines a convex contour inside the triangle. For small values of η the contour is nearly triangular, and for η close to 1 the curve approaches circular shape. Equation (2) also generates three open branches, located symmetrically inside the exterior corner regions of the lines $s_j = -d$, $j = 1, 2, 3$. These branches are not part of the constitutive model, but it is important to know their existence when developing numerical integration algorithms, where the stress point may temporarily be located outside the yield surface.

The cubic equation (2) is conveniently written in terms of the deviatoric stress invariants

$$J_2 = \tfrac{1}{2} s_{ij} s_{ij} = \tfrac{1}{2}(s_1^2 + s_2^2 + s_3^2) = -(s_2 s_3 + s_3 s_1 + s_1 s_2) \tag{3}$$

and

$$J_3 = \tfrac{1}{3} s_{ij} s_{jk} s_{ki} = s_1 s_2 s_3 \tag{4}$$

When using the fact that the sum of the deviator stresses is zero, equation (2) takes the form

$$J_3 - d\, J_2 + (1 - \eta)\, d^3 = 0 \tag{5}$$

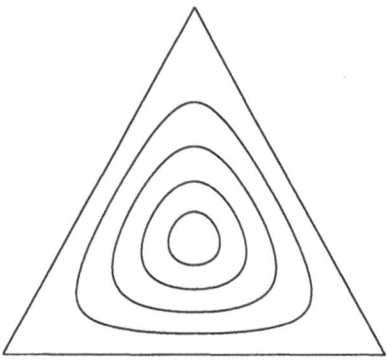

Figure 2.   Octahedral stress contours, $\eta = 0, 0.2 \cdots, 1$.

Thus, in any octahedral plane the contour is given in terms of a size parameter d, defining the circumscribing triangle, and a shape parameter η, determining the *relative* size, and thereby the shape of the contour.

2.2. THE MERIDIAN CURVE

The meridian curves correspond to stress states in which two stress components are equal. In the present notation the compression meridian shown in Fig. 1b corresponds to $\sigma_1 = \sigma_2 < \sigma_3$, and the tension meridian to $\sigma_1 = \sigma_2 > \sigma_3$. The two meridian curves can be described directly, (Krenk, 1996), or by specifying the size of the circumscribing triangle $d(p)$ and the relative size $\eta(p)$ as functions of the mean stress p. In the present theory the size of the circumscribing triangle is assumed to be proportional to the mean stress p, and the shape changes from triangular at $p = 0$ to circular at $p = p_0$ according to a power function, i.e.

$$d(p) = \alpha p \quad , \quad \eta(p) = (p/p_0)^m \tag{6}$$

This leads to the three-parameter yield function

$$f(\sigma) = -J_3 + \alpha p J_2 - (1 - \eta)\alpha^3 p^3 \tag{7}$$

The parameter p_0 is the maximum value of the mean stress, and determines the current size of the yield function. The shape is determined by the non-dimensional parameters $0 < \alpha \le 1$, giving the relative opening at $p = 0$, and the exponent m, determining the relative 'diameter' of the surface. The influence of the two shape parameters is illustrated in Fig. 3.

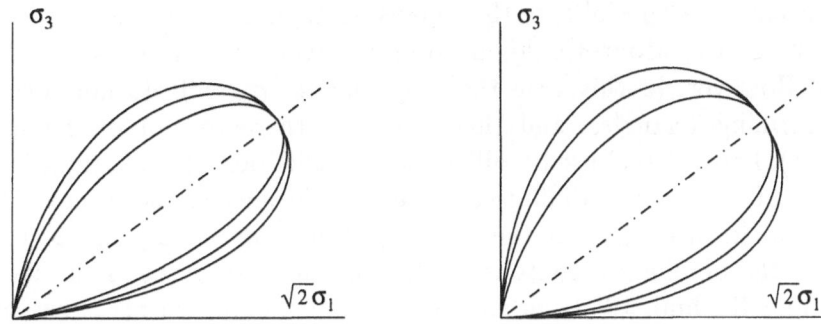

Figure 3. Meridian curves for $m = 0.4, 0.7, 1.0$. a) $\alpha = 0.8$, b) $\alpha = 1.0$.

2.3. THE YIELD SURFACE

A parametric representation of the yield surface $f(\sigma_j) = 0$ is obtained by introducing the angle

$$\cos\theta = \frac{3\sqrt{3}}{2} \frac{J_3}{J_2^{3/2}} \tag{8}$$

The yield condition can then be solved for the radius in the octahedral plane, $s = (s_{ij}s_{ij})^{1/2}$, (Krenk, 1996),

$$s = \frac{\sqrt{\tfrac{3}{2}}\,\gamma\,\alpha\,p}{\cos(\tfrac{1}{3}\arccos(\gamma\cos 3\theta))} \tag{9}$$

where $\gamma^2 = 1 - \eta$. Hereby the principal stresses become

$$\begin{bmatrix} \sigma_1 \\ \sigma_2 \\ \sigma_3 \end{bmatrix} = p \begin{bmatrix} 1 \\ 1 \\ 1 \end{bmatrix} + s\sqrt{\tfrac{2}{3}} \begin{bmatrix} \cos(\theta + \tfrac{1}{3}\pi) \\ \cos(\theta - \tfrac{1}{3}\pi) \\ -\cos\theta \end{bmatrix} \tag{10}$$

Figure 1 corresponds to the parameter values $m = 1$ and $\alpha = 0.8$.

There is evidence that the failure envelope corresponds well to a circumscribing triangle generated by the intersection of the octahedral plane with the coordinate planes $\sigma_j = 0$, (Lade and Duncan, 1975) and (Lade, 1977). In the present model with isotropic hardening this implies $\alpha = 1$. For this case the yield function can be written in terms of the third total stress invariant $I_3 = \det(\sigma_{ij}) = \sigma_1\sigma_2\sigma_3$ as

$$f(\sigma) = -I_3 + \eta\,p^3 \tag{11}$$

This form was used by Lade and Duncan (1975) with $\eta = const$, generating a smooth triangular cone in principal stress space. The general format (7) also contains as a special case the smooth triangular cone interpolating the meridians of the Mohr–Coulomb criterion given by Matsuoka and Nakai (1985). However, in this case the stress planes $\sigma_j = 0$ do not form the circumscribing triangles, and this surface is therefore not of the special form (11). Lade (1977) used a different normalization of the format (11) to describe a family of open failure surfaces. While the general format (7) and the special form (11) cover the full transition from triangular to circular shape of the octahedral contour, some other formats in common use, e.g. (Desai and Hashmi, 1989), become concave at a finite 'corner' curvature.

Kinematic hardening models can be developed from the format (7) with $\alpha < 1$ by replacing the point on the axis, $\sigma_{ij} = p\,\delta_{ij}$ with the point $\sigma_{ij} = \alpha_{ij} + p\,\delta_{ij}$, where α_{ij} is a deviator 'back stress'. Here, we shall concentrate on the isotropic format with $\alpha = 1$.

3. Stress-strain relations

In associated plasticity theory the plastic strain increment is proportional to the gradient of the yield function,

$$d\varepsilon^p = d\chi \frac{\partial f}{\partial \sigma} \tag{12}$$

When the incremental elastic stiffness tensor is denoted $\mathbf{C}$, the total strain increment is

$$d\varepsilon = \mathbf{C}^{-1} d\sigma + \frac{\partial f}{\partial \sigma} d\chi \tag{13}$$

The factor $d\chi$ is determined by plastic hardening via the consistency condition,

$$df = \frac{\partial f}{\partial \sigma} d\sigma - H\, d\chi = 0 \tag{14}$$

where the hardening parameter H is defined as

$$H = -\frac{\partial f}{\partial p_0} \frac{\partial p_0}{\partial \chi} = H_1 H_2 \tag{15}$$

The factor H_1 gives the dependence of the yield surface on the size paramter p_0, while H_2 decribes the plastic hardening via the hardening rule developed in Section 3.1.

The relation between the stress increment $d\sigma$ and the total strain increment $d\varepsilon$ is determined form (13) and (14). The factor $d\chi$ follows from multiplication of (13) with $\partial f/\partial \sigma \mathbf{C}$ and subtraction of (14). The elastoplastic stiffness tensor is then determined from (13) as

$$\mathbf{C}_{ep} = \mathbf{C} - \frac{\mathbf{C}\,\partial f/\partial \sigma\, \partial f/\partial \sigma\, \mathbf{C}}{H + \partial f/\partial \sigma\, \mathbf{C}\, \partial f/\partial \sigma} \tag{16}$$

In the following the material properties are described in tensor format, separating volumetric and deviatoric parts, while the elasto-plastic stiffnes is derived from (16) in six-component vector format.

3.1. STIFFNESS AND HARDENING PARAMETERS

The material stiffness and hardening parameters are formulated in terms of volume and deviator strains,

$$\varepsilon_v = \varepsilon_{ii} \quad , \quad e_{ij} = \varepsilon_{ij} - \tfrac{1}{3}\varepsilon_v \delta_{ij} \tag{17}$$

The elastic part of the deviator strains is assumed to be proportional to the deviator stresses, $s_{ij} = 2Ge^e_{ij}$, with constant shear modulus G.

The volumetric strain of a granular material can be expressed in terms of the specific volume v. When compression is positive the volumetric strain increment is $d\varepsilon_v = -dv/v$. A linear relation between the specific volume and the logarithm of the mean stress is assumed in the elastic and the elasto-plastic states, i.e.

$$d\varepsilon^e_v = \frac{\kappa}{v\,p}dp \quad , \quad d\varepsilon_v = \frac{\lambda}{v\,p}dp \tag{18}$$

where the two non-dimensional flexibility parameters κ and λ are the negative inclination of the v-ln p line in the elastic and the elasto-plastic state, respectively.

The plastic volume strain increment $d\varepsilon^p_v = d\varepsilon_v - d\varepsilon^e_v$ in isostatic elastic-plastic loading is then related to the mean stress increment by

$$dp = \frac{v\,p}{\lambda - \kappa}d\varepsilon^p_v \tag{19}$$

In the case of elasto-plastic isostatic loading p may be replaced by p_0, and thus relation (19) defines the hardening in this specific case.

In the clasical critical state theory hardening is derived from volume changes alone. Thus, hardening stops and unlimited plastic deformtion can take place once the critical state of zero dilatation is reached. In the present model the deviator stresses and strains give an additive contribution to the hardening. Thus, for a general stress path relation (19) is generalized to give the increment of p_0 in terms of a weighted sum of the volumetric and deviatoric parts of the plastic work,

$$dp_0 = \frac{v}{\lambda - \kappa}\left(p\,d\varepsilon^p_v + w\,s_{ij}\,de^p_{ij}\right) \tag{20}$$

where w is a non-dimensional weight parameter. At the ultimate state $dp_0 = 0$, and deformation proceeds without further hardening. In this state the plastic work of the deviator stresses $s_{ij}de_{ij}^p$ is positive, and for $w > 0$ the ultimate state therefore corresponds to a negative value of $d\varepsilon_v^p$, i.e. to dilatation. The ultimate state and the weight parameter w are discussed further in Section 4.2.

The factor H_1 from (15) follows directly from the yield function (7) by differentiation,

$$H_1 = -\frac{\partial f}{\partial p_0} = m\,p^2\,\eta^{(m+1)/m} \tag{21}$$

The hardening factor H_2 is found by inserting the plastic strain increment (12) into the hardening rule (20), and substituting the yield function (7).

$$H_2 = \frac{\partial p_0}{\partial \chi} = \frac{v}{\lambda - \kappa}\left(p\frac{\partial f}{\partial p} + w\,s_{ij}\frac{\partial f}{\partial s_{ij}}\right)$$

$$= \frac{v\,p}{\lambda - \kappa}\left[(1-w)\left(J_2 - 3(1-\eta)p^2\right) + m\,\eta\,p^2\right] \tag{22}$$

The hardening parameter H has now been determined as a function of the current specific volume and the stress invariants p and J_2.

3.2. STIFFNESS MATRICES

For numerical computations it is convenient to express the incremental stress-strain relations in the six-component format with the stress vector

$$\boldsymbol{\tau} = (\sigma_x, \sigma_y, \sigma_z, \tau_{yz}, \tau_{zx}, \tau_{xy}) \tag{23}$$

and the strain vector

$$\boldsymbol{\gamma} = (\varepsilon_x, \varepsilon_y, \varepsilon_z, \gamma_{yz}, \gamma_{zx}, \gamma_{xy}) \tag{24}$$

where the shear strains $\gamma_{yz} \ldots$ contain the factor two. The corresponding non-linear elastic stiffness matrix is

$$\mathbf{C} = G\begin{bmatrix} a & b & b & & & \\ b & a & b & & & \\ b & b & a & & & \\ & & & 1 & & \\ & & & & 1 & \\ & & & & & 1 \end{bmatrix} \tag{25}$$

with the non-dimensional parameters

$$a = \frac{v\,p}{\kappa\,G} + \frac{4}{3} \quad , \quad b = \frac{v\,p}{\kappa\,G} - \frac{2}{3} \tag{26}$$

In the six-component format the stress invariant I_3 used in the yield function (11) is

$$I_3 = \sigma_x \sigma_y \sigma_z + 2\tau_{yz}\tau_{zx}\tau_{xy} - (\sigma_x \tau_{yz}^2 + \sigma_y \tau_{zx}^2 + \sigma_z \tau_{xy}^2) \qquad (27)$$

whereby the gradient of the yield function (11) takes the form

$$\frac{\partial f(\tau)}{\partial \tau} = \begin{bmatrix} -(\sigma_y \sigma_z - \tau_{yz}^2) + (1 + \frac{1}{3}m)\,\eta\,p^2 \\ -(\sigma_z \sigma_x - \tau_{zx}^2) + (1 + \frac{1}{3}m)\,\eta\,p^2 \\ -(\sigma_x \sigma_y - \tau_{xy}^2) + (1 + \frac{1}{3}m)\,\eta\,p^2 \\ -2(\tau_{zx}\tau_{xy} - \sigma_x \tau_{yz}) \\ -2(\tau_{xy}\tau_{yz} - \sigma_y \tau_{zx}) \\ -2(\tau_{yz}\tau_{zx} - \sigma_z \tau_{xy}) \end{bmatrix} \qquad (28)$$

The elasto-plastic stiffness retains the format (16) when the elastic stiffness matrix (25) is used together with the yield function gradient (28).

4. The model parameters

The model requires five parameters: two elastic stiffness parameters G and κ, the elasto-plastic stiffness λ, the yield surface shape parameter m, and the non-dimensional weight parameter w in the plastic work. The three stiffness parameters are obtained routinely from standard tests.

Most experimental data are available from triaxial tests in which only the axial stress σ_3 and the cell pressure $\sigma_1 = \sigma_2$ are varied. It is therefore convenient to express the parameters m and w in a form where they can be determined explicitly from triaxial tests, using the stress variables p and $q = \sigma_3 - \sigma_1$. In terms of these variables the deviator stress invariants are $J_2 = \frac{1}{3}q^2$ and $J_3 = \frac{2}{27}q^3$, where q is positive for triaxial compression and negative for triaxial tension. For these two special stress conditions the yield function (7) with $\alpha = 1$ is expressed as

$$f(p,q) = -\tfrac{2}{27}q^3 + \tfrac{1}{3}q^2 p - (1 - \eta)p^3 \qquad (29)$$

with $\eta(p) = (p/p_0)^m$. Apart from the function $\eta(p)$ the yield function is homogeneous of degree three in the stresses. This gives the differential relation

$$p\frac{\partial f}{\partial p} + q\frac{\partial f}{\partial q} = 3f + \frac{d\eta}{dp}p^4 = m\,\eta\,p^3 \qquad (30)$$

where the yield condition $f = 0$ has been used. This relation leads to simple explicit formulae for the parameters m and w.

4.1. SHAPE PARAMETER AND CHARACTERISTIC STATE

The charactersitic state is described by the zero dilatation condition $d\varepsilon_v = 0$. In the following this condition is used for the plastic volumetric strain. Thus $\partial f/\partial p = 0$ at the intersection of the yield surface with the characteristic line, and by the differenital relation (30) this gives the exponent m as

$$m = \frac{q\,\partial f/\partial q}{\eta\,p^3} \tag{31}$$

When η is eliminated by use of the condition $f(p,q) = 0$, this expression reduces to

$$m = m(M_c) = \frac{6\,M_c^2}{(3 - M_c)(3 + 2M_c)} \tag{32}$$

where $M_c = (q/p)_c$ is the inclination of the characteristic line in a p-q plot.

4.2. WEIGHT PARAMETER AND ULTIMATE STATE

In the ultimate state plastic deformation can proceed without further plastic hardening, and thus the ultimate state is determined by the condition of vanishing increment of the weighted plastic work (20). In a triaxial test this in turn defines the weight factor w as

$$w = -\frac{p}{q}\frac{\partial f/\partial p}{\partial f/\partial q} = 1 - \frac{p\,\partial f/\partial p + q\,\partial f/\partial q}{q\,\partial f/\partial q} \tag{33}$$

where (p,q) is a point on the ultimate state line. When using the differential relation (30) this reduces to

$$w = 1 - m\frac{\eta\,p^3}{q\,\partial f/\partial q} = 1 - \frac{m(M_c)}{m(M_u)} \tag{34}$$

where $m(M_u)$ is expression (32) with the inclination of the ultimate state line $M_u = (q/p)_u$ substituted for M_c.

5. Representative results

Typical model results for 'drained' triaxial compression tests are shown in Fig. 4 corresponding to the material data for clay in Table 1. Each test starts from isostatic pressure $p = 100$ kPa. The figure shows normalized shear stress $q = \sigma_3 - \sigma_1$ and volumetric strain as functions of the equivalent shear strain $\varepsilon_q = \frac{2}{3}(\varepsilon_3 - \varepsilon_1)$ and illustrates the effect of pre-consolidation, represented by p_0 at the beginning of the test. Note in particular the effect on the transition from compression to dilatation in Fig. 4b.

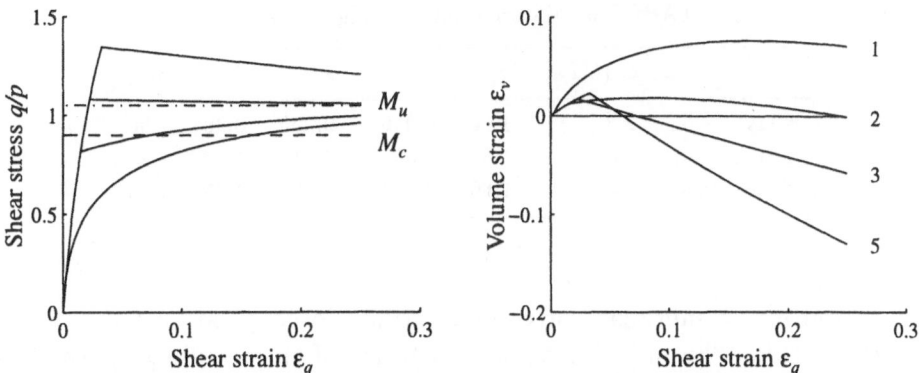

Figure 4. Stress and strain curves with $M_c = 0.9$, $M_u = 1.05$ for pre-consolidation $(p_0/p)_i = 1, 2, 3, 5$.

Table 1 also contains representative data for loose and dense sand obtained from drained triaxial compression tests, (Borup and Hedegaard, 1995). A corresponding set of theoretical undrained triaxial stress paths is shown in Fig. 5 together with the characteristic and ultimate lines. The stress path of the loose sand is more curved below the characteristic line, and exhibits a sharper bend when passing the characteristic state. This behaviour corresponds closely to that of experimental results and theoretical predictions obtained by Lade and Kim (1995) by a more elaborate model.

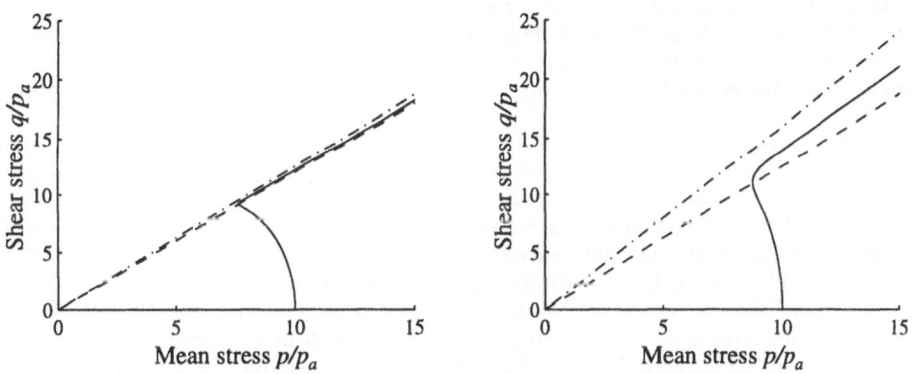

Figure 5. Theoretical stress paths for undrained triaxial compression, a) loose sand, b) dense sand.

6. Conclusions

A triaxial isotropic hardening associated plasticity model for granular materials has been developed. The model consists of three parts: an octahedral

TABLE 1. Material data for Figs. 4 and 5.

	G [MPa]	v	κ	λ	M_c	M_u
Clay	2.5	1.60	0.060	0.200	0.90	1.05
Loose sand	10	1.85	0.0045	0.010	1.20	1.25
Dense sand	20	1.60	0.004	0.006	1.25	1.60

stress contour, a meridian curve generating closed yield surfaces, and a hardening rule. The model is defined in terms of three stiffness parmeters, G, v/κ, and v/λ, and the inclination M_c and M_u of the characteristic and ultimate lines. Qualitative features like differentiation between triaxial compression and tension, dilatation at failure, and the effect of pre-consolidation are predicted well, but preliminary studies, (Krenk et al., 1995) and (Borup and Hedegaard, 1995) indicate that further improvement can be obtained by introducing a non-associated yield potential, e.g. by scaling of the volumetric strain component.

ACKNOWLEDGMENTS: Support from the Danish and Swedish Technical Research Councils is gratefully acknowledged.

References

Borup, M. and Hedegaard, J. (1995). *Characteristic State Modelling of Friction Materials.* M.Sc. Thesis, Aalborg University, Aalborg, Denmark.

Desai, C.S. and Hashmi, Q.S.E. (1989). Analysis, evaluation, and implementation of a nonassociative model for geologic materials. *Journal of Plasticity,* **5**, pp. 397-420.

Krenk, S. (1996). A family of invariant stress surfaces. *Journal of Engineering Mechanics,* **122**, pp. 201–208.

Krenk, S., Borup, M. and Hedegaard, J. (1995). A triaxial characteristic state model for sand. *Proceedings of the Eleventh European Conference on Soil Mechanics and Foundation Engineering,* Copenhagen, May 29 - June 1, **6**, pp. 89-94.

Lade, P.V. (1977). Elasto-plastic stress-strain theory for cohesionless soil with curved yield surfaces. *International Journal of Solids and Structures,* **13**, pp. 1019-1035.

Lade, P.V. and Duncan (1975). Elasto-plastic stress-strain theory for cohesionless soil. *Journal of the Geotechnical Engineering Division, ASCE,* **101**, pp. 1037-1053.

Lade, P.V. and Kim M.K. (1995). Single hardening constitutive model for soil, rock and concrete. *International Journal of Solids and Structures,* **32**, pp. 1963-1978.

Matsuoka, H. and Nakai, T. (1985). Relationship among Tresca, Mohr-Coulomb and Matsuoka-Nakai failure criteria. *Soils and Foundations,* **25**, pp. 123-128.

Schofield, A.N. and Wroth, C.P. (1968). *Critical State Soil Mechanics.* McGraw-Hill, New York.

Weidner, J. (1990). *Vergleich von Stoffgesetzen granularer Schüttgüter zur Silodruckermittlung.* Ph.D. Thesis, University of Karlsruhe, Germany.

ON THE CONSTITUTIVE POTENTIAL FOR POWDER AND POROUS BODIES

M.Shtern,

Institute for Problems of Material Sciences, Krgiganovsky
str. 3a, Kiev, 252180, Ukraine

This paper examines the general structure of constitutive laws for compressible bodies. Porous (sintered and ductile) bodies are considered. The simple cylindrical unit cell model is used to evaluate the rate-dependence of the constitutive law for isotropic porous bodies. The special choice of the stress-strain rate relation for the matrix material has provided the exact analytical expression for the given law which has been obtained for a wide spectrum of matrix rheological properties (from rigid plasticity to linear viscosity). It has been established that the macroscopic potential for the exact model does not exist. The connection between the condition for the existence of the macroscopic potential and the constitutive law for the matrix are considered. The rate dependence of the pressure transmission coefficient is established. The upper bound of the developed theory has a form which is inherent to creep models based on the elliptical macroscopic potential, including the plastic behaviour.

1. Introduction

The first theoretical estimations of various yield points for porous materials was carried out by Skorokhod (1965). Green's version of the plasticity theory for porous bodies (Green,1971) laid the foundations of the non-linear micromechanics of compressible bodies. In the middle of the 1970's Gurson obtained the exact expression for the yield condition. Further investigations(Duva,1986)) were targeted on the extension of Gurson's model for rate dependent porous bodies. An elegant asymptotic estimation obtained by Cocks (1989) as well as numerical analyses carried out by Sofronis and McMeeking (1992) clarified the main features of non-linear creep of porous bodies.

At the same time some important problems concerning the structure of the macroscopic governing equations remain to be solved. Among these there is a problem of the existence of the strain rate or stress potential. It should be remembered that the existence of these potentials is usually supposed in experimentally based models as well as in models based on numerical simulations. Also, this assumption is rather convenient for application of the models, although it is not deduced from general principals. Another problem is the rate sensitivity of the macroscopic material response of a porous body, particularly the rate dependence of Poisson's ratio and the pressure transmission coefficient. These two problems return us to the necessity for continuation of the search for the exact solution to the governing equations describing the creep of porous bodies. The experience of the authors mentioned above indicates that this cannot be achieved on the basis of a power-law. To avoid this obstacle a particular approximation of the stress-strain rate diagram for the matrix material is used in the present paper and a unit cell analysis for a cylindrical unit cell provides the macroscopic constitutive equations for non-linear creep of porous bodies in terms of ordinary functions. The problem of the existence of a potential as well as the problem of the rate sensitivity of the pressure transmission coefficient are solved on the basis of the developed model. Upper bound theories for creep is suggested.

2. Strain rate field in the cylindrical unit cell

It is supposed that the porous material considered below is an isotropic and third invariant in - dependent body. The constitutive equations are obtained using a unit cell analysis. To make this procedure more evident a cylindrical unit cell model is used (Fig.1). Assuming the matrix to be incompressible we find the strain rate field which satisfies the incompressibility equation

95

N. A. Fleck and A. C. F. Cocks (eds.), IUTAM Symposium on Mechanics of Granular and Porous Materials, 95–103.
© 1997 *Kluwer Academic Publishers.*

$$\frac{\partial V_z}{\partial z} + \frac{\partial V_r}{\partial r} + \frac{V_r}{r} = 0 \tag{1}$$

This field can be expressed by the equations

$$\dot{\varepsilon}_z = \frac{\partial V_z}{\partial z} = \pm\sqrt{\frac{2}{3}}\,\dot{\Gamma} + \frac{1}{3}\dot{E}$$

$$\dot{\varepsilon}_r = \frac{\partial V_r}{\partial r} = \mp\frac{1}{\sqrt{6}}\,\dot{\Gamma} - \left(\frac{1}{6} + \frac{1}{2}\frac{R_1^2}{r^2}\right)\dot{E} \tag{2}$$

$$\dot{\varepsilon}_\varphi = \frac{V_r}{r} = \mp\frac{1}{\sqrt{6}}\,\dot{\Gamma} - \left(\frac{1}{6} - \frac{1}{2}\frac{R_1^2}{r^2}\right)\dot{E}$$

which contain the remote strain rate components E_{ij}. So the deviatoric strain-rate for the cell can be written in terms of the remote strain-rate invariants.

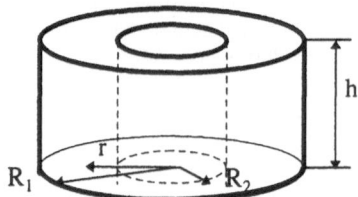

Fig. 1. The cylindrical unit cell

$$\dot{E} = \dot{E}_{kk} \qquad \dot{\Gamma} = \left(\left(\dot{E}_{ij} - \frac{1}{3}\dot{E}\,\delta_{ij}\right)\left(\dot{E}_{ij} - \frac{1}{3}\dot{E}\,\delta_{ij}\right)\right)^{1/2} \tag{3}$$

$$\dot{\gamma} = \sqrt{\dot{\varepsilon}_{ij}\,\dot{\varepsilon}_{ij}} = \sqrt{\frac{1}{2}\frac{R_1^4}{r^4}\dot{E}^2 + \dot{\Gamma}^2} \tag{4}$$

The expression for the energy dissipation rate for an isotropic, incompressible and third invariant independent matrix material can be presented in the form

$$\dot{A} = \sigma_{ij}\,\dot{\varepsilon}_{ij} = F\left(\dot{\gamma}\right) \tag{5}$$

$$\left\langle \dot{A}\right\rangle = \frac{1}{V}\int_V \dot{A}\,dV = \int_\Theta^1 F\left(\sqrt{\frac{1}{2x^2}\dot{E}^2 + \dot{\Gamma}^2}\right)dx \tag{6}$$

where the substitutions

$$\Theta = \frac{R_1^2}{R_2^2}, \qquad x = \frac{r^2}{R_2^2} \tag{7}$$

are used. The given expression transforms to Cokcs's integral (Cocks, 1989) for a power-law material. Supposing all the dissipation rate is connected with matrix deformation only we can determine the macroscopic energy dissipation rate (Skorokhod, 1965) :

$$\dot{N} = (1 - \Theta)\cdot\langle\dot{A}\rangle \tag{8}$$

3. General structure of constitutive equations

The given value is supposed to be a smooth function of the strain-rate components and defined by equation

$$\dot{N} = \dot{N}\left(\dot{E}_{ij}\right) = \Sigma_{ij}\,\dot{E}_{ij} \tag{9}$$

This equation is satisfied if we adopt the following definition for the stress tensor components (Ziegler, 1963)

$$\Sigma_{ij} = \frac{\dot{N}}{\dfrac{\partial \dot{N}}{\partial \dot{E}_{ij}}\,\dot{E}_{ij}} \cdot \frac{\partial \dot{N}}{\partial \dot{E}_{ij}} \tag{10}$$

or for the stress invariants

$$\Sigma_m = \frac{1}{3}\Sigma_{kk} = \frac{\dot{N}}{\dfrac{\partial \dot{N}}{\partial \dot{\Gamma}}\dot{\Gamma} + \dfrac{\partial \dot{N}}{\partial \dot{E}}\dot{E}} \cdot \frac{\partial \dot{N}}{\partial \dot{E}} \tag{11}$$

$$\Sigma_0 = \left(\left(\Sigma_{ij} - \Sigma \delta_{ij}\right)\left(\Sigma_{ij} - \Sigma \delta_{ij}\right)\right)^{1/2} = \frac{\dot{N}}{\dfrac{\partial \dot{N}}{\partial \dot{\Gamma}}\dot{\Gamma} + \dfrac{\partial \dot{N}}{\partial \dot{E}}\dot{E}} \cdot \frac{\partial \dot{N}}{\partial \dot{\Gamma}} \tag{12}$$

Using the result of a unit cell analysis we can link the macroscopic stress - strain rate relation with the creep law or flow mechanism of the matrix phase

$$\Sigma_0 = \frac{\displaystyle\int_\Theta^1 F\left(\dot{\gamma}\right)dx}{\displaystyle\int_\Theta^1 F^\nabla\left(\dot{\gamma}\right)\dot{\gamma}\,dx}\int_\Theta^1 F^\nabla\left(\dot{\gamma}\right)\frac{\dot{\Gamma}}{\dot{\gamma}}\,dx, \tag{13}$$

$$\Sigma_m = \frac{\displaystyle\int_\Theta^1 F\left(\dot{\gamma}\right)dx}{\displaystyle\int_\Theta^1 F^\nabla\left(\dot{\gamma}\right)\dot{\gamma}\,dx}\int_\Theta^1 F^\nabla\left(\dot{\gamma}\right)\frac{\dot{E}}{2x^2\,\dot{\gamma}}\,dx, \tag{14}$$

4. Creep law for matrix

As mentioned above, a power-law does not permit the exact solution for integrals (13) and (14) to be obtained. Here we consider a special type of matrix material for which the stress-strain rate diagram is of the "saturation type" (Fig.2). This equation describes materials which are rate independent for large values of strain rate. It may be fitted by the function

$$F\left(\dot{\gamma}\right) = \sqrt{1 + 4n^2}\;\sigma_0\left(\sqrt{\dot{\gamma}^2 + n^2\,\dot{\varepsilon}_0^2} - n\dot{\varepsilon}_0\right) \tag{15}$$

which reduces to

$$F\left(\dot{\gamma}\right) = \sigma_0 \, \varepsilon_0 \left(\frac{\dot{\gamma}}{\dot{\varepsilon}_0}\right)^2 \quad \text{and} \quad F\left(\dot{\gamma}\right) = \sigma_0 \, \dot{\varepsilon}_0 \left(\frac{\dot{\gamma}}{\dot{\varepsilon}_0}\right) \tag{16}$$

for $n \to \infty$ and $n = 0$, respectivy, which correspond to linear viscous and perfectly pastic behaviour
Now

$$\tau = \sqrt{s_{ij}s_{ij}} = \frac{F\left(\dot{\gamma}\right)}{\dot{\gamma}} = \sqrt{1+4n^2} \; \sigma_0 \frac{\sqrt{\dot{\gamma}^2 + n^2 \, \dot{\varepsilon}_0^{\,2}} - n\dot{\varepsilon}_0}{\dot{\gamma}} \tag{17}$$

5. Exact creep model for porous materials

The given approximation allows all the expressions contained in (13) and (14) to be integrated exactly. Each of these integrals can be expressed in terms of two auxiliary functions **u** and **v**:

$$\left\langle \dot{A} \right\rangle = \int_\Theta^1 F\left(\dot{\gamma}\right) dx = \sigma_0 u \, \dot{E} + \sigma_0 v \frac{\dot{\Gamma} + n^2 \dot{\varepsilon}_0}{\dot{\Gamma}} - \sigma_0 \sqrt{1+4n^2} \; n\dot{\varepsilon}_0(1-\Theta)$$

$$\int_\Theta^1 F^\nabla\left(\dot{\gamma}\right) dx = \sigma_0 u \, \dot{E} + \sigma_0 v \, \dot{\Gamma}$$

$$\int_\Theta^1 F^\nabla\left(\dot{\gamma}\right) \frac{\dot{\Gamma}}{\dot{\gamma}} dx = \sigma_0 v \tag{18}$$

$$\int_\Theta^1 F^\nabla\left(\dot{\gamma}\right) \frac{1}{2x^2} \frac{\dot{E}}{\dot{\gamma}} dx = \sigma_0 u$$

$$u = \frac{\sqrt{1+4n^2}}{\sqrt{2}} \left(A \sinh \frac{\dot{E}}{\sqrt{2}\Theta\sqrt{\dot{\Gamma}^2 + n^2 \dot{\varepsilon}_0^{\,2}}} - A \sinh \frac{\dot{E}}{\sqrt{2}\sqrt{\dot{\Gamma}^2 + n^2 \dot{\varepsilon}_0^{\,2}}} \right) \tag{19}$$

$$v = \frac{\sqrt{1\, 4n^2}}{\sqrt{\dot{\Gamma}^2 + n^2 \dot{\varepsilon}_0^{\,2}}} \dot{\Gamma} \left(\left(\frac{1}{2} \frac{\dot{E}}{\sqrt{\dot{\Gamma}^2 + n^2 \dot{\varepsilon}_0^{\,2}}} + 1 \right)^{1/2} - \left(\frac{1}{2} \frac{\dot{E}}{\sqrt{\dot{\Gamma}^2 + n^2 \dot{\varepsilon}_0^{\,2}}} + \Theta \right)^{1/2} \right), \tag{20}$$

$$\Sigma_m = (1-\Theta) \frac{\left\langle \dot{A} \right\rangle}{u \, \dot{E} + v \, \dot{\Gamma}} u, \quad \Sigma_0 = (1-\Theta) \frac{\left\langle \dot{A} \right\rangle}{u \, \dot{E} + v \, \dot{\Gamma}} v \tag{21}$$

6. Equivalent strain rate and equivalent stress - existence of potentials.

A convenient expression for the macroscopic constitutive equations may be obtained after introducing the special scalar measures of stress and strain rate. Since **u** has the same sign as the volume strain rate $\dot{E}$ and **v** and $\dot{\Gamma}$ are nonnegative by definition,

$$W = \frac{1}{\sqrt{1-\Theta}} \left(u \, \dot{E} + v \, \dot{\Gamma} \right) \tag{22}$$

is not negative. It is equal zero only when all strain-rate components are equal to zero. We shall to this quantity as the equivalent strain rate.

to this quantity as the equivalent strain rate.

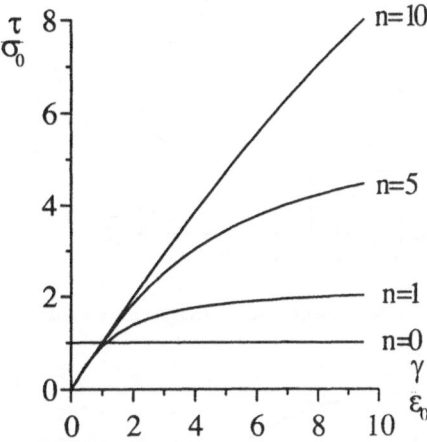

Fig.2. Relation between deviatoric stress and strain rate for the matrix material

The equivalent stress can now be defined as

$$\Sigma = \sqrt{1 - \Theta} \frac{\langle \dot{A} \rangle}{W} \tag{23}$$

It should be emphasised that each of these measures cannot be considered as a potential for a given model. Moreover, the analysis of equations (21) indicates that the potential does not exist. This feature is stipulated by the special form of the energy dissipation rate. Unlike other widely used models, the energy dissipation rate predicted by the model developed above is not a homogeneous function of the strain rate components. At the same time these values help to simplify the scalar constitutive equations and allow us to present them in the form

$$\Sigma_m = \Sigma u , \ \Sigma_0 = \Sigma v \tag{24}$$

7. Connection with Gurson's model. Surfaces of constant equivalent stress and energy dissipation rate.

Excluding the term $\dfrac{\dot{E}}{\sqrt{\dot{\Gamma}^2 + n^2 \dot{\varepsilon}_0^2}}$ from (19) and (20) one can obtain an interesting result expressed by

$$\frac{1}{1 + 4n^2} \cdot \frac{\dot{\Gamma}^2 + n^2 \dot{\varepsilon}_0^2}{\dot{\Gamma}^2} \left(\frac{\Sigma_0}{\Sigma} \right)^2 + 2\Theta \cdot \cosh\left(\frac{\sqrt{2}}{\sqrt{1 + 4n^2}} \cdot \frac{\Sigma_m}{\Sigma} \right) = 1 + \Theta^2 \tag{25}$$

It is evidently that this equation reduces to the Gurson yield condition when $n = 0$. At the same time equation (25) cannot be considered as an extension of the correlation between equivalent stress and stress invariants for a creeping material because it contains the effective deviatoric strain-rate $\dot{\Gamma}$. The correlation has a more complex form. Surfaces in Stress space for which the equivalent stress is constant are plotted in Fig.3.
Fig.4 shows surfaces of constant energy dissipation rate in the same space. Both plots are similar to those obtained by Sofronis and McMeeking (1992) and Cocks (1989) respectively.

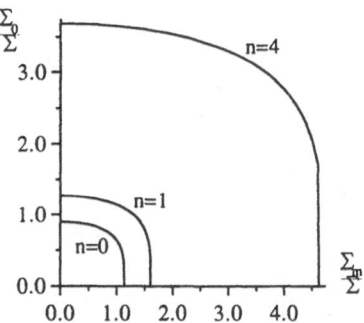

Fig.3. Surfaces in Stress space for which equivalent stress is constant

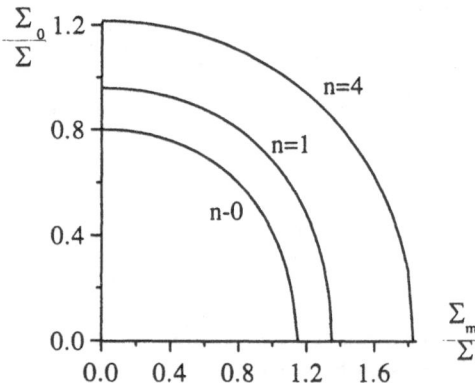

Fig.4. Surfaces in Stress space for which macroscopic energy dissipation rate is constant

8. Rate dependence of pressure transmission coefficient.

The model developed above is used to estimate the pressure transmission coefficient ζ (the ratio of stress components in die compaction when external friction is negligible). Models based on elliptical yield surfaces or elliptical creep potentials give different dependencies of this coefficient on the porosity, although they are similar in form. According to the model developed by Sofronis and McMeeking(1992) it depends also upon the rate sensitivity exponent. The model developed in the present paper predicts that the pressure transmission coefficient depends also on axial strain rate (the radial strain rate is equal to zero for die compaction).
Employing the current model we find

$$\zeta = \frac{\Sigma_r}{\Sigma_z} = \frac{3\Sigma_m}{3\Sigma_m + 2\Sigma_0} \frac{\Sigma_0}{} , \tag{26}$$

Where

$$\frac{\Sigma_m}{\Sigma_0} = \frac{\Sigma_z + 2\Sigma_r}{\sqrt{6}\,|\Sigma_z - \Sigma_r|} = \frac{\sqrt{3}}{2}\frac{1}{t}\frac{\mathrm{Arsh}\dfrac{t}{\Theta} - \mathrm{Arsh}\,t}{\sqrt{t^2 + 1} - \sqrt{t^2 + \Theta}} , \tag{27}$$

and

$$t = \frac{\dfrac{\dot{E}_z}{\dot{\varepsilon}_0}}{\sqrt{2}\sqrt{\dfrac{2}{3}\dfrac{\dot{E}_z^2}{\dot{\varepsilon}_z^2} + n^2}} . \tag{28}$$

Eqn(27) reduces to

$$\frac{\Sigma_m}{\Sigma_0} = \frac{\text{Arsh}\frac{\sqrt{3}}{2\Theta} - \text{Arsh}\frac{\sqrt{3}}{2}}{\sqrt{\frac{7}{4}} - \sqrt{\frac{3}{4} + \Theta}} \quad \text{and} \quad \frac{\Sigma_m}{\Sigma_0} = \frac{\sqrt{3}}{2}\frac{1}{\Theta}$$

for **n=0** and **n** $\to \infty$, respectively.

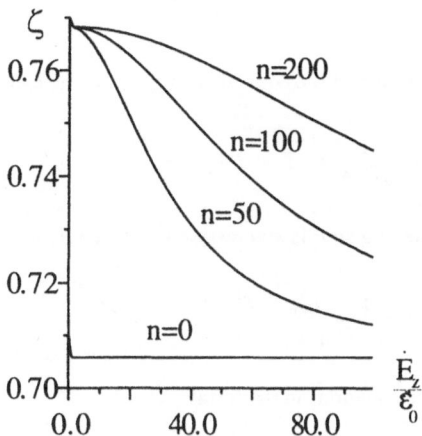

Fig.5. Rate dependence of the pressure transmission coefficient

Eqn(27) is plotted in Fig.5.

Roman (1971) obtained the similar results from his experiment on iron powder. Despite the difference between theoretically predicted and experimental results it can be seen that the current non-linear creep model is sensitive to strain rate.

9. Upper bound theory for non-linear creep.

To understand the connection between the above creep model and more widely used models with elliptical potentials we return to the expression for the energy dissipation rate (3). The simple analysis shows that this expression may be considered as a convex function of argument $a(x)$

$$a(x) = \frac{1}{2x^2}\dot{E}^2 + \dot{\Gamma}^2 \tag{29}$$

$$F(\sqrt{a(x)}\,) = \sigma_0\sqrt{1 + 4n^2}\left(\sqrt{a(x) + n^2\,\dot{\varepsilon}_0^2} - n\dot{\varepsilon}_0\right) \tag{30}$$

Then using Jensen's inequality we obtain the bound

$$A \le F\left(\sqrt{\frac{1-\Theta}{2\Theta}\dot{E}^2 + (1-\Theta)\dot{\Gamma}^2}\right) \tag{31}$$

The creep model where

$$N = F\left(\sqrt{\frac{1-\Theta}{2\Theta}\dot{E}^2 + (1-\Theta)\dot{\Gamma}^2}\right) \tag{32}$$

may be considered as an upper bound theory for the model developed above.

The general upper bound creep model can be formulated in the form

$$\dot{N} = F\left(\frac{1}{\sqrt{1-\Theta}}\sqrt{\varphi\,\dot{\Gamma}^2 + \psi\,\dot{E}^2}\right) \qquad (33)$$

Where functions φ and ψ satisfy the requirements

$$\lim_{\Theta\to 0}\varphi = 1, \ \lim_{\Theta\to 0}\psi = \infty, \ \int_0^\Theta \psi\,d\Theta < \infty \ .$$

According to the general definition of the equivalent strain rate one can write the equation

$$W = \frac{1}{\sqrt{1-\Theta}}\sqrt{\varphi\,\dot{\Gamma}^2 + \psi\,\dot{E}^2} \qquad (34)$$

Hence we can obtain the following simple expression for the equivalent stress

$$\Sigma = \frac{1}{\sqrt{1-\Theta}}\sqrt{\frac{\Sigma_m^2}{\psi} + \frac{\Sigma_0^2}{\varphi}} \qquad (35)$$

and present scalar constitutive equations in the form

$$\Sigma_m = \psi\,\frac{\Sigma}{W}\,\dot{E}, \qquad \Sigma_0 = \varphi\,\frac{\Sigma}{W}\,\dot{\Gamma}, \qquad (36)$$

where Σ and W are related to each other by the same relationship as the deviatoric stress and strain rate for the matrix phase.

Contrary to the exact creep model the constitutive equations obtained from the creep upper bound theory can be presented in the potential form

$$\Sigma_{ij} = \frac{\partial}{\partial E_{ij}}\int\frac{\dot{N}(W)}{W}\,dW \qquad (37)$$

10. Conclusions

An exact constitutive model for non-linear creep of porous materials is obtained. The main equations are directly deduced from the unit cell model by use of an averaging procedure. The special choice of the analytical approximation of the stress-strain rate diagram for the matrix material provides the opportunity to express the main equations of the developed model in terms of ordinary functions. The wellknown Gurson model is a limit of the model presented here when the matrix material is the rate-independent. For a linear viscous matrix the developed model also reduces to the known result. Unlike these two models strain-rate or stress potentials do not exist for intermediate cases. The pressure transmission coefficient depends on axial strain rate for non-linear and non-plastic behaviour. The upper bound of the developed models reduces to the set of non-linear models based on the elliptical stress or strain-rate potential.

Acknowledgements-The author expresses gratitude to Prof.V.Skorokhod and to Prof.A.Cocks for helpful discussions and to Dr. Ju.Panfilov for help in carring out the calculations.

References

Skorokhod,V.(1965) Meansquare stress and strain rate for porous materials, *Soviet powder metallurgy and metal ceramics*, **12**, pp.31-35.
Green,R.J.(1971) Plasticity theory for porous solids, *Int.J.Mech.Sci.*, **14**, pp.215-224.

Gurson,A.L.(1977) Continuum theory for ductile rupture by void nucleation and growth: Part 1-Yield criteria and flow rules for porous media, *J.Eng.Mater.Technol.*, **99**, pp.2-15.

Duva,P.(1986) A constitutive description of non-linear materials containing voids, *Mechanics of materials*, **5**, pp.137-149.

Cocks,A. (1989) Inelastic deformations of porous materials, *J.Mech. and Phys.Solids*, Vol.37, **6**, pp.693-715.

Sofronis,P. and McMeeking,R.M. (1992) Creep of power-law material containing spherical voids, *J.appl.Mech.*, Vol. 59, pp. 88-101

Ziegler,H. (1963) Some extremum principles in irreversible thermodynamics with application to continuum mechanics, *North-Holland publishing company,* Amsterdam.

Roman,O.V. et al (1971) Impact compaction of metal powder, *Soviet Powder Metallurgy and Metal Ceramics,* **9**, pp.12-17.

SCALING LAWS IN THE CONSOLIDATION OF POWDER COMPACTS

A. Casagranda* and P. Sofronis
Department of Theoretical and Applied Mechanics
University of Illinois at Urbana-Champaign
104 South Wright Street
Urbana, Illinois, 61801

Abstract

Powder consolidation under load is a common means of forming ceramic or intermetallic components for advanced structural applications. Micrometer/nanometer sized powders are used as a means for creating new composite microstructures on a fine scale. However, current models of the deformation processes in powders do not adequately represent the experimental observations in micrometer/nanometer sized powder compacts. In this paper, a constitutive model for the consolidation of these powders is proposed. The mechanisms considered include elasticity of the aggregate, diffusion along the interparticle contact area, and relative slip between the particles. The finite element method is used to predict the deformation of a powder compact for a range of loading conditions. Unit cell calculations are performed and scaling laws for the macroscopic response during the deformation of the powder aggregate are detected in the numerical results.

1. Introduction

Powder processing is often used for forming complex shaped components from ceramic and intermetallic materials. The advantages of using powders include control over the final microstructure (grain size and porosity) as well as allowing for mixing of powders to produce alloys or composites. In general, models to date predict the constitutive behavior of a micron-sized aggregate from calculations on a representative volume element. For stage I densification, i.e., for porosity $\rho < 0.90$, models have been proposed in which the dominant densification mechanism is either bulk deformation of the grains in the plasticity [1 - 4] or power law creep [5] regimes, or grain boundary diffusion [6 - 8] or interface reaction control [9]. In addition, a number of stage II ($\rho \geq 0.90$) models are also available in which the isolated porosity closes by power law creep deformation of the surrounding matrix [10 - 12] or by diffusional creep [13, 14] or by grain boundary

---

* Current address: Concurrent Technologies Corporation, 1450 Scalp Ave. Johnstown, PA 15904

N. A. Fleck and A. C. F. Cocks (eds.),
IUTAM Symposium on Mechanics of Granular and Porous Materials, 105–116.
© 1997 *Kluwer Academic Publishers.*

diffusion and slip [15]. Cocks [16] in a recent overview paper by using appropriate admissible stress or velocity fields calculated bounds on the form of macroscopic potentials which govern the constitutive behavior of the powder under the action of a given densification mechanism.

Experimental results on densification with cylindrical compact specimen of nanocrystalline TiO_2 under uniaxial compression [17, 18] indicate that the ratio of the radial to the axial strain rates is approximately equal to zero over a range of relative density (60% to 90%) and a wide range of applied stress. For nanocrystalline ZrO_2 (monoclinic), on the other hand, this ratio varies with load [18]. It is notable that this phenomenon, the absence of radial expansion, is also observed in the case of micron-sized Al_2O_3 powder [19]. An important and as yet unexplained feature of a densifying nanometer sized powder is that it exhibits a superplastic macroscopic behavior with a creep exponent on the order of 2 [17, 18]. A similar response has also been observed in the densification of micron-sized powder [19]. Current constitutive theories for micron-sized aggregates based on kinetic diffusion models are deficient since they predict or consider a value for the stress exponent equal to 1 [6]. It should be emphasized that such nonlinear dependence on stress is puzzling, especially in the case on nanocrystalline materials, in view of the fact that the potential mechanisms involved, namely diffusion, slip and grain elasticity due to the absence of any dislocations in the bulk [18, 20] are all linear in nature.

In this paper a finite element model for stage I densification of powder compacts is developed using a micromechanical approach [1 - 3, 5, 6, 11] under steady state, plane strain deformation conditions. As a starting point, the present model assumes the powder to be linear in nature, but can easily be extended to non-linear materials. Therefore, the present model is appropriate for nanocrystalline powder compacts (e.g., ceramic powders [17, 18, 20]), in which the grains exhibit linear behavior. Thus the mechanisms included in the model are elastic deformation of the individual particles, interparticle stress driven diffusion and relative slip. Free surface diffusion is not considered, but a parametric study on the values of the local capillary stress, σ_0, acting on the perimeter of the interparticle area is conducted. The contact region between particles is analyzed in great detail and the interaction among the mechanisms involved is considered. Scaling laws on the relationship between the macroscopic stress and strain rate quantities are sought in an effort to understand the nonlinear behavior of the aggregate as has been observed experimentally. The methodology to be outlined in this work can be extended to the case of powder compacts in which nonlinear constitutive laws govern the deformation in the bulk.

2. Material constitutive law

The interior of the particles is modeled to deform elastically. The complexity of the numerical calculations in the present case of a linear material model is markedly less than in straining situations of micron-sized particles with nonlinear bulk response. Thus, the rate deformation of the grains is assumed to be characterized by the standard linear isotropic elasticity law

$$\dot{\sigma}_{ij} = \frac{E}{1+v}\dot{\varepsilon}_{ij} + \frac{vE}{(1+v)(1-2v)}\dot{\varepsilon}_{kk}\delta_{ij} \tag{1}$$

where σ_{ij} denotes the stress, a superposed dot denotes differentiation with respect to time, strain rate $\dot{\varepsilon}_{ij}$ is defined through the velocity v_i by

$$\dot{\varepsilon}_{ij} = \left(v_{i,j} + v_{j,i}\right)/2, \tag{2}$$

where $(\ )_{,i} = \partial(\ )/\partial x_i$, E and v are respectively Young's modulus and Poisson's ratio, Latin indices i and j are in the range 1, 2, 3, δ_{ij} is the Kronecker delta, and the standard summation convention over the range is implied on a repeated index.

Stress motivated diffusional mass transport along the interface between two particles in contact is characterized by [16, 21]

$$j_\alpha = \mathcal{D}\partial\sigma_n/\partial x_\alpha, \tag{3}$$

where j_α denotes volumetric flux per unit length crossing the interface, σ_n is the interface normal stress (see Figure 1), α is in the range 1, 2 and refers to a local interface coordinate system. The interface diffusion constant whose dimensions are volume divided by stress per unit time is defined by

$$\mathcal{D} = D_b\delta_b\Omega/kT, \tag{4}$$

where D_b is the interface diffusion coefficient, δ_b is an effective interface thickness, k is Boltzmann's constant, T is the absolute temperature, and Ω is the atomic volume. A repeated Greek index is summed over the range 1, 2. Matter conservation along the interparticle contact requires

$$\partial j_a/\partial x_a + \dot{h} = 0, \tag{5}$$

where $\dot{h}$ denotes the normal interpenetration velocity had the particles continued to deform penetrating into each other under the action of the external load. The value of the local normal stress, σ_0, acting at the boundary of the interparticle area is given by [16, 22 - 24]

$$\sigma_0 = \gamma_s\left(k_1 + k_2\right), \tag{6}$$

where γ_s is the free energy per unit area on the surface of the pore, and k_1 and k_2 are the principal curvatures of the adjoining pore surfaces. Relative slip between the particles along the interface is assumed to be described phenomenologically by a linear viscous rheology. Thus,

$$v_t = \sigma_t/\mu, \tag{7}$$

where v_t is the local tangential relative slip velocity between the particles along the interparticle area, σ_t is the corresponding tangential (or shear) stress (see Figure 1) and μ is the viscosity of the interface.

3. Formulation of the boundary value problem in plane strain

Assuming a cubic array of cylinders, one can analyze a unit cell containing one contact zone (see Figure 1). Following Hill [25], Needleman and Rice [22], McMeeking and Kuhn [6], and Cocks [16], one can state the governing equations for the deformation of the cell in the form of the principle of virtual power as

$$\int_{S_T} T_i \delta v_i \, ds = \int_V \sigma_{ij} \delta \dot{\varepsilon}_{ij} \, dA + \int_{S_d} \sigma_n \delta v_n \, ds + \int_{S_d} \sigma_t \delta v_t \, ds, \qquad (8)$$

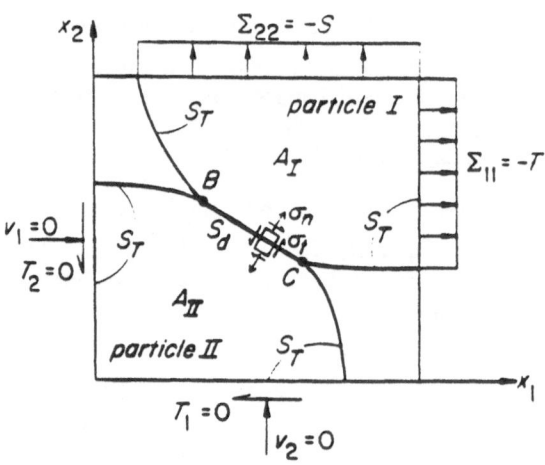

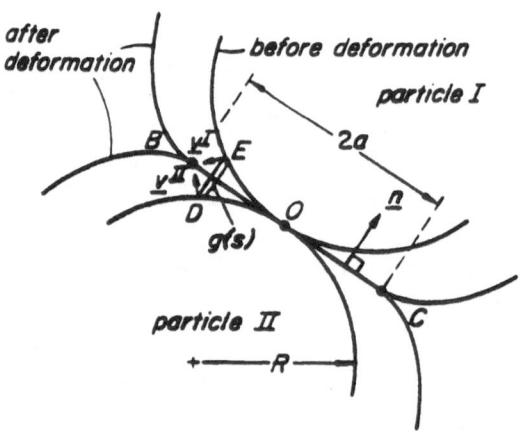

Figure 1. The unit cell under plane strain deformation with a detailed drawing of the contact area S_d.

where T_i is the specified traction on the boundary, S_T, of the cell over which tractions are prescribed. The pore surface is a traction free part of the boundary S_T (see Figure 1), σ_{ij}

is the microscopic stress within the region V occupied by the particles, σ_n and σ_t are respectively the normal and tangential stresses along the interparticle contact area S_d, δv_i is an arbitrary virtual variation of the velocity v_i in V such that δv_i vanishes on the part of the surface where velocities are prescribed, $\delta \dot{e}_{ij}$ is the corresponding virtual variation of the strain rate, and δv_n and δv_t are respectively virtual variations of the normal and tangential velocity difference along the interparticle contact area S_d. It should be noted that any normal velocity difference v_n along the interparticle area, S_d, is accommodated by diffusion.

In this paper, in which the particles are modeled to deform elastically, the local interparticle contact area, S_d, and the associated stresses as they develop during deformation are so determined that global equilibrium as expressed by eq. (8) is maintained. Thus the size, $BC = 2a$, and orientation of the contact area S_d (see Figure 1) are calculated as part of the solution procedure which involves iteration [26]. For the treatment of the contact problem in the presence of interface diffusion and slip a methodology similar to that of Simo et al. [27] was used [26]. Initially, a contact larger than the expected is assumed and the problem is solved. The tractions along the contact surface are then calculated and checked for consistency (i.e., all tractions along the contact zone should be compressive and diffusive nodes should be in contact). If the solution does not satisfy global equilibrium then a smaller contact area S_d is assumed and the process is continued.

The interaction between the various mechanisms involved in the deformation process was investigated through the following dimensionless parameters [28]:

$$\chi = E\mathcal{D}t/R^3 \tag{9}$$

for the effects of interparticle diffusion,

$$\xi = Et/R\mu \tag{10}$$

for the effects of interface slip between the particles, and

$$\zeta = \sigma_0/E \tag{11}$$

for the effect of the free energy of the pore surface. Here the time t shows explicitly in the definition of the dimensionless parameters of eqs. (9 - 10) because of its explicit appearance in the determination of the elastic microscopic strain rate, $\dot{e}_{ij}$. It should be emphasized that the present paper addresses only the steady state response of the system under constant loads S and T. Therefore parameters such as the contact zone size are determined by keeping the macroscopic loads constant at their values S and T while time is not varied during the calculation. The macroscopic velocities, from which the macroscopic strain rates are calculated, are determined by dividing the macroscopic displacements at loads S and T by the time t. The purpose is to characterize the constitutive response of the unit cell by predicting the dependence of the macroscopic strain rate, $\dot{E}_{ij}$, on the applied macroscopic stress Σ_{ij}.

4.   Results

A given set of loading conditions, T, S was measured in terms of the in-plane hydrostatic stress, $\Sigma_m = (\Sigma_{11} + \Sigma_{22})/2$, in-plane deviatoric stress, $\Sigma_e = \sqrt{3}|\Sigma_{22} - \Sigma_{11}|/2$, and the angle of loading, $\theta = \tan^{-1}(\Sigma_e/\Sigma_m)$. Similarly the homogeneous macroscopic strain rate was measured in terms of the in-plane hydrostatic and deviatoric strain rate components, $\dot{E}_m = (\dot{E}_{11} + \dot{E}_{22})/2$ and $\dot{E}_e = \sqrt{3}|\dot{E}_{22} - \dot{E}_{11}|/2$ respectively, where $\dot{E}_{11} = v_1/(\sqrt{2}R)$ and $\dot{E}_{22} = v_2/(\sqrt{2}R)$. Numerical calculations were carried out for the range of non-dimensionlized parameters. The limits of the ranges were values beyond which the results of the calculation either did not change significantly or the model broke down and no convergent solution could be obtained.

A power law relationship of the form

$$\dot{E}_m \propto \Sigma_m^{n_m}, \qquad \dot{E}_e \propto \Sigma_e^{n_e} \qquad\qquad (12)$$

is commonly used to characterize rate dependent deformation, where n_m and n_e are material constants. This assumption is consistent with reported experimental observations [17 - 19]. In the powder consolidation models to date, exponents n_m and n_e are taken equal, i.e. $n_m = n_e = n$, where n is the nonlinearity exponent for the bulk deformation within the particles. Figure 2 shows numerical predictions for the mean strain rate, $\dot{E}_m$, plotted versus the normalized mean stress, Σ_m/E,

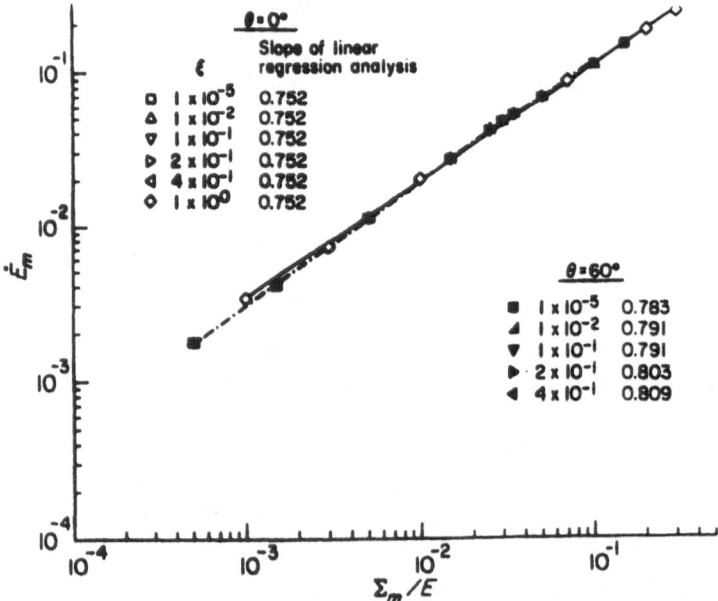

Figure 2. Plot of the mean strain rate, $\dot{E}_m$, versus the normalized mean stress, Σ_m/E, for the full range of the slip parameter ξ in the absence of interparticle diffusion, $\chi = 10^{-5}$, capillary stress parameter, $\zeta = 0$, and loading angles $\theta = 0^0$, 60^0.

in the absence of interparticle diffusion for the full range of the slip parameter ξ under both hydrostatic, $\theta = 0$, and uniaxial loading, $\theta = 60^0$. It appears that for both angles, the parameter n_m is clearly independent of the slip parameter ξ. However, a weak dependence on the loading angle can be deduced. In Figure 3 the effective strain rate, $\dot{E}_e$, is plotted against the normalized effective stress, Σ_e/E, in the presence of interparticle diffusion under no-slip conditions.

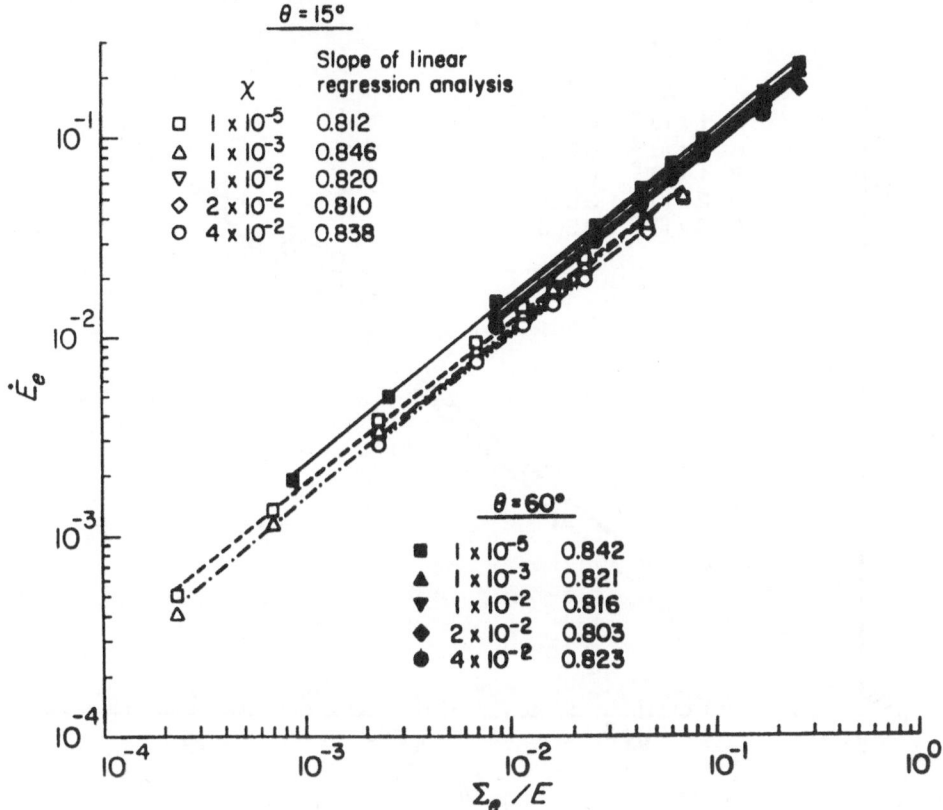

Figure 3. Plot of the effective strain rate, $\dot{E}_e$, versus the normalized effective stress, Σ_e/E, for the full range of the diffusion parameter χ, capillary stress parameter $\zeta = 0$, sticking interface, $\xi = 10^{-5}$, and loading angles $\theta = 15^0$, 60^0.

The calculations were carried out with the diffusion coefficient spanning the entire range from the no diffusion limit to very fast diffusion. For both loading angles, $\theta = 15^0$ and $\theta = 60^0$, it can be inferred from the regression analysis on the finite element results that the slope n_e depends rather weakly on the diffusion parameter χ, and exhibits a similar dependence on the loading angle θ to that of n_m. These results motivate the following form of the stress exponents

$$n_m = f_m(\theta, \chi, \zeta), \qquad n_e = f_e(\theta, \xi, \zeta). \tag{13}$$

In order to understand the form of the stress exponent functions, f_m and f_e, a comprehensive set of calculations was carried out over the full range of the arguments. For each set of calculations the stress exponent was determined using linear regression analysis. Typical results for the dependence of n_m on the diffusion parameter χ are shown in Figure 4 at loading angle $\theta = 30^0$, and capillary stress parameter $\zeta = 0$.

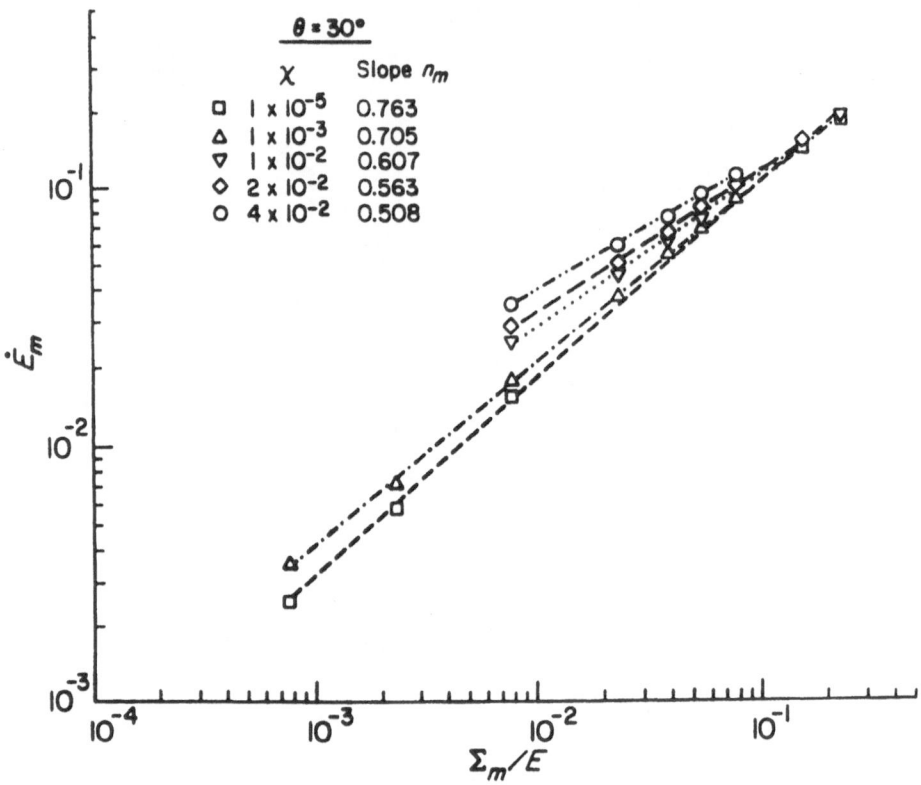

Figure 4. Dependence of the stress exponent n_m on the diffusion parameter χ for a loading angle $\theta = 30^0$ when the interface is sticking, $\xi = 10^{-5}$, and the capillary stress parameter ζ is zero.

Similarly, the dependence of n_e on the slip parameter ξ, at loading angle $\theta = 0$ and capillary stress parameter $\zeta = 0.02$ is shown in Figure 5.

The results of the finite element calculations are summarized in Tables 1 and 2. Both n_m and n_e are clearly decreasing functions of χ and ξ respectively. From the tables one can also see that there is a slight dependence of the stress exponents on the loading angle as well as the surface capillary stress parameter, ζ. Table 2 demonstrates that n_e is an increasing function of the loading angle if $\xi < 0.02$ or $\zeta \geq 0.02$. However, no similar dependence can be inferred from Table 1 for the exponent n_m.

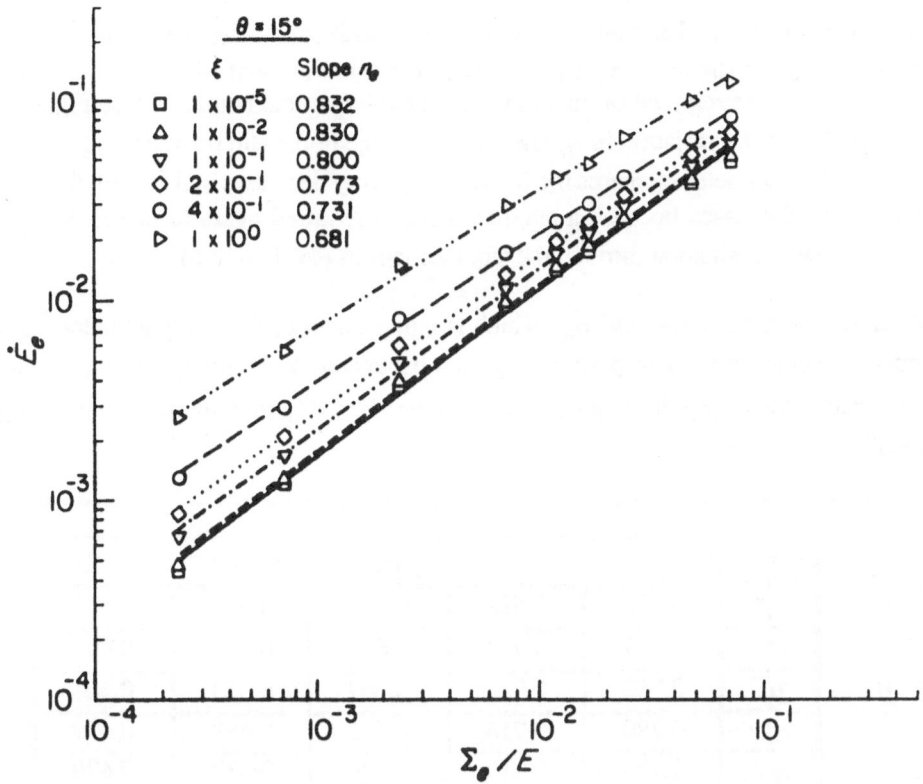

Figure 5. Dependence of the stress exponent n_e on the slip parameter ξ for a loading angle $\theta = 15^0$ in the absence of interparticle diffusion, $\chi = 10^{-5}$, and capillary stress parameter $\zeta = 0.02$.

5. Discussion

The form of the constitutive relation expressed by eqs. (12) and (13) reflects the natural separation observed in the results of the numerical calculations. The diffusion parameter is associated with the mean stress, while the slip parameter is related to the effective stress. In view of these relationships the following remarks are supported by the finite element calculations: a) The effective stress does not affect the size of the contact zone; b) the mean strain rate and the contact zone size are independent of the slip parameter. Physically, this indicates that the size of the contact is not affected by the viscosity of the interface under given applied mean stress; c) The effective strain rate is independent of the diffusion parameter. Again, this indicates that the amount of interfacial slip is not affected by the diffusive properties of the interface under a given applied effective stress. In view of these observations, a separation of the mean and effective stress as shown in eqs. (12) and (13) is justified.

The effect of the slip parameter and diffusion parameter can be seen in the change in the stress exponents n_m and n_e. Table 1 shows the variation in n_m over the range of the diffusion parameter χ. The value of n_m decreases with increasing diffusion parameter, indicating that less stress is required to cause the same amount of power dissipation through the diffusive removal of material from the interparticle boundary (see Figure 4). Table 2 illustrates the variation in n_e over the full range of the slip parameter, ξ. In this case as the slip parameter increases the interface viscosity decreases and in turn, the stress exponent, n_e, decreases because a smaller stress is required again to effect the same amount of power dissipation through the slip mechanism (see Figure 5).

TABLE 1. The stress exponent n_m, which is independent of the slip parameter ξ, for the range of values of the interparticle diffusion parameter, χ, capillary stress parameter ζ and loads which vary from purely hydrostatic, $\theta = 0^0$, to uniaxial compression, $\theta = 60^0$.

ζ	θ	χ				
		10^{-5}	10^{-3}	10^{-2}	2×10^{-2}	4×10^{-2}
0	0^0	0.752	0.698	0.613	0.593	0.552
	15^0	0.757	0.704	0.633	0.553	0.496
	30^0	0.763	0.705	0.607	0.563	0.508
	45^0	0.780	0.716	0.585	0.489	0.439
	60^0	0.783	0.718	0.576	0.538	0.479
1×10^{-2}	0^0	0.750	0.726	0.630	0.577	0.559
	15^0	0.779	0.732	0.638	0.585	0.563
	30^0	0.760	0.740	0.635	0.609	0.505
	45^0	0.778	0.719	0.640	0.568	0.533
	60^0	0.778	0.734	0.652	0.595	0.497
2×10^{-2}	0^0	0.752	0.749	0.637	0.634	0.570
	15^0	0.758	0.759	0.655	0.604	0.605
	30^0	0.787	0.744	0.681	0.607	0.524
	45^0	0.802	0.763	0.644	0.629	0.594
	60^0	0.752	0.747	0.631	0.590	0.549

An interesting new result of this investigation is the form of the dependencies in the stress exponents as stated by eqs. (13). In the work to date, the stress exponent for both hydrostatic and deviatoric components of the deformation was taken equal to that for the bulk deformation of the particle. Therefore, if that approach had been adopted in the present case of linearly deforming particles such an exponent for the hydrostatic part of the constitutive law would have been $n = 0.75$ in view of the plane deformation assumption in the model under investigation (see Table 1). However, as is deduced from the numerical results, this is not the case, and instead exponents n_m and n_e dependent on the slip and diffusion characteristics of the system should be used in the constitutive description of the material. In the case of n_m, the deviation of its numerically calculated value from 0.75 can be viewed as arising from the presence of the interparticle diffusion (see Figure 4) and its interaction with the bulk deformation of the particle in the process of development of the contact zone. The emergence of such a source for nonlinearity in the overall macroscopic behavior may be the rationale behind the experimentally observed feature of the consolidation law that the densification rate is proportional to the applied stress raised to the second power. Of course, a better 3-D model which also includes the free surface diffusion may be needed in order to quantitatively understand such a law.

TABLE 2. The stress exponent n_e, which is independent of the diffusion parameter χ, for the range of values of the interparticle slip parameter, ξ, capillary stress parameter ζ and loads which vary from $\theta = 15^0$, to uniaxial compression, $\theta = 60^0$.

ζ	θ	ξ					
		10^{-5}	10^{-2}	10^{-1}	2×10^{-1}	4×10^{-1}	1.0
0	15^0	0.812	0.808	0.752	0.709	0.654	0.573
	30^0	0.824	0.817	0.761	0.739	0.722	0.666
	45^0	0.842	0.838	0.773	0.730	0.668	
	60^0	0.842	0.850	0.793	0.748	0.701	
1×10^{-2}	15^0	0.832	0.832	0.801	0.773	0.733	0.685
	30^0	0.846	0.841	0.811	0.785	0.759	0.732
	45^0	0.856	0.856	0.810	0.788	0.758	
	60^0	0.856	0.868	0.835	0.822	0.823	
2×10^{-2}	15^0	0.832	0.830	0.800	0.773	0.731	0.681
	30^0	0.846	0.841	0.811	0.784	0.758	0.731
	45^0	0.863	0.865	0.822	0.807	0.785	
	60^0	0.856	0.865	0.833	0.817	0.822	

6. Conclusions

A model has been presented for the consolidation of powder compacts incorporating elasticity, interparticle diffusion and interface slip effects. A number of calculations over a wide range of conditions have been performed. The results agree with the experimental observation that the stress exponent for the macroscopic densification law is different from that for the bulk deformation in the powder particles. It can be inferred from the developed scaling laws under plain strain deformation conditions, that the aggregate behaves nonlinearly despite the linearity feature present in all mechanisms involved. It has been clearly demonstrated that the nonlinear nature of the macroscopic law arises from the interaction of the diffusion and relative slip mechanisms with the particle deformation along the interparticle contact area.

Acknowledgments

The research was supported by the Department of Energy under grant DEFGO2-91ER45439.

References

1.   N. A. Fleck, L. T. Kuhn and R. M. McMeeking, *J. Mech. Phys. Solids* **40**, 1139 (1992).
2.   A. R. Akisanya and A. C. F. Cocks, *J. Mech. Phys. Solids* **43**, 605 (1995).
3.   A. R. Akisanya, A. C. F. Cocks and N. A. Fleck, *J. Mech. Phys. Solids* **42**, 1067 (1994).
4.   N. A. Fleck, *Acta* Metall. **43**, 3177 (1995).
5.   L. T. Kuhn and R. M. McMeeking, *Int. J. Mech. Sci.* **34**, 563 (1992).
6.   R. M. McMeeking and L. T. Kuhn, *Acta Metall.* **40**, 961 (1992).
7.   J. Pan and A. C. F. Cocks, *Acta Metall.* **43**, 1395 (1995).
8.   D. Bouvard and R. M. McMeeking, *J. Am. Ceram. Soc.* In press.
9.   A. C. F. Cocks, *Mech. Mater.* **13**, 165 (1992).
10.   A. C. F. Cocks, *J. Mech. Phys. Solids* **37**, 693 (1989).
11.   P. Sofronis and R. M. McMeeking, *J. Appl Mech.* **59**, 88 (1992).
12.   J. M. Duva and P. D. Crow, *Acta Metall.* **40**, 31 (1992).
13.   H. Riedel, H. Zipse, and J. Svoboda, *Acta Metall.* **42**, 445 (1994).
14.   J. Pan and A. C. F. Cocks, *Acta Metall.* **42**, 1215 (1994).
15.   H. Riedel, V. Kozak, and J. Svoboda, *Acta Metall.* **42**, 3093 (1994).
16.   A. C. F. Cocks, *Acta Metall.* **42**, 2191 (1994).
17.   M. Uchic, H. J. Hofler, W. J. Flick, R. Tao, P. Kurath and R. S. Averback, *Scr. Met.* **26**, 791 (1992).
18.   R. S. Averback, private communication
19.   A. Casagranda, J. Xu, R. M. McMeeking and A. G. Evans, *J. Am. Ceram. Soc.* In press.
20.   V.G. Gryaznov, A.M. Kaprelov, and A.E. Romanov, *Scr. Met.* **23**, 1443 (1989).
21.   C. Herring, *J. Appl. Phys.* **21**, 437 (1950).
22.   A. Needleman and J. R. Rice, *Acta Metall.* **28**, 1315 (1980).
23.   T. J. Chuang, K. I. Kagawa, J. R. Rice and L. B. Sills, *Acta Metall.* **27**, 265 (1979).
24.   J. R. Rice and T. J. Chuang, *J. Am. Ceram. Soc.* **64**, 46 (1981).
25.   R. Hill, *J. Mech. Phys. Solids* **15**, 79 (1967).
26.   A. Casagranda and P. Sofronis, submitted to *Acta Metall.*
27.   J. C. Simo, P. Wriggers and R. L. Taylor, *Comput. Meth. Appl. Mech. Engng* **50**, 163 (1985).
28.   P. Sofronis and R. M. McMeeking, *Mech. Mater.* **18**, 55 (1994).

CONSTITUTIVE EQUATION FOR COMPACTION OF CERAMIC POWDERS

N.D. Cristescu, O. Cazacu and J. Jin
Department of Aerospace Engineering,
Mechanics & Engineering Science
University of Florida, Gainesville, Fl. USA.

1. Introduction

Die compaction is a widely applied technique for consolidation of ceramic powders. However, closed-die operations result in inhomogeneous densification. Die wall friction and complicated die shapes contribute to this. To reduce production costs, optimum processing is required. Thus, there is a need for a rational characterization and prediction of powder behavior.

A great number of empirical laws for the evolution of the green density as a function of external pressures have been proposed (see Heckel 1961, Kawachita and Ludde 1970, Shinohara 1984, Rong 1991, Glass *et al.*,1995). However, empirical laws are system specific. The use of these laws for conditions different than that for which they have been determined, would generally lead to erroneous results.

Microscopical models for consolidation have also been developed. Nevertheless, it seems difficult to apply such models to realistic complicated situations, such as the collapse of powder particles during densification or, the effect of irregularly shaped particles. From an engineering viewpoint a continuum approach appears to be the most appropriate.

Generally, continuum models for powders are rate-independent. It is assumed that at temperatures below 0.3 of the absolute melting temperature, compaction is by low temperature plasticity; at higher temperatures power law creep and diffusional flow aid densification (Fleck, 1995). For a wide range of powder morphologies and relative densities, powder yield behavior resembles that of granular materials. Thus, an important effort has been undertaken to adapt models developed for granular materials to powders. Puri and co-workers (see for example Puri *et al.*, 1995) applied a Cam-clay elastoplastic model to predict the stress-strain behavior of wheat flour. From the results of their investigation it can be concluded that this model does not accomodate the increase in stiffness with pressure, nor does it allow any dependence of its parameters on stress (Tripodi *et al.*, 1995). Also, it cannot describe both compressibility and dilatancy behavior. However, experimental data on powders reported by various authors (see for example, Shinohara 1984) suggest that there exists, for each confining pressure, a stress value for which the slope of the octahedral shear stress *versus* volumetric strain curve changes sign, i.e., above which the volume increases. Thus, the behavior of powders is not described correctly for undrained conditions by the critical state class of models. It appears then necessary to develop a constitutive model that properly accounts for the complex volumetric behavior of powders.

All the models discussed so far neglect the time effect on cold compaction. However,

N. A. Fleck and A. C. F. Cocks (eds.),
IUTAM Symposium on Mechanics of Granular and Porous Materials, 117–128.
© 1997 *Kluwer Academic Publishers.*

even at low temperatures, the strain-rate sensitivity of powders cannot be neglected (see Es-Saheb, 1993).

In the present paper an attempt is made to formulate a nonassociated elastic/viscoplastic constitutive equation for powder compaction. First, we present the experimental results concerning the compaction behavior of alumina powder. Then, based on the data, we formulate a 3-D elastic/viscoplastic model which describes the behavior of the material under general loading conditions. Finally, data from conventional triaxial compression tests are used to demonstrate the capability of the model to reproduce the commonly observed features of a powder's response.

2. Experimental Studies

Two types of alumina powder have been tested. One type was A10 as furnished by ALCOA, with the particle size within the range 40 - 200 μm. Another type was A16-SG, with particle size in the range 0.5 - 1.0 μm. A10 has an initial relative density of 0.36, while A16-SG has an initial relative density of 0.41. The theoretical density of alumina is 3.93 g/cm³.

The alumina powder was tested in a triaxial testing device of the kind used in soil mechanics. The first stage of the test is hydrostatic: the specimen is subjected to a pressure increased in successive steps. After each such increase the pressure is held constant for about 20 minutes. During this time interval the volume continues to decrease by creep. Fig.1 shows a typical volumetric creep curve for alumina A16-SG. The mean stress versus volumetric strain curve for A10 alumina is presented in Fig.2. The plateaux correspond to volumetric creep under constant mean stress (for 10 minutes). At the end of this time interval, when the volumetric strain rate is quite small, a small unloading-reloading cycle was applied in order

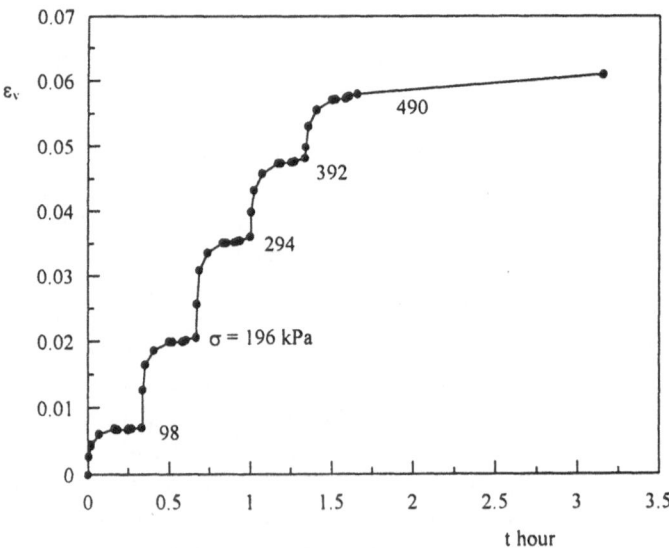

Figure 1.   Volumetric creep curves in hydrostatic tests of A16-SG.

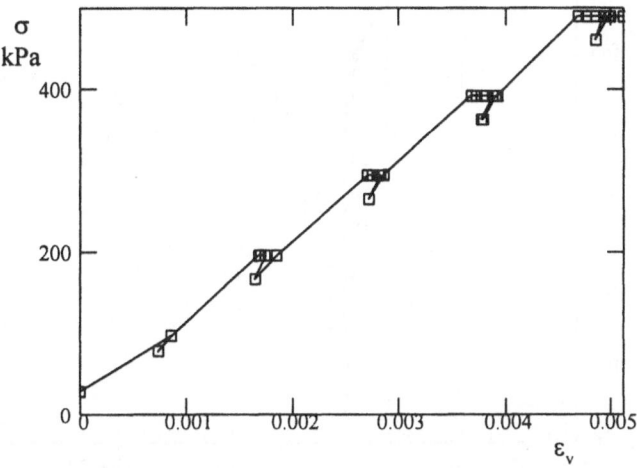

Figure 2. Mean stress - volumetric strain curves for dense A10 obtained in hydrostatic tests with small unloading following a short period of creep.

to determine the value of the bulk modulus. This procedure to determine K allows the rheological properties to be separated from the unloading characteristics (Cristescu, 1989). As expected, K increases with increasing mean stress σ. Since the pores close as the pressure is raised, at sufficiently high pressure, when most of the pores are closed, one can expect that further increase in pressure does not cause additional increase of K. This dependency can.be described by an exponential law of the form

$$K(\sigma) = K^{\infty} - p_a \exp\left(-b\frac{\sigma}{p_a}\right) \qquad (1)$$

where $K^{\infty} = 1 \times 10^7$ kPa is the constant value towards which the bulk modulus tends for high pressures, $\alpha = 10^7$, $b = -1.219 \times 10^{-4}$ and $p_a = 1$ kPa is a reference pressure (Cazacu et al., 1996). For small intervals of variation of σ, this law can be approximated by a linear one (Jin et al., 1997).

After reaching a hydrostatic predetermined pressure, in the second stage of the triaxial test the confining (lateral) pressure is kept constant, while the axial stress σ_1 is further increased. Fig.3 shows a typical octahedral shear stress τ versus strain curve for alumina A10. The material exhibits volumetric compressibility, for small values of τ, and dilatancy for higher values. At various stress levels the axial strain was kept constant, producing a significant stress-relaxation within minutes. When the stress rate reached a small enough value during relaxation(i.e., the material reached a quasi-stable state), a small unloading was performed. In Fig.3, about half of the stress decrease shown is due to relaxation, and half to unloading. By unloading after a short period of relaxation, the rheological influence becomes negligibly small, and thus one can accurately determine the elastic moduli (Cristescu, 1989).

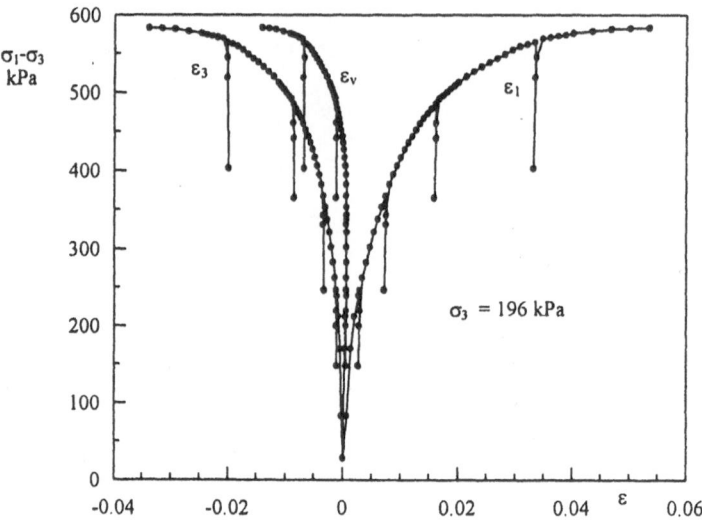

Figure 3. Stress-strain curves for alumina A10 in deviatoric tests. Half of the stress decreases shown correspond to relaxation and half to unloading.

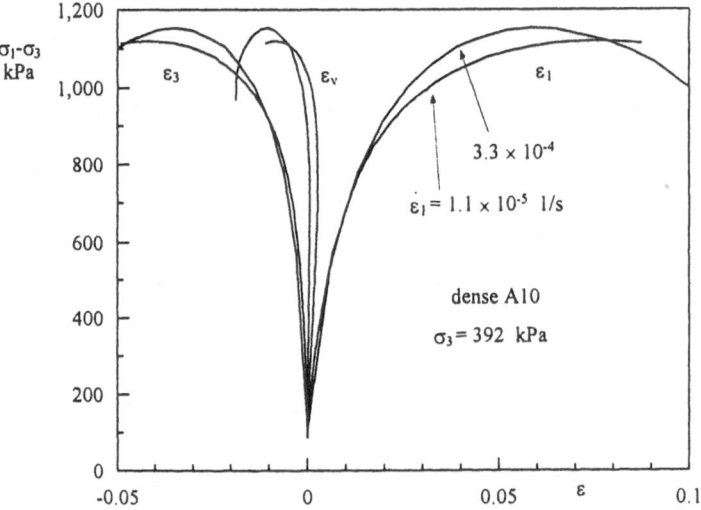

Figure 4. Strain rate influence on the mechanical response of alumina in triaxial tests.

Alumina powder also exhibits rate effects. Fig.4 shows stress-strain curves obtained for two different strain rates. Though the two strain rates are not significantly different, a marked influence of the strain rate is observed. For practical applications, it is important to point out that a lower loading rate produces more compressibility and more dilatancy. It is clearly seen (Fig.3 and Fig.4) that additional compaction can be obtained by superposing a certain amount of shearing stress on the hydrostatic pressure. The magnitude of shearing stress that has to be added to get the minimum volume depends on the confining pressure. This is shown in Fig.5. Each curve starts from the value of ε_v at the end of the hydrostatic stage of the triaxial test. A black dot indicates when alumina passes from compressibility to dilatancy in the deviatoric part of the test. It is clearly seen that the stress value for which the slope of the $\tau - \varepsilon_v$ curve changes sign, depends on the confining pressure. These critical points can be used to define the boundary between compressible and dilatant domains in the $\sigma\tau$ - plane (Cristescu, 1989,1994a). For A10 this boundary is

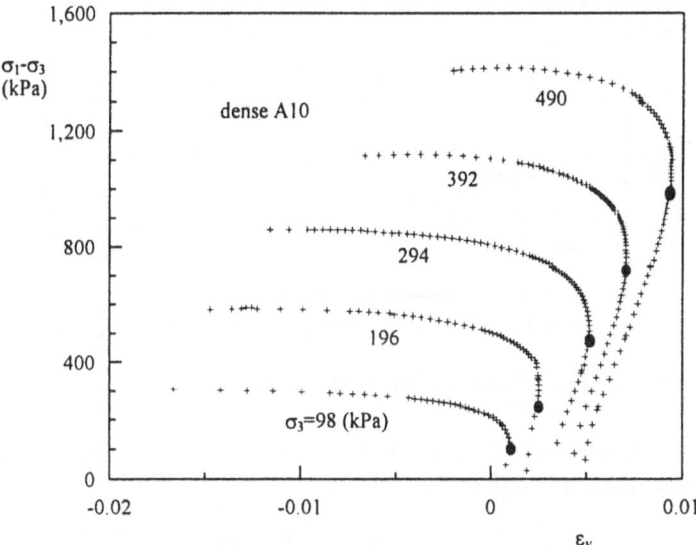

Figure 5. Deviatoric stress versus volumetric strain relationship for various confining pressures for dense A10 alumina.

$$X(\sigma,\tau) = 0.555\,\sigma - \tau . \tag{2}$$

For the other alumina tested the coefficients are slightly different.

It is interesting to mention that not all alumina powders exhibit compressible followed by dilatant behavior in deviatoric tests. Fig.6 shows the volumetric behavior of fine particle size alumina A16-SG. Comparing Fig.5 and Fig.6 one can see that the particle size has a fundamental influence on the mechanical characteristics of alumina powder: alumina powder with a large particle size is first compressible and afterwards dilatant in deviatoric tests, while alumina of small particle size is compressible throughout the test for any stress state.

3. The elastic/viscoplastic model for alumina

Since the experiments have shown that alumina powder exhibits time effects and that unloading is elastic, the model formulated is of **elastic/viscoplastic nonassociated type**. Two variants have been formulated for A10. The first variant of the constitutive equation is presented below (Cristescu 1989, 1994a, 1994b; Cazacu 1995).

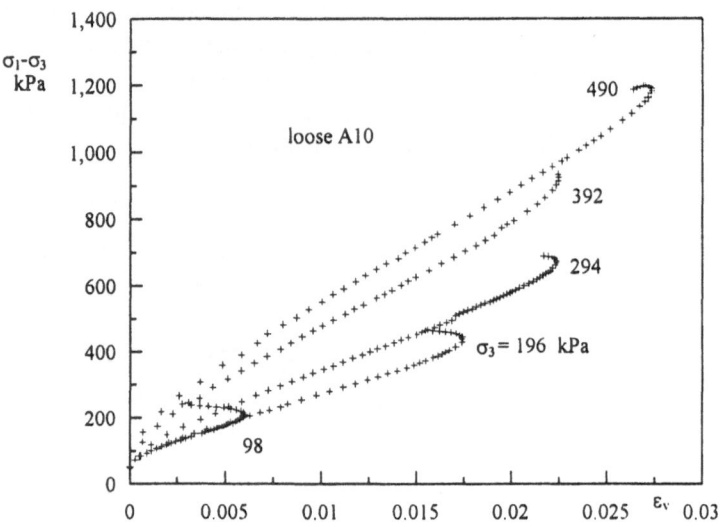

Figure 6. Same as in Figure 5 but for A16-SG alumina.

The strain rate $\dot{\varepsilon}$ is written as a superposition of the elastic response $\dot{\varepsilon}^E$ and the viscoplastic (irreversible) $\dot{\varepsilon}^I$ response, as follows:

$$\dot{\varepsilon} = \dot{\varepsilon}^E + \dot{\varepsilon}^I = \frac{\dot{\sigma}}{2G} + \left(\frac{1}{3K} - \frac{1}{2G}\right)\dot{\sigma}\mathbf{1} + k\left\langle 1 - \frac{W(t)}{H(\sigma)}\right\rangle N(\sigma).$$ (3)

Here, K and G are the bulk modulus and shear modulus (not necessarily constant), respectively, $\sigma = (1/3)Tr(\sigma)$ the mean stress, and $\mathbf{1}$ the identity tensor. The last term corresponds to the irreversible strain rate $\dot{\varepsilon}^I$, with $H(\sigma)$ the yield function, $N(\sigma)$ an irreversible stress orientation tensor, and k the viscosity. The hardening parameter is the irreversible stress work per unit volume $W(t)$ (energy stored by the body if compressibility takes place, or energy released if the body is dilatant), defined as:

$$W(T) = \int_0^T \sigma(t)\dot{\varepsilon}_v^I(t)\,dt + \int_0^T \sigma'(t):\dot{\varepsilon}^{\prime I}\,dt$$ (4)

where $\dot{\varepsilon}_v^I$ is the volumetric part of $\dot{\varepsilon}^I$, and "prime" stands for deviator. The second term on the right-hand side is always positive and represents the energy related to the change in shape. The first term on the right-hand side is the energy related to volume change. It increases in the compressibility domain and decreases in the dilatancy domain. Thus the irreversible volumetric work can be thought of as a damage indicator. The procedure used to determine the constitutive coefficients and constitutive functions is described by Cazacu *et al.* (1996). Let us summarize the main steps. First, the **yield function** $H(\sigma)$ is determined as a sum of two terms:

$$H(\sigma) = H_H(\sigma) + H_D(\sigma,\tau) \tag{5}$$

such that $H_D(\sigma,0) \equiv 0$. Thus, for hydrostatic conditions the yield function reduces to H_H. For the determination of H_H the irreversible volumetric work is determined from hydrostatic creep tests (first term of (4)). This work increases monotonically as the pressure increases. A second order polynomial provides a good fit to the data for A10 alumina. Assuming that $H_H(\sigma) = W(t)$ is the stabilization boundary, for hydrostatic tests we have

$$H_H(\sigma) = a_h \left(\frac{\sigma}{p_a} \right)^2 + b_h \frac{\sigma}{p_a} \tag{6}$$

where $a_h = 5.833 \times 10^{-6}$, $b_h = 1 \times 10^{-6}$, and $p_a = 1\,kPa$ is a reference pressure.

Further, $H_D(\sigma,\tau)$ can be computed from the deviatoric part of the compression tests. The energy dissipated is obtained using the formula

$$W(T) = \int_{T_H}^{T} \left(\frac{3}{\sqrt{2}} \tau \right) \dot{\varepsilon}_1^I(t)\,dt + \int_{T_H}^{T} \sigma_3\, \dot{\varepsilon}_v^I\,dt \tag{7}$$

where T_H represents the beginning of the deviatoric part of the test and $\dot{\varepsilon}_1^I$ is the irreversible part of axial rate of deformation. Figure 7 shows the energy dissipated obtained from 6 triaxial compression tests on alumina with confining pressures $\sigma_2 = \sigma_3 = 98; 196; 294; 343; 392$ and $490\,kPa$, respectively. A possible function that approximates the energy dissipation contours is

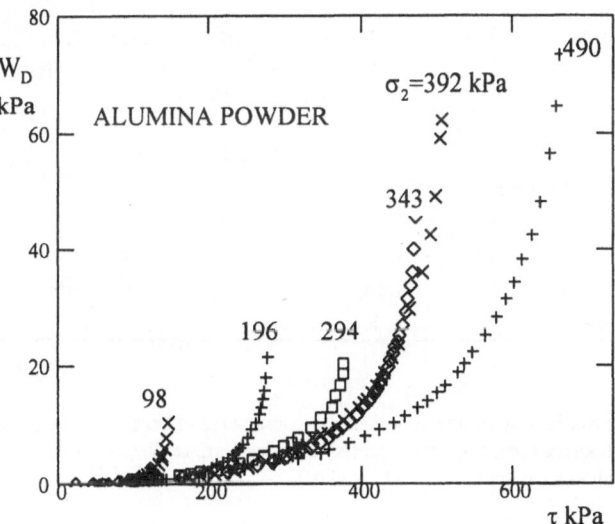

Figure 7. Variation with τ of irreversible stress work for six confining pressures.

$$H_D(\sigma,\tau) = A\left(\frac{\tau}{p_a}\right)^9 + B\left(\frac{\tau}{p_a}\right)^2. \tag{8}$$

It was found that B can be considered constant $B = 4.6 \times 10^{-5}$ and A depends on the confining pressure:

$$A(\sigma_3) = a_1\left(\frac{\sigma_3}{p_a} + a_2\right)^{-m} \tag{9}$$

with $a_1 = 0.27$, $a_2 = 37$ and $m = 8.5$. Since $\sigma_3 = \sigma - \tau/\sqrt{2}$, it follows that the expression for the yield function in terms of invariants is:

$$H(\sigma,\tau) = a_h\left(\frac{\sigma}{p_a}\right)^2 + b_h\frac{\sigma}{p_a} + a_1\left(\frac{\sigma - \tau/\sqrt{2}}{p_a}\right)^{-m}\left(\frac{\tau}{p_a}\right)^9 + B\left(\frac{\tau}{p_a}\right)^2. \tag{10}$$

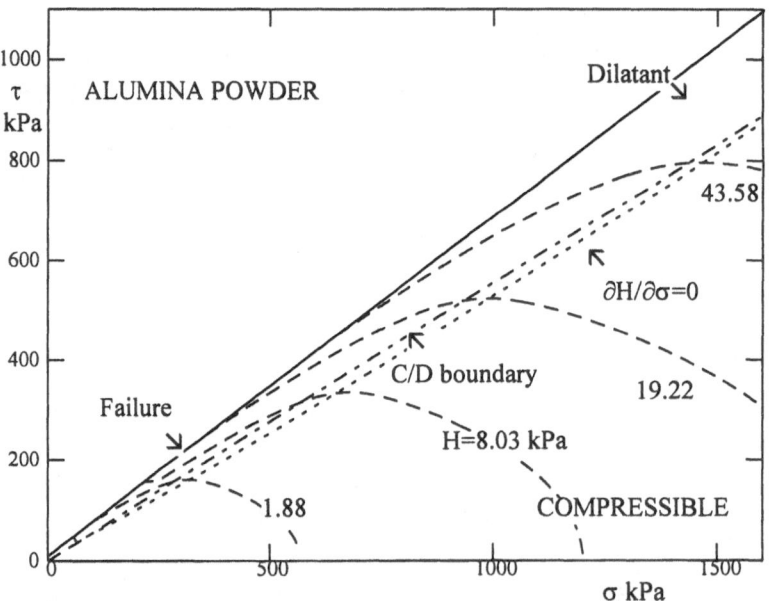

Figure 8. Shape of several yield surfaces H=constant (solid lines), compressibility/dilatancy boundary (dash-dot line), stability boundary surface $\partial H/\partial\sigma$=0 (dotted line), and short-term failure surface (solid line).

Fig.8 shows the shape of several surfaces of constant H in the $\tau\sigma$- plane. The short-term failure surface is shown as a full line while the compressibility/dilatancy boundary as a dash-dot line. The line $\partial H/\partial\sigma = 0$ is shown as an interrupted line. In an associated model this last line would coincide with the compressibility/dilatancy boundary. For a saturated powder the domain situated between the compressibility/dilatancy boundary and the line $\partial H/\partial\sigma = 0$ is a domain of possible loss of stability (Cristescu, 1991). The domain between the compressibility/dilatancy boundary and the short-term failure surface is the dilatancy domain. All the stress states under the compressibility/dilatancy boundary produce compaction. From this figure one can conclude again that a shear (i.e., τ) superposed on a hydrostatic pressure (i.e., σ) produces additional compaction. In fact, for each value of σ the maximal compaction is obtained for the corresponding value of τ located on the compressibility/dilatancy boundary (see also Fig.5).

Further, the **viscoplastic strain rate orientation tensor** $N(\sigma)$ is determined. As the material is isotropic, N must satisfy the invariance requirement

$$N(Q\sigma Q^{T}) = QN(\sigma)Q^{T} \tag{11}$$

for any orthogonal transformation Q. Using classical theorems of representation of isotropic tensor functions, and disregarding terms involving σ^2, $N(\sigma)$ can be defined as:

$$N(\sigma) = N_1 1 + N_2\frac{\sigma'}{\tau} \tag{12}$$

with $N_i = N_i(\sigma,\tau)$. Since

$$\dot{\varepsilon}^I = k\left\langle 1 - \frac{W(t)}{H(\sigma)}\right\rangle\left\{N_1 1 + N_2\frac{\sigma'}{\tau}\right\} \tag{13}$$

it follows that

$$\dot{\varepsilon}_v^I = 3k\left\langle 1 - \frac{W(t)}{H(\sigma)}\right\rangle N_1. \tag{14}$$

Using this equation and the data from the hydrostatic part of the test, we determine $kN_1|_{\tau=0}$. A second order polynomial in σ seems to fit the data well:

$$kN_1|_{\tau=0} = \varphi(\sigma) = a_k\left(\frac{\sigma}{p_a}\right)^2 + b_k\left(\frac{\sigma}{p_a}\right) \tag{15}$$

where $a_k = -8\times10^{-11}$ and $b_k = 1.64\times10^{-6}$. Next we must determine $N_1(\sigma,\tau)$ for $\tau\neq0$. From eqn.(14) it follows that this function must be positive in the compressibility domain and negative in the dilatancy one. A possible simple function satisfying these properties is

$$kN_1 = \varphi(\sigma) + \frac{\tau}{P_a} sgn X(\sigma,\tau)\, \psi(\sigma,\tau)\,, \tag{16}$$

where $X(\sigma,\tau)$ is defined by eqn.(1). Using the data from the deviatoric part of the triaxial test, ψ can be determined from eqns. (14) and (16):

$$\psi(\sigma,\tau) = \frac{\dfrac{\dot{\varepsilon}_v^I}{3\left(1 - \dfrac{W(t)}{H(\sigma)}\right)} - \varphi(\sigma)}{\dfrac{\tau}{P_a} sgn X(\sigma,\tau)} \qquad where \quad sgn X(\sigma,\tau) = \begin{cases} 1 & for \;\; X(\sigma,\tau) \ge 0 \\ 0 & otherwise \end{cases} \tag{17}$$

For A10 alumina a possible expression for $\psi(\sigma,\tau)$ is

$$\psi(\sigma,\tau) = z_a \left[\frac{\tau}{P_a} - y_a \left(\frac{\sigma - \tau/\sqrt{2}}{P_a} \right) - y_c \right]^5 \exp\left(z_b \frac{\sigma - \tau/\sqrt{2}}{P_a} \right) \tag{18}$$

where $y_a = 0.84$, $y_c = -56.06$, $z_b = -0.01$ and $z_c = 1.127 \times 10^{-15}$. From the flow rule (13) it follows that

$$kN_2(\sigma,\tau) = \frac{3}{\sqrt{2}} \frac{\dot{\varepsilon}_1^I - \dot{\varepsilon}_3^I}{\left(1 - \dfrac{W(t)}{H(\sigma)}\right)} \tag{19}$$

From the deviatoric part of the triaxial tests performed at $\sigma_3 = 294$; 392; and 490 kPa, in conjunction with eqn.(19) we found that kN_2 can be approximated by

$$kN_2(\sigma,\tau) = m_1 \left(\frac{\tau}{P_a} \right)^{14} \exp\left[m_2 \left(\frac{\sigma - \tau/\sqrt{2}}{P_a} \right) \right] + n_1 \left(\frac{\tau}{P_a} \right)^2 \exp\left[n_2 \left(\frac{\sigma - \tau/\sqrt{2}}{P_a} \right) \right] + r \tag{20}$$

where $m_1 = 2.38 \times 10^{-33}$, $m_2 = -0.038$, $n_1 = 4 \times 10^{-7}$, $n_2 = -0.005$ and $r = 8 \times 10^{-4}$.

4. Comparison with experimental data

The accuracy of the elastic/viscoplastic nonassociated model described previously was evaluated by comparing predicted and measured strains in drained compression tests at different confining pressures. Fig.9 shows the stress-strain curves obtained in triaxial tests

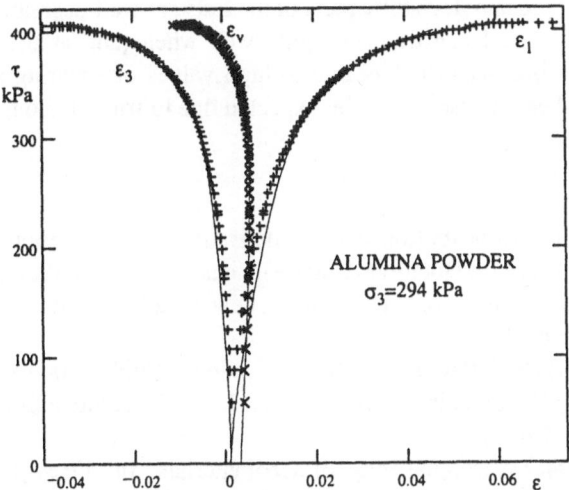

Figure 9. Stress-strain curves in deviatoric test with $\sigma_3 = 294$ kPa: symbols - experimental data, solid lines - model prediction.

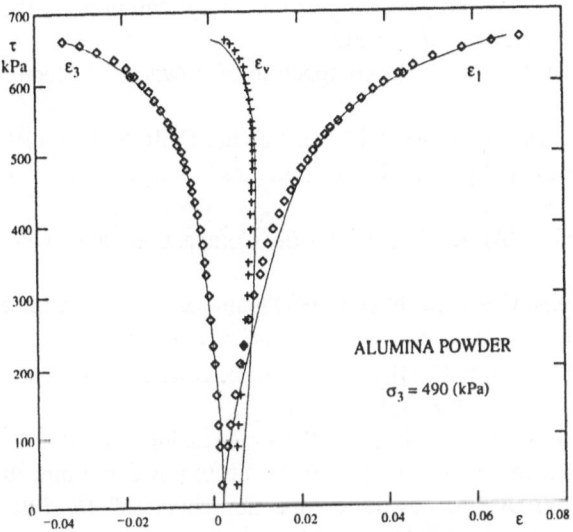

Figure 10. Same as in Fig.9 but for $\sigma_3 = 490$ kPa. The strains obtained in the first, hydrostatic, stage of the test have also been accounted for.

with the confining pressure $\sigma_3 = 294$ kPa. The symbols are the experimental data and the solid lines the model prediction. The model predicts reasonable well both volumetric compressibility and dilatancy. The strains obtained in the first, hydrostatic, stage of the test have not been accounted for. Figure 10 shows similar results for $\sigma_3 = 490$ kPa. In this case the values of the strains at the end of the first, hydrostatic, stage of the test have been taken into account and shown by the starting point of these curves on the strain axis. The results shown in this figure are obtained with another, more involved variant of the model. Fig.10 shows a good correlation of the model prediction with experimental data. Compressibility obtained in both the hydrostatic and deviatoric parts of the tests are well described. Fig.9 and Fig.10 also show that the model can describe quite well when (and at which stress state) the compacted powder becomes dilatant due to high values of shearing stresses. In powder compaction large shear stresses are to be expected due to friction along the die walls.

5. References

Cazacu, O. (1995)   Contribution à la modelisation élasto-viscoplastique d'une roche anisotrope. Thèse de Doctorat, Université des Sciences et Technologie de Lille.

Cazacu, O., Jin, J. and Cristescu, N.D. (1996) A new model for alumina powder compaction. KONA (in press).

Cristescu, N. (1989) *Rock Rheology*, Kluwer Academic Publishers, Dordrecht.

Cristescu, N. (1991) Nonassociated Elastic/Viscoplastic Constitutive Equations for Sand. *Int. J. Plasticity*, **7**,41-64.

Cristescu, N. (1994a) Viscoplasticity of Geomaterials. In: *Visco-plastic Behaviour of Geomaterials*. (Eds.N.D.Cristescu and G. Gioda) Springer Verlag, Wien-New York, 103-207.

Cristescu, N. (1994b) A procedure to determine nonassociated constitutive equations for geomaterials. *Int.J. Plasticity*, **10**, 103-131.

Es-Saheb, M.H.H. (1993) "Powder compaction interpretation using the power law"*Journal of Materials Science*. **28.** p 1269.

Fleck, N. A.(1995) "On the cold compaction of powders" J.Mech.Phys.Solids, **43**. No. 9, p 1409.

Glass, H.J., With,G.De, Graaf, M. J. M.De, Van der Drift, R. J. A. (1995) "Compaction of homogeneous (Mn, Zn)-ferrite potcores" *Journal of Materials Science*. **30**. p 3162.

Heckel, R. W. (1961) ."An Analysis of Powder Compaction Phenomena" Trans AIME, **221.**, p 1001.

Jin, J., Cazacu, O. and Cristescu, N.D. (1997) Compaction of ceramic powders. *Proc. Fifth American Congress of Applied Mechanics* (in press).

Kawachita, K. and Ludde, K.H. (1970/1971)"Some considerations on Powder Compression Equations" *Powder Technol.* **4**. p 61.

Puri, V. M ,Tripodi,M. A., Manbeck, H.B., and Messing G.L (1995) ."Constitutive model for dry cohesive powders with applications to powder compaction" *Kona*. **13.**

Rong, Ge-de (1991) "A new powder compaction equation" *The International Journal of Powder Metallurgy*. **27**. No. 3.p 211.

Shinohara,K. (1984) "Rheological properties of particulate solids." In: *"Handbook of Powder Science and Technology"* (Eds.M.E. Fayed, and L. Otten.) Van Nostrand Reinhold Comp., New York. pp.129-169.

Tripodi,M.A., Puri,V. M., Manbeck, H.B., and.Messing, G.L (1995) "Elastoplastic finite element model development and validation for low uniaxial compaction of dry cohesive powders", *Powder Technol*. **85**.p 241.

A CONSTITUTIVE MODEL FOR THE SINTERING OF FINE GRAINED ALUMINA

JAN MA
Cambridge University Engineering Department
Trumpington Street, Cambridge CB2 1PZ, UK

ALAN C.F. COCKS
Department of Engineering, Leicester
University, Leicester LE1 7RH, UK

1. Introduction

In recent years a number of constitutive laws have been proposed for the compaction and sintering of a range of engineering materials. These models can be divided into two groups, namely micromechanical and empirical models. Micromechanical constitutive laws have been developed from simple models of the way in which the microstructure evolves during the compaction process. These models provide detailed forms of relationships which can then be compared and calibrated against experimental data. Different forms of constitutive law have been proposed for different mechanisms and different stages of the compaction process. In the empirical models the basic form of the constitutive law is assumed and suitable experiments are devised in order to determine any unknown functional relationships. The choice of basic structure of the constitutive law might be guided by certain assumptions about the underlying mechanisms of deformation and densification, but no detailed modelling of the mechanisms is undertaken. In this paper we critically examine the constitutive relationships that have been proposed for the sintering of ceramic components and determine the most appropriate structure of constitutive law for the sintering of fine grained alumina. We limit our attention here to Stage 1 of the sintering process, ie when the porosity is in the form of interconnected channels. Stage 2 sintering, ie when the channels have pinched-off to leave isolated pores, is considered by Ma (1997). The transition from Stage 1 to Stage 2 sintering occurs at a relative density of about 0.95.

During the sintering of a ceramic component the material densifies and the mean grain size of the material generally increases. It is then appropriate to express the material response in terms of two state variables; the relative density, ρ , and the mean grain size, L. In ceramic components it is generally accepted that diffusional mechanisms dominate the response. The rate of deformation can then be either controlled by the rate of diffusional transport between the sources and sinks or the rate

129

N. A. Fleck and A. C. F. Cocks (eds.),
IUTAM Symposium on Mechanics of Granular and Porous Materials, 129–138.
© 1997 Kluwer Academic Publishers.

at which the sources and sinks can provide, or accept, material for the diffusional process. We refer to the situation where the rate of operation of the sources and sinks is controlling as interface reaction controlled sintering. In the following section we examine the general structure of constitutive model proposed by Cocks (1994), in which the strain-rate and the rate of change of the state variable ρ are determined from a scalar potential. Mechanistic and empirical forms of the potential are presented for grain-boundary diffusion and interface reaction controlled sintering. The full model also requires equations for the rate of change of L. We bypass the need to completely specify these relationships by expressing the material response in terms of grain-size normalised quantities. The procedures for doing this are fully described in section 3. Constitutive laws for the grain-growth process are described by Ashby (1990) and Du and Cocks (1992). Two different methods of determining the sintering potential are employed for a given assumed constitutive law. It is demonstrated through these procedures that the micromechanical interface reaction controlled model of Cocks (1994) adequately describes the sintering response of fine grained alumina.

2. Constitutive Laws for the Sintering of Engineering Materials

Cocks (1994) has demonstrated that the deformation of a porous material can be expressed in terms of a scalar strain-rate potential

$$\phi = \phi\left(\sigma_{ij}, \sigma_s, \rho, L\right)$$

which is a function of stress, σ_{ij}, sintering potential, σ_s , and the two state variables ρ and L, such that

$$\dot{\varepsilon}_{ij} = \frac{\partial \phi}{\partial \sigma_{ij}} \quad \text{and} \quad \dot{\rho} = \rho \frac{\partial \phi}{\partial \sigma_m} \tag{1}$$

where $\sigma_m = \frac{1}{3}\sigma_{kk}$ is the mean stress.

2.1 GRAIN-BOUNDARY DIFFUSION CONTROLLED SINTERING

Micromechanical and empirical constitutive laws for grain-boundary diffusion controlled sintering can be expressed in the same general form:

$$\phi_b = \frac{1}{2}\dot{\varepsilon}_{ob}\sigma_o \left(\frac{L_o}{L}\right)^3 \left[c_b(\rho)\left(\frac{\sigma_e}{\sigma_o}\right)^2 + 9f_b(\rho)\left(\frac{\sigma_m - \sigma_s}{\sigma_o}\right)^2 \right] \tag{2}$$

where the subscript 'b' indicates boundary controlled diffusion, $\dot{\varepsilon}_{ob}$ is the uniaxial strain-rate under an applied stress σ_o for a fully dense material of grain size L_o , $f_b(\rho)$ and $c_b(\rho)$ are dimensionless functions of the relative density and $\sigma_e = \sqrt{\frac{3}{2} s_{ij} s_{ij}}$, where s_{ij} are the deviatoric components of stress, is the von Mises effective stress.

The only difference between the mechanistic and empirical models is the detailed form of the functions $f_b(\rho)$, $c_b(\rho)$ and σ_s , which is a function of the state of the material, (Du and Cocks, 1992).

2.2 INTERFACE REACTION CONTROLLED SINTERING

Unlike grain-boundary diffusion controlled sintering, different forms of strain-rate potential have been proposed in the literature for interface reaction controlled sintering. Here we consider the micromechanical model developed by Cocks (1994) and an empirical model proposed by Besson and Abouaf (1992). The strain-rate potential obtained by Cocks (1994) is:

$$\phi_r = \tfrac{1}{3}\dot{\varepsilon}_{or}\sigma_o \frac{L_o}{L}\left[c_r(\rho)\left(\frac{\sigma_e}{\sigma_o}\right)^{3/2} + f_r(\rho)\left(\frac{\left|3(\sigma_m - \sigma_s)\right|}{\sigma_o}\right)^{3/2} \right]^2 \tag{3}$$

where the subscript 'r' indicates interface reaction controlled sintering and, as before, $\dot{\varepsilon}_{or}$ is the strain-rate in a uniaxial test at a stress σ_o on a fully dense material of mean grain size L_o . The dimensionless functions $c_r(\rho)$ and $f_r(\rho)$ proposed by Cocks (1994) and Du and Cocks (1994) are

$$f_r(\rho) = \frac{1}{3}\frac{(1-\rho_o)}{\rho^{3/2}\rho_o^{1/2}(\rho-\rho_o)} = \frac{1}{2}c_r(\rho) \tag{4}$$

where ρ_o is the initial relative density of the compact.

The strain-rate potential obtained from the model of Besson and Abouaf (1992) is

$$\phi_r = \tfrac{1}{3}\dot{\varepsilon}_{or}\sigma_o \frac{L_o}{L}\left[c_r(\rho)\left(\frac{\sigma_e}{\sigma_o}\right)^2 + f_r(\rho)\left(\frac{3(\sigma_m - \sigma_s)}{\sigma_o}\right)^2 \right]^{3/2} \tag{5}$$

where now

$$f_r(\rho) = \left[20.19(1-\rho)\right]^{1.572} \quad \text{and} \quad c_r(\rho) = 1 + \left[28.55(1-\rho)\right]^{1.678} \tag{6}$$

2.3 THE SINTERING POTENTIAL

It was noted earlier that the sintering potential σ_s is a function of the state of the material, which we measure in terms of the relative density and mean grain-size. Theoretical studies of sintering (Ashby, 1990; McMeeking and Kuhn, 1992; Cocks, 1994; and Svoboda et al, 1994) suggest a functional relationship of the form

$$\sigma_s = \frac{\gamma_s}{L} h(\rho) \qquad (7)$$

where γ_s is the surface energy and $h(\rho)$ is a dimensionless function of ρ and γ_s/γ_b, where γ_b is the grain-boundary energy, which is constant for a given material.

If we know, or assume, the general structure of the constitutive law, we can devise an experimental programme to determine any functional relationships contained within it, such as $f_b(\rho)$, $c_b(\rho)$ of (2), $c_r(\rho)$ and $f_r(\rho)$ of (3) and (5) and $h(\rho)$ of (7) and compare the experimental values with the theoretical predictions. In the following section we initially concentrate on the determination of $h(\rho)$. In this section we have presented three different constitutive models. If we were to assume one form of model it would be possible to determine a functional form for $h(\rho)$ that provides a reasonable fit of the chosen model to the data. This does not, however, mean that the assumed model is the most appropriate nor that it adequately reflects the underlying mechanism of compaction. In the following, we therefore present two independent methods of determining $h(\rho)$. If the two methods produce completely different values of $h(\rho)$ for a given constitutive relationship, then it is evident that the assumed form of model is inappropriate. The most appropriate model is that which produces similar functional forms for $h(\rho)$, although it is not possible at this stage to categorically state that this is the most exact model.

3 Experimental Determination of Sintering Potential

In this section we examine how the results of free sintering and uniaxial forging experiments can be used to determine the function $h(\rho)$ in the sintering potential expression of (7). The form of the constitutive law suggests how the material data should be manipulated and plotted. We consider two strategies here. In the first, which we refer to as method I, we determine the ratio of the volumetric to effective strain-rate at a given density. In the second, method II, we compare the volumetric strain rate in the forge tests with that in the free sintering experiments at the same relative density. Experiments were conducted on 99.99% pure AKP30 Sumitomo alumina powders with an average initial grain size of 0.3 μm. Full experimental details are given by Ma (1997).

3.1 GRAIN-BOUNDARY DIFFUSION CONTROLLED SINTERING

Here we assume that the sintering response is controlled by the rate of grain-boundary diffusion and employ the strain-rate potential of (2) to model the material response. Differentiating (2) with respect to the effective stress σ_e and mean stress σ_m to obtain the effective and volumetric strain-rates, $\dot{\varepsilon}_e$ and $\dot{\varepsilon}_v$, gives

$$\frac{\dot{\varepsilon}_v}{\dot{\varepsilon}_e} = 9\frac{\overline{\sigma}_s}{\sigma}\frac{L_o}{L}\frac{f(\rho)}{c(\rho)} + 3\frac{f(\rho)}{c(\rho)} \tag{8}$$

where σ is the uniaxial stress employed in the forging experiments,

$$\overline{\sigma}_s = \frac{\gamma_s}{L_o}h(\rho) \tag{9}$$

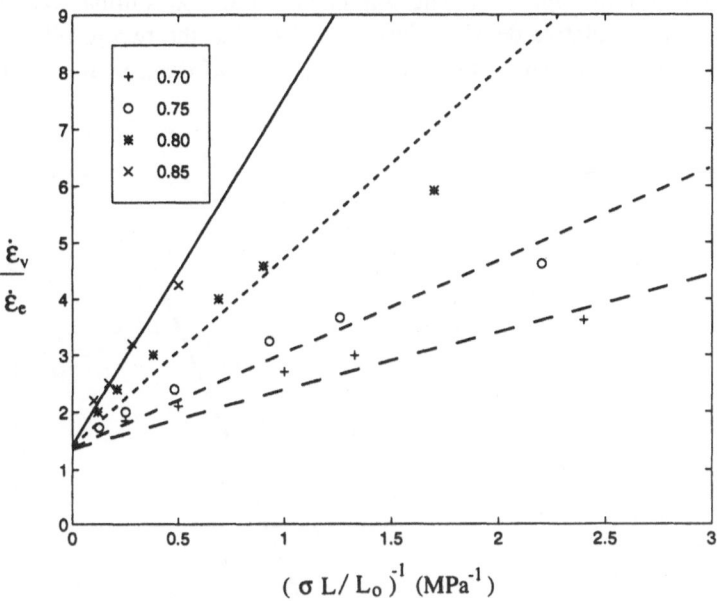

Figure 1 The ratio of volumetric to effective strain-rate determined from the forged sintering experiments plotted as a function of $(\sigma L/l_v)^{-1}$ for a range of relative densities.

and L_o is interpreted as the initial mean grain size of the material. From (2) it is evident that the grain-size compensated sintering potential of (9) and the ratio $f(\rho)/c(\rho)$ can be determined from a plot of $\dot{\varepsilon}_v/\dot{\varepsilon}_e$ against $(\sigma L/L_o)^{-1}$. Figure 1 shows plots for a range of values of ρ, together with the best linear fits through the data. The resulting variation of $\overline{\sigma}_s$ with ρ is labelled I in Figure 2.

Alternatively, we can compare the volumetric strain-rates obtained in the forged and freely sintered sintered samples at the same relative density. Again, differentiating (2) with respect to σ_m and setting σ_m to $\sigma/3$, to determine the forged volumetric strain-rate, $\dot{\varepsilon}_{v\sigma}$, and to zero to determine the freely sintered strain rate, $\dot{\varepsilon}_{vF}$, we obtain

$$\frac{\dot{\varepsilon}_{v\sigma}}{\dot{\varepsilon}_{vF}}\left(\frac{L_\sigma}{L_F}\right)^4 = \frac{\sigma}{3\bar{\sigma}_s}\frac{L_\sigma}{L_o}+1 \qquad (10)$$

where the subscripts 'σ' and 'o' refer to the forged and free sintering results respectively. The grain-size compensated sintering potential can be determined from

the slope of grain size compensated volumetric strain-rate ratio $\frac{\dot{\varepsilon}_{v\sigma}}{\dot{\varepsilon}_{vF}}\left(\frac{L_\sigma}{L_F}\right)^4$ against

$\sigma L_\sigma / L_o$. The resulting variation of $\bar{\sigma}_s$ with ρ is compared with that obtained from the previous method in Figure 2. The approaches produce similar values of the sintering potential for relative densities less than 0.75, but the two results diverge as the relative density is increased further, with a factor of 4 difference when ρ=0.85. We

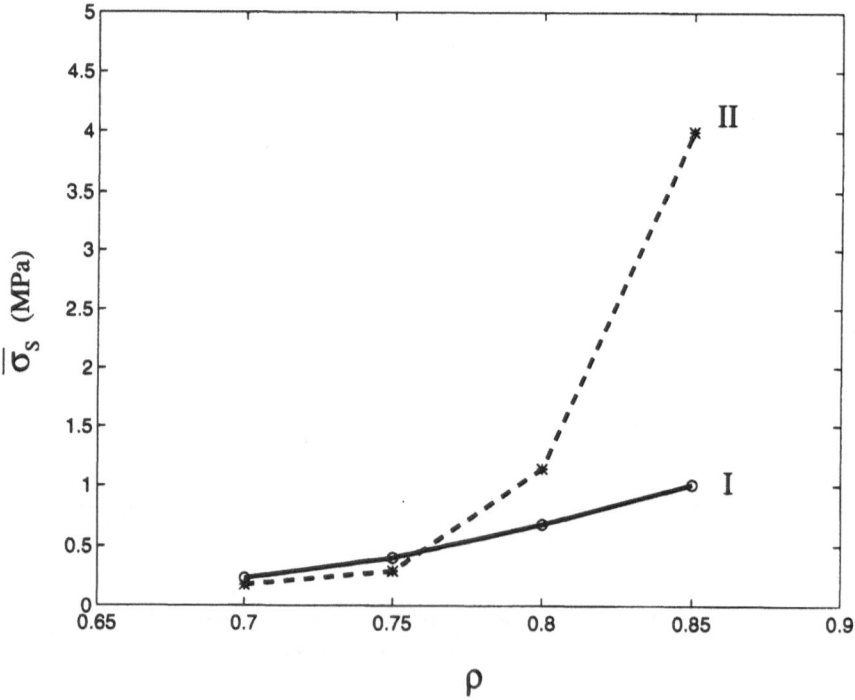

Figure 2 Grain-size normalised sintering potential as a function of density obtained assuming the strain-rate potential of (3) for grain-boundary diffusion controlled sintering for methods I and II of Section 3.1.

reserve further consideration of this result until after we have examined the predictions obtained using the interface reaction controlled constitutive laws.

3.2 INTERFACE REACTION CONTROLLED SINTERING

Here we employ the same two methods as discussed in section 3.1, but base the manipulation of material data on the strain-rate potentials of (3) and (5). Employing method I and the strain-rate potential of (3) for the micromechanical model we find that the ratio of volumetric to effective strain-rate is

$$\frac{\dot{\varepsilon}_v}{\dot{\varepsilon}_e} = 3\frac{f_r(\rho)}{c_r(\rho)}\left[1 + \frac{3\overline{\sigma}_s}{\sigma}\frac{L_o}{L}\right]^{\frac{1}{2}} \tag{11}$$

Equation (11) suggests that the grain-size compensated sintering potential, $\overline{\sigma}_s$, and the ratio $f_r(\rho)/c_r(\rho)$ can be obtained from a plot of $(\dot{\varepsilon}_v/\dot{\varepsilon}_e)^2$ against $(\sigma L/L_o)^{-1}$. A plot of the data for fine grained alumina is shown in Figure 3 over a range of densities, together with the best linear fits to the data. The resulting values of $\overline{\sigma}_s$ determined from the slope of these lines are plotted in Figure 4 as a function of the relative density. This is compared with the variation obtained using method II, in which we compare the grain-size compensated volumetric strain-rates during forging and free sintering. The procedures for doing this are fully described by Ma (1997).

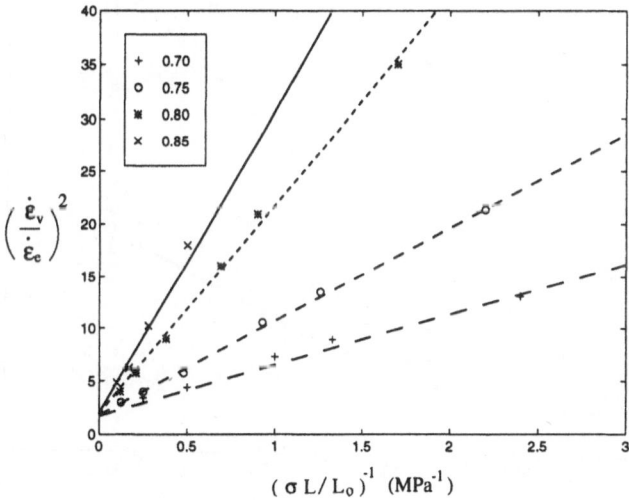

Figure 3 The ratio of volumetric to effective strain-rate squared determined from the forged sintering experiments plotted as a function of $(\sigma L/L_o)^{-1}$ for a range of relative densities.

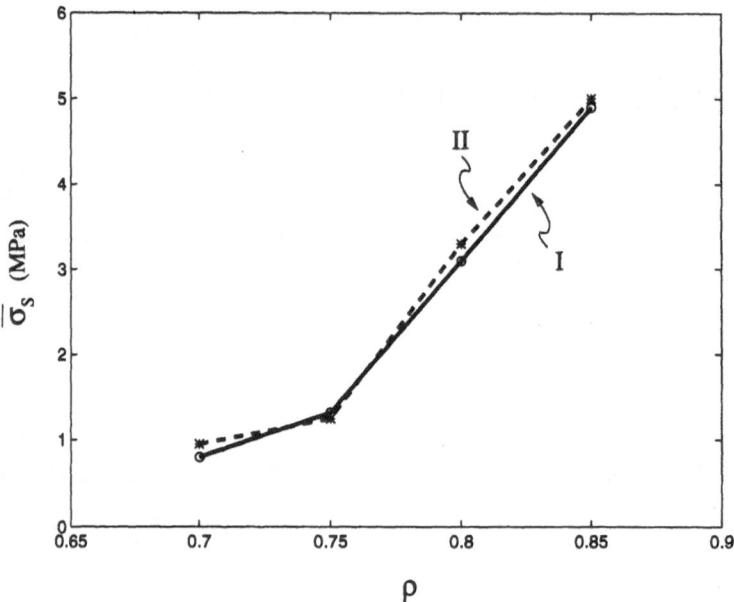

Figure 4 Grain-size normalised sintering potential assuming the strain-rate potential of (3) plotted as a function of relative density for methods I and II of Section 3.2

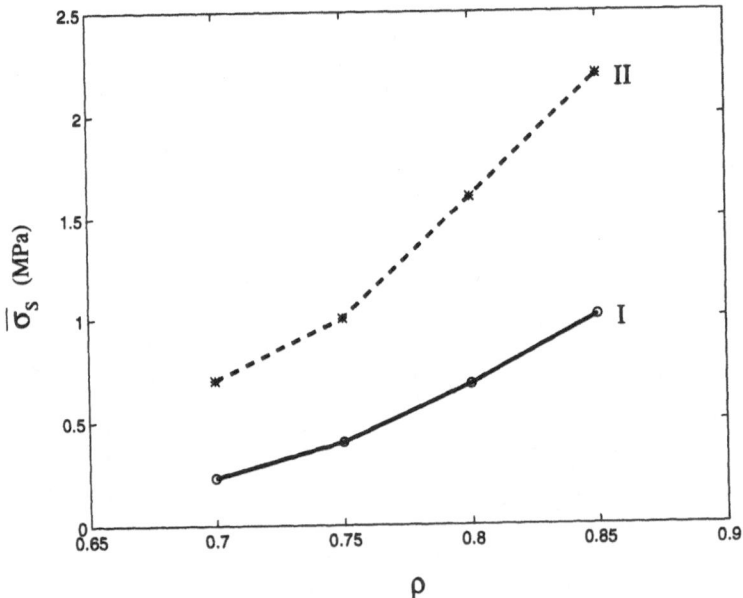

Figure 5 Grain-size normalised sintering potential assuming the strain-rate potential of (5) plotted as a function of relative density for methods I and II of Section 3.2

Employing the strain-rate potential of (5) in method I again gives the strain-rate ratio of (8) obtained for grain-boundary diffusion controlled sintering, thus $\overline{\sigma}_s$ is again determined from the slope of Figure 1. It can readily be demonstrated that this method of determining $\overline{\sigma}_s$ is valid for any constitutive law in which the strain-rate potential is a function of $\left[c(\rho)\sigma_e^2 + f(\rho)(\sigma_m - \sigma_s)^2 \right]$, we refer to such a potential as a quadratic potential. The resulting variation of the sintering potential with relative density is compared with that obtained from method II in Figure 5.

It is evident from Figures 2, 4 and 5 that the mechanistic interface reaction controlled constitutive law provides the most consistent result for the sintering potential. Further examination of Figures 1 and 3 shows that when $(\dot{\varepsilon}_v / \dot{\varepsilon}_e)^2$ is plotted against $(\sigma L / L_o)^{-1}$ the data almost exactly satisfies a linear relationship, but the quality of the linear fit is less good when $(\dot{\varepsilon}_v / \dot{\varepsilon}_e)$ is plotted against $(\sigma L / L_o)^{-1}$, as suggested by a quadratic potential, particularly at low stress levels.

As well as providing information about the sintering potential the above analyses and manipulation of data also provides information about the ratio $f_r(\rho)/c_r(\rho)$, which is determined from the intercept of the linear plot with the strain-rate ratio axis. Over the full range of densities represented by Figure 3, we find that $f_r(\rho)/c_r(\rho) = 0.5$. This result is consistent with forging studies at uniaxial stresses much greater than the sintering potential (Sofronis and Casagranda, 1997), and is indicative of the fact that the transverse strain-rate in these tests is zero during

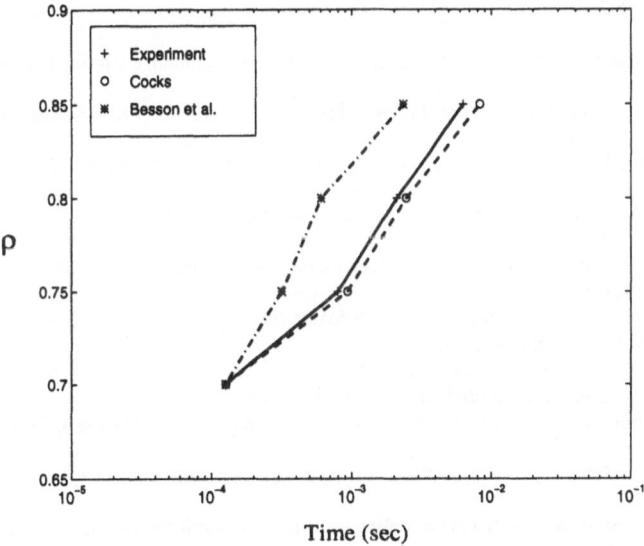

Figure 6 The variation of relative density as a function of time compared with the predictions employing $f(\rho)$ of (4) and (6)

the early stages of compaction. This observation was made use of in the determination of the dimensionless functions $c_r(\rho)$ and $f_r(\rho)$ proposed by Du and Cocks (1994) as given in (4). Using this relationship we can compute the variation of density as a function of time for any assumed form of the relationship $f_r(\rho)$. Figure 6 compares the experimental results with predictions obtained using the functions of (4) and (6) for a forging stress of 0.75MPa. The kinetic constants were determined by ensuring that the computations agreed with the theoretical predictions at a relative density of 0.7. In order to avoid the added complications associated with the choice of grain growth law the experimental results for the variation of grain-size with relative density was employed in the computations. It is evident that the mechanistic model of Cocks (1994) provides an accurate representation of the experimental data, while the empirical model of Besson and Abouaf (1992) provides a less good fit. Similar trends are observed for other values of applied stress (Ma, 1997).

4. Conclusions

The series of plots presented in section 3 all support the conclusion that the mechanistic interface reaction controlled constitutive law proposed by Cocks (1994) accurately describes the stage 1 sintering of fine grained alumina. This observation is consistent with the creep studies on fine grained alumina reported by Cannon et al (1980) and Chen and Xue (1990) who observed that the creep index in a power-law creep expression gradually increases from 1 towards 2 as the grain-size is reduced below 3μm, which is indicative of a change to interface reaction controlled creep.

5. References

Ashby, M.F. (1990) Background reading HIP 6.0, *Technical Report*, Cambridge University Engineering Department.

Besson, J. and Abouaf, M. (1992) Rheology of porous alumina and simulation of hot isostatic pressing, *J. Am. Ceram. Soc.*, **75**, 2165-2172.

Cannon, R.M., Rhodes, W.H. and Heuer, A.H. (1980) Plastic deformation of fine grained alumina, I. interface-controlled diffusional creep, *J. Am. Ceram. Soc.*, **63**, 46-58.

Chen, I. and Xue, L.A. (1990) Development of superplastic structural ceramics, *J. Am. Ceram. Soc.*, **73**, 2585-2609.

Cocks, A.C.F. (1994) The structure of constitutive laws for the sintering of fine grained materials, *Acta Met. et Mat.*, **42**, 2191-2210.

Du, Z-Z and Cocks, A.C.F. (1992) Constitutive models for the sintering of ceramic components, I- material models, *Acta Met. et Mat.*, **40**, 1969.

Du, Z-Z and Cocks, A.C.F. (1994) Sintering of fine grained materials by interface reaction controlled grain-boundary diffusion, *Int. J. Solids Structures*, **31**, 1429-1445.

Ma, J (1997) Constitutive modelling of the sintering of fine grained alumina, *PhD thesis*, University of Cambridge.

McMeeking, R.M. and Kuhn, L.T. (1992) A diffusional creep law for powder compacts, *Acta Met. et Mat.*, **40**, 961-969.

Sofronis, P. and Casagranda, A. (1997) On the scaling laws for the consolidation of nanocrystalline powder compacts, IUTAM Symposium on Mechanics of Granular and Porous Materials, this volume.

Svoboda,J., Riedel, H. and Zipse, H. (1994) Equilibrium pore surfaces, sintering stresses and constitutive equations for the intermediate and late stages of sintering, I- computation of equilibrium surfaces, *Acta Met. et Mat.*, **42**, 435-443.

IMPORTANCE OF INCREMENTAL NONLINEARITY IN THE DEFORMATION OF GRANULAR MATERIALS

YOSHIO TOBITA
Tohoku-Gakuin University, Department of Civil Engineering
Tagajyo, Miyagi, 985 JAPAN

1.Introduction

The property of incremental nonlinearity indicating that the tangent stiffness/ compliance depends on the direction of strain/ stress rate has received much attention in recent years. Experimental results by the triaxial and hollow cylindrical triaxial tests among others have clearly shown that the incremental nonlinearity is a common feature of deformation of granular materials like sand under complex loading paths including the rotation of principal stress axes. The incremental nonlinearity may be considered as a natural consequence from the view of micromechanisms for inelastic deformation of granular materials. Responding to experimental observations, many types of constitutive equations have already been proposed with different concepts and mathematical structures.

The incremental nonlinearity with noncoaxiality between the principal directions of plastic strain rates and stresses, has been considered crucial in the bifurcation analysis of shear banding, and provides results which are consistent with experimental observations. Even in a simple but practically important case such as the one dimensional liquefaction analysis, incremental nonlinearity plays a decisive role in order to get numerical results showing good correspondences with experimental and field observations.

In this paper, we summarize and discuss the fundamentals of the incremental nonlinearity: micromechanical background and constitutive modeling. The importance of this property for practical geotechnical problems is discussed with a particular emphasis on the liquefaction analysis. We also point out the importance of incremental nonlinearity in the discussion of the stability condition of granular materials.

2. Micromechanical Foundations for the Incremental Nonlinearity

The overall (averaged) constitutive behavior of any material is obtained as an integration of micromechnisms occurring at grain levels over a representative volume element. If there are two or more micromechanisms activated in different fashions, the consequent

139

N. A. Fleck and A. C. F. Cocks (eds.),
IUTAM Symposium on Mechanics of Granular and Porous Materials, 139–150.
© 1997 *Kluwer Academic Publishers.*

overall inelastic deformation will yield the incremental nonlinearity. Even if one mechanism such as slip between grains is recognized as the source of inelastic deformation, the degree of activation of each microslip associated with the referential plane defined by its unit normal depends on the direction of stress increments. It is thus understood that the overall deformation, as a suitably averaged quantity, becomes incrementally nonlinear.

In order to gain more insight into the nature of the incremental nonlinearity of the inelastic behavior of granular materials, we briefly discuss the micromechanically derived constitutive model including the fabric change and microslips between grain contacts as the two sources of inelastic deformation (see Fig.1 for thekinematic concept). The micromechanical model discussed in this chapter is not well founded by micromechanical observations, but is a hypothetical model. It is however claimed that most inelastic deformation behavior can be explained by this micromechanical model at least from a qualitative point of view (Tobita, 1987).

One source of inelastic deformation is attributed to fabric changes which occur in order to support further applied tractions and displacements on the surfaces. The kinematical changes between the previous fabric and the new one are recognized as the incremental displacements due to fabric changes. The other is microslip between grain contacts, which is triggered by the fabric change. The distribution of the referential slip plane may be assumed to exist in any directions. In order to evaluate the state of the fabric transmitting interparticle forces in a granular assembly, the fabric tensor $\mathbf{H}$ with second order tensor being averaged over a representative volume element is introduced. The incremental stress -fabric strain, being closely related to the redistribution of interparticle forces, have the following mathematical form:

$$\dot{\varepsilon}^* = \mathbf{C}^* : \dot{\sigma} \tag{1}$$

where $\mathbf{C}^*$ denotes the fourth order compliance tensor and is a function of the stress tensor σ and the fabric tensor $\mathbf{H}$. This relationship may exhibit an incremental nonlinearity since the redistribution of interparticle forces leading to fabric changes occurs separately to the external changes. For simplicity, Eq.(1) is supposed to be an incrementally linear relationship. The fabric strain rate may be decomposed into two parts: elastic ε^e and fabric change ε^f. Whether the fabric change occurs or not may depend on the stress and the direction of stress increments, since the elastic range is expected in which the existing fabric can sustain further external changes without fabric changes. The easiest way to formulate this behavior into the constitutive model is to introduce a kind of loading function as in classical elasto-plasticity. By assuming the fabric potential G^f in order to determine the direction of the strain rate due to fabric changes in addition to the loading function F^f for fabric changes, the strain rate due to fabric changes may be mathematically formulated such that

$$\dot{\varepsilon}^* = \dot{\varepsilon}^e + \dot{\varepsilon}^f = \mathbf{C}^e : \dot{\sigma} + \{(\mathbf{M} \otimes \mathbf{N}) : \dot{\sigma}\} / h^f \tag{2}$$

with $\mathbf{M} = (\partial G^f / \partial \sigma)$, $\mathbf{N} = (\partial F^f / \partial \sigma)$, and fabric strain rate occurs as

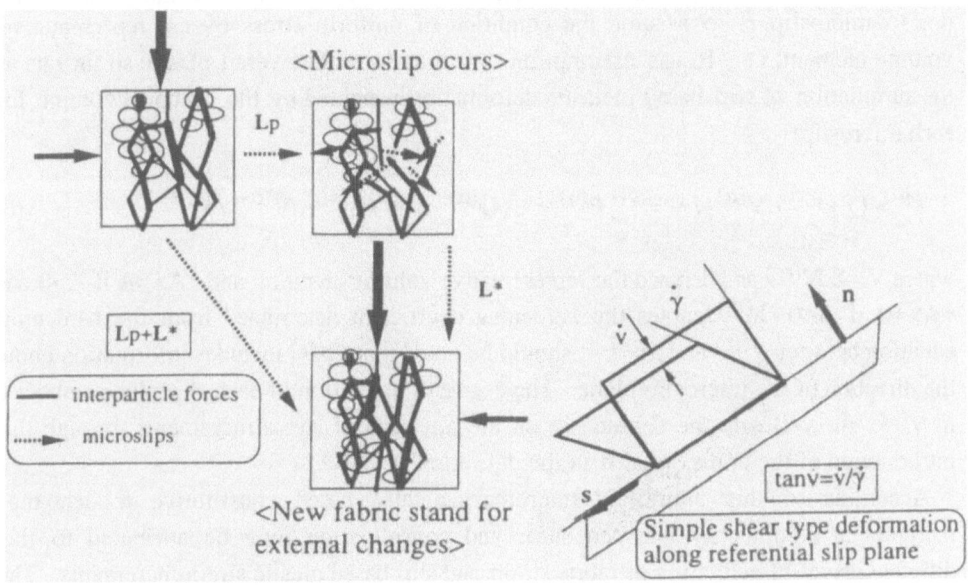

Fig.1 Kinematic in fabric change and microslip model

Fig.2 Simple shear type deformation of each microslip along the referential slip plane wtih unit normal **n**

$$\dot{\varepsilon}^{f} \neq 0 \text{ when } F^{f} = 0 \text{ and } tr(\mathbf{N} \cdot \dot{\sigma}) > 0; \quad \dot{\varepsilon}^{f} = 0 \text{ in other cases} \tag{3}$$

In order to formulate the inelastic deformation stemming from microslip between grains, we introduce the referential slip plane for each microslip identified by its unit normal **n** and the unit vector defining the slip direction **s** as shown in Fig. 2. It is supposed that microslip takes place along the referential slip plane in the form of simple shear: shear strain rate $\dot{\gamma}^{(\alpha)}$ in the direction of **s** and the normal strain $v^{(\alpha)}$ in the direction of the unit normal **n**. By transforming the simple shear type deformation into the background rectangular Cartesian coordinate system, we have the following mathematical description for each microslip.

$$\dot{\varepsilon}^{p} = \mathbf{P}^{(\alpha)}\gamma^{(\alpha)}; \quad \text{with } \mathbf{P} = (\mathbf{s} \otimes \mathbf{n} + \mathbf{n} \otimes \mathbf{s} + 2 \tan v \, \mathbf{n} \otimes \mathbf{n}) / 2 \tag{4}$$

The magnitude of $\dot{\gamma}^{(\alpha)}$ depends on the state of stress and fabrics and on the direction of stress increments. If one can assume the existence of a loading function $f^{p(\alpha)}$ for each microslip, $\dot{\gamma}^{(\alpha)}$ can be formulated in such a manner that

$$\dot{\gamma}^{(\alpha)} = tr(\mathbf{Q}^{(\alpha)} \cdot \dot{\sigma}) / h^{(\alpha)}; \quad \mathbf{Q} = (\mathbf{s} \otimes \mathbf{n} - \mu \mathbf{n} \otimes \mathbf{n}) \tag{5}$$

with $f^{p(\alpha)} = s \cdot \sigma \cdot n - \mu n \cdot \sigma \cdot n = 0$. The simplest way to get an overall plastic strain due to microslip is to assume the condition of uniform stress over a representative volume element, i.e., Reuss' assumption. We thus have the overall plastic strain rate as the summation of slip based inelastic deformation weighted by the occupied volume for each microslip:

$$\dot{\varepsilon}^p = \{\sum_\alpha \dot{\varepsilon}^{p(\alpha)} dV^{(\alpha)}\} / V = \{\sum_\alpha P^{(\alpha)} < tr(Q^{(\alpha)} \cdot \dot{\sigma}) > / h^{p(\alpha)} dV^{(\alpha)}\} / V \qquad (6)$$

where $V = \Sigma dV^{(\alpha)}$ and denoted the representative volume element, and <A>=A if A>0 and <A>=0 if A<0. $h^{(\alpha)}$denotes the hardening coefficient determined from the hardening equation between $\gamma^{(\alpha)}$ and $\mu^{(\alpha)}$. It should be noted that $Q^{(\alpha)}$ includes information about the direction of the microslip plane. The degree of activation of each slip plane embodied in $\dot{\gamma}^{(\alpha)}$ show clearly the dependence on the direction of stress increments through the introduction of the unit normal n in the definition of $Q^{(\alpha)}$.

According to this simplified micromechanically based constitutive model, the incremental nonlinearity between shear and consolidation may be attributed to the different degree of activation in fabric strain and slip based plastic strain increments. The incremental nonlinearity in shear deformation during the general stress path associated with the rotation of principal stress axes may be attributed to the different degree of activation of each microslip, which is a consequence of the relative direction of microslip to that of the stress increments.

We may suppose that most engineering materials will show the incremental nonlinearity when inelastic deformation prevails. The problem is whether a sophisticated constitutive model being consistent with micromechanisms is really necessary for the good prediction of practical problems. If the simple constitutive model can yield sufficient results to engineering problems, no sophisticated constitutive model is required. It is however understood that for geotechnical problems the introduction of the nonlinearity is of essence for many important problems.

3. Incrementally nonlinear constitutive models

Many types of incrementally nonlinear constitutive models have been proposed. In this chapter we briefly review and discuss the principal aims and the fundamental mathematical features of the proposed models by classifying them into, for convenience, three categories: (1)multimechanism models; (2)modification of the classical elasto-plasticity; and (3) mathematically well defined models.

3.1 MULTIMECHANISM MODELS

As was discussed in chapter 2, when we assume many slips and other micro-mechanisms to which directional quantities are attached, the resultant inelastic deformation after using the suitable averaging method becomes incrementally nonlinear. By assuming a relatively simple deformation pattern and hardening relation, the overall response was found to be in accordance with experimental observations under complex loading paths including the rotation of principal stress axes. The models proposed by Miura et al. (1986), Tobita et al.(1986), and Iai et al. (1991) among others for the deformation behavior of granular materials may be classified into this category. The multimechanism constitutive models are very attractive from a physical point of view. It is, however, pointed out that these models are not well formulated from a mathematical point of view (e.g., the violation of the objectivity) and have problems in application to numerical analysis due to the fact that they require extensive memory. The Double shearing (slip) model (e.g., Mehrabadi and Cowin, 1980, and Tobita, 1993) showing the incremental nonlinearity may be classified into this category.

3.2 MODIFICATION OF THE ELASTO-PLASTIC MODEL

The pioneering work in this category for geomaterials may be attributed to Rudnicki and Rice(1975), which shows that the non-coaxiality between the plastic strain rate and the stress tensors and the non-associativity between the plastic potential and the loading function are of primary importance in the bifurcation analysis of shear bands in geomaterials. The noncoaxial property in itself may not be an important factor for the bifurcation analysis, since the non-coaxiality obtained after the introduction of an anisotropic loading function does not change the qualitative feature of the bifurcation analysis compared with the original isotropic coaxial model. One may say that the incremental nonlinearity rather than the noncoaxiality changes the nature of the bifurcation analysis and leads to better results. The yield vertex model leading to incremental nonlinearity was discussed in detail for non-frictional materials. The tangential plasticity model, being mathematically well formulated, was proposed by Hashiguchi (1993). Some models based on the bounding surface concept can account for the incrementally nonlinear response by adopting the proper definition of the conjugate stress point on the bounding surface at which the direction of the plastic strain rate is evaluated. The direction of the stress rate as well as the state quantities, such as the stress tensor and the internal variables, is formulated to influence the conjugate stress point (Wang and Dafalias, 1990, Gutierrez et al.1991).

The advantage of the incremental nonlinear models obtained by the modification of the classical elastoplastic theory over the other type of models can be attributed to the fact that the inelastic constitutive models for geomaterials have been developed along the line of the theory of elasto-plasticity. The modification appears to be slight in many cases, it should, however, be remembered that the mathematical structure of the resultant models changes considerably from the original linear one. This fact may be easily understood by noting that many useful theorems associated with the theory of elasto-plasticity, like the maximum dissipation theorem are no longer available to the modified theory.

3.3 MATHEMATICALLY WELL DEFINED MODEL

Rather than following the line of the theory of elasto-plasticity, completely different formulations were proposed by Darve (1984) and Kolymbas & Wu (e.g., 1993) among others. They considered the incrementally nonlinear feature as the utmost important feature of inelastic deformation of granular materials.

The multilinear constitutive model assuming many tensorial zones in an incremental space in which the incrementally linear response is taking place was proposed by Darve (1984). In this context, the elasto-plastic formulation is considered as a multi-linear constitutive model having two tensorial zones: elastic and elasto-plastic space. The hypoplastic model proposed and developed by Kolymbas and his colleagues has also been applied to the bifurcation analysis and its mathematical features relating to the uniqueness and stability of solution have also been discussed.

The common feature of these formulations may be found in the rigorous mathematical formulation satisfying all the requirements of the principles of continuum mechanics.

3.4 BRIEF COMMENTS ON INCREMENTALLY NONLINEAR MODELS

The introduction of incremental nonlinearity into the constitutive models seems to be of essence in order to explain and evaluate experimental observations. It is, however, recognized that this property makes it almost impossible to evaluate analytically the global mathematical structures such as the uniqueness and the stability of solutions to the boundary/initial value problems, since the very powerful mathematical method such as the eigen-value analysis of the constitutive matrix is of no use for the incrementally nonlinear constitutive models.

The shortcomings of this lack of an elegant mathematical feature of the incrementally nonlinear constitutive model may be overcome by the advent of digital computers, since uniqueness and stability can be evaluated by cumbersome numerical methods.

4. Incremental Nonlinearity in Liquefaction Analysis

A simple but most important liquefaction analysis is found in the saturated horizontally layered sand deposit being subjected to upward traveling shear wave. The initial stress state is in most cases anisotropic with principal vertical stress σ_V and the principal horizontal stress σ_H. The ratio defined as (σ_H/σ_V) is called the Ko-value and in most cases Ko< 1. The existence of initial shear stress due to the anisotropic static stress complicates the analysis when the classical constitutive model is used. In order to make the discussion explicit, let us consider the elasto-plastic constitutive model for which an isotropic medium is assumed. The existence of the initial static shear stress makes the stress point occupy A in the shear stress- shear strain, shear strain-dilatancy diagrams as

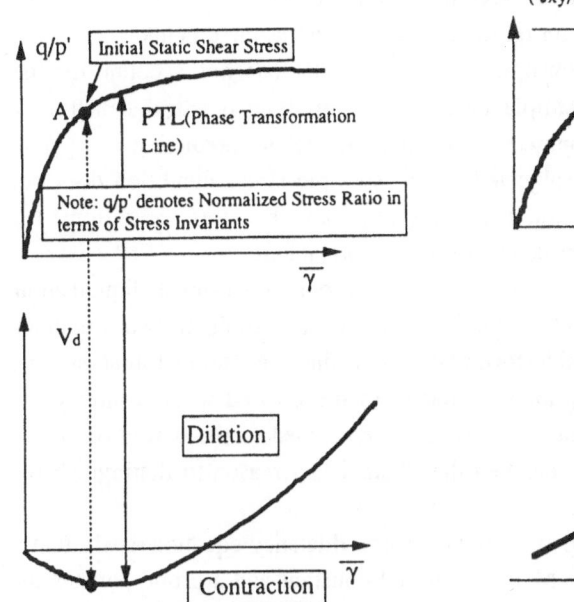

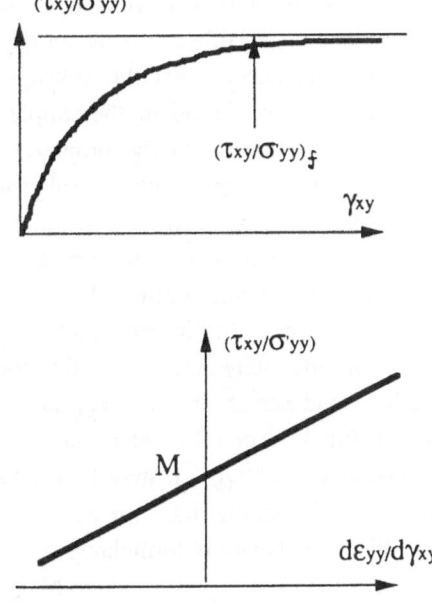

Fig.3 Effect of initial static shear stress on the normalized stress strain relationship

Fig.4 A conventional one dimensional analysis for liquefaction process

shown in Fig.3. The excitation of sand deposits due to the propagation of a shear wave yields the dynamic shear stress $d\tau_{xy}$, which leads to $d\gamma_{xy}$ and the negative dilatancy (contraction in volume) in the solid phase and increases the pore-water pressure due to the constraint of the undrained condition. The induced pore water pressure leads to the increase in the effective stress ratio even if the magnitude of dynamic shear stress is constant, which results in the successive decrease of the stiffness and further increase of the pore water pressure.

The use of an isotropic hardening model for this liquefaction process yields a different behavior from experimental observations; the increase of the pore water pressure is remarkably small compared with experiments (see, Tobita and Yoshida(1994) for more detail). This trend inherent in the isotropic hardening model can be understood easily by referring to Fig.3. The initial static stress condition is already close to the maximum density point (marked as PTL(Phase Transformation Line)) and the dilatancy behavior changes from negative (contraction) to positive (dilation) after a small dynamic shear stress is applied. In actual experiments to simulate the liquefaction process mentioned above, the magnitude of negative dilatancy is rather high, and is almost independent of the existence of initial static shear stress.

This experimentally observed trend may be considered as the incrementally nonlinear property. The dynamically induced shear stress increment $d\tau_{xy}$ is different by 45 degrees in direction from the initial shear stress condition, and in this case the resultant shear behavior is almost independent of the initial shear stress. The isotropic hardening model in which a unique relationship is assumed between the effective stress ratio and the effective (normalized) shear strain can not account for this property: the dependency of the shear deformation and dilatant behavior on the direction of stress increments.

It is thus understood that the isotropic hardening model can not account for the important behavior of sand in the simple case of an important practical bench-mark condition. The success of the proposed constitutive model in predicting the cyclic loading behavior of sands initially subjected to an isotropic stress condition does not indicate that it is also applicable to the practically important Ko-stress conditions; the isotropic stress condition may be considered as an exceptional case.

It seems worthwhile noting that the conventional one-dimensional liquefaction analysis accounts for the incrementally nonlinear character in a straightforward manner, which is unfortunately an unreasonable formulation. In the conventional analysis, the dynamically induced shear stress τ_{xy} alone is related to the induced shear strain γ_{xy} as shown in Fig.4. The effect of initial shear stress is embodied in the definition of the failure shear stress τ^f_{xy}. It may be claimed that there is no reason to distinguish the dynamic stress from the static stress.

The dilatant behavior is formulated as a kind of stress- dilatancy equation such that : $(d\varepsilon_{yy}/d\gamma_{xy}) = (\tau_{xy}/\sigma'_{yy})-M$, where M is a material constant and a prime denotes the effective stress. Such a relationship is meaningless from a continuum mechanical point of view. The horizontally constrained strain inherent in the problem is not considered at all (The horizontally layered problem is in fact a two dimensional problem in a rigorous sense).

The conventional dynamic analysis yields better results compared with those from more reasonable constitutive models and problem specification, since the model behavior closely follows the actual behavior. It is, however, claimed that the extension of the conventional analysis to multi-dimensional problems is almost impossible, since the analysis for such cases requires the reasonable formulation of problems. The development of constitutive models accounting for the incremental nonlinearity seems to be of primary importance, even for the purpose of better numerical simulations of practical engineering problems. There are many situations in which better constitutive models are required in order to capture the fundamental features of the problem.

5 Importance of Incremental Nonlinearity on Stability Condition

The introduction of the incremental nonlinearity to the constitutive modeling influences much of the uniqueness and stability conditions for solutions of boundary/initial value problems. The incremental nonlinearity renders the analytical investigation of the conditions almost impossible and the numerical investigation for each increment successively seems to be inevitable. In the investigation of reasonable and accurate conditions of stability of frictional granular materials, the introduction of incremental nonlinearity will be helpful.

5.1 BRIEF REVIEW OF STABILITY CONDITION OF GRANULAR MATERIALS

Drucker's stability condition expressed by the inner product of stress and *plastic* strain increments: $\text{tr}(\dot{\sigma}\dot{\varepsilon}^p) > 0$, is well known in the branch of geomechanics. Hill's stability condition with respect to the inner product of stress and *total* strain increments: $\text{tr}(\dot{\sigma}\dot{\varepsilon}) > 0$ is more tractable for the theoretical discussion. Note that Drucker's postulate is sufficient for Hill's postulate so long as the elastic stiffness matrix is positive definite.

As general remarks on the investigation of stability condition, we should pay due attention to the following facts:
(1) The investigation of the stability condition deals with *the stability of fundamental (unperturbed) solutions of mathematically formulated problems,* which are simplified situations obtained by dropping some factors from the actual problems;
(2) The stability condition is only a sufficient condition and never a necessary condition in any investigation of stability;
(3) The stability criterion using some scalar functional like the potential energy in elastostatics is principally applicable to conservative mechanical systems.

The feature of (1) indicates that the stability condition obtained by numerical calculation becomes meaningless if the mathematical formulation of actual problems is poor, which necessarily leads to a poor correspondence between numerical results and actual observations. It is also understood that the stability condition does not deal with real material behavior but with the fundamental solution of the mathematical problem. The feature of (2) indicates that there is no reasonable theoretical method to asses the extent of the gap between the stability condition and the actual behavior of mathematical systems. The feature of (3) indicates that Hill's stability condition which was originally developed for conservative mechanical systems can not be used as the most proper condition for other mechanical systems; e.g., when we use the non-associated flow model leading to unsymmetric mathematical structures.

The introduction and discussion of the proper stability condition for granular materials seems to be of utmost importance when we want to numerically simulate the progressive failure and flow type failure of geostructures. By restricting our discussion to the stability of uniform deformation (an element behavior assumed for a specimen) of granular materials, we review and investigate the interesting discussion between Lade and Drucker.

5.2 EXPERIMENTAL OBSERVATION VERSUS THEORETICAL PREDICTION

Lade et al. (e.g., 1987) carried out interesting experimental observations on the stability condition of uniform deformation of granular materials by using the triaxial testing apparatus. The experimental results may be summarized as follows:
(1) In drained deformation, Hill's stability condition as well as Drucker's show poor

correspondence with experimental observations; i.e., no sign of instability of uniform deformation appears even when tr($\overset{..}{\sigma}\overset{.}{\varepsilon}$)< 0;

(2) In undrained deformation of saturated sands, Hill's stability condition gives good predictions of experimental observations;

(3) For unsaturated sands, Hill's condition becomes poor with decreasing degree of saturation.

It is also reported that the non-associated formulation is essential for the deformation behavior of granular materials. After investigating the comprehensive experimental observations, they proposed a conventional stability condition peculiar to granular materials, which includes the trend of dilatancy and the degree of saturation as well as Hill's stability condition. This condition seems to be appealing from an engineering point of view; however, it is not linked to any rational methods in stability theory.

Drucker and his colleagues (e.g., Drucker and Li, 1993) tried to explain the feature (1) mentioned above based on Drucker's stability condition. They show theoretically that the incidence of shear band type bifurcation becomes possible just after Drucker's stability condition fails; however, due to the strong boundary constraints imposed on the triaxial testing specimen, the shear band cannot develop to the extent which can be observed experimentally. They show that Drucker's stability condition: tr($\overset{.}{\sigma}\overset{..}{\varepsilon}{}^{P}$) >0 is still useful for the stability of uniform deformation of granular materials.

One may notice the fact that Lade et al. investigated experimentally whether the bifurcated modes other than uniform (if observed, it indicates the instability of uniform deformation) develop markedly or not in a specimen; while Drucker et al. principally discuss the onset condition of the bifurcation mode, in other words, the possibility of bifurcation (possibility of loss of stability of uniform deformation).

5.3 INCREMENTAL NONLINEARITY AND STABILITY CONDITION

When we can assume that plastic loading continues everywhere in a specimen, the elasto-plastic constitutive model, irrespective of associated or non-associated flow, can be converted into the linear comparison material showing always the elasto-plastic loading behavior. We thus reduce the problem to an incrementally linear problem. In this case, Hill's stability condition is coincident with the uniqueness condition of solutions and this condition can be judged by the loss of positive value of the lowest eigenvalue of the constitutive matrix $\mathbf{E}$ (in the case of non-associated flow, the symmetrized constitutive matrix: $\mathbf{E}^{S}$=($\mathbf{E}$ + $\mathbf{E}^{T}$]/2, where T denotes the transpose of the matrix, should be used), when uniform deformation is considered. We thus conclude that the stability of uniform deformation should be satisfied so long as no bifurcated solution becomes possible. A unique solution should be stable if a solution exists.

From a mathematical point of view, Hill's and Drucker's stability conditions for the fundamental solution under static equilibrium are always sufficient conditions. However, their conditions may not necessarily lead to good predictions of the stability of real behavior. After the violation of Hill's stability condition, a bifurcated solution becomes possible at least from a mathematical point of view. In order to investigate the actual behavior of the bifurcated solution of mathematically formulated problems, we should carry out very complicated dynamic analysis of stability (e.g., Leroy 1991) for every possible bifurcation mode. This is not yet available for practical problems.

Considering this stage of the development of the stability condition for granular materials, it seems a clever choice to adopt Hill's stability condition as a conventional stability condition to granular materials and to develop an incremental constitutive model satisfying Hill's stability condition up to the peak of the stress-strain curve to be in good agreement with experimental observations. An associated elasto-plastic constitutive model is always a candidate for this purpose as was emphasized by Drucker. However, the associated model fails to show good simulation with the undrained behavior of granular materials and can not be applicable to the case in which both the mean pressure and shear magnitude decreases but the stress ratio is increasing. It is therefore recommended that the deformation features as a natural consequence of the non-associated flow model should be embodied in the constitutive formulation. The non-associated flow model in itself predicts too early loss of stability condition if Hill's stability condition is employed in drained deformation. The introduction of incremental nonlinearity may be expected to remedy this deficiency of the non-associated flow model. The mathematical framework of the incrementally nonlinear constitutive model for this purpose and the application to liquefaction analysis is now being studied.

6 Concluding Remarks

The importance of incremental nonlinearity in the inelastic deformation of granular materials was reviewed and discussed in this paper. The micromechanically developed constitutive model is briefly introduced in order to show that incremental nonlinearity is a natural consequence from a micromechanical point of view. The incrementally nonlinear models proposed up to now are reviewed and discussed. The importance of this property from a practical geotechnical engineering point of view was emphasized with particular emphasis on the liquefaction analysis of horizontally layered sand deposits. The stability condition of granular materials was discussed; it was shown that Hill's stability condition is principally not applicable to non-symmetric mechanical systems, which requires more elaborate dynamic method. It was, however, argued that Hill's stability condition is equivalent to the uniqueness condition and thus a sufficient condition of the stability for the fundamental solution of structures in equilibrium. When one wants to use HIll's stability condition as a convenient criterion, the introduction of incremental nonlinearity into the constitutive modeling seems to be useful.

7 References

Darve, F.(1984): An incrementally nonlinear constitutive law of second order and its application to localization, *Mechanics of Engineering Materials*(Desai and Gallagher eds.), John Wiley and Sons, 179-196

Drucker D.C. and Li, M.(1993): Triaxial test instability of a non-associated flow rule model, J. Engng. Mech. ASCE, **119**, 1188-1204

Gutierrez, M. , Ishihara, K. and Towhata, I.(1991):Flow theory for sand during rotation of principal stress direction, Soils and Foundations, **31**(4), 121-132

Hashiguchi, K.(1993): Fundamental requirements and formulation of elastoplastic constitutive equation with tangential plasticity, J. Plasticity, **9**(5), 525-549.

Iai, S. Matunaga,Y. and Kameoka, T.(1992): Strain space plasticity model for cyclic mobility, Soils and Foundations, **32**(2), 1-15

Kolymbas, D. and Wu, W.(1993): Introduction to hypoplasticity, *Modern approaches to Plasticity*, (edt. Kolymbas), Elsevier Sc. Pub. 213-223

Lade,P.V., Nelson, R.B. and Ito, Y.M.(1987): Nonassociated flow and stability of granular materials, J. Eng. Mech. ASCE, **113**(9), 1302-1318

Leroy,Y.M.(1991):Linear stability analysis of rate dependent discrete systems, Int. Solids and Structures, **27**(6), 781-808

Mehrabadi, M. and Cowin, S.C.(1980): Prefailure and post-failure soil plasticity models, J. Eng. Mech. **106**(5), 991-1003

Miura, K., Miura, S. and Toki, S.(1986): Deformation prediction for anisotropic sand during the rotation of principal stress axes, Soils and Foundations, **26**(3), 36-52

Rudnicki, J.W. and Rice, J.R.(1975): Conditions for the localization of deformation in pressure sensitive dilatant material, J. Mech. Phys. Solids, **12**, 371-394

Tobita, Y.(1987) A micromechanical consideration on constitutive model for granular materials, Dr. thesis to Tohoku University

Tobita, Y.(1993): Modified double slip model for anisotropic hardening behavior of granular materials, Mechanics of Materials, **16**, 91-100

Tobita, Y. , Kato, N., and Yanagisawa, E.(1986): Formulation and applicability of slip models as a constitutive model for granular materials, Journal of JSCE, Geotechnical Division, No.370, 57-66(in Japanese)

Tobita, Y. and Yoshida, N.(1994):An isotropic bounding surface model for undrained cyclic behavior:Limitation and Modification, *Prefailure Deformation of Geomaterials* (eds. Shibuya et al.) Balkema , 457-462

Wang, Z.L. and Dafalias, Y.F.(1990): Bounding surface hypoplasticity model for sand, J. Engng. ASCE, **116**(5), 983-1001

THE USE OF LEGENDRE TRANSFORMATIONS IN DEVELOPING THE CONSTITUTIVE LAWS OF GEOMECHANICS FROM THERMODYNAMIC PRINCIPLES

I.F. COLLINS
The University of Auckland
School of Engineering, Private Bag 92019, Auckland,
New Zealand

1 Introduction

In recent years there has been much activity in the systematic development of theories of continua which start from the laws of thermodynamics. The books by Ziegler (1983), Lemaitre and Chaboche (1990) and Maugin (1992), and the articles by Germain, Nguyen and Suquet (1983) and Ziegler and Wehrli (1987) provide general accounts of these developments. Whilst these approaches provide a very elegant unified development of the theory of many of the model materials used in engineering including the so called "standard" elastic/plastic solids, they fail, in general, to include the type of elastic/plastic models commonly used to describe the behaviour of geomaterials - specifically materials which exhibit non-associated flow rules and which violate Drucker's postulate. Building upon earlier work by Houlsby (1981, 1982) Collins and Houlsby (1996) have recently demonstrated that these "non-standard" models can be included in the general thermodynamic develpoment, provided the constitutive potentials and dissipation functions are suitably chosen - see also the paper by Houlsby (1996) in this volume.

A key feature of this development is the use of the Gibbs potential function, which depends on stress, instead of the more usual Helmholtz free energy, which is a function of strain. The major mathematical tool needed for this theory is the Legendre Transformation, which not only provides the method of transferring between potentials, but also establishes the relation between the dissipation function and the yield condition plus flow rule. The object of this paper is to review the salient properties of Legendre Transformations and to indicate briefly the arguments which lead to the predictions of non-associated elastic/plastic behaviour for geomaterials.

N. A. Fleck and A. C. F. Cocks (eds.),
IUTAM Symposium on Mechanics of Granular and Porous Materials, 151–159.
© 1997 *Kluwer Academic Publishers.*

2. The Legendre Transformation

The Legendre transformation is one of the basic tools in generating duality
structures in mechanics. It plays a central role in the general theory of
complementary variational and extremum principles, and Sewell (1987)
has given a comprehensive account of the theory in this context. The
transformation is widely used in modern thermomechanical treatments of
the theory of continua e.g. Mauguin (1992), Coussy (1995) and Lemaitre
and Chaboche (1990).

The transformation can be given several different geometric
interpretations, but a particularly illuminating account can be found in the
book by Callen (1960). Consider firstly the differentiable function $Z = X(x)$ of a single independent variable. This can be represented by a graph
Σ in the (x, Z) plane (Figure 1), which has slope $y = X'(x)$ at the point x.
This graph can also be generated as the envelope of tangent lines. The
position of the tangent line with slope y can be determined by specifying
the point, $- Y$ say, where this line cuts the Z - axis. (The inclusion of the
minus sign is a matter of choice, but as will be seen it makes the
transformation symmetric.) Thus the family of tangent lines is specified by
the function $Z = Y(y)$, which can be viewed as an alternative way of
defining the curve Σ in a 'linewise' as opposed to a 'pointwise' manner. It
follows from simple geometry (Figure 1) that $X(x) + Y(y) = xy$, and
hence that $x = Y'(y)$ by differentiation.

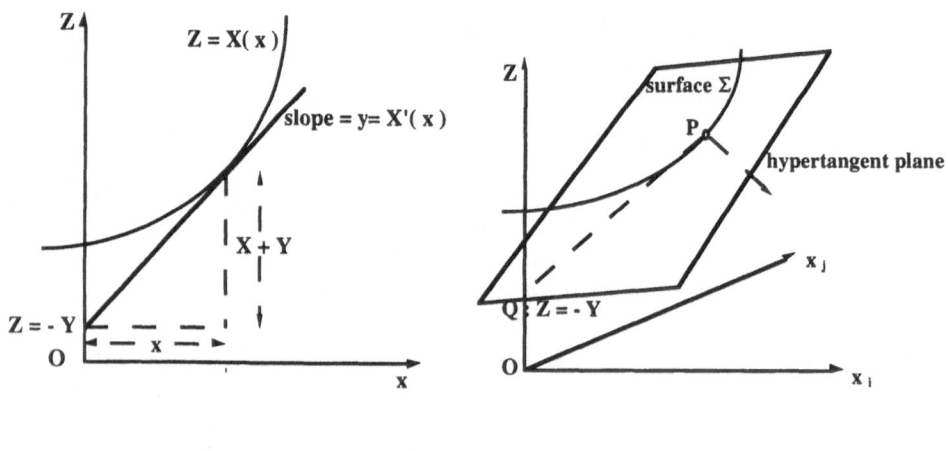

Figure 1   Figure 2

The Legendre transformation $X(x) \rightarrow Y(y)$ is hence a mapping which
replaces a 'pointwise' description of a curve by a 'linewise' description in

which the curve is viewed as an envelope of a family of straight lines. The main properties of the transformation can be summarised by the relations.

$$y = X'(x) \text{ and } x = Y'(y) \tag{1}$$

and

$$X(x) + Y(y) = xy \tag{2}$$

which clearly demonstrates the symmetry of the transformation. There is a dual graph Σ' say, whose 'pointwise' equation is $Z = Y(y)$ in the (y, Z) plane, and whose tangent lines have equation $Z = X(x)$, where $x = Y'(y)$.

The transformation generalises to a function $Z = X(x_i)$ of n variables $x_1, x_2....x_n$, which represents the $(n + 1)$ dimensional surface Σ and has gradient $y_i = \partial X/\partial x_i$. If the point at which a tangent hyperplane cuts the Z-axis is again denoted by $-Y$ (Figure 2), the family of enveloping hypertangents are defined by the function $Z = Y(y_i)$. The generalisations of (1) and (2) are hence

$$y_i = \partial X/\partial x_i \text{ and } x_i = \partial Y/\partial y_i \tag{3}$$

and

$$X(x_i) + Y(y_i) = x_i \, y_i \,. \tag{4}$$

The latter relation follows by taking the inner product of the orthogonal vectors: QP ie. $(X + Y, x_i)$ and the normal $(-1, y_i)$ to Σ at P; whilst the second relation (3) follows by differentiation of (4), provided the transformation Hessian matrix $\partial Z/\partial x_i \partial x_j$ is non-singular (see Sewell (1987) for a full discussion of singular cases). Again the symmetric self-duality of the transformation is evident for (3) and (4).

The transformation is frequently algebraically complex. An exception is where the original function is homogeneous of degree 2, so that $X(x_i) = 1/2 \, A_{ij} \, x_i x_j$ say (summation convection), for then $y_i = A_{ij} \, x_j$ which inverts to give $x_i = A_{ij}^{-1} y_j$, so that from (4) it follows that $Y(y_i) = 1/2 \, A_{ij}^{-1} y_i y_j$.

More generally if $X(x_i)$ is a homogeneous function of degree m, so that $x_i \, \partial X/\partial x_i \equiv x_i y_i = m \, X(x_i)$ by Euler's theorem for homogeneous functions, it follows for (4) that $Y(y_i)$ is homogeneous of degree n in y_i, where

$$\frac{1}{m} + \frac{1}{n} = 1 \tag{5}$$

In the above example $m = n = 2$.

An important special case, which is of prime importance in the theory of rate-independent materials, is when $m = 1$ and X is homogeneous of degree one in which case n becomes infinite. In order to understand what this means it is helpful to return to the above geometric interpretation of the Legendre transformation. A homogeneous function $Z = X(x_i)$ of degree one has the basic property that

$X(\alpha x_i) = \alpha X(x_i)$ for all α and so is represented by a cone with the origin as vertex in the $(n + 1)$ dimensional (Z, x_i) space (Figure 3). The tangent hyperplanes all pass through the origin, so that the value of the dual function $Y(y_i)$ is identically zero. This can be seen alternatively from (4) since with m = 1, Euler's theorem requires $x_i \partial X/\partial x_i \equiv x_i \, y_i = X(x_i)$.

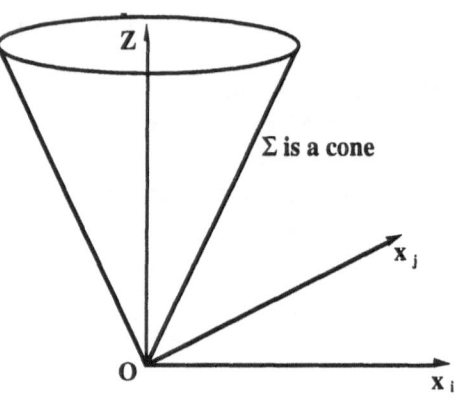

Figure 3: Singular Transformation

Moreover since each tangent hyperplane now has an infinite number of contact points with the cone, the inverse relation is not now unique and relations (3) and (4) must be replaced by

$$y_i = \partial X/\partial x_i \ \text{ and } \ x_i = \lambda \, \partial Y/\partial y_i \qquad (6)$$

and

$$X(x_i) = x_i \, y_i \ \text{ and } \ Y(y_i) = 0 \qquad (7)$$

where λ is an undetermined scalar.

In applications in continuum mechanics the X and Y functions are frequently potentials which depend on more than one set of variables. Suppose, for example the original function is $Z = X(x_i, \alpha_i)$ where x_i and α_i are vectors or tensors. Performing the Legendre transformation with respect to the x_i yields the dual potential $Y(y_i, \alpha_i)$ where $y_i = \partial X/\partial x_i$. The α_i are 'passive' variables in the transformation, and by virtue of (4) satisfy the relations:

$$\beta_i \equiv \partial X/\partial \alpha_i = -\partial Y/\partial \alpha_i . \tag{8}$$

3. Representation of Rate Independent Elastic/Plastic Solids

The stress power in an isothermal deformation defined by the small strain tensor ε_{ij} can be written

$$\sigma_{ij} \, \dot\varepsilon_{ij} = \dot F + D \tag{9}$$

where σ_{ij} is the Cauchy stress tensor, F is the Helmholtz free energy function and D the dissipation function. The Second law of thermodynamics (Clausius-Duhelm inequality) is satisfied if the value of D is non-negative. In an elastic/plastic material the "state" of a material element undergoing isothermal deformations is defined by the total (small) strain ε_{ij} and the plastic strain α_{ij}, which plays the role of an internal variable. Thus the free energy is assumed to be a function of ε_{ij} and α_{ij}, whilst the dissipation function also depends on $\dot\alpha_{ij}$. Viscous behaviour can be modelled by allowing D to also depend on $\dot\varepsilon_{ij}$, the rate of change of total strain (Ziegler (1983), Maugin (1992)). Such effects will be ignored here. Moreover, since the material is assumed rate-independent the dissipation function must be homogeneous of degree one in the $\dot\alpha_{ij}$ components, so that

$$D = \dot\alpha_{ij} \, (\partial D/\partial \dot\alpha_{ij}) \tag{10}$$

from Euler's theorem. Equation (11) can hence be rewritten:

$$\sigma_{ij} \, \dot\varepsilon_{ij} = (\partial F/\partial \varepsilon_{ij}) \, \dot\varepsilon_{ij} + (\partial F/\partial \alpha_{ij}) \, \dot\alpha_{ij} + (\partial D/\partial \dot\alpha_{ij}) \, \dot\alpha_{ij} \tag{11}$$

from which we can deduce that

$$\sigma_{ij} = \partial F/\partial \varepsilon_{ij} \tag{12}$$

and

$$\tau_{ij} \equiv -\partial F/\partial \alpha_{ij} = \partial D/\partial \dot\alpha_{ij} \tag{13}$$

Equation (12 follows from (11) since $\dot{\varepsilon}_{ij}$ and $\dot{\alpha}_{ij}$ are independent and neither term in (12 depends on these rates; however equation (13) is not a strict consequence of (11) since $\partial D / \partial \dot{\alpha}_{ij}$ is not, in general, independent of $\dot{\alpha}_{ij}$. Instead (13) has the status of a "constitutive postulate" (an example of Ziegler's postulate- Ziegler(1983)), which defines a very broad class of materials. This class is much wider than the "standard materials" defined by Drucker's postulate, but excludes incrementally nonlinear models such as hypoplastic materials.

The tensor τ_{ij} defined in (13) has the dimensions of stress and here will be referred to as a "generalised" stress. Note that from (10) and (13) the dissipation function is the product of the generalised stress, not true stress, with the plastic strain-rates.

By virtue of (11) we can perform a Legendre transformation on F with respect to total strain and define a Gibbs free energy function $G(\sigma_{ij}, \alpha_{ij})$ which depends on total stress and plastic strain which is such that

$$\varepsilon_{ij} = \partial G / \partial \sigma_{ij} \qquad (14)$$

This, of course, is the standard transformation between strain-energy and complementary strain-energy in purely elastic materials. Here, however both F and G depend on the "passive variable" α_{ij}, which by virtue of (8) and (13) satisfies

$$\tau_{ij} = \partial G / \partial \alpha_{ij} \qquad (15)$$

For decoupled materials, ie ones in which the elastic modulus is independent of plastic strain, it can be shown (Lubliner (1972)) that the Gibbs function must take the form

$$G(\sigma_{ij}, \alpha_{ij}) = G_1(\sigma_{ij}) + \sigma_{ij}\alpha_{ij} + G_2(\alpha_{ij}) \qquad (16)$$

Hence from (14) we see that the gradient of G_1 defines the elastic strain, whilst from (15) we deduce that

$$\tau_{ij} = \sigma_{ij} - \rho_{ij} \text{ , where } \rho_{ij} = - \partial G_2 / \partial \alpha_{ij} \qquad (17)$$

is the "shift " stress. The G_2 part of the Gibbs function hence determines that part of the plastic deformation which is non-dissipative.

Consider now the Legendre transformation of the dissipation function $D(\varepsilon_{ij}, \alpha_{ij}, \dot{\alpha}_{ij})$. By virtue of (13) we can define a dual function in which $\dot{\alpha}_{ij}$ is replaced by τ_{ij} and the two strain variables are treated as passive. However since D is homogeneous of degree one in the $\dot{\alpha}_{ij}$, $Z = D(\dot{\alpha}_{ij})$ represents a cone in $(n + 1)$ dimensional $(Z, \dot{\alpha}_{ij})$ space and the Legendre Transform is singular as discussed in the previous section. The dual function is hence **identically zero:**

$$f(\varepsilon_{ij}, \alpha_{ij}\, \tau_{ij}) = 0 \qquad (18)$$

and the non-unique inverse relation is:

$$\dot{\alpha}_{ij} = \dot{\lambda}\, \partial f/\partial \tau_{ij} \qquad (19)$$

where $\dot{\lambda}$ is an undetermined multiplier. These are of course the familiar yield condition and flow rule for rate-independent plastic materials, but formulated in **generalised stress space and not true stress space**.

Since both the total strain ε_{ij} and the generalised stress τ_{ij} can, in general, be expressed in terms of σ_{ij} and α_{ij}, using the Gibbs function and equations (14) and (15) we can eliminate these two tensor variables from (18) and define a yield function in **true** stress space by:

$$\hat{f}\,(\alpha_{ij}, \sigma_{ij}) \equiv f(\partial G/\partial \sigma_{ij}\,,\, \alpha_{ij}\,,\, \partial G/\,\alpha_{ij}) \qquad (20)$$

Using (16) and (19) the flow rule in true stress space becomes

$$\dot{\alpha}_{ij} = \dot{\lambda}\, (\partial \hat{f}\,/\partial \sigma_{ij} + M_{ijkl}\, \partial f/\partial \varepsilon_{kl}) \qquad (21)$$

where $M_{ijkl} = \partial^2 G_1/\partial \sigma_{ij}\, \partial \sigma_{kl}$ are the instantaneous elastic moduli, or

$$\dot{\alpha}_{ij} = \dot{\lambda}\, (\partial \hat{f}\,/\partial \sigma_{ij}) - M_{ijkl}\, \partial D/\partial \varepsilon_{kl} \qquad (22)$$

which follows from the transformation properties for passive variables.

We are thus led to the important conclusion that whenever the dissipative function depends on **total** strain in addition to the plastic strain, normality is lost in true stress space, but that the form of the flow rule, and hence of the plastic potential, is still uniquely determined, from knowledge of the two basic constitutive functions: F and D. If the dissipation function depends only on the **volumetric** component of the total strain. as though the voids ratio for example, and the elastic behaviour is isotropic, then it follows from (21) or (22) that the deviatoric (distortional) components of $\dot{\alpha}_{ij}$ are given by the associated flow rule, and it is only the volumetric component of the plastic strain rate which is non-associated. For example equation (22) would simplify to:

$$\dot{\alpha}_{ij} = \dot{\lambda} \, (\partial \hat{f} / \partial \sigma_{ij}) - K \, \delta_{ij} \, \partial D / \partial \varepsilon_v \qquad (23)$$

where K is the instantaneous elastic bulk modulus and ε_v is the total volumetric strain. This result is consistent with many observations and is a common feature of many constitutive models.

Collins and Houlsby (1996) have adopted a slightly different approach in which ε_{ij} is eliminated from D initially using (14) and D is regarded from the outset as being of the form $D(\sigma_{ij}, \alpha_{ij}, \dot{\alpha}_{ij})$. This approach is of direct relevance to geomaterials, since on dimensional grounds, purely frictional materials, which possess no characterising material stress parameters, must have dissipation functions which depend explicitly on stress. Applications to standard friction models of the Coulomb, Drucker-Prager type as well to more sophisticated critical-state models can be found in this paper.

4. Conclusions

In conventional metal plasticity the dissipation function is normally regarded as depending only on the plastic part of the total strain. With this assumption the thermomechanical theory predicts the existence of a yield surface, with a normal flow rule in true stress space - ie. the "standard" elastic/plastic model. However soils, sands and other geomaterials undergo significant volume changes. If the dissipation is assumed to depend on the total volumetric strain aswell as possibly on the plastic shear and volumetric strains, then it has been shown that application of the standard thermomechanical arguments and of Legendre transformations predicts the existence of a yield surface in true stress space with a flow rule for the plastic strain rates which is associated in the deviatoric components but non-associated in the volumetric part.

One of the main aims of this paper has been to demonstrate that the Legendre Transformation plays a central role in the development of thermomechical constitutive models. This has been demonstrated here by the introduction of the Gibbs potential and of the singular transformation between the dissipation function for a rate-independent material and the correspondig yield condition and flow rule.

5. References

Callen, H.B. (1960) *Thermodynamics,* John Wiley and Sons, New York.
Collins, I.F. and Houlsby, G.T. (1966) Application of thermomechanical principles to the modelling of geotechnical materials, in preparation.
Coussy,O. (1995) *Mechanics of Porous Continua.* John Wiley and Sons, New York.
Germain, P., Nguyen, Q.S. and Suquet, P. (1983) Continuum thermodynamics, *J. Appl. Mechanics,* **50**, 1010-1020.
Houlsby, G.T. (1981) A study of plasticity theories and their applicability to soils, PhD thesis, University of Cambridge.
Houlsby, G.T. (1982) A derivation of the small-strain incremental theory of plasticity from thermodynamics. *Proc. IUTAM conference on deformation and failure of granular materials,* Delft, 109-118.
Houlsby,G.T. (1996) Derivation of incremental stress-strain response for plasticity models based on thermodynamic functions. *Proc. IUTAM Symposium on Mechanics of Granular and Porous Materials,* Cambridge.
Lemaitre, J. and Chaboche, J.L. (1990) *Mechanics of solid materials,* Cambridge University Press, UK.
Lubliner, J. (1972) On the thermodynamic foundations of non-linear solid mechanics, *CISM - Courses and lectures,* **54**, Springer-Verlag, Wien.
Maugin, G.A. (1992) *The thermodynamics of plasticity and fracture,* Cambridge University Press, UK.
Sewell, M.J. (1987) *Maximum and minimum principles.* Cambridge University Press, UK.
Ziegler, H. (1983) *An introduction to thermomechanics,* North-Holland, Amsterdam (second edition).
Ziegler, H. and Wehrli, C. (1987) The derivation of constitutive relations from the free energy and the dissipation function. *Advances in Applied Mechanics,* **25**, 183-238.

DERIVATION OF INCREMENTAL STRESS-STRAIN RESPONSE FOR PLASTICITY MODELS BASED ON THERMODYNAMIC FUNCTIONS

G.T. HOULSBY
Oxford University
Department of Engineering Science, Parks Road, Oxford

1. Introduction

The constitutive behaviour of soils and other granular materials is often usefully described within the framework of plasticity theory. The existence of a yield surface in stress space, enclosing a region in which the strains are substantially elastic, is well established empirically. The validity of the concepts of a flow rule and hardening law are also well established for many granular materials. Within this broad framework many constitutive models, of varying complexity, have been proposed. Recent research on the non-linearity of soils at small strains has raised important questions about the range of applicability of these models, but the basic framework remains unchallenged.

One of the most important steps in the development of plasticity theory was the development of a standard approach which allows the incremental stress-strain response to be derived, given the functional forms of (a) the elastic response, (b) the yield surface, (c) the plastic potential and (d) a hardening law. A model must be formulated incrementally if it is to be used in non-linear finite element analysis, which is undoubtedly the most important technique currently available for the analysis of engineering problems involving materials with complex constitutive behaviour. For a standard derivation of incremental response for conventional plasticity models see for example Zienkiewicz (1977).

A feature of most plasticity models for granular materials is that, unlike models for metals, they employ "non-associated" flow rules, that is the plastic potential is not the same as the yield surface. This is essential if the dilative properties of granular materials are to be modelled realistically. Unfortunately it is possible to devise non-associated plasticity models which violate thermodynamic principles, and no simple procedure has been devised to apply retrospective checks to plasticity models to check that they are thermodynamically acceptable.

The above problem has led to an alternative approach to plasticity modelling of granular materials (Houlsby 1981, 1982). In this approach the entire constitutive behaviour is determined from two thermodynamic functions: the Helmholtz Free Energy and a Dissipation Function. By adopting the orthogonality principle (Ziegler, 1977),

161

N. A. Fleck and A. C. F. Cocks (eds.),
IUTAM Symposium on Mechanics of Granular and Porous Materials, 161–172.
© 1997 *Kluwer Academic Publishers.*

which may be regarded as a more restrictive form of the Second Law of Thermodynamics, the entire constitutive behaviour may then be determined.

An important observation is that, for certain forms of the thermodynamic functions, the phenomena of elastic behaviour, a yield surface, flow rule and hardening law all appear as consequences of the above formulation and do not need to be imposed as separate features. The models derived are, however, guaranteed not to violate thermodynamic principles. An obstacle to use of this approach has been that, although incremental stress-strain response has been determined for specific models, no general technique has been established to derive the incremental response directly from the thermodynamic functions. The purpose of the paper is to set out a general technique (subject to one minor restriction on the form of the dissipation function), so that incremental response may be derived. This paper clears the way for the use of the new method, guaranteeing thermodynamic acceptability, in non-linear finite element analysis.

2. Functions of Kinematic Variables

The most important feature of the approach to constitutive modelling adopted here is that the entire constitutive behaviour is determined by the specification of only two functions: the Helmholtz Free Energy and the Dissipation Function. Both of these are functions only of kinematic variables, *i.e.* the strains and internal kinematic variables such as the plastic strain. Once the form of the functions is determined then the entire constitutive behaviour may be derived purely by manipulation of these functions, and no further assumptions need to be made.

Let the strains be denoted by ε_i and the internal variables by α_i. For convenience a single subscript notation is used here, with stresses and strains treated as vectors, as is usual in development of finite element code. In general the Helmholtz Free Energy may be a function of both the strains and the internal variables $F = F(\varepsilon_i, \alpha_i)$. The Dissipation Function may not only be a function of these variables, but also their rates, thus $D = D(\varepsilon_i, \alpha_i, \dot{\varepsilon}_i, \dot{\alpha}_i)$. If the material is rate-independent, then it can easily be shown that the Dissipation Function must be homogeneous and of first order in the rates, so that $D(\varepsilon_i, \alpha_i, n\dot{\varepsilon}_i, n\dot{\alpha}_i) = nD(\varepsilon_i, \alpha_i, \dot{\varepsilon}_i, \dot{\alpha}_i)$.

Applying the standard procedures developed for this approach (Ziegler (1977), Houlsby (1981, 1982)), the constitutive behaviour is then entirely determined by the pair of equations:

$$\sigma_i = \frac{\partial F}{\partial \varepsilon_i} + \frac{\partial D}{\partial \dot{\varepsilon}_i} \tag{1}$$

$$0 = \frac{\partial F}{\partial \alpha_i} + \frac{\partial D}{\partial \dot{\alpha}_i} \tag{2}$$

The proof of the above equations involves detailed consideration of the application of the First and Second Laws of Thermodynamics to constitutive models employing internal kinematic variables. They embody Ziegler's "orthogonality principle", which is in effect a stronger statement than the Second Law. A discussion of the general validity of this principle is not possible here. Instead we may simply regarded as a classifying principle: there is a class of materials for which the constitutive behaviour can be defined using equations (1) and (2). This class automatically obeys the thermodynamic laws, and is sufficiently broad to encompass realistic models for granular materials.

Unfortunately, however, the above equations are not immediately in an appropriate form for expressing the constitutive behaviour for application in numerical computation, and considerable manipulation (see Houlsby (1981, 1982)) is required to achieve this.

Whilst it has always proved possible to derive the constitutive behaviour from the above expressions, this has hitherto been by application of *ad hoc* procedures, and no standard method has been developed. A particularly important form of the constitutive equations is an incremental relationship between the stress and strain. This form is used, for example, in non-linear finite element analysis to obtain the stiffness matrix. The main purpose of this paper is to develop a standard analysis procedure (for slightly restricted cases) by which the incremental response may be derived.

Firstly it is useful to re-cast the forms of the Helmholtz Free Energy and Dissipation Function. It is often convenient to consider internal variables which correspond to the conventional plastic strains $\alpha_i \equiv \varepsilon_i^p$. The elastic strain is then defined through the standard decomposition of strain into elastic and plastic components $\varepsilon_i = \varepsilon_i^e + \varepsilon_i^p$. It is then convenient to redefine the Helmholtz Free Energy as a function of elastic and plastic strains $F = F\left(\varepsilon_i^e, \varepsilon_i^p\right)$, since this often gives a simple functional form. This change is purely for convenience, and does not restrict in any way the form of F.

The Dissipation Function is also re-written in terms of elastic and plastic strains. In addition the restriction is made that the Dissipation Function is not a function of the elastic strain rates. This simplification is not unduly restrictive, but it eliminates the possibility of materials which exhibit changes of stress with no change of strain (*i.e.* rigid responses). In most developments of constitutive models based on thermodynamics this restriction is assumed without question or discussion. The form of the Dissipation Function therefore becomes $D = D\left(\varepsilon_i^e, \varepsilon_i^p, \dot{\varepsilon}_i^p\right)$.

The standard approach for defining the constitutive behaviour needs to be modified slightly to account for the introduction of elastic and plastic strains. A simple way is to introduce a set of constraint functions $C_i = \dot{\varepsilon}_i - \dot{\varepsilon}_i^e - \dot{\varepsilon}_i^p = 0$, and to modify the formulation to:

$$\sigma_i = \frac{\partial F}{\partial \varepsilon_i} + \frac{\partial D}{\partial \dot{\varepsilon}_i} + \lambda_j \frac{\partial C_j}{\partial \dot{\varepsilon}_i} \tag{3}$$

$$0 = \frac{\partial F}{\partial \varepsilon_i^e} + \frac{\partial D}{\partial \dot{\varepsilon}_i^e} + \lambda_j \frac{\partial C_j}{\partial \dot{\varepsilon}_i^e} \qquad (4)$$

$$0 = \frac{\partial F}{\partial \varepsilon_i^p} + \frac{\partial D}{\partial \dot{\varepsilon}_i^p} + \lambda_j \frac{\partial C_j}{\partial \dot{\varepsilon}_i^p} \qquad (5)$$

where λ_i is a Lagrangean multiplier. Note that throughout this paper the summation convention is used for a repeated index. From equation (3) it rapidly follows that $\lambda_i = \sigma_i$, so that (4) and (5) can be rewritten:

$$\sigma_i = \frac{\partial F}{\partial \varepsilon_i^e} \qquad (6)$$

$$\sigma_i = \frac{\partial F}{\partial \varepsilon_i^p} + \frac{\partial D}{\partial \dot{\varepsilon}_i^p} \qquad (7)$$

In this paper the constitutive behaviour will be derived entirely from the thermodynamic functions in their original form. An alternative approach is to use the powerful technique of Legendre Transformations to change from the original functions of kinematic variables to new functions of stress variables. This alternative approach (Collins and Houlsby (1996), Collins (1996)) proves to be valuable in that it provides useful insights into the relationships between the forms of the Dissipation Function, the yield surface and the plastic potential.

3. Derivation of Incremental Response

In the following development it will be useful to use the shorthand $F_i^e \equiv \frac{\partial F}{\partial \varepsilon_i^e}$, $F_{ij}^{ep} \equiv \frac{\partial^2 F}{\partial \varepsilon_i^e \partial \varepsilon_j^p}$ etc.. Note then that the differential of (6) gives:

$$\dot{\sigma}_i = F_{ij}^{ee} \dot{\varepsilon}_j^e + F_{ij}^{ep} \dot{\varepsilon}_j^p = F_{ij}^{ee} \dot{\varepsilon}_j + \left(F_{ij}^{ep} - F_{ij}^{ee} \right) \dot{\varepsilon}_j^p \qquad (8)$$

Consider now the form of D. Although D is first order in the plastic strain rates, it is not usually linear, but D^2 is not only second order, but for many models of material behaviour it is quadratic (see Section 4 below for examples). This restriction will be applied in the following development. For this case it is useful to define:

$$q_{ij} = \frac{1}{2} \frac{\partial^2 \left(D^2 \right)}{\partial \dot{\varepsilon}_i^p \partial \dot{\varepsilon}_j^p} \tag{9}$$

where $q_{ij} = q_{ij}\left(\varepsilon_i^e, \varepsilon_i^p\right)$, *i.e.* q_{ij} is not a function of the plastic strain rates. Define also the inverse of q_{ij} as p_{ij}:

$$p_{ij} q_{jk} = \delta_{ik} \tag{10}$$

In some cases it will be convenient to determine the inverse analytically, in other cases it can be determined numerically. Note that q_{ij} and p_{ij} are of course symmetric.

Clearly the existence of the inverse depends on the assumption that the determinant of q_{ij} is non-zero. In fact certain important forms of D result in a zero determinant, with these being cases where there is a mode of plastic deformation which would induce no dissipation. This deformation mode is then suppressed by a side constraint that it must be zero. Models involving either zero or constant plastic dilation come into this category. One approach to their treatment is to reduce the number of internal variables by one. In this case the internal variables no longer correspond precisely to the conventional plastic strains (although they are closely related to them) and the modified formulation described in Appendix A is appropriate.

Multiplying (7) by $\dot{\varepsilon}_i^p$ gives:

$$\left(\sigma_i - F_i^p\right)\dot{\varepsilon}_i^p = \frac{\partial D}{\partial \dot{\varepsilon}_i^p} \dot{\varepsilon}_i^p = D \tag{11}$$

where the second part of equation (11) follows from the fact that D is first order in the strain rates. Note now that because $\dfrac{\partial \left(D^2 \right)}{\partial \dot{\varepsilon}_i^p}$ is also first order in the rates it follows that:

$$q_{ij} \dot{\varepsilon}_j^p = \frac{1}{2} \frac{\partial^2 \left(D^2 \right)}{\partial \dot{\varepsilon}_i^p \partial \dot{\varepsilon}_j^p} \dot{\varepsilon}_j^p = \frac{1}{2} \frac{\partial \left(D^2 \right)}{\partial \dot{\varepsilon}_i^p} = D \frac{\partial D}{\partial \dot{\varepsilon}_i^p} \tag{12}$$

Multiplying (7) by D therefore gives:

$$\left(\sigma_i - F_i^p\right)D = \frac{\partial D}{\partial \dot{\varepsilon}_i^p} D = q_{ij}\dot{\varepsilon}_j^p \tag{13}$$

or, defining $r_j = p_{ij}\left(\sigma_i - F_i^P\right)$:

$$p_{ij}\left(\sigma_i - F_i^P\right)D = r_j D = \dot{\varepsilon}_j^P \tag{14}$$

Equations (8), (11) and (14) are $2n+1$ equations in the $3n+1$ variables $\dot{\sigma}_i$, $\dot{\varepsilon}_i$, $\dot{\varepsilon}_i^P$ and D (where n is the number of strain variables), and at first sight $\dot{\varepsilon}_i^P$ and D could be eliminated to give n equations representing the stiffness relationship between $\dot{\sigma}_i$ and $\dot{\varepsilon}_i$. However, (11) and (14) are not linearly independent equations in the rates, and they can be combined to give either $D = 0$ (which is the case of no dissipation, when all plastic strains are zero and elastic behaviour occurs) or:

$$p_{ij}\left(\sigma_i - F_i^P\right)\left(\sigma_j - F_j^P\right) = 1 \tag{15}$$

which is simply the yield condition. Thus it is seen that the existence of the yield surface arises as a natural consequence of the formulation.

To obtain $2n+1$ useful equations in the rates it is necessary to drop equation (11) and introduce instead the consistency condition that the stress point remains on the yield surface during plastic deformation. This is given by the differential of (15):

$$\dot{p}_{ij}\left(\sigma_i - F_i^P\right)\left(\sigma_j - F_j^P\right) + 2p_{ij}\left(\sigma_i - F_i^P\right)\left(\dot{\sigma}_j - F_{jk}^{pe}\dot{\varepsilon}_k^e - F_{jk}^{pp}\dot{\varepsilon}_k^P\right) = 0 \tag{16}$$

Noting that $\dot{p}_{ij} = -p_{ik}\,p_{jk}\,\dot{q}_{kl}$, this can be re-arranged to:

$$r_k r_l \dot{q}_{kl} = 2r_j\left(\dot{\sigma}_j - F_{jk}^{pe}\dot{\varepsilon}_k - \left(F_{jk}^{pp} - F_{jk}^{pe}\right)\dot{\varepsilon}_k^P\right) \tag{17}$$

Defining $q_{ijk}^e = \dfrac{\partial q_{ij}}{\partial \varepsilon_k^e}$ etc., and substituting for $\dot{\varepsilon}_i^P$, (17) can further be re-written as:

$$r_k r_l\left(q_{klm}^e \dot{\varepsilon}_m + \left(q_{klm}^P - q_{klm}^e\right)r_m D\right) = 2r_j\left(\dot{\sigma}_j - F_{jk}^{pe}\dot{\varepsilon}_k - \left(F_{jk}^{pp} - F_{jk}^{pe}\right)r_k D\right) \tag{18}$$

Equation (8) may be rewritten in the form:

$$\dot{\sigma}_i = F_{ij}^{ee}\dot{\varepsilon}_j + r_j\left(F_{ij}^{ep} - F_{ij}^{ee}\right)D \tag{19}$$

which can be substituted into (18) to give, on rearrangement:

$$r_k r_l \left(r_m \left(q_{klm}^p - q_{klm}^e \right) - 2 \left(F_{kl}^{ep} - F_{kl}^{ee} - F_{kl}^{pp} + F_{kl}^{pe} \right) \right) D =$$

$$r_i \left(-r_l q_{ilj}^e + 2 \left(F_{ij}^{ee} - F_{ij}^{pe} \right) \right) \dot{\epsilon}_j \qquad (20)$$

This is in effect an equation of the form $AD = B_j \dot{\epsilon}_j$ which can be used to substitute for D into (19), giving after some minor manipulation:

$$\dot{\sigma}_i = \left(F_{ij}^{ee} + r_k \left(F_{ik}^{ep} - F_{ik}^{ee} \right) \frac{B_j}{A} \right) \dot{\epsilon}_j = d_{ij} \dot{\epsilon}_j \qquad (21)$$

where d_{ij} is the required stiffness matrix. All the terms are easily determined from the original functions. Furthermore, equations (20) and (21) have been derived for very general forms of the thermodynamic functions, and in many simple cases a number of the terms are zero.

4. Examples

Two examples are pursued here: (a) a very simple analogue of elastic-perfectly plastic (cohesive) behaviour, (b) a critical state model appropriate for soft clays. These cases both involve associated flow, a third example, that of a simple frictional plasticity model involving non-associated flow, is given by Houlsby (1996).

4.1 COHESIVE ELASTIC PERFECTLY-PLASTIC MODEL

Consider first a simple 2-dimensional analogue of plasticity in which the strain vector is $\gamma_i = \begin{bmatrix} \gamma_1 & \gamma_2 \end{bmatrix}^T$ and stress vector $\tau_i = \begin{bmatrix} \tau_1 & \tau_2 \end{bmatrix}^T$. The use of γ and τ in place of ϵ and σ is simply to follow convention for shear components. The elastic behaviour is given by $\tau_i = G\gamma_i^e$, and the yield surface (and plastic potential) is $\tau_i \tau_i = c^2$. The thermodynamic functions required to derive this model are:

$$F = \frac{G}{2} \gamma_i^e \gamma_i^e \qquad (22a)$$

$$D = c \sqrt{\dot{\gamma}_i^p \dot{\gamma}_i^p} \qquad (22b)$$

From these expressions the standard manipulation rapidly gives $F_i^e = G\gamma_i$, $F_i^p = 0$, $F_{ij}^{ee} = G\delta_{ij}$, $F_{ij}^{ep} = 0$, $F_{ij}^{pe} = 0$, $F_{ij}^{pp} = 0$, $q_{ij} = c\delta_{ij}$, $p_{ij} = \frac{1}{c}\delta_{ij}$, $r_i = \frac{\sigma_i}{c}$, $q_{ijk}^e = 0$ and

$q_{ijk}^p = 0$. Equation (8) gives the required elastic behaviour and equation (15) gives the required form of the yield locus. The factors in equation (21) are determined as $A = 2G\dfrac{\tau_i \tau_i}{c^2} = 2G$ and $B_j = \dfrac{2G}{c}\tau_j$, so that $d_{ij} = G\left(\delta_{ij} - \dfrac{\tau_i \tau_j}{c^2}\right)$, which can easily be verified as the correct plastic stiffness matrix.

4.2 CRITICAL STATE MODEL

The second example is that of the "Modified Cam-Clay" critical sate model which is often used to model the work-hardening behaviour of normally consolidated and lightly overconsolidated clays. Houlsby (1981) demonstrates how this model can be derived from the thermodynamic functions. For triaxial stress states it is convenient to express the model in terms of the stress and strain vectors $[p \quad q]^T$ and $[v \quad \varepsilon]^T$ commonly used in critical state soil mechanics. Using a slightly simplified form of the functions used by Houlsby (1981), with the simplification being related to the datum chosen for plastic strain, the required functions are:

$$F = p_r \kappa \exp\left(\frac{v_e}{\kappa}\right) + \frac{3G\varepsilon_e^2}{2} + p_r(\lambda - \kappa)\exp\left(\frac{v_p}{\lambda - \kappa}\right) \tag{23}$$

$$D = p_r \exp\left(\frac{v_p}{\lambda - \kappa}\right)\sqrt{\dot{v}_p^2 + M^2 \dot{\varepsilon}_p^2} \tag{24}$$

where p_r is a reference pressure, M, λ and κ are material parameters commonly used in critical state soil mechanics (except that λ and κ refer to slopes in $(\ln p, \ln V)$ space rather than in $(\ln p, V)$ space) and G is the shear modulus. Note that the model defines a bulk modulus proportional to pressure, but a constant shear modulus. It follows that $F_i^e = [p \quad q]^T = \left[p_r \exp\left(\dfrac{v_e}{\kappa}\right) \quad 3G\varepsilon_e\right]^T$, $F_i^p = \left[p_r \exp\left(\dfrac{v_p}{\lambda - \kappa}\right) \quad 0\right]^T$. It is convenient to introduce the shorthand $p_x = p_r \exp\left(\dfrac{v_p}{\lambda - \kappa}\right)$. It then follows that

$F_{ij}^{ee} = \begin{bmatrix} \dfrac{p}{\kappa} & 0 \\ 0 & 3G \end{bmatrix}$, $F_{ij}^{ep} = 0$, $F_{ij}^{pe} = 0$, $F_{ij}^{pp} = \begin{bmatrix} \dfrac{p_x}{\lambda - \kappa} & 0 \\ 0 & 0 \end{bmatrix}$. Differentiation of D^2 gives

$q_{ij} = p_x^2 \begin{bmatrix} 1 & 0 \\ 0 & M^2 \end{bmatrix}$ and $p_{ij} = p_x^{-2} \begin{bmatrix} 1 & 0 \\ 0 & M^{-2} \end{bmatrix}$, so that equation 15 gives the yield locus as the ellipse $\dfrac{(p-p_x)^2}{p_x^2}+\dfrac{q^2}{M^2 p_x^2}=1$. One can further derive the relations

$r_i = p_x^{-2}\left[(p-p_x)\quad \dfrac{q}{M^2}\right]^T$, $q_{ijk}^e = 0$, $q_{ij1}^p = \dfrac{2p_x^2}{\lambda-\kappa}\begin{bmatrix}1 & 0 \\ 0 & M^2\end{bmatrix}$ and $q_{ij2}^p = 0$. Noting that

$r_k q_{ijk}^p = \dfrac{2(p-p_x)}{\lambda-\kappa}\begin{bmatrix}1 & 0 \\ 0 & M^2\end{bmatrix}$, the factors in equation (20) are determined as

$A = \dfrac{2}{p_x^4}\left((p-p_x)^2\dfrac{p\lambda}{\kappa(\lambda-\kappa)}+\dfrac{q^2}{M^4}\left(\dfrac{M^2(p-p_x)}{(\lambda-\kappa)}+3G\right)\right)$ and

$B_j = \dfrac{2}{p_x^2}\left[\dfrac{p(p-p_x)}{\kappa}\quad \dfrac{3Gq}{M^2}\right]^T$. Finally the stiffness matrix can be determined as:

$$d_{ij} = \begin{bmatrix}\dfrac{p}{\kappa} & 0 \\ 0 & 3G\end{bmatrix} - \dfrac{2}{Ap_x^4}\begin{bmatrix}\dfrac{p^2(p-p_x)^2}{\kappa^2} & \dfrac{3Gqp(p-p_x)}{M^2\kappa} \\ \dfrac{3Gqp(p-p_x)}{M^2\kappa} & \dfrac{9G^2q^2}{M^4}\end{bmatrix} \quad (25)$$

Equation (25) can be manipulated into a variety of alternative forms. It has been verified that equation (25) is correct by comparing it with values from a totally different approach used to derive a compliance matrix in Houlsby (1981).

5. Conclusions

Plasticity models for modelling the behaviour of granular materials can be derived from thermodynamic functions. This approach has the advantage that the models are guaranteed to obey the laws of thermodynamics, and retrospective criteria need not be applied.

Within this framework a standard approach has been developed (subject only to a minor restriction on the form of the Dissipation Function) which allows the incremental stiffness matrix to be derived solely by manipulation of the original thermodynamic functions. This is an essential step if this approach is to be used to derive models appropriate for non-linear finite element analysis. Whilst the new approach needs further refinement, it can now readily be used for the development and implementation of realistic models for granular materials.

6. Acknowledgement

The Author is grateful to the University of Auckland for support as a Foundation Visitor for the period during which the work reported here was started.

7. References

Collins, I.F. (1996) The use of Legendre transformations in developing the constitutive laws of geomechanics from thermodynamic principles, *Proc. IUTAM Symposium on Mechanics of Granular and Porous Materials*, Cambridge

Collins, I.F. and Houlsby, G.T. (1996) Application of thermomechanical principles to the modelling of geotechnical materials, Report No. OUEL 2100/96, Department of Engineering Science, Oxford University

Houlsby, G.T. (1981) *A Study of Plasticity Theories and Their Applicability to Soils*, Ph.D. Thesis, Cambridge University

Houlsby, G.T. (1982) A derivation of the small strain incremental theory of plasticity from thermomechanics, *Proc. IUTAM Symp. on Deformation and Failure of Granular Materials*, Delft, 109-118

Houlsby, G.T. (1992) Interpretation of dilation as a kinematic constraint, *Proc. Workshop on Modern Approaches to Plasticity*, Horton, Greece, 19-38

Houlsby, G.T. (1996) Derivation of Incremental Stress-Strain Response for Plasticity Models Based on Thermodynamic Functions, Report No. OUEL 2108/96, Department of Engineering Science, Oxford University

Matsuoka, H. and Nakai, T. (1974) Stress-deformation and strength characteristics of soil under three different principal stresses, *Proc. JSCE*, **232**, 59-70

Ziegler, H. (1977) *An Introduction to Thermomechanics*, North Holland, Amsterdam (see also 2nd Edition, 1983).

Zienciewicz, O.C. (1977) *The Finite Element Method*, 3rd Edition, McGraw Hill, London

8. Appendix A: An Alternative Formulation Using Internal Variables

In this Appendix the incremental response is derived from the original forms of the Helmholtz Free Energy and Dissipation Function, without introducing the definitions of elastic and plastic strains. On some occasions this approach proves to be more useful, since the internal variables are not restricted to be the plastic components of the strains. The numbering of the equations in this Appendix is chosen to correspond as closely as possible to the equations in the original text, so some numbers are omitted.

As in the main text, the dissipation function is not a function of the strain rates, so that equations (1) and (2) become:

$$\sigma_i = F_i^{\varepsilon} \tag{A1}$$

$$- F_i^\alpha = \frac{\partial D}{\partial \dot{\alpha}_i} \tag{A2}$$

Where the shorthand $F_i^\varepsilon \equiv \dfrac{\partial F}{\partial \varepsilon_i}$, $F_{ij}^{\varepsilon\alpha} \equiv \dfrac{\partial^2 F}{\partial \varepsilon_i \partial \alpha_j}$ etc. is introduced. Note that the dimension of the internal variable α_i need not be the same as that of the strain ε_i. The differential of (A1) gives:

$$\dot{\sigma}_i = F_{ij}^{\varepsilon\varepsilon} \dot{\varepsilon}_j + F_{ij}^{\varepsilon\alpha} \dot{\alpha}_j \tag{A8}$$

Again assuming that D^2 is quadratic, define:

$$q_{ij} = \frac{1}{2} \frac{\partial^2 (D^2)}{\partial \dot{\alpha}_i \partial \dot{\alpha}_j} \tag{A9}$$

Define also p_{ij}, the inverse of q_{ij} as in the main text. Multiplying (A2) by $\dot{\alpha}_i$ gives:

$$- F_i^\alpha \dot{\alpha}_i = \frac{\partial D}{\partial \dot{\alpha}_i} \dot{\alpha}_i = D \tag{A11}$$

Multiplying (A2) by D gives:

$$- F_i^\alpha D = \frac{\partial D}{\partial \dot{\alpha}_i} D = q_{ij} \dot{\alpha}_j \tag{A13}$$

or, defining $r_j = -p_{ij} F_i^\alpha$:

$$- p_{ij} F_i^\alpha D = r_j D = \dot{\alpha}_j \tag{A14}$$

The yield surface can therefore be derived as:

$$p_{ij} F_i^\alpha F_j^\alpha = 1 \tag{A15}$$

and the differential of the yield surface is:

$$\dot{p}_{ij} F_i^\alpha F_j^\alpha + 2 p_{ij} F_i^\alpha \left(F_{jk}^{\alpha\varepsilon} \dot{\varepsilon}_k + F_{jk}^{\alpha\alpha} \dot{\alpha}_k \right) = 0 \tag{A16}$$

This can be re-arranged to:

$$r_k r_l \dot{q}_{kl} = -2r_j \left(F_{jk}^{\alpha\varepsilon} \dot{\varepsilon}_k + F_{jk}^{\alpha\alpha} \dot{\alpha}_k \right) \tag{A17}$$

Which can be further be re-written as:

$$r_k r_l \left(q_{klm}^{\varepsilon} \dot{\varepsilon}_m + q_{klm}^{\alpha} \dot{\alpha}_m \right) = -2r_j \left(F_{jk}^{\alpha\varepsilon} \dot{\varepsilon}_k + F_{jk}^{\alpha\alpha} \dot{\alpha}_k \right) \tag{A18}$$

where $q_{klm}^{\varepsilon} = \dfrac{\partial q_{kl}}{\partial \varepsilon_m}$ etc.. This allows a solution for D:

$$r_k r_l \left(r_m q_{klm}^{\alpha} + 2F_{kl}^{\alpha\alpha} \right) D = -r_i \left(r_l q_{ilj}^{\varepsilon} + 2F_{ij}^{\alpha\varepsilon} \right) \dot{\varepsilon}_j \tag{A20}$$

which is again an equation of the form $AD = B_j \dot{\varepsilon}_j$, and one can derive:

$$\dot{\sigma}_i = \left(F_{ij}^{\varepsilon\varepsilon} + r_k F_{ik}^{\varepsilon\alpha} \frac{B_j}{A} \right) \dot{\varepsilon}_j = d_{ij} \dot{\varepsilon}_j \tag{A21}$$

A minor variation is that it often more convenient to express D in the form $D = D(\sigma_i, \alpha_i, \dot{\alpha}_i)$, where the substitution of stress for strain variables is allowable because equation (A1) gives $\sigma_i = \sigma_i(\varepsilon_i, \alpha_i)$. The rest of the analysis proceeds exactly as above, except that (A18) becomes:

$$r_k r_l \left(q_{klm}^{\sigma} \dot{\sigma}_m + q_{klm}^{\alpha} \dot{\alpha}_m \right) = -2r_j \left(F_{jk}^{\alpha\varepsilon} \dot{\varepsilon}_k + F_{jk}^{\alpha\alpha} \dot{\alpha}_k \right) \tag{A18a}$$

leading to:

$$r_k r_l \left(q_{klm}^{\alpha} r_m + q_{klm}^{\sigma} F_{mn}^{\varepsilon\alpha} r_n + 2F_{kl}^{\alpha\alpha} \right) D = -\left(r_k r_l q_{klm}^{\sigma} F_{mj}^{\varepsilon\varepsilon} + 2r_k F_{kj}^{\alpha\varepsilon} \right) \dot{\varepsilon}_j \tag{A20a}$$

Which is again of the form $AD = B_j \dot{\varepsilon}_j$, and can be substituted into equation (A21).

LOCAL CONTACT BEHAVIOUR OF VISCOPLASTIC PARTICLES

BERTIL STORÅKERS
Department of Solid Mechanics
Royal Institute of Technology
S-100 44 Stockholm, Sweden

Abstract

In the past models for the analysis of powder compaction have involved arbitrary assumptions regarding contact kinematics. Recently, some fundamental solutions to indentation problems have made it possible to generate results for powder compaction from first micromechanical principles with well defined physical interpretations. Here an intermediate problem is considered where two spheres are compressed with different strength and size. A background is first given to the basic indentation problem and it is shown how a bimaterial solution may be obtained for general power law viscoplastic behaviour. Rigorous and invariant results of high accuracy are derived for the evolution of flattening and contact size under prescribed loading. Besides being of intrinsic interest the findings serve as useful tools to determine the macroscopic properties of mixed powders. Universal but simple formulae are presented for a combination of soft and hard particles where strain-hardening plastic flow, creep and rigid behaviour arise as special cases. A comparison is made between the predictions of the proposed model and experimental data from observations of soft and hard materials.

1. Introduction

Investigations of the compaction of granular and powder materials have made steady progress through combined efforts in materials science and mechanics. As the issues involved are complex a balanced view has to taken where phenomenological assumptions and a more rigorous basis have to be drawn upon in parallel. As to powder compaction the general mechanisms have been reviewed by Ashby (1990) and methods of powder processing have been summarized at an ASME Congress (1995).

A mechanically inclined analysis of the densification of powders under compression rests heavily on the local contact properties of particles. The matter has recently been investigated by Larsson et al (1996) based on first micromechanical principles in order to obtain densification formulae for both cold and hot compaction. The local analysis by Larsson et al (1996) rests heavily on earlier models of Brinell indentation of power law materials i.e. impression of a rigid sphere into a deformable half-space. In this respect to

N. A. Fleck and A. C. F. Cocks (eds.),
IUTAM Symposium on Mechanics of Granular and Porous Materials, 173–184.
© *1997 Kluwer Academic Publishers.*

simulate the behaviour of specific materials Larsson et al adopted plastic flow theory as analysed by Biwa and Storåkers (1995) for the cold case and Norton creep theory by Storåkers and Larsson (1994) for the hot case.

The investigation by Larsson et al (1996) was based on random packing of identical spherical particles. In real compounds, however, particles may be of different sizes and in composite powders also of different strength as investigated e.g. by Lange et al (1991) for for the case of lead and steel spheres. Accordingly, it is desirable to obtain a general solution to the compression of two dissimilar spheres.

In the classical investigations by Hertz, (1882), it was shown in linear elasticity that when two curved surfaces come into contact the resulting boundary value problem may be reduced to indentation of a half-space. When it comes to nonlinear material behaviour the corresponding problem seems to have received little attention so far. Some results are available from Storåkers (1989) who analysed the contact between two spheres of different sizes for nonlinear elasticity or alternatively plastic deformation theory behaviour. It is the present purpose to examine in detail the contact properties for a wide variety of dissimilar nonlinear materials based on a general viscoplastic constitutive equation as recently applied to Brinell indentation by Storåkers et al (1996).

2. Spherical indentation of viscoplastic solids

A useful viscoplastic constitutive equation, essentially to reproduce the behaviour of metals at ambient and elevated temperatures, reads

$$\sigma = \sigma_0 \varepsilon^{1/m} \dot{\varepsilon}^{1/n} , \tag{1}$$

in one dimensional form and obvious notation.

In the general case eq. (1) reproduces primary creep or strain-rate dependent plasticity. In the limit $n \to \infty$, eq. (1) corresponds to strain-hardening plasticity and in the extreme when $m \to \infty$ also, it reduces to perfect plasticity. On the other hand for general n-values and $m \to \infty$, eq. (1) corresponds to power law (Norton) creep.

In a recent contact analysis, Storåkers et al (1996), have condensed some appropriate three-dimensional forms of eq. (1) proposed by earlier writers. In the most simple case when isotropy and von Mises behaviour is assumed, eq. (1) may be generalized to read

$$\sigma_e = \sigma_0 \varepsilon_e^{1/m} \dot{\varepsilon}_e^{1/n} , \tag{2}$$

$$\sigma_e^2 = \frac{3}{2} s_{ij} s_{ij}, \quad s_{ij} = \sigma_{ij} - \frac{1}{3} \sigma_{kk} \delta_{ij} , \tag{3}$$

$$\dot{\varepsilon}_e^2 = \frac{2}{3}\dot{\varepsilon}_{ij}\dot{\varepsilon}_{ij} \qquad (4)$$

and

$$\sigma_{ij} = \sigma_e\frac{\partial\dot{\varepsilon}_e}{\partial\dot{\varepsilon}_{ij}} \qquad (5)$$

in common notation.

Consider first the fundamental Brinell problem as shown in Fig. 1a. The field equations as to equilibrium and kinematic compatibility are

$$\frac{\partial\sigma_{ij}}{\partial x_i} = 0 \qquad (6)$$

and

$$\varepsilon_{ij} = \frac{1}{2}\left(\frac{\partial u_i}{\partial x_j} + \frac{\partial u_j}{\partial x_i}\right) \quad . \qquad (7)$$

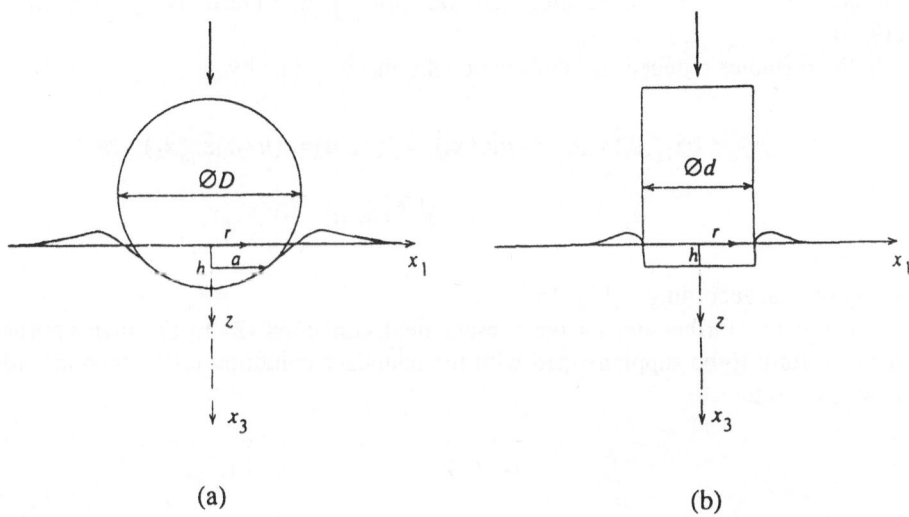

Fig. 1 (a) Spherical Brinell indentation and (b) the related Boussinesq problem.

The boundary conditions for the surface of a half-space indented by a rigid sphere are

$$u_3 = h - r^2/D, \quad \sigma_{31} = \sigma_{32} = 0 \qquad r \leq a, \; x_3 = 0 \tag{8}$$

and

$$\sigma_{3i} = 0 \qquad r > a, \; x_3 = 0 \tag{9}$$

with the notation illustrated in Fig. 1a. For simplicity, but not by necessity, frictionless indentation is assumed.

Contemplating the structure of the constitutive equations (2) to (5) a rate formulation of the boundary value problem is natural in particular as hereditary behaviour is present in general. To this end to effect the rate forms of eqs. (5) and (6) is obvious and the remaining inhomogeneous boundary condition (8) accordingly reduces to

$$\dot{u}_3 = \dot{h} \qquad r \leq a, \; x_3 = 0. \tag{10}$$

As a primary observation by eq. (10) it is clear that the rate form of the problem posed corresponds to that of indentation by a flat die as illustrated in Fig. 1b. Once this intermediate problem is solved a full solution can be obtained by supplementary cumulative superposition. This was proved to be so in a Brinell analysis for creep by Storåkers and Larsson (1994) and subsequently also for strain-hardening plasticity by Biwa and Storåkers (1995) and for the more general form of viscoplasticity by Storåkers et al. (1996).

In these studies reduced variables were introduced ad hoc by

$$x_i = a\tilde{x}_i, \quad \dot{u}_i(x_j, a) = \dot{h}\tilde{u}_i(\tilde{x}_j), \quad \dot{\varepsilon}_{ij}(x_k, a) = (\dot{h}/a)\tilde{\varepsilon}_{ij}(\tilde{x}_k) \quad ,$$

$$\sigma_{ij}(x_i, a) = \sigma_o(\dot{h}/a)^{1/m}(\dot{h}/a)^{1/n}\tilde{\sigma}_{ij}(\tilde{x}_k) \tag{11}$$

with notation according to Fig. 1a.

Leaving out further details the present field equations (2) to (7) then appear in a reduced form to be supplemented with the boundary condition (10) which accordingly now reduces to

$$\tilde{u}_3 = 1 \qquad \tilde{r} \leq 1, \; \tilde{x}_3 = 0. \tag{12}$$

The intermediate boundary value problem so posed is a stationary one independent of the contact radius and of natural time. The reduced variables depend only on the power law exponents m and n. It remains, however, to satisfy the prescribed contact condition (8) and also to solve the complete problem expressed in the original variables.

The boundary condition (8) can be rewritten as

$$\int_0^t \tilde{u}_3(r/a)\dot{h}dt = h(a) - r^2/D \quad . \tag{13}$$

Using a variable transformation from t to a, eq. (13) reduces to an ordinary Volterra integral equation. Its solution,

$$h(a) = a^2/(c^2 D) \tag{14}$$

with

$$c^2 = 1 - 2\int_1^\infty \frac{\tilde{u}_3}{\tilde{r}^3}d\tilde{r} \quad , \tag{15}$$

then determines the physical value of the contact radius a. The corresponding field variables may accordingly be determined in the same spirit of cumulative superposition.

The eigenvalue $c^2(m, n)$ appearing in eqs. (14), (15) is an invariant and depends only on m and n. It may also be shown to give the relative indentation height above the contour. Once its value has been found the present interest focuses on the relation between total load, L, and the indentation depth, h. Using the contact radius, a, as an intermediate variable by eq. (14), it is then clear from the reduced variables introduced in (11) that the total load may be expressed in a separable form as

$$L = \sigma_0 a^2 (a/D)^{1/m}(\dot{a}/D)^{1/n}\tilde{L}(m, n) \tag{16}$$

where

$$\tilde{L} = -\frac{2^{1 + 1/n}\pi}{c^{2/m + 2/n}}\int_0^1 \tilde{\sigma}_{33}\tilde{r}d\tilde{r}. \tag{17}$$

In various formulations the values of the invariants c^2, eq. (15), and $\tilde{L}$, eq (17), have been determined by different writers based on computational methods. It suffices here to mention that Hill et al (1989) adopted plastic deformation theory and used a finite element method based on total strains to solve the Brinell problem for different hardening exponents m. In contrast, subsequent writers, Storåkers et al (1994), (1995), (1996), designed a finite element procedure to apply to the intermediate flat die problem supplemented by cumulative superposition. The constitutive properties used were stationary creep (n), plastic flow theory (m) and viscoplasticity (m,n) respectively.

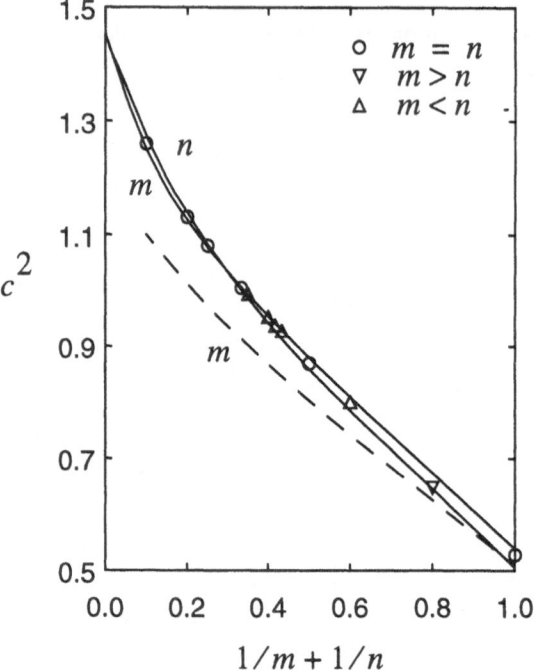

Fig. 2 The invariant $c^2(m, n)$ as function of $1/m + 1/n$ at spherical indentation of viscoplastic materials. General viscoplasticity, (o) $m = n$, (∇) $m > n$, (Δ) $m < n$, Storåkers et al. (1996), (—— m) plastic flow theory, Biwa and Storåkers (1995), (—— n) creep theory, Storåkers and Larsson (1994), (- - - m) deformation theory of plasticity, Hill et al. (1989).

All the results for c^2 are shown in Fig. 2. It may be first observed that the creep results by Storåkers and Larsson (1994) and the plastic flow theory results by Biwa and Storåkers (1995) do indeed come close as functions of n and m respectively. They should be identical at $m,n \rightarrow \infty$, that is for perfect plasticity, but not otherwise as the plastic material behaviour is hereditary for finite m-values. Based on these findings it was found rational, however, to plot the viscoplastic results as a function of $1/m + 1/n$ and as may be seen in Fig. 2, within a very good accuracy

$$c^2(m, n) = c^2(1/m + 1/n) \ . \tag{18}$$

Eq. (18) being approximate in general was shown by Storåkers et al (1996) to be exact for a certain reduced form of proportional straining.

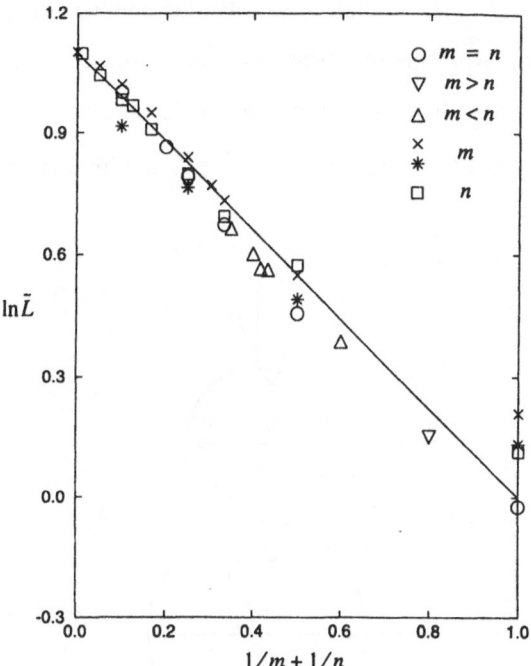

Fig. 3 Reduced mean pressure $ln\tilde{L}$ as function of $1/m + 1/n$ at spherical indentation of viscoplastic materials. General viscoplasticity, (o) $m = n$, (∇) $m > n$, (Δ) $m < n$, Storåkers et al. (1996), (x) plastic flow theory $n \to \infty$, Biwa and Storåkers (1995), ($\square$) creep theory $m \to \infty$, Storåkers and Larsson (1994), (*) deformation theory of plasticity $n \to \infty$, Hill et al. (1989), (———) fit, eq. (19), for general viscoplasticity, Storåkers et al. (1996).

In Fig. 3 the reduced load, $\tilde{L}$, is shown as a function of the combined exponents $1/m + 1/n$. By introducing a factor $(n+2)/n$, Storåkers et al (1996), all results can be combined in a universal hardness formula, in the spirit of Tabor (1951), as

$$L = 3[(n+2)/n]\pi a^2 \sigma_0 (a/3D)^{1/m} (\dot{a}/3D)^{1/n} . \qquad (19)$$

Its reduced form is shown in Fig. 3 and seen to confirm well compared with the corresponding computational results.

3. The two-particle problem

The present problem concerns two spherical but dissimilar particles symmetrically compressed as shown in Fig. 4. The objective is to solve the combined contact problem by extracting the results, σ_{ij}^o, u_i^o, from the Brinell problem outlined above for given values of m and n. The single strength parameter σ_0 in the viscoplastic constitutive equation, (1), (2), is now replaced for two spheres by σ_k, $k = 1,2$, and likewise the two diameters are written D_k.

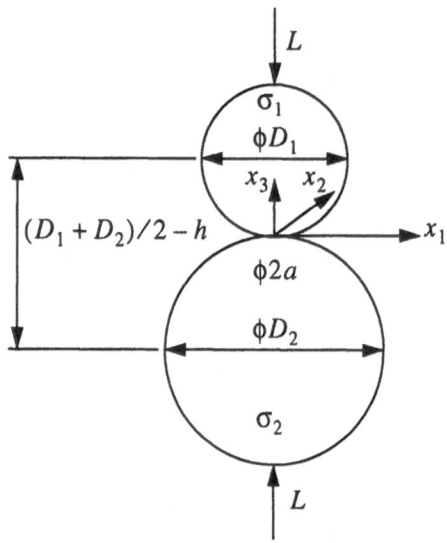

Fig. 4 Compression of two dissimilar spheres.

For a combined solution to be achieved it is to be expected that the reduced field equations must be essentially unaffected and it appears natural to start by the basic assumption

$$\sigma_{ij}^{(k)}(x_l) = \sigma_{ij}^o(x_l) \qquad k = 1, 2 \tag{20}$$

as continuity of traction is required at the contact region. As a consequence of (20) local equilibrium is also satisfied.

By further scaling displacements as

$$u_i^{(k)}(x_j) = (\sigma_o/\sigma_k)^q u_i^o(x_j) \qquad k = 1, 2 \tag{21}$$

where

$$1/m + 1/n = 1/q \qquad (22)$$

it is readily seen that the complete field equations (2) to (7) are satisfied.

It remains to meet with the inhomogeneous boundary condition (8) which is now to be replaced by

$$u_3^{(1)} + u_3^{(2)} = h - r^2/(1/D_1 + 1/D_2) \qquad r \le a, \ x_3 = 0 \qquad (23)$$

By introducing eq. (21) into (23) and subsequently choosing

$$1/\sigma_1^q + 1/\sigma_2^q = 1/\sigma_0^q \qquad (24)$$

and

$$1/D_1 + 1/D_2 = 1/D \quad , \qquad (25)$$

equation (8) is fulfilled and as a consequence the combined boundary value problem is equivalent to the single one posed above. Accordingly, stresses and displacements may be directly read off from equations (20) and (21).

For clarity and explicitness it is of further interest to elaborate on the invariant $c^2 \, (m,n)$ for the relation between the loading, L, and the contact depth, h, or alternatively contact radius, a. Upon setting $D_1 = \alpha D_2$, the change in distance between the centre of two particles follows from (14) and (25) as

$$h = (1 + \alpha)a^2/(c^2 D_1) \quad . \qquad (26)$$

By setting $\sigma_1 = \beta \sigma_2$ and adopting the hardness formula (19), the contact load is

$$L = 3[(n+2)/n][(1+\alpha)/(1+\beta^q)]^{1/q} \pi a^2 \sigma_1 (a/3D_1)^{1/m} (\dot{a}/3D_1)^{1/n} \quad . \qquad (27)$$

Thus, the fundamental solution $L_0 = L_0(a, t, m, n, \sigma_1)$ given by equation (19) provides the solution for the two-particle problem as

$$L(a, t, m, n) = [(1+\alpha)/(1+\beta^q)]^{1/q} L_0 \qquad (28)$$

On the other hand when $L(a) = L_o(a_o)$, then

$$a = [(1 + \beta^q)/(1 + \alpha)]^{\frac{1}{2q+1}} a_o \quad . \tag{29}$$

Some special cases are of particular interest. First for the fully symmetric case, $\alpha = \beta = 1$, (28) and (29) obviously reduce to $L = L_o$ and $a = a_o$ respectively.

In an intermediate case when $\alpha = 1$ and $\beta = 0$, i.e. two spheres of equal size, one of them being rigid, then by (28) and (29)

$$L = 2^{1/q} L_o \tag{30}$$

and

$$a = 2^{-\frac{1}{2q+1}} a_0 \tag{31}$$

respectively. In the asymptotic limit when $\alpha = 1, \beta = 0, q \to \infty$, (30) and (31) reduce to $L = L_o$ and $a = a_o$ which might not be immediately foreseen.

As regards the indentation depth, h, the corresponding fundamental loading relation reads

$$L_0 = 3[(n + 2)/n]\pi c^{2+1/q} \sigma_0 h_0 D(h_0/9D)^{1/2m}\left[\frac{d}{dt}(h_0/9D)^{1/2}\right]^{1/n} \quad . \tag{32}$$

Thus in analogy with (28), when $L_0 = L_0(h, t, m, n, \sigma_1)$,

$$L(h, t, m, n) = (1 + \alpha)^{\frac{1}{2q}-1}(1 + \beta^q)^{-\frac{1}{q}} L_0 \quad . \tag{33}$$

Alternatively when $L = L_o$

$$h = (1 + \alpha)[(1 + \beta^q)/(1 + \alpha)]^{\frac{2}{2q+1}} h_0 \quad . \tag{34}$$

For the symmetric case when $\alpha = \beta = 1$

$$L = 2^{-1-\frac{1}{2q}} L_o \qquad (35)$$

or equivalently

$$L = L_o(h/2) \qquad (36)$$

and

$$h = 2h_o \ . \qquad (37)$$

Likewise, when $\alpha = 1, \beta = 0$

$$L = 2^{-1+\frac{1}{2q}} L_o \qquad (38)$$

and

$$h = 2^{\frac{2q-1}{2q+1}} h_o . \qquad (39)$$

Finally for the perfectly plastic case, when $\alpha = 1, \beta = 0, q \to \infty$, (38) and (39) reduce to (35) and (37).

4. Comparison with experimental observations

In an extensive study of isostatic pressing of composite powders by Turner and Ashby, (1995), for one thing these writers performed some discriminating experiments of bimodal spheres pertinent to the problem analysed above. Balls of equal sizes, made of Plasticine and steel, were subjected to compression in accordance with Fig. 4. Centre-to-centre displacements were measured as a function of the loading for compression of soft spheres and the combination of soft and hard spheres. The resulting loading characteristics are shown in Fig. 5.

The normal grade, black Plasticine was chosen to resemble a plastic material having pronounced work-hardening with no time dependence. Spheres were prepared by rolling Plasticine slugs to a diameter of 12.7 mm equal to that of the steel spheres. Under the assumption that the steel spheres remain essentially rigid and Plasticine admits power law behaviour, the compression characteristics correspond directly to those predicted by eq. (39). Adopting monolithic compression as a basis, the sought power law exponent, m, was readily determined by aid of the load-displacement relation (32). It was found that for $m = 1.35$ a fit of high accuracy was obtained for the plasticine-plasticine behaviour in Fig. 5.

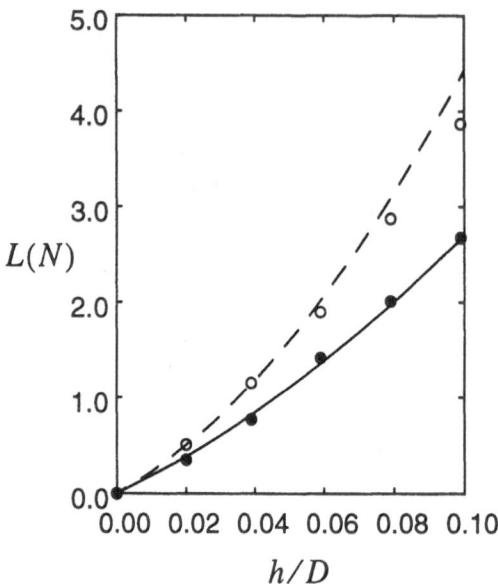

Fig. 5 Contact force, L, as function of relative centre-to-centre displacement (h/D). Plasticine-Plasticine (•) and Plasticine-Steel (o) indentation experiments by Turner and Ashby (1995), with corresponding predictions by eqs. (32) and (39).

The corresponding predicted curve based on eq. (39) for the bi-modal experiment is also shown in Fig. 5 and found to be favourably reproducable.

References

American Society of Mechanical Engineers (1995) Net Shape Processing of Powder Materials, AMD-Vol. 216, (eds. S. Krishnasawami, R. M. McMeeking and J. R. Trassoras), New York.

Ashby, M. F. (1990) HIP 6.0, Engineering Department Cambridge University, Cambridge, U.K.

Biwa, S. and Storåkers, B. (1995) An Analysis of Fully Plastic Brinell Indentation, *J. Mech. Phys. Solids* **43**, 1303-1333.

Hertz, H. (1882) Über die Berührung fester elastischer Körper, *J. reine angewandte Mathematik* **92**, 156-171.

Hill, R., Storåkers, B. and Zdunek, A. (1989) A Theoretical Study of the Brinell Hardness Test, *Proc. R. Soc. Lond.* **A423**, 301-330.

Lange, F. F., Atteraas, L., Zok, F. and Porter, J. P. (1991) Deformation Consolidation of Metal Powders Containing Steel Inclusions, *Acta Metall. Mater.*, **39**, 209-219.

Larsson, P.-L., Biwa, S. and Storåkers, B. (1996) Analysis of Cold and Hot Isostatic Compaction of Spherical Particles, *Acta materialia* **44**, 3655-3666.

Storåkers, B. (1989) Some Self-Similarity Features of Hertzian Contact of Nonlinear Solids, Report **112**, Dept. of Solid Mechanics, Royal Institute of Technology, Stockholm.

Storåkers, B. and Larsson, P.-L. (1994) On Brinell and Boussinesq Indentation of Creeping Solids, *J. Mech. Phys. Solids* **42**, 307-332.

Storåkers, B., Biwa, S. and Larsson, P.-L. (1996) Similarity Analysis of Inelastic Contact, *Int. J. Solids Structures* (to appear).

Tabor, D. (1951) *Hardness of Metals*, Clarendon Press, Oxford.

Turner, C. D. and Ashby, M. F., (1995) The Cold Isostatic Pressing of Composite Powders, Technical Report CUEDIC-MATS/TR 227, Cambridge University, Cambridge, U.K.

CEMENT AMONG GRAINS IN ROCKS

JACK DVORKIN
Department of Geophysics, Stanford University
Stanford, CA 94305-2215, USA

1. Introduction

An important contact mechanism in a cemented granular medium is stress transmission between deformable particles through deformable interparticle cement bonds. The results presented here are based on a solution to the problem of normal and shear deformation of two elastic cemented spherical particles bonded by elastic cement. Once a two-particle problem is solved, the effective deformational characteristics of a granular assembly are derived using standard statistical approaches. Contact cement, even if it is soft, acts to dramatically increase the effective elastic stiffness of a granular medium.

We apply this theoretical result to predicting elastic wave velocities in granular rocks. If the elastic wave velocities are measured, and the porosity of a rock is established, it is possible to solve the inverse problem, i.e., to determine the amount of contact cement and its type.

Not all diagenetic cement in rock is contact cement. Part of the cement can be deposited in the pore space away from contacts. This non-contact cement, in contrast to the contact cement, only weakly affects the stiffness of the rock. We provide an ad-hoc effective medium model to calculate the elastic constants of rocks with non-contact cement. This model can also be used to predict elastic wave velocities in rocks, and to diagnose rocks for the amount of the non-contact cement.

It is remarkable that the position of diagenetic cement -- whether it is contact cement, or non-contact cement -- influences the permeability of rocks. At the same porosity, a rock with non-contact cement has smaller permeability than rock with contact cement. This fact leads to the possibility of inferring permeability from elastic wave velocities.

2. Contact Problem for Two Spheres and Implications

The solution to the problem of normal and shear deformation of two elastic cemented spherical particles bonded by elastic cement is given in Dvorkin et al. (1994). The results reveal a peculiar picture of normal stress distribution at the cemented grain contacts: the stresses are maximum at the center of the contact region when the cement is soft relative to the grain, and are maximum at the periphery of the contact region when the cement is stiff. Stress distribution shape gradually varies between these two extremes as the cement's stiffness increases (Figure 1). These results allow for the following simple physical interpretation. When a soft cement layer is confined between two rigid grains, its maximum compression occurs at its thinnest part -- near the point of direct grain-to-grain contact. If the cement is stiff, the problem approaches that of a rigid punch (the cement) penetrating an elastic half-space (the grain). In the latter case, normal stress becomes infinite at the edge of the punch (Johnson, 1985) which explains the observed maximum of the normal stresses at the periphery of the contact region.

The predicted stress distribution patterns have been validated in photoelasticity experiments of Sienkiewicz et al. (1996). The computer-generated fringe contours were very close to those observed in two transparent plastic disks with aluminum cement between them.

N. A. Fleck and A. C. F. Cocks (eds.),
IUTAM Symposium on Mechanics of Granular and Porous Materials, 185–192.
© 1997 *Kluwer Academic Publishers.*

An important result is that even for relatively compliant elastic cement, stress concentration at the center of a contact is not large. The normal and shear stresses are almost uniformly distributed along the contact zone. Therefore, even soft cement is load-bearing. It acts to diffuse stress concentrations inevitable at uncemented contacts and thus prevents grains from breaking.

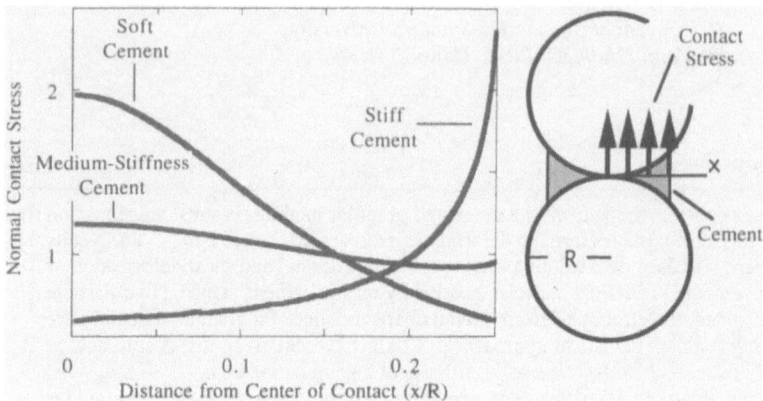

Figure 1. Normal stresses (normalized by the average stress) at the cemented interface of two elastic spheres.

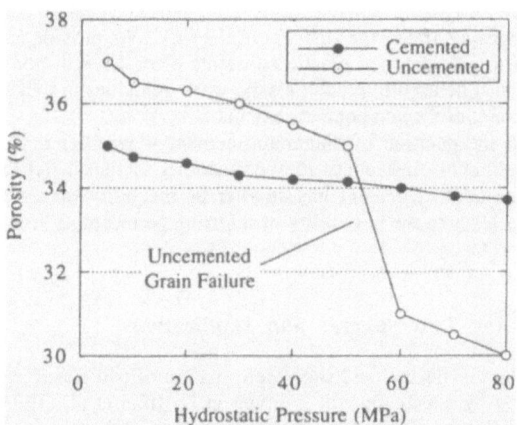

Figure 2. Porosity of the cemented sample and uncemented one versus hydrostatic confining pressure. The sharp porosity reduction in the uncemented sample indicates sudden grain crushing. The cemented sample stayed intact. This effect was confirmed by visual inspection of the samples.

We observed this effect in compaction tests (isotropic drained loading) on randomly packed glass beads that were a) uncemented and b) cemented by epoxy at their contacts. In the latter case, the volume of the epoxy accounted for only 10 percent of the pore space. Intensive crushing of grains was observed in the first case at about 50 MPa (Figure 2). In the second case, the cemented grains stayed intact, the failure being localized within the epoxy. Notice that epoxy is much softer than glass: the bulk and shear moduli of glass are 50.0 GPa and 26.2 GPa, respectively, whereas those of the epoxy are 6.8 GPa and 2.0 GPa, respectively.

3. Elastic Constants of Aggregates with Contact Cement

We can model a granular aggregate as a random pack of identical spherical particles with cement added on the surface of the spheres. This cement may be either contact or non-contact (Figure 3a). For simplicity, we consider two patterns of contact cement deposition: Scheme 1 where all cement is deposited at grain contacts (Figure 3b) and Scheme 2 where cement is evenly deposited on the grain surface (Figure 3c). If porosity reduction in an aggregate is due to cement deposition, then, at the same stiffness, an aggregate with Scheme 1 has higher porosity than one with Scheme 2.

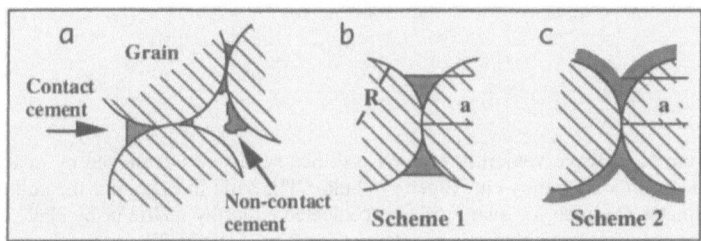

Figure 3. Cement among grains. a. Contact and non-contact cement. b. Scheme 1 of cement deposition. c. Scheme 2 of cement deposition.

In Figure 4 we show examples of predicting the elastic wave velocities in artificial cemented materials using the contact cement theory. Clearly, our theoretical predictions are well supported by experimental results. The experiments also support the theoretical conclusion that even soft contact cement is load-bearing. Even small amounts of soft cement, if deposited precisely at grain contacts, act to dramatically increase the stiffness of a particulate aggregate.

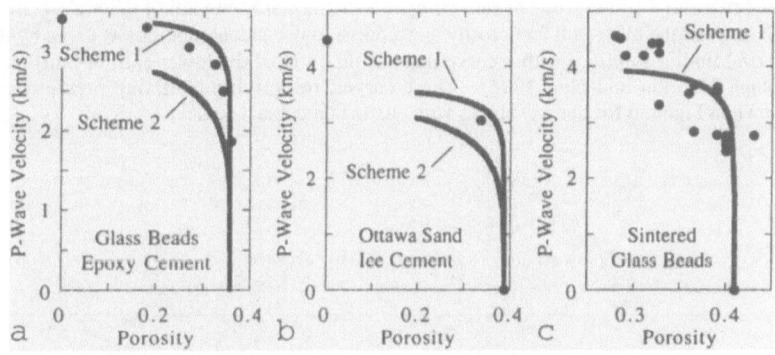

Figure 4. Examples of the theoretical and experimental values of P-wave velocities in cemented granular materials. Circles are from experiments, solid lines are our theoretical predictions. From left to right: epoxy-cemented glass beads (data from Yin, 1993); ice-cemented Ottawa sand (data from Jacoby et al., 1995); and sintered glass beads (data from Berge et al., 1993).

The cementation theory has good predictive power for elastic wave velocities not only in artificial, but also in natural cemented aggregates (Figure 5). It allows one, by forward modeling, to determine the cementing mineral (quartz or clay in this example).

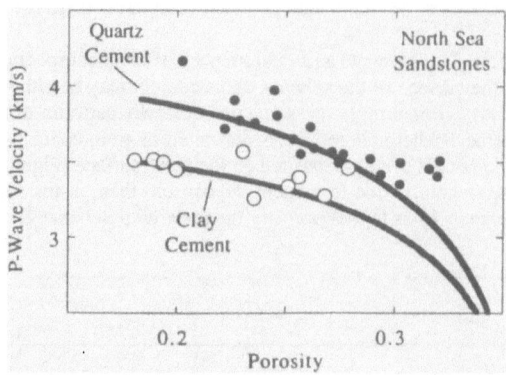

Figure 5. P-wave velocities in dry North Sea sandstones cemented by quartz (filled symbols) and by clay (open symbols). The solid lines are our theoretical estimates obtained for quartz spheres cemented either by quartz or by clay. In these estimates cement evenly envelopes the grains (Scheme 2).

4.   Elastic Constants of Aggregates with Non-Contact Cement

A granular aggregate with non-contact cement is also modeled as a random pack of identical spherical particles. However, in this case the contact law of particle interaction is given by the Hertz-Mindlin solution (Mindlin, 1949). This law is used to calculate the effective elastic constants of the aggregate at 0.36 porosity -- the porosity of a dense random pack of identical spheres. As porosity decreases, the elastic moduli slightly increase, and, finally, reach those of the non-porous solid that is the mixture of 0.64 volumetric parts of the grain material and 0.36 parts of the cement material. We connect these two end members, one at zero porosity with the elastic moduli of the grain-cement mixture and the other at 0.36 porosity and a pressure-dependent modulus as given by the Hertz-Mindlin solution, with a curve that has the form of the lower Hashin-Shtrikman bound (Dvorkin and Nur, 1996). These curves, for varying confining pressure, are shown in Figure 6 for quartz spheres with quartz non-contact cement.

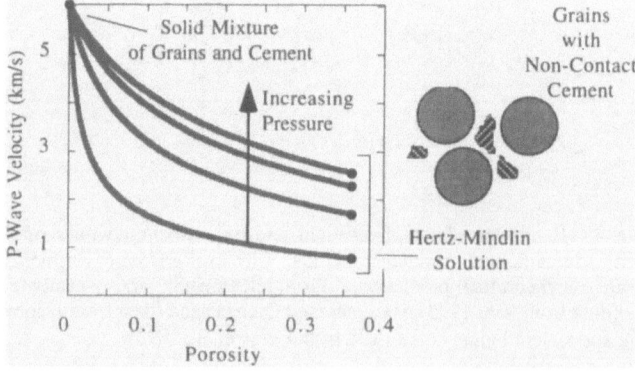

Figure 6. P-wave velocity versus porosity at varying confining pressure, as given by the non-contact cement theory.

This heuristic theoretical model can be used to accurately estimate and bound elastic-wave velocities in loose high-porosity sands, such as found in the North Sea Troll field (Figure 7). The prediction quality increases with the increasing confining pressure.

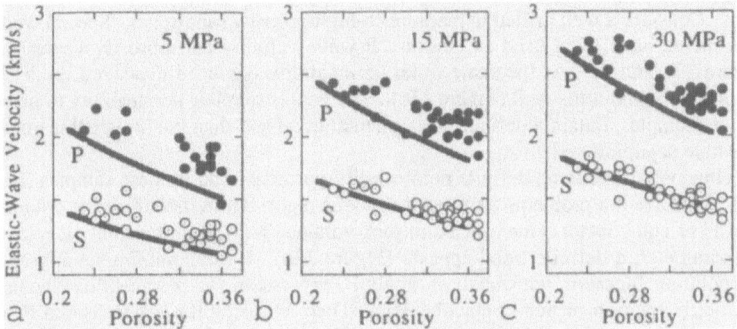

Figure 7. P- (filled symbols) and S-wave (open symbols) velocities in dry North Sea sandstones with non-contact cement (mica). The solid lines are our theoretical estimates. The confining pressure is 5 (a), 15 (b), and 30 (c) MPa. Data from Blangy (1992).

5. Diagnosing High-Porosity Rock

The above theoretical models can be used to diagnose granular rocks for the amount and type of the contact and non-contact cement. Of course, in order to do so, the following axiom has to be accepted. If in the (elastic wave) velocity-porosity plane an experimental datapoint falls near a theoretical trajectory then the internal structure of the rock is such as used to theoretically generate this trajectory. An example of such diagnostic is given in Figure 8.

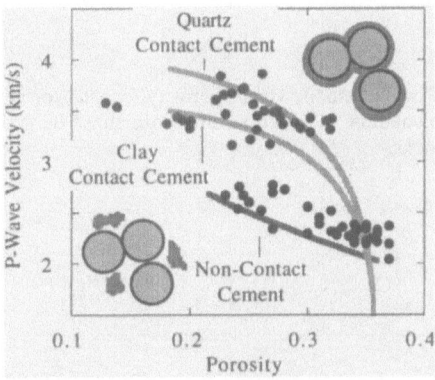

Figure 8. Diagnosing rocks for the type and position of cement.

Such diagnostic becomes desirable if we consider how permeability depends on the position of cement among grains. At the same porosity, the permeability of sandstones increases with the increasing amount of contact cement (Figure 9). Therefore, by identifying the position of cement, one can increase the accuracy of determining permeability. For high-porosity granular rocks a general rule is that at a fixed porosity, normalized permeability increases with the increasing elastic-wave velocity (our numerical experiments, unpublished). Resulting from this law are interesting relations between elastic-wave velocity and permeability (Figure 10).

Finally, we give an example of a practical application of the described diagnostic of rocks. Consider a well drilled in medium-to-high-porosity sandstones. Special well log instruments have been used to measure P-wave velocity and porosity versus depth (Figure 11a and b). At the same time, permeability has been measured on selected samples throughout the well (Figure 11c). The goal is to relate permeability to porosity of these samples, obtain a definite experimental trend and then interpolate this trend for the whole depth interval.

However, there is no definite permeability-porosity trend in these samples (Figure 12a). To solve the problem, we run a diagnostic algorithm to determine the volumetric fraction of non-contact cement in a unit rock volume. Now, if permeability is related to this parameter, a definite trend appears (Figure 12b). By determining an appropriate interpolation function, we obtain an analytic expression for permeability versus the volumetric fraction of non-contact cement. Then, by using the log-measured P-wave velocity and porosity values, we determine this parameter for the whole well (Figure 13a). And, finally, using this result and the derived interpolation function, we arrive at permeability estimate for the whole well (Figure 13b).

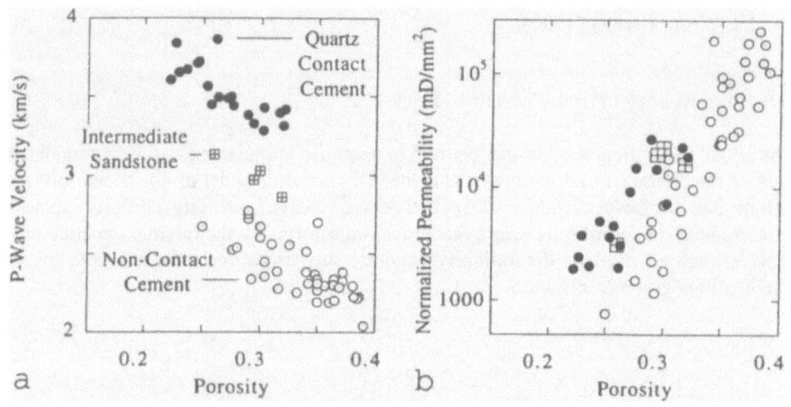

Figure 9. a. P-wave velocity versus porosity for sandstones with varying degree of contact cementation. b. Permeability normalized by grain diameter squared for the same rocks.

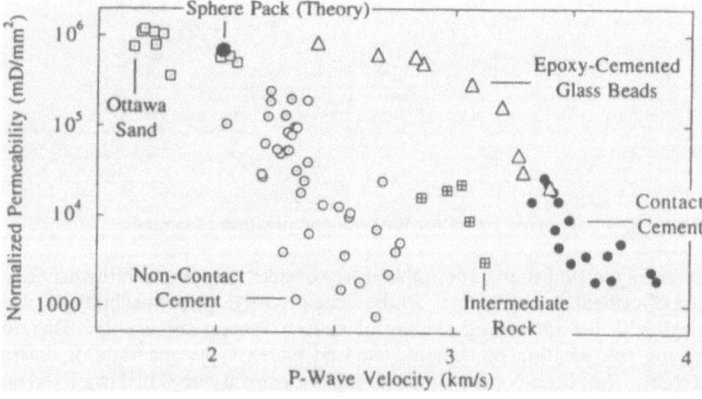

Figure 10. Granular materials with varying position of cement among grains. Permeability normalized by grain diameter squared versus P-wave velocity.

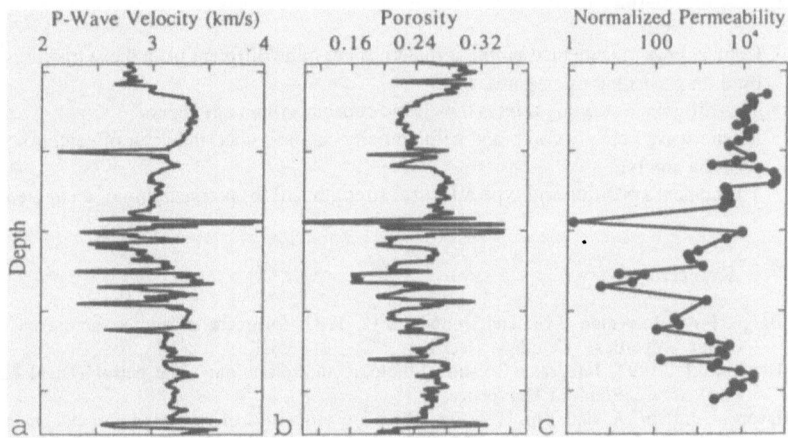

Figure 11. Well-log values of P-wave velocity (a) and porosity (b). Sample-measured permeability values (c). Permeability is normalized by grain diameter squared. Depth values concealed for confidentiality.

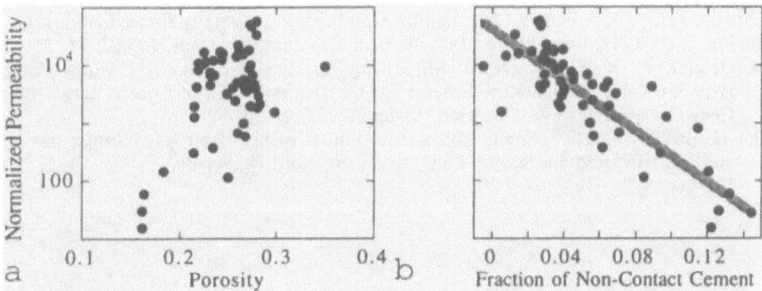

Figure 12. Permeability normalized by grain diameter squared versus (a) porosity and (b) volumetric fraction of non-contact cement. Permeability is in mD/mm^2.

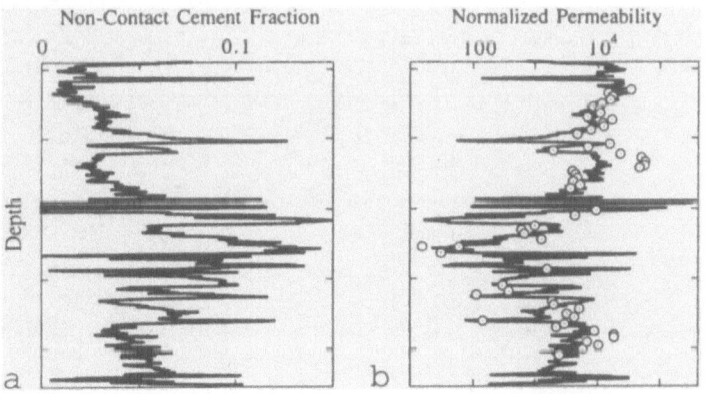

Figure 13. Non-contact cement fraction (a) and normalized permeability (b) versus depth. Open symbols are from sample measurements.

6.   Conclusions

* Contact laws in cemented granular materials are quite different from those traditionally used for particulate aggregates.
* This difference strongly affects the elastic constants of an aggregate.
* Elastic-wave velocity combined with porosity can be a good indicator of cement's position and type.
* The cement's position and type affect (a) strength and (b) permeability of an aggregate.

7.   References

Berge, P.A., Berryman, J.G., and Bonner, B.P., 1993, Influence of microstructure on rock elastic properties: Geophys. Res. Lett., **20**, 2619-2622.

Blangy, J.P., 1992, Integrated seismic lithologic interpretation: The petrophysical basis: Ph.D. thesis, Stanford University.

Dvorkin, J., Nur, A. and Yin, H., 1994, Effective properties of cemented granular material: Mechanics of Materials, **18**, 351-366.

Dvorkin, J., and Nur, A., 1996, Elasticity of High-Porosity Sandstones: Theory for Two North Sea Datasets: Geophysics, **61**.

Hashin, Z., and Shtrikman, S., 1963, A variational approach to the elastic behavior of multiphase materials: J. Mech. Phys. Solids, **11**, 127-140.

Jacoby, M., Dvorkin, J., and Liu, F., 1996, Elasticity of Partially Saturated Frozen Sand: Geophysics, **61**, 288-293.

Johnson, K.L., 1985, Contact Mechanics, Cambridge University Press, Cambridge.

Mindlin, R.D., 1949, Compliance of elastic bodies in contact: Trans. ASME, **71**, 259-268.

Sienkiewicz, F., Shukla, A., Sadd, M., Zhang, Z., and Dvorkin, J., 1996, A Combined Experimental and Numerical Scheme for the Determination of Contact Loads Between Cemented Particles: Mechanics of Materials, **22**, 43-50.

Yin, H., 1993, Acoustic velocity and attenuation of rocks: Isotropy, intrinsic anisotropy, and stress induced anisotropy, Ph.D. thesis, Stanford University.

THREE-DIMENSIONAL DISCRETE MECHANICS OF GRANULAR MATERIALS

Masao Satake
Tohoku Gakuin University
Chuo 1-13-1, Tagajo 985, Japan

1. Introduction

This paper intends to summarize the three-dimensional discrete mechanics of granular materials. Since the granular material is an assembly of granular particles, the discrete-mechanical formulation is considered to be natural and fundamental, and such approaches have been proposed by using graph theory [1-4]. Although the approach was two-dimensional and limited to the case of circular discs in the preliminary studies, the extension to the case of arbitrarily-shaped particles and to the three-dimensional case have been made. It is noted, however, that there is some problem in the treatment of voids in the 3D analysis, and this paper will also discuss this point.

2. Formulation of Discrete Mechanics

We begin with the explanation of 2D discrete mechanics of granular materials. As is seen in Fig.1, an assembly of particles is replaced by a graph called the *particle graph*. The three elements of the graph,-- point, branch and loop--, correspond to three elements of the assembly, --particle, contact and void--, respectively. If we consider a void as a *dual particle*, as is shown by the dotted line in Fig.2 (a), we can introduce the *void graph* shown in Fig.2 (b). The particle and void graphs have a geometrical duality with each other as shown in Table 1.

N. A. Fleck and A. C. F. Cocks (eds.),
IUTAM Symposium on Mechanics of Granular and Porous Materials, 193–202.
© 1997 *Kluwer Academic Publishers.*

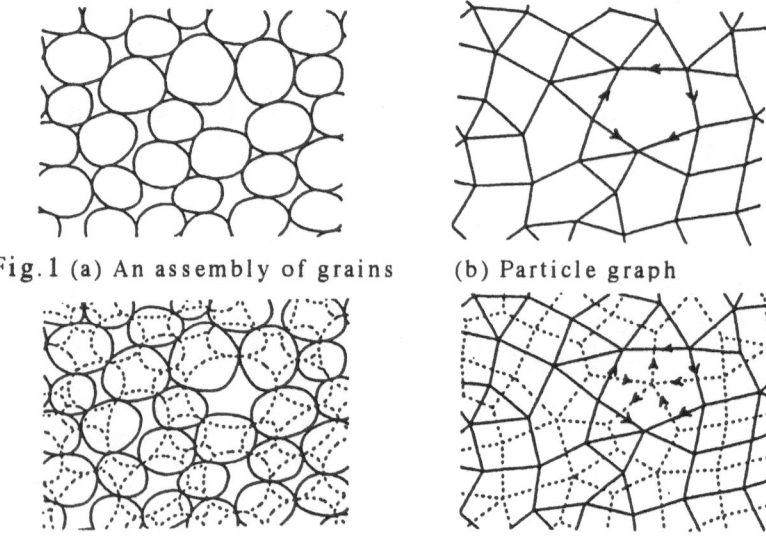

Fig.1 (a) An assembly of grains   (b) Particle graph

Fig.2 (a) Voids and dual particles   (b) Void graph

Table 1   Correspondence between a 2D granular assembly and graphs

Granular assembly	Particle	Contact	Void
Particle graph	Point	Branch	Loop
Void graph	Loop	Branch	Point
(Dual particle graph)			

For the convenience of analysis, we introduce an orientation to every branch in both graphs. The orientation of branches in a particle graph is given arbitrarily, and that of branches in a void graph (called dual branches) is determined by clockwise rotation of the direction of the corresponding branch of the particle graph, as is shown in Fig.3. All loops in both graphs are given clockwise orientation.

For all elements in an assembly of granular particles explained above, we define mechanical quantities, as shown in Table 2, where subscripts P, C and V indicate an ordered number of particles, contacts and voids, respectively. Particle force is the body force in common terminology, and contact displacement and rotation mean the relative displacement and rotation at a contact point. All quantities of contact are to be defined with respect to a *positively-connected* particle for the concerned contact, where the term of "positively-(or negatively-) connected" means that the branch corresponding to the contact is oriented away from (or towards) the point corresponding to the particle (Fig.4). The meaning of void quantities will be clarified later.

Table 2 Mechanical quantities in 2D discrete mechanics

		Particle	Contact	Void
Force	$F = (f, m)^{\mathrm{T}}$	F_P	F_C	F_V
	(force, moment)			
Displacement	$U = (u, w)^{\mathrm{T}}$	U_P	U_C	U_V
	(displacement, rotation)			

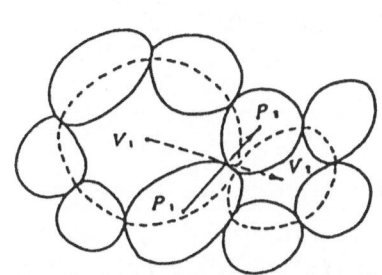 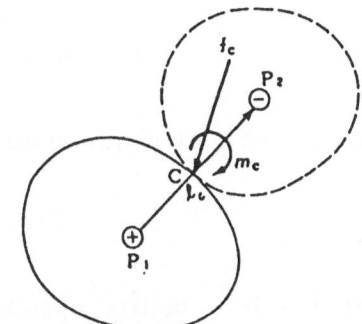

Fig.3 Orientation of dual branch   Fig.4 Positive and negative connection

For particle and contact quantities, we have a relation written as

$$F_P = - \widetilde{D}_{PC} F_C, \qquad U_C = - \widetilde{D}_{CP} U_P \qquad (1)$$

where D_{PC} is a matrix operator called the *extended incidence matrix* and expressed as

$$\widetilde{D}_{PC} = \begin{pmatrix} D_{PC} & 0 \\ D_{PC} \, r_{PC} \times & D_{PC} \end{pmatrix} \qquad (2)$$

where D_{PC} denotes the incidence matrix in graph theory and r_{PC} is a radius vector from the centroid of particle P to the concerned contact C

(Fig.5). Eq.(1_1) is the equilibrium equation for particle P, and Eq.(1) defines the generation of F_P and U_C from F_C and U_P, respectively.

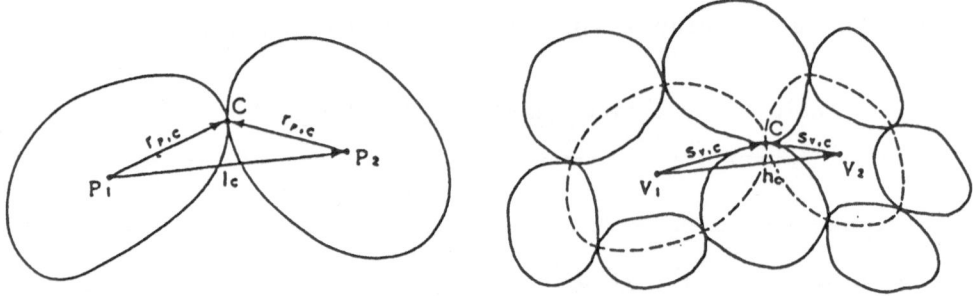

Fig.5 Two particles in contact   Fig.6 Two voids in contact

As we have the following identity for loop matrix L_{CV}:

$$L_{VC} \, D_{CP} = 0, \qquad D_{PC} \, L_{CV} = 0 \qquad\qquad (3)$$

we define the void displacement as

$$U_V = - \widetilde{L}_{VC} \, U_C \qquad\qquad (4)$$

where $\widetilde{L}_{CV}$ is a matrix operator called the *extended loop matrix* and expressed as

$$\widetilde{L}_{VC} = \begin{pmatrix} L_{VC} & L_{VC} \, s_{VC} \times \\ 0 & L_{VC} \end{pmatrix} \qquad\qquad (5)$$

where s_{VC} is a radius vector from the centroid of a void (dual particle) V to the contact point C (see Fig.6). As we have the following identity, similar to Eq. (3):

$$\widetilde{L}_{VC} \, \widetilde{D}_{CP} = 0, \qquad \widetilde{D}_{PC} \, \widetilde{L}_{CV} = 0 \qquad\qquad (6)$$

we have $U_V = 0$ for U_C which satisfies Eq.(1_2). However, in the general case, where Eq.(1_2) does not hold, we must consider a non-zero U_V. In such a meaning, the void displacement U_V is considered to correspond with *incompatibility* in continuum mechanics.

If F_C is generated from the void force F_V in the manner

$$F_C = - \widetilde{L}_{CV} F_V \tag{7}$$

we have $F_P = 0$, from Eq.(6). In this meaning, the void force F_V corresponds with the *stress function* in continuum mechanics.

From the above consideration, it is seen that F_C and U_C, expressed by Eqs.(1_1) and (4), respectively, are indeterminate as F_C and U_C, expressed by Eqs.(7) and (1_2), respectively. As the void graph is regarded as a graph of dual particles, L_{VC} and D_{CP} for a particle graph are considered as the incident and loop matrices for the dual particle graph, respectively, and we have also a duality that displacement and force quantities in a particle graph correspond with force and displacement quantities in the graph of dual particles, respectively. It is noted that the relations expressed by Eqs.(1),(4) and (7) are readily applied in computer simulation techniques.

Now, we proceed to the three-dimensional case. In 3D, duality between particle and void (dual particle) graphs is similar to the 2D situation shown in Table 1. However, the number of elements becomes four, and we must add the *void cells* (see Fig.7) as a new element in a particle graph. A void cell is surrounded by loops and considered to represent a unit of void in 3D, although the void in 3D is connected throughout the assembly. In 2D, a contact of particles is also a contact of voids and corresponds with a branch of the particle and void graphs, respectively. In 3D, a particle contact corresponds with a branch, whereas a contact of void cells corresponds with a loop in a particle graph. For this reason, such a loop is called the *dual contact*. Table 3 explains the correspondence between an assembly of granular particle and graphs and also the duality between two graphs in 3D.

Table 3 Correspondence between a 3D granular assembly and graphs

Granular assembly	Particle	Contact	Dual contact	Void
Particle graph	Point	Branch	Loop	Cell
Void graph	Cell	Loop	Branch	Point

In 3D, we have some problems in the above definition. The definition of a loop is very clear in 2D, but is not in 3D, so that the partition of a particle graph by void cells is not unique. We will discuss this problem in **4**.

Mechanical quantities in 3D are defined similarly to the way they are described in 2D, as shown in Table 4, where subscripts D and V are indices corresponding to dual contacts and void cells, respectively. It is noted here that the orientation of branches and loops in a particle graph are determined arbitrarily and that of corresponding loops and branches in a void graph is to be determined by the rule of right-handed twining.

Table 4 Mechanical quantities in 3D discrete mechanics

		Particle	Contact	Dual contact	Void
Force	F	F_P	F_C	F_D	F_V
Displacement	U	U_P	U_C	U_D	U_V

In 3D, Eqs.(4) and (7) are replaced by the following Eqs.:

$$F_C = - \widetilde{L}_{CD} F_D, \qquad U_D = - \widetilde{L}_{DC} U_C \qquad (8)$$

Here we introduce a new matrix, the *cell matrix* defined as

$$C_{VD} = \begin{cases} 1 & \text{if the void cell is positively-connected} \\ & \text{at the dual contact} \\ -1 & \text{if the void cell is negatively-connected} \\ & \text{at the dual contact} \\ 0 & \text{otherwise} \end{cases} \qquad (9)$$

where the term of "positively-(or negatively-)connected" means that the branch in a void graph corresponding to the dual contact is oriented away from (or towards) the point in a void graph corresponding to the void cell. We also define the *extended cell matrix* in the following form:

$$\widetilde{C}_{VD} = \begin{pmatrix} C_{VD} & C_{VD} \, \boldsymbol{t}_{VD} \times \\ 0 & C_{VD} \end{pmatrix} \tag{10}$$

where $\boldsymbol{t}_{VD}$ is a radius vector from the centroid of a void cell V to that of the dual contact, i.e. the contact point of dual particles. For mechanical quantities of dual contacts and void cells, we have generative relations

$$\boldsymbol{F}_D = - \widetilde{C}_{DV} \, \boldsymbol{F}_V, \qquad \boldsymbol{U}_V = - \widetilde{C}_{VD} \, \boldsymbol{U}_D \tag{11}$$

Similarly, as in 2D, we have identities of the form

$$C_{VD} \, L_{DC} = 0, \qquad L_{CD} \, C_{DV} = 0 \tag{12}$$

and

$$\widetilde{C}_{VD} \, \widetilde{L}_{DC} = 0, \qquad \widetilde{L}_{CD} \, \widetilde{C}_{DV} = 0 \tag{13}$$

Hence we can state that

$$\boldsymbol{F}_C = - \widetilde{L}_{CD} \, \boldsymbol{F}_D = 0, \qquad \text{if} \quad \boldsymbol{F}_D = - \widetilde{C}_{DV} \, \boldsymbol{F}_V \tag{14}$$
$$\boldsymbol{U}_V = - \widetilde{C}_{VD} \boldsymbol{U}_D = 0, \qquad \text{if} \quad \boldsymbol{U}_D = - \widetilde{L}_{DC} \, \boldsymbol{U}_C \tag{15}$$

3. Constitutive Relations

Here we consider simple constitutive relations of mechanical quantities in 3D discrete mechanics of granular materials.

Assume a simple constitutive relation between contact force and displacement expressed as

$$\boldsymbol{F}_C = S_C \, \boldsymbol{U}_C \tag{16}$$

where S_C is a stiffness matrix (non-singular) at contact C and is called the *contact stiffness*.

When U_C is generated from U_P, we have, from Eq.(1), a constitutive relation for particle P, written as

$$F_P = S_P\, U_P, \qquad \text{where} \qquad S_P = \overset{\frown}{D}_{PC}\, S_C\, \widetilde{D}_{CP} \qquad (17)$$

and when F_C is obtained from F_V in 2D, we have, using Eqs.(4) and (7), a constitutive relation for void V, written as

$$U_V = C_V\, F_V, \qquad \text{where} \qquad C_V = \widetilde{L}_{VC}\, S_C^{-1}\, \widetilde{L}_{CV} \qquad (18)$$

In 3D, subscripts V in Eq. (18) should be replaced by D, which indicates dual contacts. S_P and C_V (C_D) are called *particle stiffness* and *void compliance* (*dual contact compliance*), respectively.

In 3D, we can assume another constitutive relation between dual contact force and displacement expressed as

$$F_D = S_D\, U_D \qquad\qquad (19)$$

where S_D is a stiffness matrix (non-singular) for the dual contact, called the *dual contact stiffness*. Similarly, as in 2D, when U_D is obtained from U_C, we have, using Eq.(8), a constitutive relation for contact C written as

$$F_C = S'_C\, U_C \qquad \text{where} \qquad S'_C = \overset{\frown}{L}_{CD}\, S_D\, L_{DC} \qquad (20)$$

and when F_D is obtained from F_V, we have, using Eq.(11), a constitutive relation for void V written as

$$U_V = C_V\, F_V \qquad \text{where} \qquad C_V = C_{VD}\, S_D^{-1}\, \widetilde{C}_{DV} \qquad (21)$$

S'_C and C_V are called *second contact stiffness* and *void compliance*, respectively. It is noted here that the analysis for a void cell and dual contact explained above is considered to be applicable to the analysis of cellular and porous materials rather than granular materials.

4. Partition of Void in 3D

In 3D, we introduce the void cell as a unit of void. However, as is mentioned in the previous section, the partition of a 3D particle graph by loops is not uniquely defined. This point is very different to the 2D case, where the partition of a 2D particle graph by loops is unique and very clear. In this section, we discuss this problem further.

The following two ways are considered to be appropriate for the solution of this problem.

The first way is one in which the stiffness of a dual contact is taken into consideration. As explained in **3**, we can introduce some stiffness to the dual contacts. In this case, it may be natural that such a stiffness is introduced not to all loops but to some selected loops, which do not intersect with each other. Using these selected loops we can easily obtain a unique partition of a particle graph, which is the same as that of the whole void in a 3D assembly of particles. Since dual loops without stiffness are not necessary for the analysis explained in **3**, the above partition is considered to be an appropriate method for our analysis.

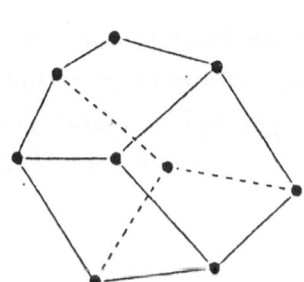
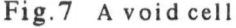
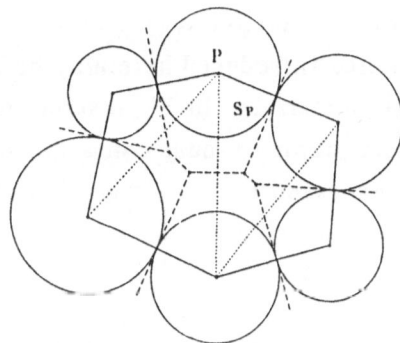

Fig.7 A void cell   Fig.8 Modified Voronoi partition in 2D

In the second way, we use the *virtual branches* in a particle graph. In 2D, we have a *modified Voronoi partition* as shown in Fig.8. In this case, the domain S_P belonging to particle P obtained by the modified Voronoi partition is defined as

$$S_P = \{p \mid d\,(p,P)^2 - d\,(p,X)^2 \leqq r_P{}^2 - r_X{}^2\} \qquad (22)$$

where p represents a point in the whole domain, $d(x,y)$ denotes the distance between points x and y, X is an arbitrary particle and r_X is the radius of particle X (mean radius in the case of arbitrarily-shaped particles).

It is important that the above partition defines the virtual branches which correspond to virtual contacts of non-contacting neighboring particles, as is seen in Fig. 8. Further it is noteworthy that Eq. (22) is also valid in 3D. Thus the modified Voronoi partition also defines virtual branches in a 3D particle graph. All branches including virtual branches give a fundamental partition by tetrahedrons with four triangles in a 3D particle graph, similar to the partition by triangles in 2D. It is easily understood that we can obtain loops corresponding to dual contacts, by connecting such triangles, which surround the above fundamental tetrahedrons, and ignoring virtual branches.

5. Concluding Remarks

This paper explains the formulation of 3D discrete mechanics of granular materials. Many mechanical quantities are introduced for the general discussion of discrete mechanics, and as is explained, some quantities introduced here may be useful for the analysis of cellular and porous materials. In 3D discrete mechanics, the method of partition of particle graph by dual contacts becomes important and this problem is also discussed and two ways of addressing this problem are explained.

References

1. Satake, M. (1976) Constitution of mechanics of granular materials through graph representation, in Theo. Appl. Mech. **26**, University of Tokyo Press, Tokyo, 257-266.
2. Satake, M. (1992) A discrete-mechanical approach to granular materials, *Int. J. Engng Sci.* **30**-10, 1525-1533
3. Satake, M. (1993) Discrete-mechanical approach to granular media, in C. Thornton (ed.), Powders & Grains 93, A. A. Balkema, Rotterdam, 3-9
4. Satake, M. (1993) New formulation of graph-theoretical approach in the mechanics of granular materials, *Mechanics of Materials* **16**, 65-72

NUMERICAL ANALYSIS OF CLAY PARTICLE ASSEMBLIES

A. ANANDARAJAH
Department of Civil Engineering
The Johns Hopkins University
Baltimore, Maryland 21218, U.S.A.

1. Introduction

The current level of understanding of the role of fabric and compositional parameters of the microstructure of clays is inadequate for the development of physically meaningful mathematical models of the engineering behavior of soils. In this context, research was initiated several years ago, which seeks to develop methods for quantifying particle-level forces and use them to model behavior of an assembly consisting of many such particles. To date, methods have been developed for quantifying the double-layer repulsive force and the van der Waals attractive force between two nonparallel plate like particles as a function of the system variables. Two different numerical methods and an approximate analytical modeling method have thus far been pursued in analysing assemblies. The numerical methods are based on the discrete element method (DEM) and an energy minimization method. The progress made in these areas are summarized in this paper.

2. Methods of Quantifying Double-Layer Repulsive Force and van der Waals Attractive Force

2.1. DOUBLE-LAYER REPULSIVE FORCE

Due to the presence of negative charges on their faces, clay particles, when

N. A. Fleck and A. C. F. Cocks (eds.),
IUTAM Symposium on Mechanics of Granular and Porous Materials, 203–213.
© 1997 *Kluwer Academic Publishers.*

placed in an electrolyte solution, interact with each other electrostatically, leading to a net repulsion. This force depends on the intensity of the negative surface charge, valence (ν) and concentration (n_0) of ions, the static dielectric constant of pore fluid (ϵ), temperature (T), and dimensions (area=$L_p \times W_p$) and orientation of particles (θ_r) . Analytical solutions are available for systems with parallel particles. The governing equation (Poisson-Boltzmann) can not be solved analytically when clay particles are inclined to each other. Using the finite element (Anandarajah and Lu, 1991a) and the finite difference techniques (Anandarajah and Chen, 1994), the governing equation was solved for a wide range of system parameters, and from the results, an approximate method has now been developed. Using this method, the repulsive force and its location can approximately be computed for a system of two nonparallel particles such as the one shown in Fig. 1. In this figure, two clay particles (shown by lines), inclined to each other at an angle of $2\theta^r$ and separated at the closest ends by a distance of $2d^r$, are placed in a fluid of infinite extent. The surface charge is characterized by an electric potential ψ_0.

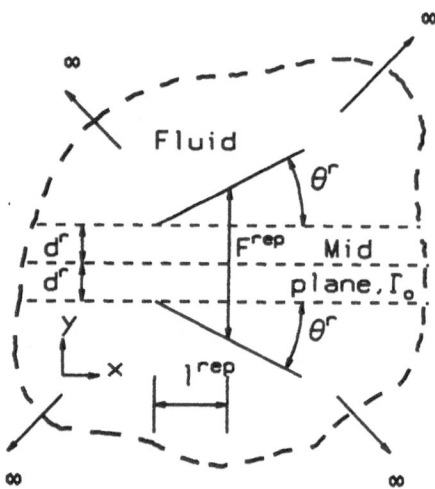

Figure 1. A System of Two Particles Immersed
in a Fluid with Repulsive Force Between Them.

The numerical results obtained from the aforementioned study revealed a number of useful features of the double-layer repulsive force. An example of this is presented in Fig. 2, which shows the decay in magnitude of the repulsive force, F^{rep}, as a function of separation distance for different values of θ^r. While the qualitative trend is implied by the existing simple theories, the quantitative effect, particularly that of θ^r, could not have been assessed

using the existing simple theories. For example, F^{rep} could decrease by an
order of magnitude as the orientation of clay particles deviates from parallel
by about 5^0; there are no simple theories to compute this effect. In view
of the fact that clay particles are never parallel to each other in a real
soil, this is very useful information. The parameters used to compute the
results are shown in the figure. While the numerical analyses (e.g., finite
element analysis) may be repeated for any specific set of system variables,
the computational effort involved is not trivial. The approximate simplified
method yields results within 10 to 15% accuracy.

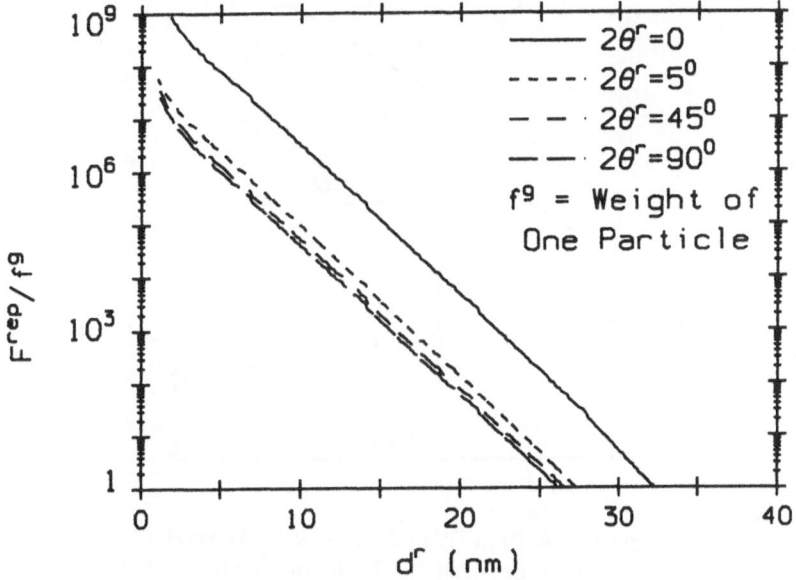

Figure 2. Variation of Repulsive Force with Separation for Different
Inclinations ($L_p = 1.2\mu m$, $h = 37nm$, $\epsilon = 80$,$A = 10 \times 10^{-20}$ J,
$n_0 = 0.01$ mole/liter, $\nu = 1$,$T = 293^0$K, $\psi_0 = 188mv$).

2.2. VAN DER WAALS ATTRACTIVE FORCE

A similar method was needed for computing the second most significant
physico-chemical force between two clay particles: the van der Waals attractive force. In this case, a closed-form solution has been developed for
the plate-fluid-wall system shown in Fig. 3. While a plate-fluid-plate system is what is needed to model a soil system, the plate-fluid-wall system
provides a good approximation, except for small edge effects. The details of
these theories may be found in Anandarajah and Chen (1995, 1996a) and
Chen and Anandarajah (1996b). The effect of the fluid filling the pores is
accounted for in the theory through a constant known as the Hamaker's
constant, A; theories have been developed for computing this. For example,
for a kaolinite-water system, $A = 10 \times 10^{-20}$ Joules. Using this value for

A, and the same values for the other relevant variables as in the example shown in Fig. 2, the results shown in Fig. 4 were computed. The figure shows the manner in which the van der Waals attractive force F^{att} decays as a function of the separation distance d^a and the interparticle angle θ^a. As in the case of F^{rep}, it is seen that θ^a has a significant effect on F^{att}. Approximate methods have now been developed for the completely general case of a three-dimensional plate-fluid-plate system.

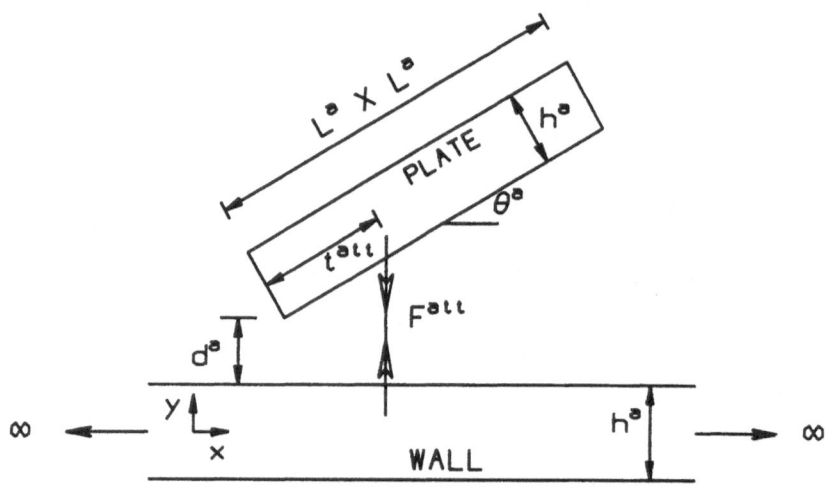

Figure 3. A Plate-Wall System with Attractive
Force Between the Plate and Wall.

3. Discrete Element Method of Assembly Analysis

The most systematic and comprehensive analytical method for determining the behavior of an assembly is perhaps the discrete element method (DEM) of Cundall and Strack (1979). The method was developed for a granular assembly, and it has thus far been applied only to such material assemblies. The first successful application of DEM to an assembly of clay particles was presented by Anandarajah (1994). In this preliminary study, where the objective was largely to demonstrate the feasibility of performing a DEM analysis on an assembly of plate like cohesive particles, the double-layer repulsive force was simulated, along with the contact mechanical forces and the bending of clay particles (Anandarajah and Lu, 1991b), but the van der Waals force was ignored. It was shown that numerical results of one-dimensional compression behavior resembled that of dispersive clays such as montmorillonite, where the repulsive force is the dominant component, owing to the large specific surface of the particles. The analysis allows the evolution of fabric anisotropy to be computed. Recent experimental studies (Anandarajah et al., 1994, 1996; Kuganenthira et al., 1996) are in good agreement with the discrete element studies.

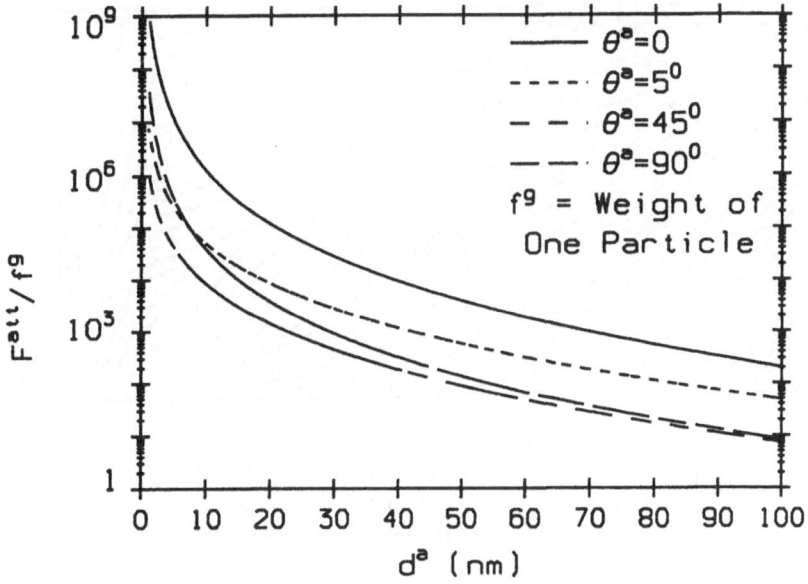

Figure 4. Variation of Attractive Force with Separation for Different Inclinations ($L_p = 1.2\mu m$, $h = 37nm$, $\epsilon = 80$, $A = 10 \times 10^{-20}$ J, $n_0 = 0.01$ mole/liter, $\nu = 1$, $T = 293^0$K, $\psi_0 = 188$mv).

The lack of a rational method for quantifying the van der Waals force was the reason at that time for ignoring it; the problem has now been addressed as described in the preceding section. Efforts are currently underway in performing DEM analyses with consideration of the attractive force. Preliminary results indicate that the method of preparing the initial numerical assembly is rather complex. For example, when the van der Waals force was ignored, the assembly shown in Fig. 5 was obtained after 50% vertical compression of an initially loose assembly. The walls were assumed to be physico-chemical; i.e., there are repulsive forces between the walls and the nearby particles. A similar initial assembly was loaded under similar conditions, but with consideration of the van der Waals attractive force. The compressed assembly at static equilibrium is shown in Fig. 6. It is seen that the assembly in Fig. 5 has uniform particle spacing across the height of the specimen, whereas particles "stick" to the walls in the assembly shown in Fig. 6. When the attractive forces are ignored, the forces between the wall and the particles are the contact mechanical and double-layer repulsive forces, both of which are repulsive in that they "push" the particles away from each other, thus permitting waves to propagate across

the height of the specimen. This process becomes localized in the studies
which include the attractive force, where particles "stick" to each other

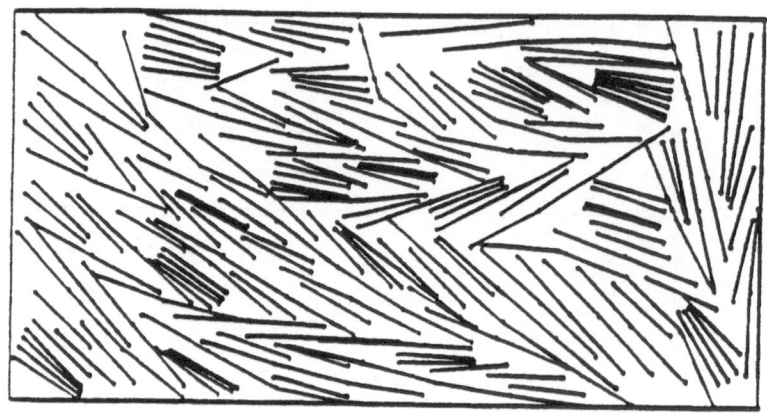

Figure 5. DEM Simulation without Consideration of
van der Waals Forces – After 50% Vertical
Compression of an Initial Loose Assembly.

Figure 6. DEM Simulation with Consideration of
van der Waals Forces – After 50% Vertical
Compression of an Initial Loose Assembly.

and to the wall, thus dissipating the energy locally near the wall. In other words, there is a severe edge effect. Simulations are currently being conducted with purely mechanical walls. Also, the gravity-consolidation method is being examined.

4. Approximate Simplified Analytical Assembly Analysis Method

A direct analytical approach has some advantages over a comprehensive numerical approach such as the DEM method. A realistic DEM analysis should use a sufficiently large number of clay particles in the assembly being analysed, which renders the analyses rather computer CPU intensive, preventing analyses to be routinely performed; an approximate analytical method which incorporates the essential microscopic features is desirable in this context. In addition, an analytical approach could eventually evolve into a simple, physically-based, micromechanical constitutive model to use in finite element analyses in place of a conventional elasto-plastic constitutive model. On this basis, an analytical approach is simultaneously being pursued.

In a preliminary analysis, a simple model was developed by extending Bolt's (1956) compressibility theory. Bolt (1956) used a parallel particle model to represent a volume of clay, with the assumption that the externally applied (one-dimensional) load is counter-balanced entirely by the double-layer repulsive force. In the study presented here, the particles are allowed to be nonparallel, as shown in Fig. 7. In addition to the repulsive force, the van der Waals attractive force is considered. Also, to analyse a soil suspension, the gravity force acting on individual particles are explicitly considered. The limitation of the model is that the contact mechanical forces are ignored, and at this time, it is applicable only for one-dimensional situations (e.g., consolidation, suspension, etc.). Owing to the first assumption, it is directly applicable only to a soil suspension, where the contact mechanical forces are either absent or insignificant. Nevertheless, it serves to verify the validity of the physico-chemical force theories employed.

Table 1 presents a comparison of the theory with experimental data. In this example, suspension tests were conducted on kaolinite suspensions in water with varying amounts of NaCl. Equilibrium void ratios were measured. The table compares the experimental void ratios with those calculated assuming that (1) the particles are parallel to each other (Bolt's model), and (2) the particles are randomly oriented ($2\theta = 45^0$). It may be observed that Bolt's model yields very poor predictions (which is in agreement with the observation made by Bolt himself in his original study in 1956), and the present model with the assumption of random particle orientation yields results very close to the observed results. The study

Table 1: **Results for Kaolinite-Water-$NaCl$ Suspensions ($2\theta = 0$ and 45^0)**

n_0 mole/liter	e Exp.	ψ_0 (mv)	Z (nm)	Theory ($\theta = 0$)			Theory ($2\theta = 45^0$)		
				e_d	e_θ	e	e_d	e_θ	e
0.002	12.1	228.3	6.85	1.70	0	1.70	1.00	11.2	12.2
0.010	11.6	187.7	3.06	0.76	0	0.76	0.60	11.2	11.8
0.100	10.5	129.8	.968	0.17	0	0.17	0.20	11.2	11.4
1.000	9.30	74.20	.306	.001	0	.001	.001	11.2	11.2

not only reveals the importance of considering the particle orientation in any assembly analysis, but provides some validity for the theories used to quantify the physico-chemical forces. Several similar comparisons have been obtained for contaminated kaolinite suspensions.

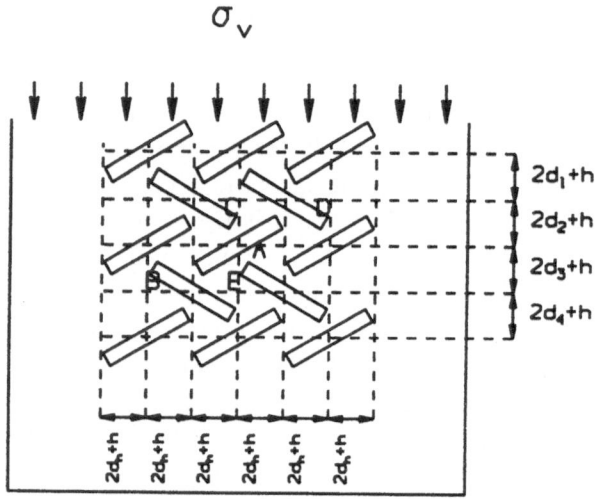

Figure 7. Layers of Nonparallel Particles
in a 1-D Consolidation/Suspension Cylinder.

5. An Energy Minimization Method of Assembly Analysis

A third approach that is being pursued is similar to DEM, but uses the energy minimization technique to establish the equilibrium configuration. Just as in DEM, Newton's law is enforced. The energy of a trial assembly depends on the coordinates (x_i, y_i) and orientations α_i of each particle i. Defining a vector $\mathbf{X}$ for the nth particle in an assembly as

$$\mathbf{X} = [x_1, y_1, \alpha_1, ..., x_i, y_i, \alpha_i, ..., x_n, y_n, \alpha_n]^T \qquad (1)$$

the energy of the system at any state can be written as.

$$U(\mathbf{X}) = nkT + E_{int} - \sum_i m_i g(y_{io} - y_i) - \int_0^{\Delta h_x} \sigma_x dh_x - \int_0^{\Delta h_y} \sigma_y dh_y \qquad (2)$$

where kT is the kinetic energy, with k denoting Boltzmann's constant and T the absolute temperature, E_{int} is the internal energy of the system including the repulsive and attractive energies, m_i is the mass of a particle i, g is the gravitational acceleration, Δh_x is the change in soil specimen width, Δh_y is the change in soil specimen height, h_x and h_y are specimen width and length respectively, and σ_x and σ_y are the horizontal and vertical stresses applied to the soil specimen respectively. Thus the first term represents kinetic energy. The second term denotes internal energy, consisting of repulsive and attractive potential energies. The third term considers loss in potential of the system due to work done by gravity and the fourth and fifth terms include the loss in potential due to the work done by externally applied surface traction.

U is minimized with respect to $\mathbf{X}$, subject to certain constraints, in determining the equilibrium configuration. A nonlinear optimization technique is used for this purpose. An initial assembly representing a suspension as shown in Fig. 8a was formed. At the end of the energy minimization process, the assembly shown in Fig. 8b was obtained, which represents the equilibrium configuration. By this procedure, analyses were carried out to simulate the suspensions tested in the laboratory. Fig. 9 presents a comparison of the predicted and measured equilibrium void ratios of kaolinite suspensions in various organic fluids, each having a specific value of the static dielectric constant, ranging between about 1.91 (heptane) to 110 (formamide). It may be seen that the trend is predicted, however, further study is needed in examining the apparent discrepancy from a quantitative view point.

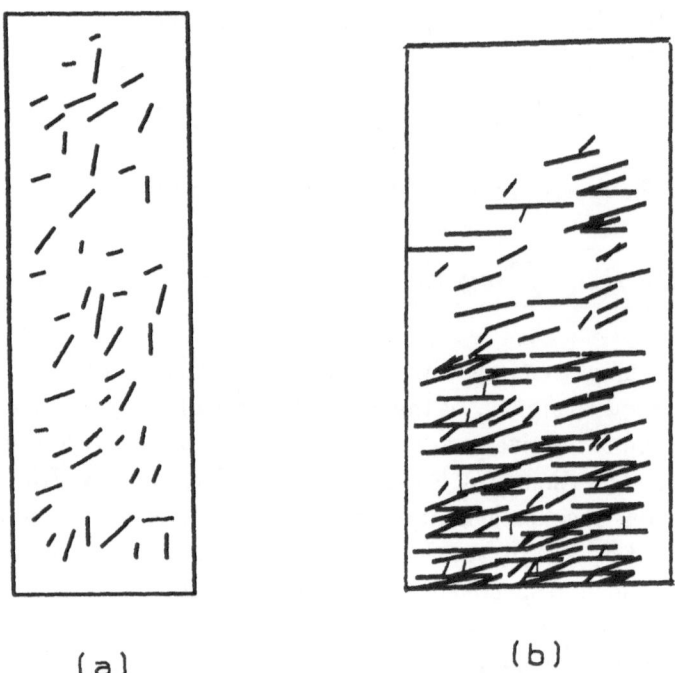

(a)   (b)

Figure 8. Simulation by Energy Minimization Technique.
(a) Initial Assembly, (b) Assembly at Equilibrium.

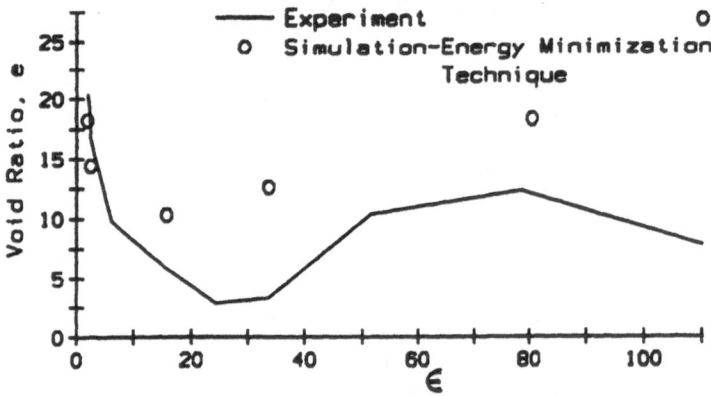

Figure 9. Comparison of Suspension Behavior Predicted by Energy
Minimization Technique and Measured Experimentally.

6. Conclusions

Theories have been developed for approximately computing the double-layer repulsive and van der Waals attractive forces between two nonparallel clay particles. Three different methods are being pursued in analysing an assembly of many particles; the discrete element method, an energy minimization technique and an approximate, simplified analytical method. These methods are found to yield reasonable results, indicating the potential of using them to develop a microstructural understanding of the engineering behavior of cohesive soils.

7. Acknowledgement

The research was partially supported by the U.S. Air Force Office of Scientific Research (AFOSR) under grant No. F49620-93-1-0265 and by the U.S. National Science Foundation (NSF) under grant No. MSS-9215847.

8. References

Anandarajah, A. (1994), "Discrete Element Method for Simulating Behavior of Cohesive Soils," *J. Geotech. Eng. Div.*, ASCE **120**, 1593-1615.

Anandarajah, A. and Chen, J. (1995), "Single Correction Function for Retarded van der Waals Attraction," *J. Coll. Inter. Sci.* **176**, 293-300.

Anandarajah, A. and Chen, J. (1996), "Van der Waals Attractive Force Between Clay Particles in Water and Contaminant," *Soils and Foundations*, JSSMFE (in press).

Anandarajah, A., Kuganenthira, N. and Zhao, D. (1996), "Variation of Fabric Anisotropy of Kaolinite in Triaxial Loading," *J. Geo. Engr. Div.*, ASCE, **122**, 633-640.

Anandarajah, A. and Chen, J. (1994), "Double-Layer Repulsive Force Between Two Inclined Platy Particles According to Gouy-Chapman Theory," *J. Coll. Inter. Sci.* **168**, 111-117.

Anandarajah, A. and Lu, N. (1991a), "Numerical Study of the Electrical Double-Layer Repulsion between Non-Parallel Clay Particles of Finite Length," *Int. J. Numer. Analy. Meth. Geomech.* **15**, 683-702.

Anandarajah, A. and Lu, N. (1991b), "Structural Analysis by the Distinct Element Method," *J. Engr. Mech. Div.*, ASCE **117**, 2156-2159.

Anandarajah, A. and Kuganenthira, N. (1995), "Some Aspects of Fabric Anisotropy of Soils," *Geotechnique*, **45**, 69-81.

Bolt, G. H. (1956). "Physico-chemical analysis of the compressibility of pure clays." *Geotechnique*, Vol. 6, pp. 86-93.

Chen, J. and Anandarajah, A. (1996), "Van der Waals Attractive Force Between Spherical Particles," *J. Coll. Inter. Sci.* **180**, 519-523.

Cundall, P. A. and Strack, O. D. L. (1979), "A Discrete Numerical Model for Granular Assemblies," *Geotechnique*, **29**, 47-65.

Kuganenthira, N., Zhao, D. and Anandarajah, A. (1996), "Measurement of Fabric Anisotropy in Triaxial Shearing," *Geotechnique* **46** (in press).

ON GRADIENT-ENHANCED DAMAGE THEORIES

R. DE BORST
Delft University of Technology, Faculty of Civil Engineering
P.O. Box 5048, 2600 GA Delft, The Netherlands

A. BENALLAL
Laboratoire de Mécanique et Technologie
61, Avenue du Président Wilson, 94235 Cachan, France

R.H.J. PEERLINGS
Eindhoven University of Technology, Faculty of Mechanical Engineering
P.O. Box 513, 5600 MB Eindhoven, The Netherlands

1. Introduction

Damage mechanics theory can be used to describe degradation and failure of structural materials and components. In its simplest form, it degrades the elastic properties, in particular Young's modulus with the accumulation of damage. Often, and this is the approach that will also be used here, the damage is linked to the maximum strain level attained during the loading history (Lemaitre and Chaboche, 1990).

A particular difficulty that is inherent in damage mechanics theory, and in all continuum mechanics models that try to capture degradation and failure of materials, is that at a certain threshold level of loading the governing differential equations locally lose ellipticity (or hyperbolicity if dynamic loading conditions are considered). Consequently, the boundary or initial value problem becomes ill-posed (Benallal *et al.*, 1988) and analytical or numerical solutions become meaningless.

A host of solutions has been suggested to remedy this deficiency of the standard continuum approach. For high-speed phenomena the inclusion of the inherent rate dependence of a material seems natural, see for instance Needleman (1988) for metals, Loret and Prévost (1990) for soils and Sluys (1992) and Sluys and de Borst (1992, 1994) for concrete. For granular materials a revival of the Cosserat continuum has been observed, since micromechanical foundations for applying such a theory exist (Mühlhaus and Vardoulakis, 1987). Numerical approaches have been elaborated that can be implemented in standard finite element codes in a straightforward fashion (de Borst, 1991,1993). For cracking in concrete and ceramics, and for describing void growth in metals nonlocal theories either in an integral format or in a differential format seem most appropriate. Within the context of a simple elastic-damaging material a fully nonlocal theory has been proposed by Pijaudier-Cabot and Bazant (1987) and Bazant and Pijaudier-Cabot (1988). Aifantis (1984, 1987, 1992), Coleman and Hodgdon (1985), Schreyer and Chen (1986), Lasry and Belytschko (1988), Vardoulakis and Aifantis (1991) and Sulem *et al.* (1995)

N. A. Fleck and A. C. F. Cocks (eds.),
IUTAM Symposium on Mechanics of Granular and Porous Materials, 215–226.
© 1997 *Kluwer Academic Publishers.*

have proposed gradient theories in a plasticity-based format, while, motivated by the work of Mühlhaus and Aifantis (1991), de Borst and Mühlhaus (1992), Sluys et al. (1993) and Pamin (1994) have derived algorithms for finite element implementations of a gradient-enhanced plasticity theory. Recently, Mühlhaus et al. (1994) have proposed a gradient theory within a damage mechanics framework. A different formalism was advocated by Peerlings et al (1996a,1996b) and was successfully implemented in a finite element code.

In this contribution we shall consider enriched damage theories in which the damage is coupled to (isotropic) elasticity. Firstly, we shall briefly summarise the existing knowledge on standard damage theories and non-local damage theories in an integral format. Based on the latter class of theories a family of gradient-enhanced damage theories will be derived. Some remarks about their numerical implementation will be made and a comparison will be carried out between some of the gradient damage theories and a damage theory in an integral format.

2. Elasticity-Based Damage Models

Herein, we shall restrict ourselves to isotropic damage formulations. In a strain-based formulation we then have

$$\sigma = (1 - \omega)\mathbf{D}\varepsilon \tag{1}$$

with σ the stress tensor, ε the strain tensor, $\mathbf{D}$ the virgin elastic stiffness tensor, and ω a scalar-valued internal variable, which reflects the amount of damage which the material has experienced. It starts at zero (undamaged state) and grows to one (complete loss of integrity) as a function of a scalar-valued history parameter κ, which represents the most severe deformation the material has experienced: $\omega = \omega(\kappa)$. The history parameter initiates at a threshold level κ_i and damage growth is possible if the damage loading function

$$f(\varepsilon_{eq}, \kappa) = \varepsilon_{eq} - \kappa \tag{2}$$

vanishes. In particular, the damage loading function f and the growth rate of the history parameter κ have to satisfy the Kuhn-Tucker conditions

$$f \leq 0 \quad , \quad \dot{\kappa} \geq 0 \quad , \quad f\dot{\kappa} = 0 \tag{3}$$

In eq. (2) ε_{eq} is the local equivalent strain, which can be a function of strain invariants, the principal strains, or the local energy release rate due to damage (Lemaitre and Chaboche, 1990), e.g. $\varepsilon_{eq} = \frac{1}{2}\varepsilon^T\mathbf{D}\varepsilon$.

3. Non-Local Damage Models

In a non-local generalisation the equivalent strain ε_{eq} is normally replaced by a spatially averaged quantity (Pijaudier-Cabot and Bazant, 1987; Bazant and Pijaudier-Cabot, 1988):

$$f(\bar{\varepsilon}_{eq}, \kappa) = \bar{\varepsilon}_{eq} - \kappa \qquad (4)$$

where the non-local average strain $\bar{\varepsilon}_{eq}$ is computed as:

$$\bar{\varepsilon}_{eq}(\mathbf{x}) = \frac{1}{V_r(\mathbf{x})} \int_V g(\mathbf{s}) \varepsilon_{eq}(\mathbf{x}+\mathbf{s}) \, dV \, , \quad V_r(\mathbf{x}) = \int_V g(\mathbf{s}) \, dV \qquad (5)$$

with $g(\mathbf{s})$ a weight function, e.g., the error function, and $\mathbf{s}$ a relative position vector pointing to the infinitesimal volume dV. Alternatively, the locally defined history parameter κ may be replaced in the damage loading function f by a spatially averaged quantity:

$$f(\varepsilon_{eq}, \bar{\kappa}) = \varepsilon_{eq} - \bar{\kappa} \qquad (6)$$

where the non-local history parameter $\bar{\kappa}$ follows from:

$$\bar{\kappa}(\mathbf{x}) = \frac{1}{V_r(\mathbf{x})} \int_V g(\mathbf{s}) \kappa(\mathbf{x}+\mathbf{s}) \, dV \, , \quad V_r(\mathbf{x}) = \int_V g(\mathbf{s}) \, dV \qquad (7)$$

The Kuhn-Tucker conditions can now be written as:

$$f \le 0 \, , \quad \dot{\kappa} \ge 0 \, , \quad f\dot{\kappa} = 0 \qquad (8)$$

4. Gradient Damage Models

Non-local constitutive relations can be considered as a point of departure for constructing gradient models. Again, this can either be done by expanding the kernel ε_{eq} of the integral in (5) in a Taylor series, or by expanding the history parameter κ in (7) in a Taylor series. We will first consider the first-mentioned case and then we will do the same for κ. If we truncate after the second-order terms and carry out the integration implied in (5) under the assumption of isotropy, the following relation ensues:

$$\bar{\varepsilon}_{eq} = \varepsilon_{eq} + \bar{c} \nabla^2 \varepsilon_{eq} \qquad (9)$$

where $\bar{c}$ is a material parameter of the dimension length squared. It can be related to the averaging volume and then becomes dependent on the precise form of the weight function g. For instance, for a one-dimensional continuum and taking

$$g(s) = \frac{1}{\sqrt{2\pi} l} e^{-s^2/2l^2} \qquad (10)$$

we obtain $\bar{c} = \frac{1}{2} l^2$. Here, we adopt the phenomenological view that $\sqrt{\bar{c}}$ reflects the length scale of the failure process that we wish to describe macroscopically.

Formulation (9) has a severe disadvantage when applied in a finite element context, namely that it requires computation of second-order gradients of the local equivalent strain ε_{eq}. Since this quantity is a function of the strain tensor, and since the strain tensor involves first-order derivatives of the displacements, third-order derivatives of the displacements have to be computed, which necessitates C^1-continuity of the shape functions.

To obviate this problem, eq. (9) is differentiated twice and the result is substituted in eq. (9). Again neglecting fourth-order terms then leads to

$$\bar{\varepsilon}_{eq} - \bar{c}\nabla^2 \bar{\varepsilon}_{eq} = \varepsilon_{eq} \tag{11}$$

When $\bar{\varepsilon}_{eq}$ is discretised independently and use is made of the divergence theorem, a C^0-interpolation for $\bar{\varepsilon}_{eq}$ suffices (Peerlings et al., 1996a).

Higher-order continua require additional boundary conditions. With eq. (11) governing the damage process, either the averaged equivalent strain $\bar{\varepsilon}_{eq}$ itself or its normal derivative must be specified on the boundary S of the body:

$$\bar{\varepsilon}_{eq} = \bar{\varepsilon}_s \quad \text{or} \quad \mathbf{n}^T \nabla \bar{\varepsilon} = \bar{\varepsilon}_{ns} \tag{12}$$

with $\mathbf{n}$ the outward normal vector to the boundary of the body S. In the example calculations the natural boundary condition $\bar{\varepsilon}_{ns} = 0$ has been adopted.

In a fashion similar to the derivation of the gradient damage models based on the averaging of the equivalent strain ε_{eq}, we can elaborate a gradient approximation of (7), i.e., by developing κ into a Taylor series. For an isotropic, infinite medium and truncating after the second term we then have

$$\bar{\kappa} = \kappa + \bar{c}\nabla^2 \kappa \tag{13}$$

where the gradient constant $\bar{c}$ depends on the weighting function. In principle, eq. (6) could now be replaced by

$$f = \varepsilon_{eq} - \kappa - \bar{c}\nabla^2 \kappa \tag{14}$$

but we will consider κ as an independent variable in the finite element implementation to be discussed next, and therefore we shall retain the form (6) for the damage loading function, where $\bar{\kappa}$ is given by (13).

5. Finite Element Aspects

As an example we shall elaborate the finite element implementation of the damage model with gradient enhancement according to (13). We consider equilibrium at iteration $n + 1$:

$$\mathbf{L}^T \sigma_{n+1} + \rho \mathbf{g} = 0 \tag{15}$$

with $\mathbf{g}$ the gravity acceleration vector and $\mathbf{L}$ a differential operator matrix. Multiplying the equilibrium equation with $\delta \mathbf{u}$, where $\mathbf{u}$ is the continuous displacement field vector and the symbol δ denotes the variation of a quantity, and integrating over the entire volume occupied by the body, one obtains the corresponding weak form:

$$\int_V \delta \mathbf{u}^T (\mathbf{L}^T \sigma_{n+1} + \rho \mathbf{g}) \, dV = 0 \tag{16}$$

Similarly, the weak form of the Helmholtz equation for the distribution of the local history parameter κ, eq. (13), can be derived as:

$$\int_V \delta\kappa(\kappa_{n+1} + \bar{c}\nabla^2\kappa_{n+1} - \bar{\kappa}_{n+1})dV = 0 \tag{17}$$

We now introduce the decompositions

$$\sigma_{n+1} = \sigma_n + d\sigma \tag{18}$$

and

$$\kappa_{n+1} = \kappa_n + d\kappa \tag{19}$$

for the stress σ and the history parameter κ, respectively. The d-symbol signifies the iterative improvement of a quantity between two successive iterations. With aid of these decompositions and applying the divergence theorem to eqs (16)-(17) one obtains

$$\int_V (\mathbf{L}\delta\mathbf{u})^T d\sigma dV = \int_V \rho\delta\mathbf{u}^T \mathbf{g}dV + \int_S \delta\mathbf{u}^T \mathbf{t}dS - \int_V (\mathbf{L}\delta\mathbf{u})^T \sigma_n dV \tag{20}$$

where $\mathbf{t}$ is the boundary-traction vector and

$$\int_V (\delta\kappa d\kappa - \bar{c}\nabla\delta\kappa \cdot \nabla d\kappa)dV - \int_V \delta\kappa d\bar{\kappa}dV =$$

$$\int_V \delta\kappa\bar{\kappa}_n dV - \int_V (\delta\kappa\kappa_n - \bar{c}\nabla\delta\kappa \cdot \nabla\kappa_n)dV \tag{21}$$

where the non-standard natural boundary condition

$$\mathbf{n}^T\nabla\kappa = 0 \tag{22}$$

has been adopted, $\mathbf{n}$ being the outward normal to the body surface. Since κ can be directly related to the damage variable ω this condition can be interpreted as no damage flux through the boundary of the body, which seems physically reasonable.

Finally we interpolate displacements $\mathbf{u}$ and the history parameter κ as

$$\mathbf{u} = \mathbf{Ha} \tag{23}$$

and

$$\kappa = \tilde{\mathbf{H}}\mathbf{k} \tag{24}$$

with $\mathbf{a}$ and $\mathbf{k}$ the vectors that contain the nodal values of $\mathbf{u}$ and κ, respectively. $\mathbf{H}$ and $\tilde{\mathbf{H}}$ contain the interpolation polynomials of $\mathbf{u}$ and κ, respectively. The gradient of κ is then computed as

$$\nabla\kappa = \tilde{\mathbf{B}}\mathbf{k} \quad , \quad \tilde{\mathbf{B}} = \nabla \cdot \tilde{\mathbf{H}} \tag{25}$$

Substitution of eqs (23)-(25) into eqs (20) and (21) and using the fact that the ensuing relations must hold for any admissible $\delta\mathbf{u}$ and $\delta\kappa$ then yields

$$\int_V \mathbf{B}^T d\boldsymbol{\sigma} dV = \mathbf{f}_{ext} - \mathbf{f}_{int,n} \tag{26}$$

where $\mathbf{B} = \mathbf{LH}$, and

$$\mathbf{f}_{ext} = \int_V \rho \delta \mathbf{u}^T \mathbf{g} dV + \int_S \delta \mathbf{u}^T \mathbf{t} dS \tag{27}$$

$$\mathbf{f}_{int,n} = \int_V \mathbf{B}^T \boldsymbol{\sigma}_n dV \tag{28}$$

and

$$\mathbf{K}_{kk} d\mathbf{k} - \int_V \tilde{\mathbf{H}}^T d\bar{\kappa} dV = \int_V \tilde{\mathbf{H}}^T \bar{\kappa}_n dV - \mathbf{K}_{kk} \mathbf{k}_n \tag{29}$$

where

$$\mathbf{K}_{kk} = \int_V (\tilde{\mathbf{H}}^T \tilde{\mathbf{H}} - \bar{c}\, \tilde{\mathbf{B}}^T \tilde{\mathbf{B}}) dV \tag{30}$$

We now adopt the standard elastic-damage stress-strain relation of eq. (1) and cast it into an incremental format:

$$d\boldsymbol{\sigma} = (1 - \omega_n)\mathbf{D} d\boldsymbol{\varepsilon} - \mathbf{D}\boldsymbol{\varepsilon}_n d\omega \tag{31}$$

Restricting the treatment to small displacement gradients we introduce the linear kinematic relation

$$\boldsymbol{\varepsilon} = \mathbf{Lu} \tag{32}$$

with $\mathbf{L}$ as defined in the preceding, or using eq. (23) and $\mathbf{B} = \mathbf{LH}$,

$$\boldsymbol{\varepsilon} = \mathbf{Ba} \tag{33}$$

Furthermore, the damage ω is a function of the local history parameter κ, $\omega = \omega(\kappa)$, so that

$$d\omega = \frac{\partial \omega}{\partial \kappa} d\kappa \tag{34}$$

Taking into account eqs (24), (33) and (34), eq. (31) is elaborated as:

$$d\boldsymbol{\sigma} = (1 - \omega_n)\mathbf{DB} d\mathbf{a} - \mathbf{D}\boldsymbol{\varepsilon}_n \frac{\partial \omega}{\partial \kappa} \tilde{\mathbf{H}} d\mathbf{k} \tag{35}$$

To complete the formulation the iterative improvement of the non-standard history parameter $d\bar{\kappa}$ is elaborated as

$$d\bar{\kappa} = \frac{\partial\bar{\kappa}}{\partial\varepsilon_{eq}}\frac{\partial\varepsilon_{eq}}{\partial\varepsilon}\mathbf{B}d\mathbf{a} \tag{36}$$

where $\partial\bar{\kappa}/\partial\varepsilon_{eq}=1$ for loading and $\partial\bar{\kappa}/\partial\varepsilon_{eq}=0$ otherwise. Inserting the above expressions for $d\sigma$ and $d\bar{\kappa}$ into eqs (26) and (29) yields the following set of equations that describe the incremental process in the discretised gradient-enhanced elastic-damaging continuum:

$$\begin{bmatrix} \mathbf{K}_{aa} & \mathbf{K}_{ak} \\ \mathbf{K}_{ka} & \mathbf{K}_{kk} \end{bmatrix}\begin{bmatrix} d\mathbf{a} \\ d\mathbf{k} \end{bmatrix} = \begin{bmatrix} \mathbf{f}_{ext}-\mathbf{f}_{int,n} \\ \mathbf{f}_{k,n}-\mathbf{K}_{kk}\mathbf{k}_n \end{bmatrix} \tag{37}$$

where $\mathbf{f}_{ext}$, $\mathbf{f}_{int,n}$ and $\mathbf{K}_{kk}$ are defined in accordance with eqs (27), (28) and (30), and

$$\mathbf{K}_{aa} = \int_V (1-\omega_n)\mathbf{B}^T\mathbf{D}\mathbf{B}dV \tag{38}$$

$$\mathbf{K}_{ak} = -\int_V \mathbf{B}^T\mathbf{D}\varepsilon_n \frac{\partial\omega}{\partial\kappa}\tilde{\mathbf{H}}dV \tag{39}$$

$$\mathbf{K}_{ka} = -\int_V \frac{\partial\bar{\kappa}}{\partial\varepsilon_{eq}}\tilde{\mathbf{H}}^T\frac{\partial\varepsilon_{eq}}{\partial\varepsilon}\mathbf{B}dV \tag{40}$$

and

$$\mathbf{f}_{k,n} = \int_V \tilde{\mathbf{H}}^T\bar{\kappa}_n dV \tag{41}$$

An algorithm is elaborated in Box 1.

The basic variables are differentiated only once in the above expressions and a simple C^0-continuity of the interpolation polynomials suffices. To avoid stress oscillations, the displacements should be interpolated one order higher than the history variable in order to avoid stress oscillations, cf. the Babuska-Brezzi condition for mixed finite elements in incompressible solids.

6. A Comparison of Approaches

To gain some insight into the properties of the non-local versus the gradient damage models, a dispersion analysis is now carried out for a one-dimensional bar of infinite length. Dispersion analyses can provide much insight into the wave propagation characteristics of enriched continuum models of strain-softening materials, e.g., Sluys (1992), or Huerta and Pijaudier-Cabot (1994). Here, we shall briefly summarise the main results of such an investigation that has been carried out by Peerlings et al. (1996b), which involves the non-local damage model with averaging on the equivalent strain and the two related gradient models (explicit and implicit format).

In a dispersion analysis a perturbation of the form

Box 1. Algorithm for gradient-enhanced damage model.

---

1. Update $\mathbf{K}_{aa}$, $\mathbf{K}_{ak}$, $\mathbf{K}_{ka}$

2. Solve for da and dk according to eq. (37)

3. Update $\mathbf{a}$ and $\mathbf{k}$ at the nodal points:

$\mathbf{a}_{n+1} = \mathbf{a}_n + \mathrm{da}$

$\mathbf{k}_{n+1} = \mathbf{k}_n + \mathrm{dk}$

4. Compute in the integration points:

strains: $\varepsilon_{n+1} = \mathbf{B}\mathbf{a}_{n+1}$

equivalent strain: $\varepsilon_{eq,n+1} = \varepsilon_{eq}(\varepsilon_{n+1})$

damage loading function: $f = \varepsilon_{eq,n+1} - \bar{\kappa}_n$

If $f \geq 0$: $\bar{\kappa}_{n+1} = \varepsilon_{eq,n+1}$

else: $\bar{\kappa}_{n+1} = \bar{\kappa}_n$

Interpolate: $\kappa_{n+1} = \tilde{\mathbf{H}}\mathbf{k}_{n+1}$

Compute: $\omega_{n+1} = \omega(\kappa_{n+1})$

Compute: $\sigma_{n+1} = (1 - \omega_{n+1})\mathbf{D}\varepsilon_{n+1}$

5. Update the internal forces:

$\mathbf{f}_{\mathrm{int},n+1} = \int_V \mathbf{B}^T \sigma_{n+1}\,dV$

$\mathbf{f}_{k,n+1} = \int_V \tilde{\mathbf{H}}^T \bar{\kappa}_{n+1}\,dV$

6. Check convergence criterion.

$$\delta u = \hat{u}e^{\mathrm{i}k(x - ct)} \qquad (42)$$

is introduced into the governing set of equations, namely the equation of motion, the kinematic relation and the constitutive model, which have been combined to give a single expression in terms of the axial displacement of the bar and have been linearised around a homogeneous deformation state indicated with the subscript 0. In eq. (42) k is the wave number, c is the corresponding phase velocity and $\hat{u}$ is the amplitude of the perturbation. As detailed in Peerlings *et al.* (1996b), the following expressions are obtained for the phase velocity:

❑ Non-local damage model, eq. (5):

$$c = c_e \sqrt{1 - \omega_0 - \varepsilon_0 \left(\frac{\partial \omega}{\partial \kappa} \right)_0 e^{-k^2/2 l^2}} \tag{43}$$

❑ Gradient damage model in explicit format, eq. (9):

$$c = c_e \sqrt{1 - \omega_0 - \varepsilon_0 \left(\frac{\partial \omega}{\partial \kappa} \right)_0 (1 - \tfrac{1}{2} k^2 l^2)} \tag{44}$$

❑ Gradient damage model in implicit format, eq. (11):

$$c = c_e \sqrt{1 - \omega_0 - \varepsilon_0 \left(\frac{\partial \omega}{\partial \kappa} \right)_0 (1 + \tfrac{1}{2} k^2 l^2)^{-1}} \tag{45}$$

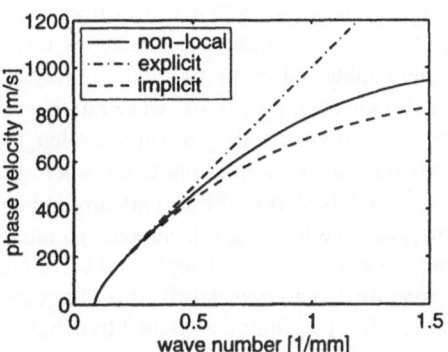

Figure 1. Wave velocity as function of wave number.

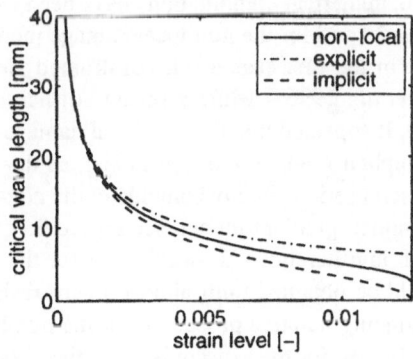

Figure 2. Cut-off wave length as function of strain level.

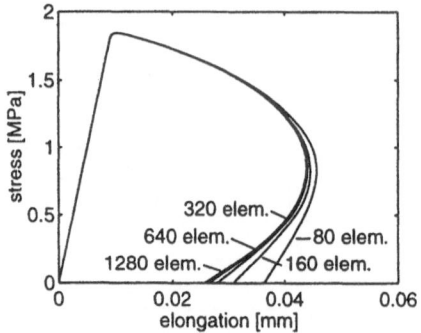

Figure 3. Load-deflection curves at mesh refinement (implicit gradient model).

In eqs (43)-(45) c_e is the elastic wave velocity, ε_0 is the existing strain level and ω_0 is the corresponding value of the damage parameter. The following material data were used: a Young's modulus $E = 20{,}000$ MPa, a density such that $c_e = 1000$ m/s, a linear degrading damage model with an initial value $\kappa_i = 0.0001$ and a value $\kappa_c = 0.0125$ (at which the local load-carrying capacity is exhausted) and an internal length scale $l = \sqrt{2}$ mm. Taking a strain level $\varepsilon_0 = \kappa_i$, the curves of Figure 1 are obtained.

We observe that all three models result in a cut-off wave number below which loading waves cannot propagate, i.e. the wave speed becomes imaginary. This phenomenon was also found for gradient-enhanced plasticity models (Sluys, 1992; Sluys et al., 1993; Sluys and de Borst, 1994). However, the cut-off wave number now depends on the existing strain level, and as seen in Figure 2, the critical wave length becomes smaller for increasing deformation, in contrast to gradient plasticity where it remains constant. Also, the three different damage models start to differ markedly at increasing strain levels. The most salient observation is that while the non-local damage model and the gradient model in an implicit format approach a zero wave length, and therefore to a physically realistic vanishing localisation zone for large strain levels, this is not so for the gradient damage model in an explicit format, thus precluding a gradual transition into a line crack.

Next, we consider a bar with a finite length, $L = 100$ mm, and take an imperfection (10% reduction of the cross sectional area A) in the centre 10 mm of the bar. Now, for reasons mentioned above, numerical computations have been carried out only for the implicit gradient damage model and for the non-local damage model, but not for the explicit gradient damage model. In the first case a full constrained Newton-Raphson procedure was adopted (Peerlings et al., 1996a) while a secant stiffness method was used for the non-local damage model. It appeared that the non-local damage model requires a less fine discretisation than the implicit gradient damage model, as the results for an 80 element and for a 160 element discretisation already coincide in the non-local approach, while this is not the case for the implicit gradient model, see Figure 3. On the other hand, convergence in terms of equilibrium iterations is much better for the gradient model, and fully converged solutions could be obtained until almost a zero residual load level, Figures 3 and 4. The equilibrium-finding iterative procedure for the non-local model fails at a non-zero residual load level. In fact, for finer discretisations the iterative procedure diverges at an earlier stage in the loading process. Figure 4 compares the load-deflection curves for both models for a discretisation of 320 elements. We observe that upon further loading, the differences between both models become more pronounced and that at a certain stage

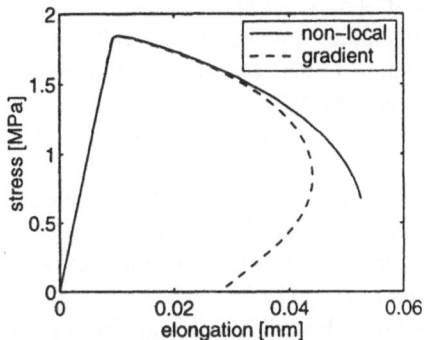

Figure 4. Comparison of the non-local and implicit gradient models.

convergence is lost for the non-local damage model.

7. Concluding Remarks

A family of gradient-enhanced damage models has been constructed where the damage is coupled to elasticity. Although they have been derived from non-local damage models in an integral format, their properties can be quite different from those of non-local integral-type damage models. Indeed, considerable differences can exist, as has been shown by dispersion analyses and by a simple one-dimensional example. For a number of cases of gradient-enhanced damage models modern, computationally efficient schemes have been constructed. This opens the way to their use in large-scale computations.

8. References

Aifantis, E.C. (1984) On the microstructural origin of certain inelastic models, *J. Engng. Mater. Technol.* **106**, 326-334.

Aifantis, E.C. (1987) The physics of plastic deformation, *Int. J. Plasticity* **3**, 211-247.

Aifantis, E.C. (1992) On the role of gradients in the localization of deformation and fracture, *Int. J. Engng. Sci.* **30**, 1279-1299.

Bazant, Z.P. and Pijaudier-Cabot, G. (1988) Nonlocal continuum damage, localization instability and convergence, *J. Appl. Mech.* **55**, 287-293.

Benallal, A., Billardon, R. and Geymonat, G. (1988) Some mathematical aspects of the damage softening rate problem, in J. Mazars and Z.P. Bazant (eds.), *Cracking and Damage*, Elsevier, Amsterdam and London, pp. 247-258.

Borst, R. de (1991) Simulation of strain localisation: A reappraisal of the Cosserat continuum, *Engng. Comput.* **8**, 317-332.

Borst, R. de and Mühlhaus, H.-B. (1992) Gradient-dependent plasticity: Formulation and algorithmic aspects, *Int. J. Num. Meth. Eng.* **35**, 521-539.

Borst, R. de (1993) A generalisation for J_2-flow theory for polar continua, *Comp. Meth. Appl. Mech. Eng.* **103**, 347-362.

Coleman, B.D. and Hodgdon, M.L. (1985) On shear bands in ductile materials, *Arch. Ra-*

tion. Mech. Anal. **90** , 219-247.

Huerta, A. and Pijaudier-Cabot, G. (1994) Discretization influence on the regularization by two localization limiters, *J. Eng. Mech.* **120** , 1198-1218.

Lasry, D. and Belytschko, T. (1988) Localization limiters in transient problems, *Int. J. Solids Structures* **24** , 581-597.

Lemaitre, J. and Chaboche, J.L. (1990) *Mechanics of solid materials,* Cambridge University Press, Cambridge.

Loret, B. and Prévost, J.H. (1990) Dynamic strain localization in elasto-(visco-)plastic solids, Part 1, *Comp. Meth. Appl. Mech. Eng.* **83** , 247-273.

Mühlhaus, H.-B. and Vardoulakis, I. (1987) The thickness of shear bands in granular materials, *Geotechnique* **37** , 271-283.

Mühlhaus, H.-B. and Aifantis, E.C. (1991) A variational principle for gradient plasticity, *Int. J. Solids Structures* **28** , 845-858.

Mühlhaus, H.-B., de Borst, R., Sluys, L.J. and Pamin, J. (1994) A thermodynamic theory for inhomogeneous damage evolution, in H.J. Siriwardane and M.M. Zaman (eds.), *Computer Methods and Advances in Geomechanics,* Balkema, Rotterdam and Boston, pp. 635-640.

Needleman, A. (1988) Material rate dependence and mesh sensitivity in localization problems, *Comp. Meth. Appl. Mech. Eng.* **67** , 69-86.

Pamin, J. (1994) *Gradient-dependent plasticity in numerical simulation of localization phenomena.* Dissertation, Delft University of Technology, Delft.

Peerlings, R.H.J., Borst, R. de, Brekelmans, W.A.M. and Vree, J.H.P. de (1996a) Gradient-enhanced damage for quasi-brittle materials, *Int. J. Num. Meth. Eng.* **39** 3391-3403.

Peerlings, R.H.J., Borst, R. de, Brekelmans, W.A.M., Vree, J.H.P. de and Spee, I. (1996b) Some observations on localisation in non-local and gradient damage models, *Eur. J. Mech./A: Solids,* in press.

Pijaudier-Cabot, G. and Bazant, Z.P. (1987) Nonlocal damage theory, *J. Engng. Mech.* **113** , 1512-1533.

Schreyer, H.L. and Chen, Z. (1986) One-dimensional softening with localization, *J. Appl. Mech.* **53** , 791-797.

Sluys, L.J. (1992) *Wave propagation, localisation and dispersion in softening solids,* Dissertation, Delft University of Technology, Delft.

Sluys, L.J. and Borst, R. de (1992) Wave propagation and localisation in a rate-dependent cracked medium - Model formulation and one-dimensional examples, *Int. J. Solids Structures* **29** , 2945-2958.

Sluys, L.J., Borst, R. de and Mühlhaus, H.-B. (1993) Wave propagation, localization and dispersion in a gradient-dependent medium, *Int. J. Solids Structures* **30** , 1153-1171.

Sluys, L.J. and Borst, R. de (1994) Dispersive properties of gradient-dependent and rate-dependent media, *Mech. Mater.* **18** , 131-149.

Sulem, J., Vardoulakis, I. and Papamichos, E. (1995) Microstructure and scale effects in granular rocks, in H.-B. Mühlhaus (ed.), *Continuum Models for Materials with Microstructure,* J. Wiley & Sons, Chichester, pp. 200-237.

Vardoulakis, I. and Aifantis, E.C. (1991) A gradient flow theory of plasticity for granular materials, *Acta Mechanica* **87** , 197-217.

REMARKS ON COAXIALITY IN FULLY DEVELOPED GRAVITY FLOWS OF DRY GRANULAR MATERIALS

A J M SPENCER
Department of Theoretical Mechanics
University of Nottingham
Nottingham NG7 2RD, UK

1. Introduction

The Coulomb-Mohr condition is quite widely accepted as a basis for determining the stress at flow or failure of a granular material, but the question of how correctly to express the equations which govern the flow behaviour still has no generally agreed answer. A brief review was given in Spencer (1982). In particular there are differing views as to whether the principal axes of the stress and the rate-of-deformation are or are not coincident. The requirement that these axes should coincide (the coaxiality assumption) is often stated as a condition for material isotropy, but this is true only under restricted constitutive assumptions. Coaxiality is a consequence of assuming that the deformation-rate is derived from a plastic potential which is an isotropic function of the stress. However coaxiality does not require the existence of a plastic potential.

To keep the analysis as simple as possible the case considered is that of a dry, cohesive or non-cohesive granular material in a fully developed quasi-static flow, so that effects such as initial dilatancy and fluid pore pressure are eliminated. The theory can be extended to include these and other complications, but it is not anticipated that such modifications would have a significant qualitative effect on the conclusions. The stress is assumed to be governed by the Coulomb-Mohr condition (again generalisations are possible but not likely to change the conclusions) and the equilibrium equations, with gravitational body force included. To determine the flow fields two alternative theories are employed. The first is the 'double-shearing' theory first formulated by Spencer (1964) and extended in Spencer (1982). The physical basis of this theory is that flow occurs by shear on the surfaces on which the critical shear traction for satisfaction of the Coulomb-Mohr

N. A. Fleck and A. C. F. Cocks (eds.),
IUTAM Symposium on Mechanics of Granular and Porous Materials, 227–238.
© *1997 Kluwer Academic Publishers.*

condition is mobilised, with the direction of shear coincident with that of the shear traction. The alternative hypothesis which is examined is the 'coaxiality' assumption (Hill, 1950, Jenike, 1964, and many others) that the principal axes of stress and deformation-rate are coincident. The governing equations for these theories are summarised in Section 2.

Four distinct problems of gravity flows are considered. Radial flows through (i) wedges and (ii) cones are described in Section 3. These were first solved, using the coaxial theory, by Jenike (1964). Bradley (1991) confirmed Jenike's results and obtained corresponding results with the double-shearing theory; these solutions are summarised in Spencer and Bradley (1996). The other problems deal with flow of material compressed between vertical walls (Spencer and Bradley, 1992) or in a contracting vertical circular cylindrical tube (Spencer and Bradley, 1997). These solutions also provide bases for approximate solutions for gravity flows in tapering channels and cylinders. They are described in Section 4. It is found in each of the four problems that the double-shearing theory gives smooth and plausible results, but the coaxial theory predicts singularities in the flow field on surfaces within the flowing region.

2. Governing Equations

2.1. STRESS FIELD

The equations that govern the stress in a granular material that flows quasi-statically and conforms to the Coulomb-Mohr yield condition are well known, and are summarised here for convenience.

2.1.1. *Plain Strain*
In terms of plane rectangular Cartesian coordinates (x, y) the relevant stress components are denoted σ_{xx}, σ_{yy}, σ_{xy}. The equations of equilibrium, with gravitational body force ρg per unit volume directed in the negative y direction, are

$$\frac{\partial \sigma_{xx}}{\partial x} + \frac{\partial \sigma_{yy}}{\partial y} = 0, \quad \frac{\partial \sigma_{xy}}{\partial x} + \frac{\partial \sigma_{yy}}{\partial y} = \rho g. \tag{2.1}$$

(Tensile stress is taken positive). The stress components can be expressed as

$$\sigma_{xx} = -p + q \cos 2\psi, \quad \sigma_{yy} = -p - q \cos 2\psi, \quad \sigma_{xy} = q \sin 2\psi, \tag{2.2}$$

where

$$p = -\tfrac{1}{2}(\sigma_1 + \sigma_3), \qquad q = \tfrac{1}{2}(\sigma_1 - \sigma_3), \tag{2.3}$$

Here σ_1, σ_2, σ_3 are the principal stress components, ordered so that $\sigma_1 \geq \sigma_2 \geq \sigma_3$, with σ_1, σ_3 the principal components whose axes lie in the (x, y) planes and ψ is the angle that the axis of the σ_1 principal stress makes with the x-axis. The Coulomb-Mohr yield condition takes the form

$$q = p \sin \phi + c \cos \phi, \tag{2.4}$$

where ϕ is the angle of internal friction, and c is the cohesion, both assumed constant. If $c = 0$ the material is cohesionless.

In terms of plane polar coordinates (r, θ), with $\theta = 0$ as the upward vertical, the stress components are denoted σ_{rr}, $\sigma_{\theta\theta}$, $\sigma_{r\theta}$, the equations of equilibrium are

$$\frac{\partial \sigma_{rr}}{\partial r} + \frac{1}{r} \frac{\partial \sigma_{r\theta}}{\partial \theta} + \frac{\sigma_{rr} - \sigma_{\theta\theta}}{r} = \rho g \cos \theta,$$

$$\frac{\partial \sigma_{r\theta}}{\partial r} + \frac{1}{r} \frac{\partial \sigma_{\theta\theta}}{\partial \theta} + \frac{2\sigma_{r\theta}}{r} = -\rho g \sin \theta, \tag{2.5}$$

and

$$\sigma_{rr} = -p + q \cos 2\chi, \quad \sigma_{\theta\theta} = -p - q \cos 2\chi, \quad \sigma_{r\theta} = q \sin 2\chi, \tag{2.6}$$

where p and q are given by (2.3), $\chi = \psi - \theta - \frac{1}{2}\pi$, and the Coulomb-Mohr condition remains as (2.4).

2.1.2. Axial Symmetry

In terms of cylindrical polar coordinates (r, Φ, z) with the z axis vertically upward, and with the relevant stress components denoted by σ_{rr}, $\sigma_{\Phi\Phi}$, σ_{zz} and σ_{rz}, the equilibrium equations are

$$\frac{\partial \sigma_{rr}}{\partial r} + \frac{\partial \sigma_{rz}}{\partial z} + \frac{\sigma_{rr} - \sigma_{\Phi\Phi}}{r} = 0, \quad \frac{\partial \sigma_{rz}}{\partial r} + \frac{\partial \sigma_{zz}}{\partial z} + \frac{\sigma_{rz}}{r} = \rho g, \tag{2.7}$$

and we have

$$\sigma_{rr} = -p + q \cos 2\Psi, \quad \sigma_{zz} = -p - q \cos 2\Psi, \quad \sigma_{rz} = q \sin 2\Psi, \tag{2.8}$$

where p and q are again given by (2.3), Ψ is the angle between the σ_1 axis and the r axis, and $\sigma_2 = \sigma_{\Phi\Phi}$. There are several possible regimes associated with the Coulomb-Mohr condition depending on the ordering of $\sigma_{\Phi\Phi}$ relative to σ_1 and σ_3; these are discussed in detail in Spencer (1982, 1986). The appropriate case here is the 'Haar-von Karman' regime in which $\sigma_{\Phi\Phi}$ is equal to either σ_1 or σ_3, and in particular the case $\sigma_{\Phi\Phi} = \sigma_3$. In that case, in addition to (2.8), there obtains

$$\sigma_{\Phi\Phi} = -p - q, \tag{2.9}$$

and again the Coulomb-Mohr condition takes the form (2.4).

In terms of spherical polar coordinates(R, Θ, Φ) with $\Theta = 0$ representing the upward vertical, the equation of equilibrium are, for axial symmetry,

$$\frac{\partial \sigma_{RR}}{\partial R} + \frac{1}{R}\frac{\partial \sigma_{R\Theta}}{\partial \Theta} + \frac{1}{R}\left(2\sigma_{RR} - \sigma_{\Theta\Theta} - \sigma_{\Phi\Phi} + \sigma_{R\Theta}\cot\Theta\right) = \rho g \cos\Theta,$$

$$\frac{\partial \sigma_{R\Theta}}{\partial R} + \frac{1}{R}\frac{\partial \sigma_{\Theta\Theta}}{\partial \Theta} + \frac{1}{R}(\sigma_{\Theta\Theta} - \sigma_{\Phi\Phi})\cot\Theta + \frac{3}{R}\sigma_{R\Theta} = -\rho g \sin\Theta,$$

(2.10)

and now

$$\sigma_{RR} = -p + q\cos 2\Lambda, \qquad \sigma_{\Theta\Theta} = -p - q\cos 2\Lambda,$$

(2.11)

$$\sigma_{R\Theta} = q\sin 2\Lambda, \qquad \sigma_{\Phi\Phi} = -p - q,$$

with p and q once more given by (2.3) and where $\Lambda = \frac{1}{2}\pi - \Theta - \Psi$.

2.2. VELOCITY FIELD

It is assumed, for simplicity, that the flows are isochoric, which is a reasonable assumption for fully developed flows of dry granular material. It is possible, though more complicated, to extend the solutions to incorporate dilatancy in the form of the dilatant double-shearing model originally proposed by Mehrabadi and Cowin (1978).

In *plain strain* velocity components are denoted by $v_x, v_y, v_z = 0$ referred to Cartesian coordinates, and by $v_r, v_\theta, v_z = 0$ referred to cylindrical coordinates. Then the incompressibility condition can be expressed in either of the forms

$$\frac{\partial v_x}{\partial x} + \frac{\partial v_y}{\partial y} = 0,$$

(2.12)

or

$$\frac{\partial v_r}{\partial r} + \frac{v_r}{r} + \frac{1}{r}\frac{\partial v_\theta}{\partial \theta} = 0.$$

(2.13)

For *axially symmetric* flows the velocity components are denoted $v_r, v_z, v_\Phi = 0$ (cylindrical polar coordinates) or $v_R, v_\Theta, v_\Phi = 0$ (spherical polar coordinates) and the incompressibility condition is

$$\frac{\partial v_r}{\partial r} + \frac{v_r}{r} + \frac{\partial v_z}{\partial z} = 0,$$

(2.14)

or

$$\frac{\partial v_R}{\partial R} + \frac{1}{R}\frac{\partial v_\Theta}{\partial \Theta} + \frac{2v_R}{R} + \cot\Theta\frac{v_\Theta}{R} = 0.$$

(2.15)

A further equation is required in each case. The double-shearing non-dilatant theory is considered first. The formulation of this is described in detail in Spencer (1964, 1982, 1986) and we simply state the equations, which are as follows:

Plane Strain, Cartesian Coordinates

$$\left(\frac{\partial v_x}{\partial y} + \frac{\partial v_y}{\partial x}\right)\cos 2\psi - \left(\frac{\partial v_x}{\partial x} - \frac{\partial v_y}{\partial y}\right)\sin 2\psi$$

$$+ \sin\phi\left(\frac{\partial v_x}{\partial y} - \frac{\partial v_y}{\partial x} + 2\Omega\right) = 0, \qquad (2.16)$$

$$\Omega = \frac{\partial\psi}{\partial t} + v_x\frac{\partial\psi}{\partial x} + v_y\frac{\partial\psi}{\partial y}. \qquad (2.17)$$

Plane Strain, Plane Polar Coordinates

$$\left(\frac{1}{r}\frac{\partial v_r}{\partial\theta} + \frac{\partial v_\theta}{\partial r} - \frac{v_\theta}{r}\right)\cos 2\chi - \left(\frac{\partial v_r}{\partial r} - \frac{v_r}{r} - \frac{1}{r}\frac{\partial v_\theta}{\partial\theta}\right)\sin 2\chi$$

$$+ \sin\phi\left(\frac{1}{r}\frac{\partial v_r}{\partial\theta} - \frac{\partial v_\theta}{\partial r} - \frac{v_\theta}{r} + 2\Omega\right) = 0, \qquad (2.18)$$

$$\Omega = \frac{\partial\chi}{\partial t} + v_r\frac{\partial\chi}{\partial r} + v_\theta\frac{1}{r}\frac{\partial\chi}{\partial\theta}. \qquad (2.19)$$

Axial Symmetry, Cylindrical Polar Coordinates

$$\left(\frac{\partial v_r}{\partial z} + \frac{\partial v_z}{\partial r}\right)\cos 2\Psi - \left(\frac{\partial v_r}{\partial r} - \frac{\partial v_z}{\partial z}\right)\sin 2\Psi$$

$$+ \sin\phi\left(\frac{\partial v_r}{\partial z} - \frac{\partial v_z}{\partial r} + 2\Omega\right) = 0, \qquad (2.20)$$

$$\Omega = \frac{\partial\Psi}{\partial t} + v_r\frac{\partial\Psi}{\partial r} + v_z\frac{\partial\Psi}{\partial z}. \qquad (2.21)$$

Axial Symmetry, Spherical Polar Coordinates

$$\left(\frac{1}{R}\frac{\partial v_R}{\partial\Theta} + \frac{\partial v_\Theta}{\partial R} - \frac{v_\Theta}{R}\right)\cos 2\Lambda - \left(\frac{\partial v_R}{\partial R} - \frac{v_R}{R} - \frac{1}{R}\frac{\partial v_\Theta}{\partial\Theta}\right)\sin 2\Lambda$$

$$+ \sin\phi\left(\frac{1}{R}\frac{\partial v_R}{\partial\Theta} - \frac{\partial v_\Theta}{\partial R} + \frac{v_\Theta}{R} + 2\Omega\right) = 0, \quad (2.22)$$

$$\Omega = \frac{\partial\Lambda}{\partial t} + v_R\frac{\partial\Lambda}{\partial R} + v_\Theta\frac{1}{R}\frac{\partial\Lambda}{\partial\Theta}. \quad (2.23)$$

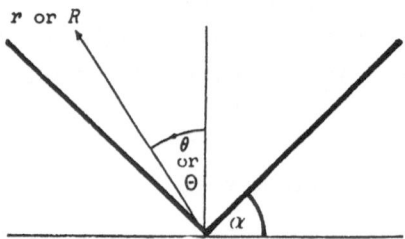

Figure 1.   Radial flow in a vertical wedge or cone.

In each case Ω represents the spin of the principal axes of stress at a material particle.

In the alternative coaxial theory it is simply assumed that the principal axes of stress coincide with the principal axes of the rate-of-deformation tensor. Then the equations in the various cases are obtained by setting $\sin \phi = 0$ in (2.16), (2.18), (2.20) and (2.22) respectively. The spin Ω does not enter the coaxial theory.

3. Steady Converging Radial Flow in Wedges and Cones

Consider the configuration illustrated in Figure 1, which represents either a wedge shaped region in plane strain, bounded by the planes $\theta = \pm \alpha$, or a conical region in axial symmetry, bounded by the cone $\Theta = \alpha$. In either case granular material flows in the radial direction towards the apex, under the influence of gravity acting towards the apex in the direction of the channel axis.

3.1. RADIAL FLOW IN A WEDGE

In the case of a wedge, Jenike (1964) showed that there exist solutions of (2.4) and (2.5) for the stress field of the form

$$\chi = \chi(\theta), \qquad q = \rho g r F(\theta). \tag{3.1}$$

When χ and q are determined, the stress is then given by (2.6).Substituting (3.1) into (2.4) and (2.5) gives, after rearrangement

$$\frac{dF}{d\theta} = \frac{F \sin 2\chi + \sin \phi \sin(2\chi + \theta)}{\cos 2\chi + \sin \phi}, \tag{3.2}$$

$$\frac{d\chi}{d\theta} = \frac{F \{\operatorname{cosec}\phi - 3 \sin \phi - 2 \cos 2\chi\} + \cos \theta + \sin \phi \cos(2\chi + \theta)}{2F(\cos 2\chi + \sin \phi)}. \tag{3.3}$$

It is assumed that the stress field is symmetrical about $\theta = 0$. The symmetry condition is

$$\chi = 0 \qquad \text{at } \theta = 0. \tag{3.4}$$

Various wall friction conditions can be treated, but for definiteness it is assumed that the wall is 'perfectly rough' so that the material slips on itself at the walls. This leads to the condition (Spencer and Bradley, 1996)

$$\chi = \tfrac{1}{4}\pi + \tfrac{1}{4}\phi \qquad \text{at } \theta = \alpha. \tag{3.5}$$

Analytical integration of (3.2) and (3.3) subject to (3.4) and (3.5) does not seem to be possible, but numerical solution is straightforward. This was done by Jenike (1964) and his results have been confirmed by Bradley (1991).

Jenike (1964) also considered the associated radial flow field, using the coaxial theory. He sought solutions of the form

$$v_r = \frac{1}{r}v(\theta), \qquad v_\theta = 0, \tag{3.6}$$

for which (2.13) is trivially satisfied and the coaxiality condition is

$$\frac{dv}{d\theta}\cos 2\chi + 2v\sin 2\chi = 0, \tag{3.7}$$

which has the solution

$$v = v_0 \exp\left\{-2\int_0^\theta \tan 2\chi \; d\theta\right\}. \tag{3.8}$$

Hence v can be calculated using the values of χ obtained in the stress solution.

If, on the other hand, the double shearing theory is employed, then (2.18) gives, in place of (3.7), for flows of the form (3.6)

$$\frac{dv}{d\theta}(\cos 2\chi + \sin \phi) + 2v\sin 2\chi = 0, \tag{3.9}$$

which has the solution

$$v = v_0 \exp\left\{-2\int_0^\theta \frac{\sin 2\chi \; d\theta}{\cos 2\chi + \sin \phi}\right\}. \tag{3.10}$$

Graphs showing the variation of $v(\theta)/v_0$ with θ, using both coaxial theory and double-shearing theory, are shown in Figure 2. Further details are given in Bradley (1991) and Spencer and Bradley (1996); the graphs shown are

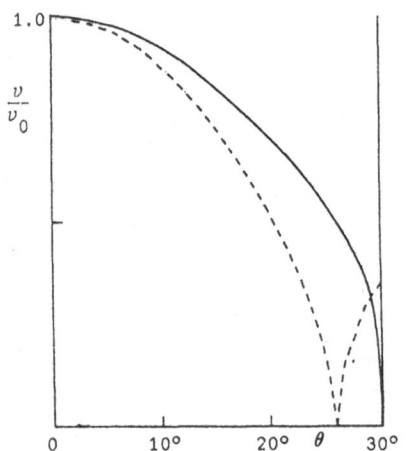
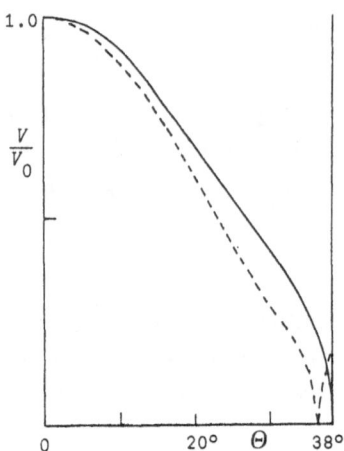

Figure 2. Profile of radial velocity in a wedge or cone with perfectly rough walls, for double shearing theory (solid line) and coaxial theory (broken line) Left: wedge, $\alpha = 30°$, $\phi = 30°$. Right: cone, $\alpha = 38°$, $\phi = 15°$.

typical and other values of ϕ and α give qualitatively similar behaviour. For the coaxial theory, similar velocity profiles were obtained by Jenike (1964). The important feature is that whereas the double-shearing theory predicts a smooth velocity profile between $\theta = 0$ and $\theta = \alpha$, the coaxial theory predicts that the radial velocity decreases to zero on a plane on which $\chi = \frac{1}{4}\pi$, and then increases. Since χ increases continuously from zero at $\theta = 0$ to $\frac{1}{4}\pi + \frac{1}{2}\phi$ at $\theta = \alpha$ when the wall is perfectly rough, there is always a radial plane within the wedge on which the velocity is zero.

3.2. RADIAL FLOW IN A CONE

Radial flow in a cone was also investigated by Jenike (1964). In this case the stress is given by (2.4) and (2.10) with Λ and q of the forms

$$\Lambda = \Lambda(\Theta), \quad q = \rho g R G(\Theta). \tag{3.11}$$

Inserting these in the equilibrium equations leads to the pair of ordinary differential equations

$$(\cos 2\Lambda + \sin \phi)\frac{dG}{d\Theta} = \sin \phi \sin(2\Lambda + \Theta)$$

$$+ 2\sin \Lambda \{\cos \Lambda - \sin \phi \operatorname{cosec} \Theta \sin(\Theta + \Lambda)\} G,$$

$$\tag{3.12}$$

$$2G(\cos 2\Lambda + \sin \phi)\frac{d\Lambda}{d\Theta} = \cos \Theta + \sin \phi \cos(2\Lambda + \Theta)$$

$$+ \{\operatorname{cosec} \phi - 4 \sin \phi - 1 - 2 \cos 2\Lambda - (1 + \sin \phi)\operatorname{cosec} \Theta \sin(2\Lambda + \Theta)\} G.$$

$$(3.13)$$

For axially symmetric solutions and perfectly rough walls the boundary conditions are

$$\Lambda = 0, \quad \Theta = 0 \quad \text{and} \quad \Lambda = \tfrac{1}{4}\pi + \tfrac{1}{2}\phi, \quad \Theta = \alpha. \qquad (3.14)$$

As in the case of a wedge, numerical solution of (3.12) and (3.13) is required, and was effected by Jenike (1964) and Bradley (1991). However it is interesting to note that (3.12) and (3.13) possess the singular solution

$$\Lambda = -\Theta + \tfrac{1}{2}\pi - \gamma, \quad G = \frac{\sin \Theta}{2 \sin 2\gamma}, \quad \text{where} \quad \cos 2\gamma = \frac{1 - \sin \phi}{2 \sin \phi}, \quad 0 < \gamma < \tfrac{1}{4}\pi.$$

$$(3.15)$$

The significance of this solution is outlined in Spencer and Bradley (1996). The associated radial flow field is of the form (Jenike, 1964)

$$v_R = \frac{V(\Theta)}{R^2}, \qquad v_\Theta = 0. \qquad (3.16)$$

For the coaxial theory, it follows that

$$\cos 2\Lambda \frac{dV}{d\Theta} + 3V \sin 2\Lambda = 0,$$

and hence

$$V = V_0 \exp\left\{-3\int_0^\Theta \tan 2\Lambda \, d\Theta\right\}, \qquad (3.17)$$

whereas, using the double-shearing theory, from (2.22),

$$(\cos 2\Lambda + \sin \phi)\frac{dV}{d\Theta} + 3V \sin 2\Lambda = 0,$$

and

$$V = V_0 \exp\left\{-3\int_0^\Theta \frac{\sin 2\Lambda \, d\Theta}{\cos 2\Lambda + \sin \phi}\right\}. \qquad (3.18)$$

Some typical velocity profiles are shown in Figure 2. Similarly to the case of a wedge, the coaxial theory predicts that the velocity is zero on the cone on which $\Lambda = \tfrac{1}{4}\pi$.

4. Compression Between Vertical Plates and Cylinders

The following two problems are considered. In plane geometry, $x = \pm a$ are vertical walls bounding granular material, which is compressed by the walls moving horizontally inwards with speed U. Details of the solution are given in Spencer and Bradley (1992). In cylindrical geometry, $r = a$ is a circular cylinder containing granular material which is compressed by contraction of the cylinder, with its surface moving inwards with speed U, and details of this solution are in Spencer and Bradley (1997). In both cases the walls are taken to be perfectly rough.

4.1. COMPRESSION BETWEEN VERTICAL WALLS

In the case of vertical walls, the relevant stress solution of (2.1) and (2.4) is

$$q = \frac{\rho g a}{\cos \phi} \left\{ 1 - \left(1 - \frac{x^2}{a^2} \right)^{\frac{1}{2}} \sin \phi \right\}, \tag{4.1}$$

$$\tan 2\psi = \frac{x \cos \phi}{a \sin \phi - (a^2 - x^2)^{\frac{1}{2}}}, \tag{4.2}$$

and for symmetrical solutions with perfectly rough walls, ψ decreases continuously from $\frac{1}{2}\pi$ at $x = 0$ to $\frac{1}{4}\pi - \frac{1}{2}\phi$ at $x = a$.

For the coaxial theory, the velocity equations are (2.12) and the coaxiality condition. With ψ given by (4.2) and the boundary condition $v_x = -U$ at $x = a$, the solution is

$$v_x = -\frac{Ux}{a},$$

$$v_y = \frac{Uy}{a} - 2U \cos \phi \left[\frac{(a^2 - x^2)^{\frac{1}{2}}}{a} - \sin \phi \ln \left\{ \sin \phi - \frac{(a^2 - x^2)^{\frac{1}{2}}}{a} \right\} \right] - V, \tag{4.3}$$

where V is a rigid-body translatory motion. Thus v_y has a logarithmic singularity at $x = a \cos \phi$, which for $\phi > 0$ lies within the body of granular material. Also, from (4.2), $x = a \cos \phi$ is the value of x at which $\psi = \frac{1}{4}\pi$.

On the other hand, using the double-shearing equation (2.16) rather than the coaxiality condition, it was shown in Spencer and Bradley (1992) that the corresponding velocity solution is

$$v_x = -\frac{Ux}{a}, \qquad v_y = \frac{Uy}{a} - \frac{2U(a^2 - x^2)^{\frac{1}{2}}}{a \cos \phi} - V, \tag{4.4}$$

and thus the double-shearing theory predicts smooth velocity fields.

Figure 3 shows typical profiles of v_y according to both coaxial and double-shearing theories.

4.2. LATERAL COMPRESSION OF A VERTICAL CYLINDER

In the problem of compression of a circular cylinder, the relevant stress solution is determined by

$$q = \tfrac{1}{2}\rho g r \,\mathrm{cosec}\, 2\Psi, \tag{4.5}$$

and $\tan \Psi$ is given implicitly by

$$\frac{r}{r_0} = \frac{\tan^{2n-1}\Psi}{\{\tan^2\Psi(1 + \sin\phi) + (1 - 3\sin\phi)\}^n}, \qquad n = \frac{1 - 2\sin\phi}{1 - 3\sin\phi}. \tag{4.6}$$

The 'perfectly rough' wall condition at $r = a$ is

$$\tan\Psi = \frac{1 - \sin\phi}{\cos\phi}, \quad \text{or} \quad \Psi = \tfrac{1}{4}\pi - \tfrac{1}{2}\phi \tag{4.7}$$

and this determines the constant r_0.

For the coaxial theory the velocity solution is (Spencer and Bradley, 1997)

$$v_r = -\frac{Ur}{a}, \qquad v_z = \frac{2Uz}{a} - \frac{3U}{a}\int \tan 2\Psi \, dr - W, \tag{4.8}$$

where W represents a rigid body motion. Explicit expressions can be obtained for v_z when n is an integer, but it is sufficient here to note that (4.8) has logarithmic singularities at $\Psi = \pm\tfrac{1}{4}\pi$, and that with the perfectly rough wall condition, Ψ decreases continuously from $\tfrac{1}{2}\pi$ at $r = 0$ to $\tfrac{1}{4}\pi - \tfrac{1}{2}\phi$ at $r = a$, and so takes the value $\tfrac{1}{4}\pi$ on a surface inside the cylinder.

In the double-shearing theory the appropriate solution is (Spencer and Bradley, 1997)

$$v_r = -\frac{Ur}{a}, \qquad v_z = \frac{2Uz}{a} - \frac{3Ur\tan\Psi}{a(1 - 2\sin\phi)} - W, \tag{4.9}$$

which is smooth and plausible for all values of r. Typical velocity profiles are shown in Figure 3.

It is concluded that, in all four of the problems discussed above, the coaxial theory leads to anomalous and physically unreasonable predictions, whereas the double-shearing yields entirely plausible results. It is suggested that one or more of these solutions might form a basis for experimental test of some of the many theories of granular material behaviour.

Acknowledgements

The author thanks the University of Canterbury for an Erskine Fellowship and the Leverhulme Foundation for an Emeritus Fellowship.

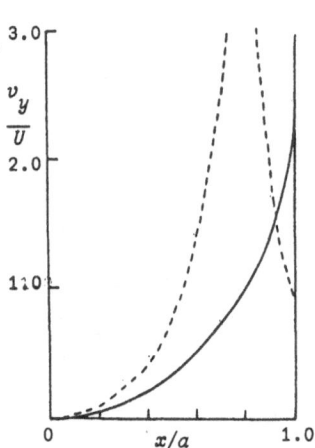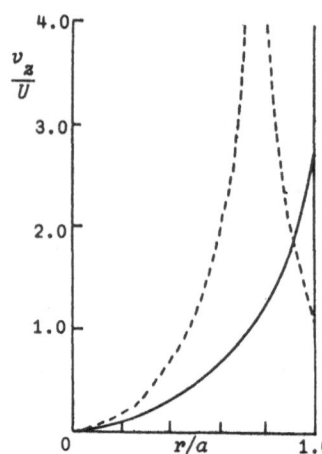

Figure 3. Profile of vertical velocity in compression between rough vertical walls or of a rough vertical cylinder, for double-shearing theory (solid line) and coaxial theory (broken line). Left: vertical walls, $\phi = 30°$. Right: cylinder, $\phi = 24°$

References

Bradley, N.J. (1991) Gravity flows of granular materials, PhD Thesis, University of Nottingham.

Hill, R. (1950), *The Mathematical Theory of Plasticity*, Clarendon Press, Oxford.

Jenike, A.W. (1964) Steady gravity flow of frictional-cohesive solids in converging channels, *J. Appl. Mech.***31**, 5-11.

Mehrabadi, M.M. and Cowin, S.C. (1978) Initial planar deformation of dilatant granular materials, *J. Mech. Phys. Solids* **26**, 269-284.

Spencer, A.J.M. (1964) A theory of the kinematics of ideal soils under plane strain conditions, *J. Mech. Phys. Solids* **12**, 337-351.

Spencer, A.J.M. (1982) Deformation of ideal granular materials, in H.G. Hopkins and M.J. Sewell (eds), *Mechanics of Solids*, Pergamon Press, Oxford and New York, pp.607-652.

Spencer, A.J.M. (1986) Axially symmetric flow of granular materials, *Solid Mechanics Archives* **11**, 185-198.

Spencer, A.J.M. and Bradley, N.J. (1992) Gravity flow of a granular material in compression between vertical walls and through a tapering vertical channel, *Q. Jl. Mech. Appl. Math.* **45**. 733-746.

Spencer, A.J.M. and Bradley, N.J. (1996) Gravity flow of granular materials in converging wedges and cones, in K.Z. Markov (ed.) *Continuum Models and Discrete Systems*, World Scientific Publishing Co., Singapore, pp.581-590.

Spencer, A.J.M. and Bradley, N.J. (1997) Gravity flow of granular material in contracting cylinders and tapering tubes. To be published.

MODELLING MATHEMATICALLY THE FLOW OF GRANULAR MATERIALS

D HARRIS
Department of Mathematics, UMIST
PO Box 88, Manchester M60 1QD

1. Introduction

In this paper a comparison is made between a discrete Newtonian many rigid body model of a confined granular material and two types of continuum plasticity model. On the basis of this comparison the rigid-body constraint in the discrete model is used to motivate and formulate a kinematic rule, proposed previously in Harris (1995a,b), which states that the relative velocity of two infinitesimally separated material particles is a superposition of dilatant shears in certain slip directions together with a local rotation of those slip directions. Continuum plasticity models, for example the plastic potential model, the double-shearing model proposed in Spencer (1964) and the double-sliding free-rotating model proposed in de Josselin de Jong (1959), may be described in terms of simultaneous slip on two such directions.

The kinematic rule may also be used to generate new models, including models of anisotropic behaviour, and examples of this are presented. The characteristic directions are established.

Two paradigms for continuum plasticity models are presented: (a) a discrete Newtonian many rigid-body model; (b) the structure of the continuum equations governing sliding on a single family of surfaces in the presence of dry Coulomb friction. We investigate existing models in the light of these paradigms. Underpinning our analysis is the notion that the closer the properties of the continuum model are to those indicated by (a) and (b) then the better the continuum model. A single slip model is proposed in Geniev (1958).

It is shown how the kinematic rule may be used as a basis for a method of validating or refuting continuum plasticity models and the models mentioned above are investigated from this point of view. It is shown that the

239

N. A. Fleck and A. C. F. Cocks (eds.),
IUTAM Symposium on Mechanics of Granular and Porous Materials, 239–250.
© 1997 *Kluwer Academic Publishers.*

widely used plastic potential model has two inadequacies: (i) for a non-associated flow rule the slip directions do not correspond to those expected by analogy with a rigid-body constraint, (ii) rotation of the slip directions is not adequately incorporated. The dilatant double-shearing model proposed in Mehrabadi and Cowin (1978), is also shown to suffer from disadvantage (i) but, for this model, a method of remedying this defect is presented. Finally, this paper together with Harris (1993) show that the plastic potential model may be transformed into a modified double-shearing model which satisfies the rigid-body constraint and allows proper rotation of the slip directions. A generalisation of the double-shearing model similar to that considered here is proposed in Anand (1983).

2. Discrete and Continuum Models

Consider a region R, occupied by granular material, of volume V in three-dimensional Euclidean space $\mathbf{R}^3$ with boundary ∂R. Let F denote an inertial frame of reference, with origin O and axes Ox_μ, $1 \leq \mu \leq 3$, and G_i, $1 \leq i \leq n$, denote a family of non-inertial frames with origins O_i and axes $O_i y_\mu$.

In the discrete Newtonian model R is occupied by $n \in \mathbf{N}$ non-overlapping rigid bodies (grains), B_i, $1 \leq i \leq n$. Let $\mathbf{r}_i$, $\mathbf{v}_i$ denote the position, velocity vector of the centre of mass of B_i, m_i its mass, $\mathbf{I}_i$ its moment of inertia tensor and let $\mathbf{h}_i = \mathbf{I}_i \boldsymbol{\omega}_i$ denote the angular momentum, where $\boldsymbol{\omega}_i$ denotes the angular velocity of grain i. Let O_i be translating with velocity $\mathbf{V}_i$ relative to O and let $O_i y_\mu$ rotate with angular velocity Ω_i relative to Ox_μ. The equations of motion of grain i expressed relative to frame G_i are

$$m_i \frac{d\mathbf{v}_i}{dt} + m_i \left(\Omega_i \times \mathbf{v}_i \right) = \mathbf{F}_i^e + \mathbf{F}_i^b + \sum_{k=1}^{c_i} \mathbf{F}_{ij_k}^c \tag{1}$$

$$\frac{d\mathbf{h}_i}{dt} + \Omega_i \times \mathbf{h}_i + m_i \left(\mathbf{V}_i \times \mathbf{v}_i \right) = \mathbf{r}_i^e \times \mathbf{F}_i^e + \mathbf{r}_i^b \times \mathbf{F}_i^b + \sum_{k=1}^{c_i} \mathbf{r}_i^c \times \mathbf{F}_{ij_k}^c \tag{2}$$

where the time derivative is relative to the frame $O_i y_\mu$, and $\mathbf{r}_i^e$, $\mathbf{r}_i^b$, $\mathbf{r}_i^c$ denote the position vectors of the points of application of the external force $\mathbf{F}_i^e$, body force $\mathbf{F}_i^b$ and contact force $\mathbf{F}_{ij_k}^c$ respectively. c_i denotes the number of contact points for grain i. The contact forces satisfy

$$\mathbf{F}_{ij_k}^c + \mathbf{F}_{j_k i}^c = 0$$

and may be decomposed parallel and perpendicular to the tangent plane at the point of contact. Let $\mathbf{n}_{ij_k}^c$ denote the outward normal to grain i at its point of contact with grain j_k then

$$\mathbf{F}_{ij_k}^c = N_{ij_k}^c \mathbf{n}_{ij_k}^c + T_{ij_k}^c \mathbf{t}_{ij_k}^c$$

where $\mathbf{t}_{ij_k}^c$ denotes the tangent vector at the contact point in the plane defined by $\mathbf{F}_{ij_k}^c$ and $\mathbf{n}_{ij_k}^c$. Assume a dry friction law $f\left(N_{ij_k}^c, T_{ij_k}^c\right) \leq 0$ at the contact points in the form of Coulomb's law,

$$|T_{ij_k}^c| \leq \mu_{ij_k} N_{ij_k}^c \tag{3}$$

where μ_{ij_k} denotes the usual coefficient of friction between grains i and j_k. Finally, let $\mathbf{v}_{ij_k}^c$, $\mathbf{r}_{ij_k}^c$ denote the contact point velocity, position vector of grain j_k relative to grain i,

$$\mathbf{v}_{ij_k}^c = \mathbf{v}_{j_k} + \boldsymbol{\omega}_{j_k} \times \mathbf{r}_{j_k i}^c - \mathbf{v}_i - \boldsymbol{\omega}_i \times \mathbf{r}_{ij_k}^c,$$

then the rigid-body constraint that grains must not overlap may be written

$$\mathbf{v}_{ij_k}^c \cdot \mathbf{n}_{ij_k}^c \geq 0, \tag{4}$$

together with the fact that $T_{ij_k}^c \mathbf{t}_{ij_k}^c$ and $\mathbf{v}_{ij_k}^c$ share the same line of action and are oppositely directed,

$$\mathbf{v}_{ij_k}^c = -\lambda T_{ij_k}^c \mathbf{t}_{ij_k}^c,$$

where λ is a positive multiplier. The discrete model is completed by specifying the initial configuration of the grains, i.e. the set of position vectors of the centres of mass of the grains together with their orientation, and the initial velocities of the centre of mass together with the initial angular velocity of the grains. Both of these must satisfy the condition of non-overlap of the grains.

In a plasticity model R is occupied by a continuum on which are defined the density $\rho(\mathbf{x},t)$, Cauchy stress $\boldsymbol{\sigma}(\mathbf{x},t)$ and Eulerian velocity vector $\mathbf{v}(\mathbf{x},t)$ fields. The equations of motion may be written

$$\nabla_{\mathbf{x}} \cdot \boldsymbol{\sigma}(\mathbf{x},t) + \mathbf{F}(\mathbf{x},t) = \rho(\mathbf{x},t)\left[\frac{\partial \mathbf{v}(\mathbf{x},t)}{\partial t} + \mathbf{v}(\mathbf{x},t) \cdot \nabla_{\mathbf{x}} \mathbf{v}(\mathbf{x},t)\right]. \tag{5}$$

$$\boldsymbol{\sigma}(\mathbf{x},t) = \boldsymbol{\sigma}^T(\mathbf{x},t) \tag{6}$$

where $\mathbf{F}$ denotes the body force and the superscript T denotes transpose. We deliberately refrain from making the usual plasticity assumption of quasi-static motion. A yield condition of the form

$$f(\boldsymbol{\sigma}(\mathbf{x},t)) \leq 0 \tag{7}$$

is assumed.

The velocity field is determined by a flow rule, constitutive equation or constraint equation

$$\mathbf{d} = g(\boldsymbol{\sigma}, \varsigma), \tag{8}$$

relating the the deformation-rate tensor $\mathbf{d}$ to the stress $\boldsymbol{\sigma}$ and fabric tensor ς. The form of g is given in later sections. The model is completed by a set of initial and boundary values, namely the initial values $\boldsymbol{\sigma}(\mathbf{x}, 0)$, $\mathbf{v}(\mathbf{x}, 0)$ and values of $\boldsymbol{\sigma}$, $\mathbf{v}$ on ∂R.

In comparing the two models equations (5), (6), (7) in the continuum model correspond to equations (1), (2), (3) respectively in the discrete. It can also be seen that the flow rule/constitutive/constraint equation (8) in the continuum model corresponds to the condition (4) of no overlap of grains in the discrete. In the next section we pursue this point further.

3. The infinitesimal constraint equation

Figure 1 shows two grains B_i, B_j such that the point R_i of B_i is in contact with the point R_j of B_j. Let P_i, P_j denote arbitrary material points in B_i, B_j respectively. For any two points P and Q let ${}_P\mathbf{v}_Q$, denote the velocity, of Q relative to P, ${}_P\mathbf{s}_Q$ the position vector of Q relative to P then

$$P_i \mathbf{v}_{P_j} = \boldsymbol{\omega}_j \times \left({}_{R_j}\mathbf{s}_{P_j}\right) + \left({}_{R_i}\mathbf{v}_{R_j}\right) + \boldsymbol{\omega}_i \times \left({}_{P_i}\mathbf{s}_{R_i}\right),$$

where $\boldsymbol{\omega}_i$, $\boldsymbol{\omega}_j$ denote the angular velocities of B_i, B_j respectively. If $R_j P_j$ and $P_i R_i$ are instantaneously collinear with the contact point dividing $P_i P_j$ internally in the ratio $\lambda_i : \lambda_j$ then

$$P_i \mathbf{v}_{P_j} = (\lambda_i \boldsymbol{\omega}_i + \lambda_j \boldsymbol{\omega}_j) \times \left({}_{P_i}\mathbf{s}_{P_j}\right) + \left({}_{R_i}\mathbf{v}_{R_j}\right),$$

where $\lambda_i + \lambda_j = 1$. The velocity vector ${}_{R_i}\mathbf{v}_{R_j}$ may be decomposed as

$$R_i \mathbf{v}_{R_j} = k_{ij}^c \mathbf{n}_{ij}^c + l_{ij}^c \mathbf{t}_{ij}^c$$

and hence

$$P_i \mathbf{v}_{P_j} = (\lambda_i \boldsymbol{\omega}_i + \lambda_j \boldsymbol{\omega}_j) \times \left({}_{P_i}\mathbf{s}_{P_j}\right) + k_{ij}^c \mathbf{n}_{ij}^c + l_{ij}^c \mathbf{t}_{ij}^c. \tag{9}$$

Now consider a continuum model and take two material points P and Q infinitesimally separated. Let $\mathbf{dv}_P^Q$, $\mathbf{ds}_P^Q$ denote the velocity and position vectors of Q relative to P. Each material point consists of many grains and hence the properties of grains are averaged in the continuum model and become pointwise defined properties which are smoother than in the discrete model. Thus we shall suppose that $\boldsymbol{\omega}_i$, $\boldsymbol{\omega}_j$ in equation (9) become $\boldsymbol{\Omega}$, $\boldsymbol{\Omega} + \mathbf{d\Omega}$, respectively. Hence,

$$\mathbf{dv}_P^Q = \boldsymbol{\Omega} \times \mathbf{ds}_P^Q + k\mathbf{n} + l\mathbf{t}, \tag{10}$$

correct to first order, where $\mathbf{n}$ is determined by the local fabric and $\mathbf{t}$ by the local slip direction. Each point of the continuum is a contact point and a slip surface passes through it. Thus the local motion consists of a rotation of the slip surface accompanied by a translational relative motion of the material either side of the slip surface. In fact, existing models of granular materials do not use the term $k\mathbf{n}$ and so we shall take $k = 0$. In a confined granular material, contacts will usually be made and lost only as grains approach and recede from each other, i.e. they meet or separate as their relative tangential velocity brings them into contact or separates them; they will not spring apart due to a relative normal velocity. However, we cannot infer from this that $k = 0$, merely that $\mathbf{\Omega} \times \mathbf{ds}_P^Q + k\mathbf{n}$ is such that contact is maintained in the ensuing motion. It does imply, however, that the term $k\mathbf{n}$ is associated with grain rotation.

4. Single slip material

Our first application of the infinitesimal constraint equation is to a fictitious material which possesses one family of slip surfaces through each point. We shall motivate the continuum model by appeal to an associated discrete model, see figure 2, in which the finite grain-size induces dilatation/compression as the grains on neighbouring surfaces ride over one another. The surfaces may be envisaged as physical surfaces with the grains attached (e.g. a "bumpy" pack of cards) or as cracks (e.g. jointed rock material). The continuum model is then obtained by letting the grain size tend to zero, the number of grains tend to infinity while maintaining geometrical and packing similarity.

4.1. KINEMATIC EQUATIONS

Let the normal to the macroscopic slip surface (i.e. that obtained by ignoring the grain-size) be denoted by $\mathbf{n}_\gamma$, the normal to the grain on grain interface be denoted by $\mathbf{n}_\beta$ and let the tangent to this interface in the direction of the motion be denoted by $\mathbf{t}_\beta$ then write the infinitesimal constraint equation (10) as

$$\mathbf{dv}_P^Q = \mathbf{\Omega} \times \mathbf{ds}_P^Q + k_\gamma \left(\mathbf{n}_\gamma \cdot \mathbf{ds}_P^Q \right) \mathbf{t}_\beta. \tag{11}$$

where the coefficient of $\mathbf{t}_\beta$ is assumed to be proportional to the projection of $\mathbf{ds}_P^Q$ onto the normal to the slip surface. Equation (11) may be written

$$\mathbf{dv}_P^Q = [\mathbf{\Omega} + k_\gamma (\mathbf{t}_\beta \otimes \mathbf{n}_\gamma)] \cdot \mathbf{ds}_P^Q. \tag{12}$$

Let v_i denote the Eulerian velocity components, define the velocity gradient $\mathbf{\Gamma}$ as the tensor with components

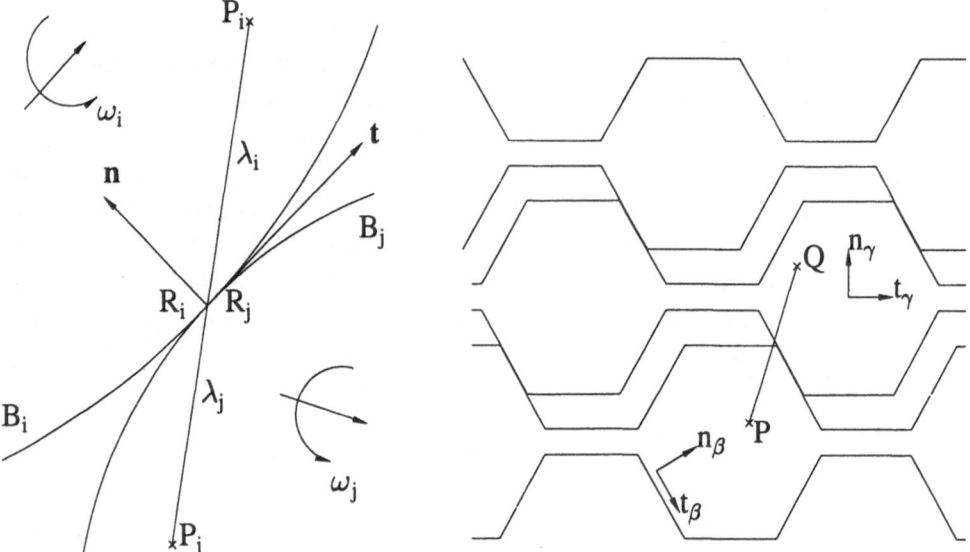

Figure 1: Rigid body constraint   Figure 2: Material with single slip family

$$\Gamma_{ij} = \frac{\partial v_i}{\partial x_j}$$

and let $\boldsymbol{\omega}$ (the spin tensor) denote the anti-symmetric part of $\boldsymbol{\Gamma}$. Then

$$\mathbf{dv}_P^Q = \boldsymbol{\Gamma} \cdot \mathbf{ds}_P^Q,$$

i.e.

$$\boldsymbol{\Gamma} = \boldsymbol{\Omega} + k_\gamma \left(\mathbf{t}_\beta \otimes \mathbf{n}_\gamma \right).$$

The dilatation rate is given by

$$tr\mathbf{d} = k_\gamma \mathbf{t}_\beta \cdot \mathbf{n}_\gamma,$$

the spin tensor by

$$\boldsymbol{\omega} = \boldsymbol{\Omega} + \frac{1}{2} k_\gamma \left[\mathbf{t}_\beta \otimes \mathbf{n}_\gamma - \mathbf{n}_\gamma \otimes \mathbf{t}_\beta \right]$$

and the deformation-rate tensor by

$$\mathbf{d} = \frac{1}{2} k_\gamma \left[\mathbf{t}_\beta \otimes \mathbf{n}_\gamma + \mathbf{n}_\gamma \otimes \mathbf{t}_\beta \right]. \tag{13}$$

Consider a plane deformation (assumed to take place in the Ox_1x_2-plane) and define angles θ_β, θ_γ by

$$\mathbf{t}_\beta = (\cos\theta_\beta, \sin\theta_\beta), \tag{14}$$

$$\mathbf{n}_\gamma = (-\sin\theta_\gamma, \cos\theta_\gamma). \tag{15}$$

From equations (13), (14) and (15) the planar equations may be written

$$(d_{11} - d_{22})\cos(\theta_\beta + \theta_\gamma) + 2d_{12}\sin(\theta_\beta + \theta_\gamma) = 0, \tag{16}$$

$$d_{11} + d_{22} = \sin(\theta_\beta - \theta_\gamma)\left[-(d_{11} - d_{22})\sin(\theta_\beta + \theta_\gamma) + 2d_{12}\cos(\theta_\beta + \theta_\gamma)\right] \tag{17}$$

$$k_\gamma = -(d_{11} - d_{22})\sin(\theta_\beta + \theta_\gamma) + 2d_{12}\cos(\theta_\beta + \theta_\gamma) \tag{18}$$

$$2(\omega_{12} + \Omega_{12}) = \cos(\theta_\beta - \theta_\gamma)\left[-(d_{11} - d_{22})\sin(\theta_\beta + \theta_\gamma) + 2d_{12}\cos(\theta_\beta + \theta_\gamma)\right] \tag{19}$$

where d_{ij}, ω_{ij} denote the components of $\mathbf{d}$ and $\boldsymbol{\omega}$. Equations (16), (17) are two equations for the two unknowns v_1, v_2 and equations (18), (19) determine k_γ, Ω_{12}.

4.2. STRESS EQUATIONS

Turning now to the stress field we suppose that a Coulomb type yield condition is satisfied on the β-surface, i.e.

$$|\tau_\beta| \leq \sigma_\beta \tan\varphi_\beta + c_\beta \tag{20}$$

where σ_β, τ_β denote the normal, tangential traction across the β-surface, and φ_β, c_β denote the angle of internal friction, cohesion associated with the β-surface. The Ox_3-axis is the intermediate principal stress direction. Define the usual planar stress variables

$$p_\sigma = -\frac{1}{2}(\sigma_{11} + \sigma_{22}), \qquad q_\sigma = \frac{1}{2}\left[(\sigma_{11} - \sigma_{22})^2 + 4\sigma_{12}^2\right]^{\frac{1}{2}},$$

$$\psi_\sigma = \frac{1}{2}\arctan\left(\frac{2\sigma_{12}}{\sigma_{11} - \sigma_{22}}\right).$$

where ψ_σ is the angle that the algebraically greater principal stress direction makes in the Ox_1x_2-plane with the x_1-axis. The yield condition (20) may be written

$$\pm q_\sigma \sin[2(\psi_\sigma - \theta_\beta) \mp \varphi_\beta] \leq p_\sigma \sin\varphi_\beta + c_\beta \cos\varphi_\beta, \tag{21}$$

where here, and in future formulae, the upper sign corresponds to $\tau_\beta > 0$, the lower sign to $\tau_\beta < 0$. The stress equations of motion are

$$
\begin{aligned}
&(a\cos 2\psi_\sigma - 1)\, p_{\sigma,1} + (a\sin 2\psi_\sigma)\, p_{\sigma,2} + (b\cos 2\psi_\sigma - 2q_\sigma \sin 2\psi_\sigma)\, \psi_{\sigma,1} \\
&+ (b\sin 2\psi_\sigma + 2q_\sigma \cos 2\psi_\sigma)\, \psi_{\sigma,2} + \rho F_1 = \rho\left[v_{1,t} + v_1\left(v_{1,1}\right) + v_2\left(v_{1,2}\right)\right],
\end{aligned}
$$

$$
\begin{aligned}
&(a\sin 2\psi_\sigma)\, p_{\sigma,1} + (a\cos 2\psi_\sigma + 1)\, p_{\sigma,2} + (2q_\sigma \cos 2\psi_\sigma + b\sin 2\psi_\sigma)\, \psi_{\sigma,1} \\
&+ (2q_\sigma \sin 2\psi_\sigma - b\cos 2\psi_\sigma)\, \psi_{\sigma,2} + \rho F_2 = \rho\left[v_{2,t} + v_1\left(v_{2,1}\right) + v_2\left(v_{2,2}\right)\right],
\end{aligned}
$$

where

$$
a = \pm \frac{\sin \varphi_\beta}{\sin\left[2\left(\psi_\sigma - \theta_\beta\right) - \varphi_\beta\right]},
$$

$$
b = \pm 2q_\sigma \cot\left[2\left(\psi_\sigma - \theta_\beta\right) - \varphi_\beta\right].
$$

4.3. CHARACTERISTIC DIRECTIONS

Using standard methods, see for example Harris (1989,1994), the velocity characteristic equation is

$$
(\sin\theta + \sin\nu)\, m^2 + (2\cos\theta)\, m - (\sin\theta - \sin\nu) = 0
$$

where

$$
\theta = \theta_\beta + \theta_\gamma, \qquad \nu = \theta_\beta - \theta_\gamma.
$$

The roots of this equation give the characteristic directions

$$
m_1 = \tan\theta_\gamma, \qquad m_2 = \tan\left(\theta_\beta + \frac{1}{2}\pi\right),
$$

i.e. the macroscopic slip direction and the perpendicular to the microscopic slip (dilatancy) direction. Similarly, the stress characteristic equation is

$$
\begin{aligned}
&(2aq_\sigma - 2q_\sigma \cos 2\psi_\sigma - b\sin 2\psi_\sigma)\, m^2 - 2\left(b\cos 2\psi_\sigma - 2q_\sigma \sin 2\psi_\sigma\right) m \\
&+ (2aq_\sigma + 2q_\sigma \cos 2\psi_\sigma + b\sin 2\psi_\sigma) = 0
\end{aligned}
$$

and the roots of this equation give the stress characteristic directions

$$
m_3 = \tan\left(\theta_\beta \pm \varphi_\beta + \frac{1}{2}\pi\right), \qquad m_4 = \tan\theta_\beta,
$$

i.e. an angle φ_β either side of the perpendicular to the microscopic slip direction and the microscopic slip direction. See figure 3(a).

Thus, the characteristic directions are the carriers of the information required to specify the model and there is a one-to-one correspondence

between the number of characteristic directions and the amount of information, namely the frictional and dilatational properties of the model and
the fixed relationship between the yielding and sliding directions.

The model may be extended to the case where there is more than one
family of slip surfaces. In order to prevent violation of the non-overlap
condition slip may occcur on at most one of these families at any given
time - this family is called the *active* family while the remaining families
are called *passive*. The equations governing the motion on each family will
be equations analogous to the above equations. In general the state of stress
will be limiting on only one family. For special cases of the stress it may
be limiting on more than one family. In such a case the deformation would
consist of a random sequence of slips on the active slip systems.

5. Continuum models and double slip materials

The β-surface in the single slip model was motivated by the intention to
represent a physical slip surface. For materials possessing more than one
family, motion takes place on each family separately. In this section we consider a class of model which is purely continuum, i.e. there are no grains.
Every direction through a point P represents the tangent to a potential
slip surface through P. The stress tensor σ at P, via the normal and tangential traction, selects the directions which give the actual slip surfaces -
there are two such directions. The continuum models considered here assume that slip takes place on *both* surfaces simultaneously, i.e. they are
concerned with double slip materials. This will not contradict the non-
overlap condition only if there are no grains or the grain-size is zero. The
continuum assumption is to replace yield and slip on a grain interface by
yield and slip on a characteristic direction. We shall here consider three
such models.

(a) The plastic potential model, M_1,

$$d_{ij} = \lambda \frac{\partial g}{\partial \sigma_{ij}},$$

which for planar deformations may be written

$$d_{11} + d_{22} = \sin \nu \left[(d_{11} - d_{22}) \cos 2\psi_\sigma + 2d_{12} \sin 2\psi_\sigma \right],$$

$$(d_{11} - d_{22}) \sin 2\psi_\sigma - 2d_{12} \cos 2\psi_\sigma = 0,$$

where ν is the angle of dilatancy.

(b) The double-sliding free-rotating model, M_2,

$$(d_{11} + d_{22}) \cos (\phi - \chi) = \sin \chi \left[(d_{11} - d_{22}) \cos 2\psi_\sigma + 2d_{12} \sin 2\psi_\sigma \right],$$

$$2\left(\Omega_{12}+\omega_{12}\right)\sin\left(\phi-\chi\right)=\cos\chi\left[(d_{11}-d_{22})\sin 2\psi_\sigma - 2d_{12}\cos 2\psi_\sigma\right].$$

where ϕ, χ denote the angle of internal friction and a dilatancy parameter, respectively.

(c) The double-shearing model, M_3, the equations for which are the same as M_2 except that Ω_{12} is replaced by $\dot{\psi}_\sigma$, the material derivative of ψ_σ.

The kinematic rule in Harris (1995b) assumes the existence of three pairs of directions: (i) the stress characteristic (yield) directions, α_j, (ii) the micro slip (dilatancy) directions, β_j, (iii) the macro slip directions, γ_j, where j=1,2 and the following is proved

Theorem : $\forall M_n$, $1\le n \le 3$, $\exists \Omega$, β_j, γ_j such that

$$\mathbf{dv}_P^Q = \Omega \times \mathbf{ds}_P^Q + \sum_j k_{\gamma_i}\left(\varepsilon_j \mathbf{n}_{\gamma_i} \cdot \mathbf{ds}_P^Q\right)\mathbf{t}_{\beta_j} \tag{22}$$

where i takes the value 1 when j is 2 and vice versa, ε_j is 1 when j is 1 and -1 when j is 2. Note that there are now five unknowns v_1, v_2, Ω_{12}, k_{γ_1}, k_{γ_2} but only four equations.

Let θ_{α_j}, θ_{β_j}, θ_{γ_j} denote the angles that the α_j, β_j, γ_j directions make with the x_1-axis and define the angles ϕ_j, ν_j and ζ_j, j=1,2.

$$\theta_{\alpha_j}=\psi_\sigma\pm\frac{\pi}{4}\pm\frac{1}{2}\phi_j, \qquad \theta_{\beta_j}=\psi_\sigma\pm\frac{\pi}{4}\pm\frac{1}{2}\nu_j, \qquad \theta_{\gamma_j}=\psi_\sigma\pm\frac{\pi}{4}\pm\frac{1}{2}\zeta_j.$$

Equation (22) may be written, Harris (1995b),

$$d_{11}+d_{22}=-R_\nu\left[k_{\beta_1}\sin\frac{1}{2}(\zeta_2-\nu_2)+k_{\beta_2}\sin\frac{1}{2}(\zeta_1-\nu_1)\right], \tag{23}$$

$$d_{11}-d_{22}=R_\nu\left[k_{\beta_1}\cos\left(2\psi_\sigma+\frac{1}{2}(\zeta_2+\nu_2)\right)+k_{\beta_2}\cos\left(2\psi_\sigma-\frac{1}{2}(\zeta_1+\nu_1)\right)\right], \tag{24}$$

$$2d_{12}=R_\nu\left[k_{\beta_1}\sin\left(2\psi_\sigma+\frac{1}{2}(\zeta_2+\nu_2)\right)+k_{\beta_2}\sin\left(2\psi_\sigma-\frac{1}{2}(\zeta_1+\nu_1)\right)\right], \tag{25}$$

$$2\left(\omega_{12}+\Omega_{12}\right)=R_\nu\left[-k_{\beta_1}\cos\frac{1}{2}(\zeta_2-\nu_2)+k_{\beta_2}\cos\frac{1}{2}(\zeta_1-\nu_1)\right]. \tag{26}$$

where $R_\nu=\sec\frac{1}{2}(\nu_1+\nu_2)$ and are valid for anistropic materials. The equations for isotropic materials are obtained by putting $\phi_1=\phi_2$, $\nu_1=\nu_2$ and $\zeta_1=\zeta_2$. We now consider each of the models M_n in turn.

Plastic potential model, M_1: choose $\nu_j=-\zeta_j$, then

$$(d_{11}-d_{22})\sin\left(2\psi_\sigma-\frac{1}{2}(\nu_1-\nu_2)\right)-2d_{12}\cos\left(2\psi_\sigma-\frac{1}{2}(\nu_1-\nu_2)\right)=0, \tag{27}$$

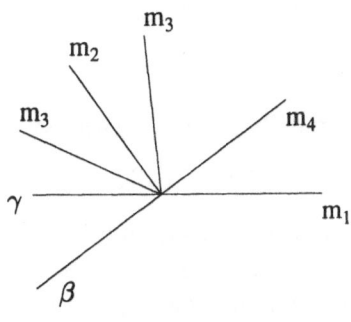

(a) Single slip model

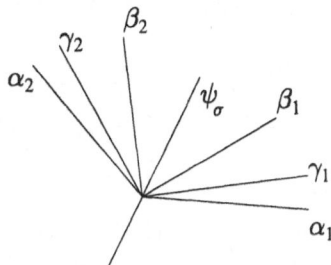

(b) Plastic potential model

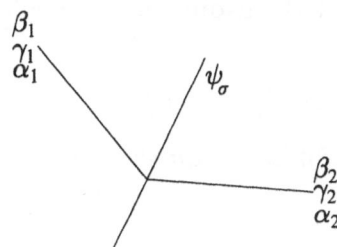

(c) Incompressible double shearing

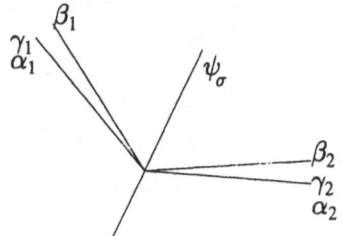

(d) Dilatant double shearing model

Figure 3: Characteristic directions

$$d_{11} + d_{22} = \sin\tfrac{1}{2}(\nu_1 + \nu_2)$$

$$\left[(d_{11} - d_{22})\cos\left(2\psi_\sigma - \frac{1}{2}(\nu_1 - \nu_2)\right) + 2d_{12}\sin\left(2\psi_\sigma - \frac{1}{2}(\nu_1 - \nu_2)\right)\right]. \quad (28)$$

Thus, the local velocity field is independent of the local rotation. The only effect of rotation on the velocity field is through its appearance in the boundary conditions. The characteristic directions are shown in figure 3(b) and it can be seen that there is no fixed relationship between the α- and γ-directions.

Double-shearing model, M_2: In this case choose $\Omega = -\dot{\psi}_\sigma$.
For the incompressible case choose $\nu_j = \phi_j = \zeta_j$, i.e. all three directions coincide, then

$$d_{11} + d_{22} = 0,$$

$$2(\omega_{12} + \dot{\psi}_\sigma)\sin(\phi_1 + \phi_2) = (d_{11} - d_{22})\left[\sin(2\psi_\sigma - \phi_1) + \sin(2\psi_\sigma + \phi_2)\right]$$
$$-2d_{12}\left[\cos(2\psi_\sigma - \phi_1) + \cos(2\psi_\sigma + \phi_2)\right].(29)$$

See figure 3(c). For the compressible case choose $\nu_j = \phi_j$, and $\zeta_j = \phi_j - 2\chi_j$, where χ_j are dilatancy parameters. Then $(d_{11} + d_{22})\sin(\phi_1 + \phi_2 - \chi_1 - \chi_2) =$

$$(d_{11} - d_{22})\left[-\sin(2\psi_\sigma - \phi_1 + \chi_1)\sin\chi_2 + \sin(2\psi_\sigma + \phi_2 - \chi_2)\sin\chi_1\right]$$

$$-2d_{12}\left[-\cos(2\psi_\sigma - \phi_1 + \chi_1)\sin\chi_2 + \cos(2\psi_\sigma + \phi_2 - \chi_2)\sin\chi_1\right], \qquad (30)$$

$$2(\omega_{12} + \dot{\psi}_\sigma)\sin(\phi_1 + \phi_2 - \chi_1 - \chi_2) =$$

$$(d_{11} - d_{22})\left[\sin(2\psi_\sigma - \phi_1 + \chi_1)\cos\chi_2 + \sin(2\psi_\sigma + \phi_2 - \chi_2)\cos\chi_1\right]$$

$$-2d_{12}\left[\cos(2\psi_\sigma - \phi_1 + \chi_1)\cos\chi_2 + \cos(2\psi_\sigma + \phi_2 - \chi_2)\cos\chi_1\right]. \qquad (31)$$

See figure 3(d). Note that there is no fixed relationship between the α- and γ-directions. Let the β-direction be the same as for the incompressible case then the γ-direction will be inclined at an angle χ_j to the other characteristic direction, i.e. $\frac{1}{2}\zeta_j = \frac{1}{2}\phi_j - \chi_j$. For isotropic materials we amend the Coulomb yield condition so that the α-direction coincides with the γ-direction. Thus,

$$q_\sigma \le p_\sigma \sin(\phi_1 - 2\chi_1) + c\cos(\phi_1 - 2\chi_1)$$

and the yield condition now contains the dilatancy direction in a manner similar to the single slip model.

6. References

1. Anand L. (1983) Plane deformations of ideal granular materials, *J. Mech. Phys. Solids*, **31**, 105-122.
2. Geniev, G.A. (1958) Problems of the dynamics of a granular medium (in Russian), *Akad. Stroit Archit. SSSR*, Moscow.
3. Harris, D. (1995a) A unified formulation for plasticity models of granular and other materials, *Proc. Roy. Soc. Series A*, **450**, 37-49.
4. Harris, D. (1995b) Equations governing plane plastic deformation in anisotropic materials, in D.F. Parker and A.H. England (eds.) *Anisotropy Inhomogeneity and Nonlinearity in Solid Mechanics* Kluwer, Dordrecht, 183-190.
5. Harris, D. TAM Report No. 32, (1994) Dept. Mathematics, UMIST 1-21.
6. Harris, D. (1993) Constitutive equations for planar deformations of rigid-plastic materials, *J. Mech. Phys. Solids*, **41**, 1515-1531.
7. de Josselin de Jong, G. (1959) *Statics and kinematics in the failable zone of a granular material.* Uitgeverij, Delft.
8. Mehrabadi, M.M. and Cowin, S.C.(1978) Initial planar deformation of dilatant granular materials, *J. Mech. Phys. Solids*, **26**, 269-284.
9. Ostrowska-Maciejewska, J. and Harris, D. (1989) Three-dimensional constitutive equations for rigid/perfectly plastic granular materials. *Math. Proc. Camb. Phil. Soc.* **108**, 153-169.
10. Spencer, A.J.M. (1964) A theory of the kinematics of ideal soils under plane strain conditions, *J. Mech. Phys. Solids*, **12**, 337-351.

THE DOUBLE-SHEARING VELOCITY EQUATIONS FOR DILATANT SHEAR-INDEX GRANULAR MATERIALS

J. M. HILL

Department of Mathematics,

University of Wollongong,

Wollongong NSW 2522

Australia

1. Introduction

Modelling the flow of a granular material constitutes one of the outstanding problems of continuum mechanics, for which there is little general agreement on the governing mathematical equations. This is because real granular materials are not readily defined and have widely varying physical characteristics, which give rise to behaviour which is so complex and variable that a single mathematical model is unlikely to account for the behaviour of all granular materials under all practical or experimental conditions. However, it is generally accepted that failure of a granular material occurs due to frictional slip between particles and that the material yields at a surface when the magnitude of the shear component of stress τ assumes a particular relationship with the normal component of stress σ, namely $|\tau| = f(\sigma)$ for some function $f(\sigma)$, where here we take the normal component of stress to be positive in tension. A large body of experimental work supports this hypothesis and it would be true to say that the resulting stress distribution, based on this assumption, is reasonably in accord with experimental evidence and generally accepted as an overall accurate model. In contrast there is very little general agreement on the associated velocity or displacement field corresponding to a given stress distribution and the topic is controversial.

The simplest and best known failure criterion is that originally proposed by Coulomb (1773) and arises by direct analogy with the laws of friction. That is, slip occurs when the shear stress overcomes frictional resistance and cohesion, so that

$$|\tau| = c - \sigma \tan \delta \,, \tag{1}$$

251

N. A. Fleck and A. C. F. Cocks (eds.),
IUTAM Symposium on Mechanics of Granular and Porous Materials, 251–262.
© 1997 *Kluwer Academic Publishers.*

where c and δ are constants known as the cohesion and angle of internal friction respectively. In many cases Coulomb's failure hypothesis provides a satisfactory basis for the determination of stress in a granular material. However, in order to more accurately accommodate experimental results a large number of authors (see Hill and Wu (1993) for detailed references) have successfully utilized failure criteria of the form

$$\left(\frac{|\tau|}{c}\right)^n = 1 - \frac{\sigma}{t} ,\tag{2}$$

where c, t and n are positive constants which are referred to as the cohesion, tensile strength and shear index respectively. This equation applies to a wide range of granular and powder-like materials and is sometimes known as the Warren-Spring equation. Clearly, the case $n = 1$ gives Coulomb's law and all other known experimentally determined numerical values of n lie between 1 and 2 (see Hill and Wu (1993)). For the general failure criterion $|\tau| = f(\sigma)$, the angle of internal friction is a stress dependent function which is defined incrementally from the equation

$$d|\tau| = -\tan \delta \, d\sigma ,\tag{3}$$

and is accordingly only constant for the linear failure hypothesis due to Coulomb.

For a given state of stress in a granular material, in order to assign a displacement or velocity field we need to speculate on the mechanics and physics underlying the interaction of granular particles at the microscopic level. Accordingly, the validity of every mathematical or numerical model of granular behaviour hinges on the assumptions underlying particle interactions. Here, we examine one particular continuum mechanical granular theory, termed the double-shearing theory, which has evolved from the work of AJM Spencer (see Spencer (1964) and (1982)). The double-shearing theory for the kinematics of soil as originally formulated by Spencer (1964) is based on the postulates of material isotropy and incompressibility and the physical idea, suggested by de Josselin de Jong (1959), that the deformation arises as a result of shear along the stress characteristics. Spencer's double-shearing theory is a properly invariant, planar, general theory for the deformation of non-dilatant granular materials, which predicts that the principal axes of stress and strain-rate will not, in general, coincide. Although incompressibility is frequently assumed in granular theory, generally a granular material flows (or is allowed to flow) as a result of an expansion of the void space, which means that the bulk density decreases. This drawback of the theory was cleverly rectified by Mehrabadi and Cowin (1978) who, based on the kinematical proposal of Butterfield and Harkness (1972), proposed an expansion of the material in a direction which is perpendicular to the plane of shearing. This expansion is measured by a third physical parameter γ which is referred to as the angle of dilatancy. In the Mehrabadi and Cowin theory, the angle of dilatancy is assumed to be constant and is such that γ zero yields Spencer's theory while the value $\gamma = \delta$ (the angle of internal friction) produces the theory arising from using the Coulomb yield condition as a plastic potential. Experimental evidence indicates that in reality the angle of dilatancy decreases to zero with increased monotonic shearing deformation.

The purpose of this paper is to present a compact derivation of the velocity equations in characteristic variables for dilatant shear index granular materials, with variable angle of internal friction. These equations are utilized by Hill and Wu (1996) to determine the indentation of a rigid punch in a shear index dilatant granular material. Their derivation and the final form of the equations are quite complicated. In applications it is clearly important to know that the correct form of the equations is being used and this is the motivation for presenting the following derivation. This derivation utilizes the formalism developed by Harris (1985) to derive the velocity equations in rectangular Cartesian coordinates and the final equations obtained here agree with those given by Harris (1993) after an error in his equations is corrected by replacing a $1/4$ by a $1/2$. In the following section we state the main details concerning the stress equations and stress characteristics. In the section thereafter we present the basic ideas underlying the double-shearing theory and we present a modification of Harris' derivation of the velocity equations, which are valid for varying angle of internal friction. In the final section of the paper we show how these equations may be transformed into characteristic variables.

2. Stress equations for plane problems

For two-dimensional states of stress, using rectangular Cartesian coordinates (x, y, z) and in which the stress components are independent of the coordinate z, with stress tensor σ such that $\sigma_{xz} = \sigma_{yz} = 0$ and σ_{zz} corresponds to the intermediate principal stress component σ_{II}. The remaining components of the stress tensor are assumed to satisfy the equilibrium equations

$$\frac{\partial \sigma_{xx}}{\partial x} + \frac{\partial \sigma_{xy}}{\partial y} = 0, \qquad \frac{\partial \sigma_{xy}}{\partial x} + \frac{\partial \sigma_{yy}}{\partial y} = 0, \qquad (4)$$

assuming also that gravitational effects may be neglected. If σ_I and σ_{III} denote the maximum and minimum principal stress components (namely $\sigma_I \geq \sigma_{II} \geq \sigma_{III}$) and if we assume the direction of σ_I makes an angle ψ with the convential x-direction, then we have the standard relations

$$\sigma_{xx} = -p + q \cos 2\psi, \qquad \sigma_{yy} = -p - q \cos 2\psi, \qquad \sigma_{xy} = q \sin 2\psi, \qquad (5)$$

where p and q are defined by

$$p = -\frac{\sigma_I + \sigma_{III}}{2}, \qquad q = \frac{\sigma_I - \sigma_{III}}{2}. \qquad (6)$$

On substitution of (5) into (4) we can deduce the stress equations in terms of the arc-lengths s_α and s_β along the α and β characteristics, obtained respectively by integration of the differential relations,

$$\frac{dy}{dx} = \tan\left(\psi - \frac{\pi}{4} - \frac{\delta}{2}\right), \quad \frac{dy}{dx} = \tan\left(\psi + \frac{\pi}{4} + \frac{\delta}{2}\right), \qquad (7)$$

which are respectively the α and β-slip lines and (4) becomes

$$\cot\delta\frac{\partial q}{\partial s_\alpha} + 2q\frac{\partial\psi}{\partial s_\alpha} = 0, \quad \cot\delta\frac{\partial q}{\partial s_\beta} - 2q\frac{\partial\psi}{\partial s_\beta} = 0. \qquad (8)$$

We observe that these equations are formally identical with those applying to the Coulomb failure condition (1) for which δ is a constant (see for example Spencer (1982)). The same equations also apply for variable angle of internal friction $\delta(\sigma)$ because in the above derivation we only need the relations

$$\frac{\partial q}{\partial x} = \csc\delta\frac{\partial q}{\partial x}, \quad \frac{\partial q}{\partial y} = \csc\delta\frac{\partial q}{\partial y},$$

or equivalently $dq = \sin\delta dp$ which can be verified as follows for any yield condition $|\tau| = f(\sigma)$. From Hill and Wu (1993) the yield function is formally deduced by elimination of δ from the equations

$$q\cos\delta = f(\zeta), \quad \tan\delta = -f'(\zeta),$$

where $\zeta = -p + q\sin\delta$ and the prime denotes differentiation with respect to the argument ζ. On taking the total derivative of the first relation and making use of the second,

$$\cos\delta dq - q\sin\delta d\delta = -\tan\delta(-dp + \sin\delta dq + q\cos\delta d\delta),$$

which simplifies to give $dq = \sin\delta dp$. Thus for any yield condition with variable angle of internal friction, the usual differential relations for α and β characteristics still apply and the stress equilibrium equations (4) become (8).

We note that the yield condition applying to (2) in principal stress space is derived in Hill and Wu (1993) and in parametric form is given by

$$\frac{\sigma_I}{t} = 1 + \frac{\beta c}{t}(\sec\delta - \tan\delta) - \beta^n, \quad \frac{\sigma_{III}}{t} = 1 - \frac{\beta c}{t}(\sec\delta + \tan\delta) - \beta^n,$$

$$\beta = \left(\frac{nt}{c}\tan\delta\right)^{\frac{1}{(1-n)}}.$$

In general the relationship between σ_I and σ_{III} is complicated except for the special cases of $n = 1$ and $n = 2$ for which we have respectively,

$$\left(1 - \frac{\sigma_I}{t}\right)^{\frac{1}{2}} = \left\{\left[1 + \left(\frac{c}{t}\right)^2\right]^{\frac{1}{2}} - \frac{c}{t}\right\}\left(1 - \frac{\sigma_{III}}{t}\right)^{\frac{1}{2}}, \quad \tan\delta = \frac{c}{t}, \quad (n = 1),$$

$$\left(1 - \frac{\sigma_I}{t}\right)^{\frac{1}{2}} = \left(1 - \frac{\sigma_{III}}{t}\right)^{\frac{1}{2}} - \frac{c}{t}, \quad \sin\delta = \left\{\frac{2t}{c}\left(1 - \frac{\sigma_{III}}{t}\right)^{\frac{1}{2}} - 1\right\}^{-1}, \quad (n = 2).$$

An assumed knowledge of the stress characteristics is critical to a development of the double-shearing kinematical theories which are the subject of the following section.

3. Double-shearing theories for dilatant granular materials

The double-shearing theory originally proposed by AJM Spencer (see Spencer (1964)) for plane problems involving incompressible granular materials, postulates that the deformation arises as a result of two shearing deformations along the α and β-slip lines and that each shear takes place by distortion only (that is, no tension or compression). The kinematical equations are formally derived as follows. We consider an arbitrary material particle located at $P(t)$ and introduce a rectangular Cartesian coordinate system (ξ, η) with origin at $P(t)$ and rotating with angular speed Ω defined by

$$\Omega = \frac{d\psi}{dt} = \frac{\partial\psi}{\partial t} + u\frac{\partial\psi}{\partial x} + v\frac{\partial\psi}{\partial y}, \tag{9}$$

so that the directions of the α and β-slip lines remain fixed in the rotating frame and (u, v) are the velocity components with respect to the fixed (x, y) Cartesian coordinates. Further, we suppose that at some time t_0 the two rectangular Cartesian axes coincide. We say that a deformation is a "shearing deformation" along the α-line if in the rotating frame the relative velocity of two adjacent points is:

(i) zero if the points lie on the same α-line,
(ii) directed along the α-line through $P(t)$ if the points do not lie on the same α-line.

The incompressible double-shearing postulate assumes that the deformation everywhere consists of two such superimposed shearing deformations on the α and β-lines. This assumption leads to the velocity equations (see Spencer (1964) and (1982)),

$$\frac{\partial u}{\partial x} + \frac{\partial v}{\partial y} = 0,$$

$$\left(\frac{\partial u}{\partial x} - \frac{\partial v}{\partial y}\right)\sin 2\psi - \left(\frac{\partial u}{\partial y} + \frac{\partial v}{\partial x}\right)\cos 2\psi = \left(\frac{\partial u}{\partial y} - \frac{\partial v}{\partial x} + 2\Omega\right)\sin\delta. \tag{10}$$

The Mehrabadi and Cowin (1978) double-shearing theory for dilatant granular materials adopts the same procedure but allows an expansion (or contraction) in directions normal to the slip lines following the kinematical proposal of Butterfield and Harkness (1972) which states:

"As successive material points along a slip line are considered, any change in velocity relative to the slip line field that occurs between one point and the next is in a direction inclined at an angle γ to the conjugate slip line", so that alternatively

$$\tan \gamma = \frac{\text{relative velocity normal to slip line}}{\text{relative velocity tangential to slip line}} . \tag{11}$$

A similar procedure as that employed to deduce (10) gives rise to the following velocity equations for dilatant granular materials.

$$\left\{ \left(\frac{\partial u}{\partial x} - \frac{\partial v}{\partial y} \right) \cos 2\psi + \left(\frac{\partial u}{\partial y} + \frac{\partial v}{\partial y} \right) \sin 2\psi \right\} \sin \gamma$$

$$= \left(\frac{\partial u}{\partial x} + \frac{\partial v}{\partial y} \right) \cos(\delta - \gamma) ,$$

$$\left\{ \left(\frac{\partial u}{\partial x} - \frac{\partial v}{\partial y} \right) \sin 2\psi - \left(\frac{\partial u}{\partial y} + \frac{\partial v}{\partial x} \right) \cos 2\psi \right\} \cos \gamma \tag{12}$$

$$= \left(\frac{\partial u}{\partial y} - \frac{\partial v}{\partial x} + 2\Omega \right) \sin(\delta - \gamma) ,$$

from which it is apparent that γ zero gives rise to (10) and it can be verified that $\gamma = \delta$ corresponds to the equations arising from using the Coulomb yield condition (1) as a plastic potential. Given that (u, v) are solutions of (12) the bulk density $\rho(x, y, t)$ is obtained from the continuity equation

$$\frac{\partial \rho}{\partial t} + u \frac{\partial \rho}{\partial x} + v \frac{\partial \rho}{\partial y} + \rho \left(\frac{\partial u}{\partial x} + \frac{\partial v}{\partial y} \right) = 0 . \tag{13}$$

Harris (1985) provides the simplest derivation of the velocity equations (12) and below we present a modification of his approach. If i and j denote the usual fixed unit vectors in the x and y directions respectively, then the unit vectors e_ξ and e_η of the rotating rectangular Cartesian coordinate system (ξ, η) are given by

$$e_\xi = i \cos(\psi - \psi_0) + j \sin(\psi - \psi_0), \quad e_\eta = -i \sin(\psi - \psi_0) + j \cos(\psi - \psi_0), \tag{14}$$

where ψ_0 denotes $\psi(x, y, t)$ evaluated at time t_0. Further the position and velocity vectors with respect to this rotating frame are given by

$$r' = \xi e_\xi + \eta e_\eta , \quad v' = v_\xi e_\xi + v_\eta e_\eta , \tag{15}$$

where the velocity components v_ξ and v_η are defined by $v_\xi = d\xi/dt$ and $v_\eta = d\eta/dt$. If R and V denote the position and velocity vectors respectively of the origin $P(t)$, then the position and velocity vectors of a particle with respect to the fixed Cartesian axes are given respectively by

$$r = R + r', \qquad v = V + \frac{d\,r'}{dt}. \tag{16}$$

Now on using

$$\frac{d\,e_\xi}{dt} = \Omega\,e_\eta, \qquad \frac{d\,e_\eta}{dt} = -\Omega\,e_\xi, \tag{17}$$

which are obtained from (14) and Ω is defined by (9), we may deduce from the above relations

$$v = V + (v_\xi - \eta\Omega)\,e_\xi + (v_\eta + \xi\Omega)\,e_\eta. \tag{18}$$

Harris (1985) exploits the fact that the relative velocity of two material particles is unaffected by the translation of the (ξ, η)-system relative to the (x, y)-system and therefore, essentially by subtraction of two vector expressions of the form (18), we have

$$d\,v = (dv_\xi - \Omega d\eta)\,e_\xi + (dv_\eta + \Omega d\xi)\,e_\eta, \tag{19}$$

from which we may deduce

$$d\,v' = d\,v + \Omega\,(\,e_\xi\,d\eta - e_\eta\,d\xi\,), \tag{20}$$

where $d\,v' = dv_\xi\,e_\xi + dv_\eta\,e_\eta$ which is the relative velocity of two material particles as viewed from the rotating frame. Similarly, by subtraction of two position vectors we have $d\,r = d\,r'$ from which we have

$$dx\,i + dy\,j = d\xi\,e_\xi + d\eta\,e_\eta, \tag{21}$$

and therefore using (14) we may obtain

$$d\xi = dx\cos(\psi - \psi_0) + dy\sin(\psi - \psi_0),\, d\eta = -dx\sin(\psi - \psi_0) + dy\cos(\psi - \psi_0), \tag{22}$$

where again ψ_0 denotes the value of ψ at time t_0. On using equations (14) and (22) we find that (20) becomes

$$d\,v' = d\,v + \Omega\,(\,i\,dy - j\,dx\,), \tag{23}$$

which is the basic equation we need to translate the Butterfield and Harkness proposal into the mathematical velocity equations.

In order to achieve this, it is convenient to use the notation first introduced by Harris (1985) and we define the following four angles

$$\theta_\alpha = \psi - \frac{\pi}{4} - \frac{\delta}{2} , \quad \theta_\beta = \psi + \frac{\pi}{4} + \frac{\delta}{2} ,$$
$$\theta_A = \psi - \frac{\pi}{4} - \frac{\delta}{2} + \gamma , \quad \theta_B = \psi + \frac{\pi}{4} + \frac{\delta}{2} - \gamma , \tag{24}$$

then from the relations

$$dx = ds_\alpha \cos \theta_\alpha + ds_\beta \cos \theta_\beta , \quad dy = ds_\alpha \sin \theta_\alpha + ds_\beta \sin \theta_\beta , \tag{25}$$

we may deduce

$$ds_\alpha = \frac{1}{\cos \delta} \left\{ dx \sin \theta_\beta - dy \cos \theta_\beta \right\} , \quad ds_\beta = -\frac{1}{\cos \delta} \left\{ dx \sin \theta_\alpha - dy \cos \theta_\alpha \right\} . \tag{26}$$

Now the kinematical proposal of Butterfield and Harkness (1972) is that the relative velocity in the rotating frame of two successive points near the origin along the α-line (β-line) is in the B-direction (A-direction) where the A-direction makes a counterclockwise angle γ with the α-line tangent at P and the B-direction makes a clockwise angle γ with the β-line tangent at P as illustrated in Figure 1. This proposal implies that

$$\frac{\partial v'}{\partial s_\alpha} = a \, e_B , \quad \frac{\partial v'}{\partial s_\beta} = b \, e_A , \tag{27}$$

for certain constants a and b and where e_A and e_B denote unit vectors in the A and B-directions respectively. Accordingly, the relative velocity of two adjacent particles, as viewed from the rotating (ξ, η) frame, is given by

$$d v' = a \, e_B \, ds_\alpha + b \, e_A \, ds_\beta . \tag{28}$$

Now from Figure 1 we have the relations

$$e_A = \cos \theta_A \, e_\xi + \sin \theta_A \, e_\eta , \quad e_B = \cos \theta_B \, e_\xi + \sin \theta_B \, e_\eta , \tag{29}$$

and from (14) these become

$$e_A = i \cos(\psi - \psi_0 + \theta_A) + j \sin(\psi - \psi_0 + \theta_A) ,$$
$$e_B = i \cos(\psi - \psi_0 + \theta_B) + j \sin(\psi - \psi_0 + \theta_B) . \tag{30}$$

Thus, altogether from (23), (26), (28) and (30) we have, on evaluating the equation at time t_0, the following result

$$dv + \Omega(i\,dy - j\,dx)$$

$$= \frac{a}{\cos\delta}\Big\{i\cos\theta_B + j\sin\theta_B\Big\}\Big\{dx\sin\theta_\beta - dy\cos\theta_\beta\Big\} \tag{31}$$

$$- \frac{b}{\cos\delta}\Big\{i\cos\theta_A + j\sin\theta_A\Big\}\Big\{dx\sin\theta_\alpha - dy\cos\theta_\alpha\Big\}\,.$$

From this equation we may deduce the following expressions,

$$\frac{\partial u}{\partial x} = \frac{1}{\cos\delta}\Big\{a\sin\theta_\beta\cos\theta_B - b\sin\theta_\alpha\cos\theta_A\Big\}\,,$$

$$\frac{\partial u}{\partial y} + \Omega = \frac{-1}{\cos\delta}\Big\{a\cos\theta_\beta\cos\theta_B - b\cos\theta_\alpha\cos\theta_A\Big\}\,,$$

$$\frac{\partial v}{\partial x} - \Omega = \frac{1}{\cos\delta}\Big\{a\sin\theta_\beta\sin\theta_B - b\sin\theta_\alpha\sin\theta_A\Big\}\,, \tag{32}$$

$$\frac{\partial v}{\partial y} = \frac{-1}{\cos\delta}\Big\{a\cos\theta_\beta\sin\theta_B - b\cos\theta_\alpha\sin\theta_A\Big\}\,,$$

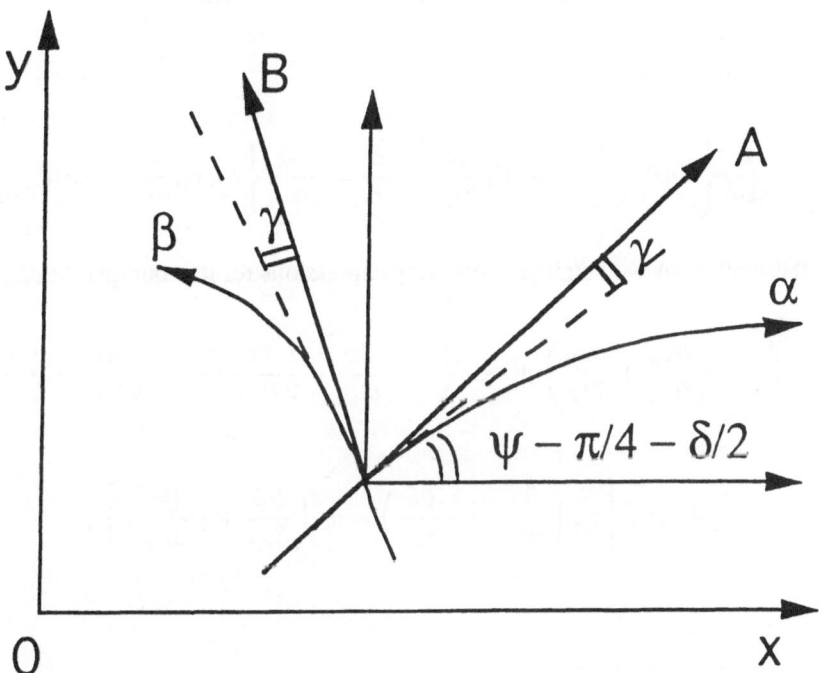

Figure 1 α and β-slip lines and A and B-directions

from which all the detailed relations given by Harris (1985) follow immediately and the velocity equations (12) arise by elimination of the constants a and b. In this derivation the equations so obtained apply for both constant and non-constant angle of internal friction. In the following section we deduce the velocity equations in characteristic variables for the case when the angle of internal friction is not constant.

4. Velocity equations in characteristic variables

We remark that in the original derivations of the velocity equations (10) and (12) given by Spencer (1964) and Mehrabadi and Cowin (1978) respectively, the equations were first obtained in characteristic variables and then transformed to the fixed rectangular Cartesian frame. However, starting from the velocity equations (12) we need the relations

$$u = v_\alpha \cos\theta_\alpha + v_\beta \cos\theta_\beta , \quad v = v_\alpha \sin\theta_\alpha + v_\beta \sin\theta_\beta , \tag{33}$$

where v_α and v_β denote velocities along the characteristic curves. We also need the relations

$$\frac{\partial}{\partial s_\alpha} = \cos\theta_\alpha \frac{\partial}{\partial x} + \sin\theta_\alpha \frac{\partial}{\partial y} , \quad \frac{\partial}{\partial s_\beta} = \cos\theta_\beta \frac{\partial}{\partial x} + \sin\theta_\beta \frac{\partial}{\partial y} , \tag{34}$$

from which we may deduce

$$\frac{\partial}{\partial x} = \frac{1}{\cos\delta}\left\{ \sin\theta_\beta \frac{\partial}{\partial s_\alpha} - \sin\theta_\alpha \frac{\partial}{\partial s_\beta} \right\}, \quad \frac{\partial}{\partial y} = \frac{-1}{\cos\delta}\left\{ \cos\theta_\beta \frac{\partial}{\partial s_\alpha} - \cos\theta_\alpha \frac{\partial}{\partial s_\beta} \right\}. \tag{35}$$

From these relations we obtain the following expressions for the four quantities:

$$\frac{\partial u}{\partial x} + \frac{\partial v}{\partial y} = \left(\frac{\partial v_\alpha}{\partial s_\alpha} + \frac{\partial v_\beta}{\partial s_\beta} \right) + \sec\delta\left\{ v_\alpha \left(\frac{\partial\psi}{\partial s_\beta} - \frac{1}{2}\frac{\partial\delta}{\partial s_\beta} \right) - v_\beta \left(\frac{\partial\psi}{\partial s_\alpha} + \frac{1}{2}\frac{\partial\delta}{\partial s_\alpha} \right) \right\}$$

$$+ \tan\delta\left\{ v_\alpha \left(\frac{\partial\psi}{\partial s_\alpha} - \frac{1}{2}\frac{\partial\delta}{\partial s_\alpha} \right) - v_\beta \left(\frac{\partial\psi}{\partial s_\beta} + \frac{1}{2}\frac{\partial\delta}{\partial s_\beta} \right) \right\}, \tag{36}$$

$$\frac{\partial u}{\partial y} - \frac{\partial v}{\partial x} = -\left\{ v_\alpha \left(\frac{\partial\psi}{\partial s_\alpha} - \frac{1}{2}\frac{\partial\delta}{\partial s_\alpha} \right) + v_\beta \left(\frac{\partial\psi}{\partial s_\beta} + \frac{1}{2}\frac{\partial\delta}{\partial s_\beta} \right) \right\}$$

$$+ \sec\delta \left(\frac{\partial v_\alpha}{\partial s_\beta} - \frac{\partial v_\beta}{\partial s_\alpha} \right) + \tan\delta \left(\frac{\partial v_\alpha}{\partial s_\alpha} - \frac{\partial v_\beta}{\partial s_\beta} \right), \tag{37}$$

$$\frac{\partial u}{\partial x} - \frac{\partial v}{\partial y} = \cos 2\psi \left[\left(\frac{\partial v_\beta}{\partial s_\alpha} + \frac{\partial v_\alpha}{\partial s_\beta} \right) \right.$$

$$+ \sec \delta \left\{ v_\alpha \left(\frac{\partial \psi}{\partial s_\alpha} - \frac{1}{2} \frac{\partial \delta}{\partial s_\alpha} \right) - v_\beta \left(\frac{\partial \psi}{\partial s_\beta} + \frac{1}{2} \frac{\partial \delta}{\partial s_\beta} \right) \right\}$$

$$\left. + \tan \delta \left\{ v_\alpha \left(\frac{\partial \psi}{\partial s_\beta} - \frac{1}{2} \frac{\partial \delta}{\partial s_\beta} \right) - v_\beta \left(\frac{\partial \psi}{\partial s_\alpha} + \frac{1}{2} \frac{\partial \delta}{\partial s_\alpha} \right) \right\} \right] \qquad (38)$$

$$+ \sin 2\psi \left[-v_\alpha \left(\frac{\partial \psi}{\partial s_\beta} - \frac{1}{2} \frac{\partial \delta}{\partial s_\beta} \right) - v_\beta \left(\frac{\partial \psi}{\partial s_\alpha} + \frac{1}{2} \frac{\partial \delta}{\partial s_\alpha} \right) \right.$$

$$\left. + \sec \delta \left(\frac{\partial v_\alpha}{\partial s_\alpha} - \frac{\partial v_\beta}{\partial s_\beta} \right) + \tan \delta \left(\frac{\partial v_\alpha}{\partial s_\beta} - \frac{\partial v_\beta}{\partial s_\alpha} \right) \right],$$

$$\frac{\partial u}{\partial y} + \frac{\partial v}{\partial x} = \sin 2\psi \left[\left(\frac{\partial v_\beta}{\partial s_\alpha} + \frac{\partial v_\alpha}{\partial s_\beta} \right) \right.$$

$$+ \sec \delta \left\{ v_\alpha \left(\frac{\partial \psi}{\partial s_\alpha} - \frac{1}{2} \frac{\partial \delta}{\partial s_\alpha} \right) - v_\beta \left(\frac{\partial \psi}{\partial s_\beta} + \frac{1}{2} \frac{\partial \delta}{\partial s_\beta} \right) \right\}$$

$$\left. + \tan \delta \left\{ v_\alpha \left(\frac{\partial \psi}{\partial s_\beta} - \frac{1}{2} \frac{\partial \delta}{\partial s_\beta} \right) - v_\beta \left(\frac{\partial \psi}{\partial s_\alpha} + \frac{1}{2} \frac{\partial \delta}{\partial s_\alpha} \right) \right\} \right] \qquad (39)$$

$$+ \cos 2\psi \left[v_\alpha \left(\frac{\partial \psi}{\partial s_\beta} - \frac{1}{2} \frac{\partial \delta}{\partial s_\beta} \right) + v_\beta \left(\frac{\partial \psi}{\partial s_\alpha} + \frac{1}{2} \frac{\partial \delta}{\partial s_\alpha} \right) \right.$$

$$\left. + \sec \delta \left(\frac{\partial v_\beta}{\partial s_\beta} - \frac{\partial v_\alpha}{\partial s_\alpha} \right) + \tan \delta \left(\frac{\partial v_\beta}{\partial s_\alpha} - \frac{\partial v_\alpha}{\partial s_\beta} \right) \right].$$

On making use of the equation

$$\Omega = \frac{d\psi}{dt} = \frac{\partial \psi}{\partial t} + v_\alpha \frac{\partial \psi}{\partial s_\alpha} + v_\beta \frac{\partial \psi}{\partial s_\beta}, \qquad (40)$$

and the above expressions we can verify that the velocity equations (12) become

$$\cos(\delta - \gamma) \frac{\partial v_\alpha}{\partial s_\alpha} - \sin(\gamma) \frac{\partial v_\beta}{\partial s_\alpha} = v_\beta \left\{ \cos(\gamma) \frac{\partial \psi}{\partial s_\alpha} + \sin(\delta - \gamma) \frac{\partial \psi}{\partial s_\beta} \right\}$$

$$+ \frac{\partial \psi}{\partial t} \sin(\delta - \gamma) + \frac{1}{2} [v_\beta \cos \gamma + v_\alpha \sin(\delta - \gamma)] \frac{\partial \delta}{\partial s_\alpha},$$

$$\qquad (41)$$

$$\cos(\delta - \gamma) \frac{\partial v_\beta}{\partial s_\beta} - \sin(\gamma) \frac{\partial v_\alpha}{\partial s_\beta} = -v_\alpha \left\{ \cos(\gamma) \frac{\partial \psi}{\partial s_\beta} + \sin(\delta - \gamma) \frac{\partial \psi}{\partial s_\alpha} \right\}$$

$$- \frac{\partial \psi}{\partial t} \sin(\delta - \gamma) + \frac{1}{2} [v_\alpha \cos \gamma + v_\beta \sin(\delta - \gamma)] \frac{\partial \delta}{\partial s_\beta}.$$

These equations apply for variable angle of internal friction and are those used by Hill and Wu (1996) for the problem of indentation of a granular material by a smooth flat rigid punch. They coincide with velocity equations given by Harris (1993) (equation (30)) on making use of the relation given in section 5 of that paper between his dilatancy angle and that introduced by Mehrabadi and Cowin (1978) and denoted here by γ. In addition the $1/4$ in Harris' equation must be replaced by $1/2$, which is an error.

References

Butterfield, R. and Harkness, R. M. (1972) The kinematics of Mohr-Coulomb materials, Proceedings of the Roscoe Memorial Symposium Cambridge, *"Stress-Strain Behaviour of Soils", edited by R. H. G. Parry, 220-233, G. T. Foulis, Henley.*

Coulomb, C. A. (1773) Essai sur une application des régles de maximis et minimis à quelques problèmes de statique, relatifs à l'architecture, *Mém. de Math. de l'Acad. Roy. des Sci.* Paris **7**, 343-382. In English translation: Heyman J. (1972) *Coulomb's Memoir on Statistics: An Essay in the History of Civil Engineering*, Cambridge University Press.

de Josselin de Jong, G. (1959) *Statics and Kinematics of the Failable Zone of the Granular Material*, Uitgeverij Waltman, Delft. Harris, D. (1985) A derivation of the Mehrabadi-Cowin equations, *J. Mech. Phys. Solids* **33**, 51-59.

Harris, D. (1993) Constitutive equations for planar deformations of rigid-plastic materials, *J. Mech. Phys. Solids* **41**, 1515-1531.

Hill, J. M. and Wu, Y.-H. (1993) Plastic flows of granular materials of shear index n. I. Yield functions, *J. Mech. Phys. Solids* **41**, 77-93.

Hill, J. M. and Wu, Y.-H. (1996) The punch problem for shear index granular materials, *Q. Jl. Mech. appl. Math.* **49**, 81-105.

Mehrabadi, M. M. and Cowin, S. C. (1978) Initial planar deformation of dilatant granular materials, *J. Mech. Phys. Solids* **26**, 269-284.

Spencer, A. J. M. (1964) A theory of the kinematics of ideal soils under plane strain conditions, *J. Mech. Phys. Solids* **12**, 337-351.

Spencer, A. J. M. (1982) *Deformation of Ideal Granular Materials*, Article in Mechanics of Solids, Rodney Hill 60th Anniversary Volume (Editors H. G. Hopkins and M. J. Sewell) Pergamon Press, Oxford, 607-652.

DISTINCT ELEMENT SIMULATIONS AND DYNAMIC MICROSTRUCTURAL IMAGING OF SLOW SHEARING GRANULAR FLOWS

U. TÜZÜN
Department of Chemical and Process Engineering

D.M. HEYES
Department of Chemistry
University of Surrey, Guildford , GU2 5XH, UK.

Abstract

Granular Dynamics simulations are presented based on the application of the Distinct Element (DE) technique to the filling and discharge events of granular materials in bins and hoppers. The results of a series of simulation case studies conducted in both plane-strain (2-D) and axial-symmetric (3-D) storage vessels respectively are compared with photographic measurements of particle trajectories in wedge-shaped and flat-bottom hoppers and with the tomographic scans of the interstitial voidage taken at different horizontal planes along the height of conical and cylindrical containers.

The simulations allow for the translational and rotational velocities of individual particles to be tracked during gravity flow by the use of a time-stepping algorithm based on Newtonian Mechanics . The interstitial fluid drag effects were also considered in the case of a radial fluid flow field in a conical hopper by calculating the local average voidage at points within a fixed Eulerian Grid superimposed on the particle flow field. The particle-particle and particle-wall interactions are modelled at the scale of the nominal particle size using a number of possible interaction laws based on Molecular Dynamics continuous potential, Hertzian elastic contact and Hooke's linear spring models respectively. These interaction laws give rise to significantly different mathematical forms and the long-range connectivity of individual pair interactions within the bulk is found to be influenced significantly both i) by the degree of variation of the contact normal compliance with contact separation distance and ii) by the arbitrarily chosen interaction cut-off separation (or force). These points are illustrated by comparing the fields of contact tangential displacements and particle velocities obtained with different interaction laws.

1. Normal Force Interaction Laws

* The simulations can use one of three options : Continuous interaction ; Hertzian interaction ; or linear spring interaction (Hooke's Law)

N. A. Fleck and A. C. F. Cocks (eds.),
IUTAM Symposium on Mechanics of Granular and Porous Materials, 263–274.
© 1997 *Kluwer Academic Publishers.*

1.1 CONTINUOUS INTERACTION

The fundamental non-cohesive CI curve is given by :

$$\phi(r) = \varepsilon \ (\ d/r\ )^{\ n} \tag{1}$$

where $\phi(r)$ is the potential energy between two particles with centre to centre distance r and diameter d is the soft sphere potential. This form has found much application in liquid state statistical mechanics. The normal force between the particles is given by :

$$F_N = \ - d\phi \ / \ dr \tag{2}$$

All length scales used in the programs are in units of particle size, d and the mass of the hopper contents is scaled in units of particle mass, m. The scaling behaviour for these interactions is made more specific for the present work than is generally the case for soft sphere fluids by setting the interaction force at a separation r = d equal to the gravitational force on the particle. The interaction parameter, ε in eqn(1) can then be expressed in terms of the gravitational constant, g as follows :

$$\varepsilon = dmg \ / \ n \tag{3}$$

In the CI model, the normal force extends beyond r > d. Hence, some cut-off is required. Initially, the particles were considered to be non-interacting when r > 1.2d. Here F_N is negligible. In some later simulations , the interaction cut-off was reduced to r = 1.064d (equivalent to a normal force minimum of 0.1mg) to observe its effect on the long-range connectivity in the bulk. One other significant feature of the CI model is that it gives rise to a separation distance dependant normal stiffness . Inserting eqn.1 into eqn.2 above and taking the partial derivative ($-\partial F_N / \partial r$) results in:-

$$k_{CI} = (n+1)\frac{F_N}{r} = \frac{\varepsilon n(n+1)}{r^2}\left(\frac{d}{r}\right)^n \tag{4}$$

The dimensionless normal force vs. dimensionless separation distance curve is shown in *Figure 1* below for values of the exponent n= 36 and n= 144 respectively. A long tail is produced with n= 36 which allows for a very soft interaction at small values of the contact normal force; F_N.

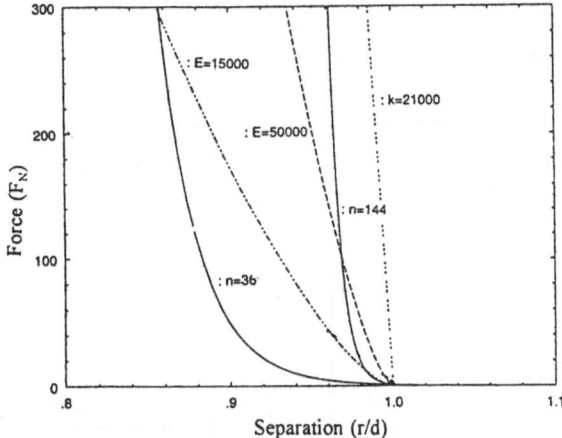

Figure 1. Continuous, Hertzian and Hooke's spring interactions

1.2 HERTZIAN INTERACTION

An analysis of the normal interaction of spheres was undertaken by Hertz from the fundamental physics of contact mechanics as described by Johnson (1985). The equation is as follows :

$$F_N = 4/3 \; E^* \; \sqrt{R} \, (\, d - r \,)^{3/2} \tag{5}$$

where $1/E^* = (1-v_1) / E_1 + (1-v_2) / E_2$ and $1/R = 1/R_1 + 1/R_2$. Here E is Young's modulus, and v is Poisson's ratio, subscripts 1 & 2 indicate particles 1 and 2. The effective Young's Modulus E^* for the granular assembly was calculated by matching the values of the continuous interaction exponent n with the equivalent value of E^* on the basis of the assembly voidage obtained at a given normal force. *Fig.2* shows a series of comparisons between the continuous and Hertzian interaction parameters on the basis of 2-D and 3-D assembly voidage respectively. Clearly , the magnitude of the assembly Young's modulus is orders of magnitude smaller than the Young's modulus of the solid particles themselves.

1.3 HOOKE'S LAW

This is the simplest relationship and is as follows :-

$$F_N = k \, (\, d - r \,) \tag{6}$$

where k is the contact stiffness. The stifness k here does not vary with the contact separation distance and it is this feature which sets Hooke's Linear Spring apart from the above two interaction laws. In both Hertzian and Hooke's interaction simulations , the interaction cut-off was chosen to be at $r = d$; i.e. at the particle boundary.It is also important to note that the value of the stiffness constant, k used in the simulations of bulk flow is required to be orders of magnitude smaller than the values measured for surface displacements of the solid particles.

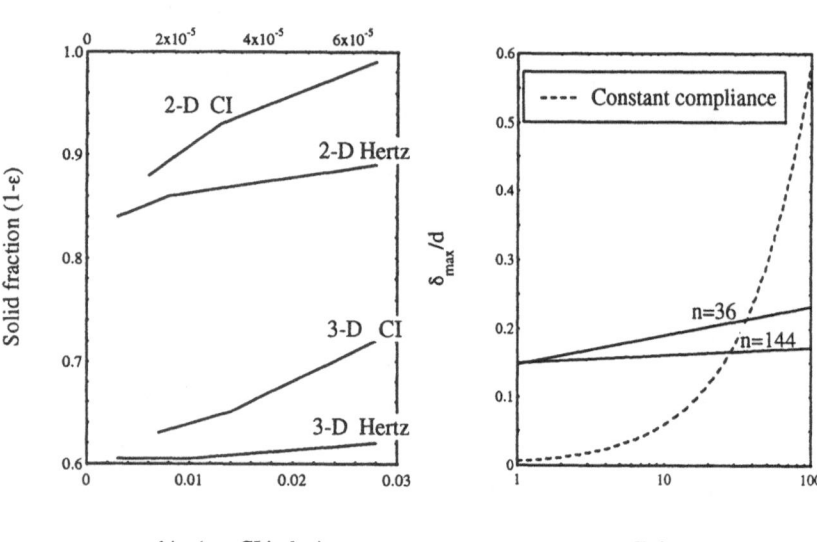

$1/E^*$ (E^* = Effective Bulk Modulus)

Figure2. Variation of asssembly voidage with Continuous and Hertzian interaction parameters

Figure3. Variation of maxiumum tangential contact displacement with contact normal force

2. Particle Tangential Friction Interaction

The model calculates the relative movement, δ, of the interacting (contacting) surfaces within each time step and accumulates the total displacement, δ_F between the surfaces from the time they initially come into contact. The variation of the frictional force, F_F prior to gross sliding (see below) between surfaces is calculated using a modification of Amonton's Law ($F_F=\mu F_N$) first proposed by Mindlin et al (1953) :

$$F_F = \mu F_N \ (\ 1 - (\ 1 - | \ \delta_F \ | \ / \ \delta_{max} \quad)^{3/2} \qquad (7)$$

$| \ \delta_F \ |$ is not allowed to exceed δ_{max}. When $| \ \delta_F \ | = \delta_{max}$, sliding occurs. It should be noted that in the model, the friction force F_F does not depend on whether δ_F is increasing or decreasing; i.e any hysteresis in the tangential loading and unloading is ignored. If the contact is broken (i.e. when $r > r_c$) δ_F is set to zero and all memory of that contact is lost. The maximum displacement before the onset of gross sliding is given by :

$$\delta_{max} \quad = \delta_R \ \delta_n \qquad (8)$$

where δ_R is a constant; Langston et al (1995). *Fig.3* compares the normalized tangential displacement versus normal force curves for constant stiffness and variable stiffness interactions. Clearly, constant contact stiffness allows little tangential displacement at small normal loads . This feature renders it inappropriate for modelling bulk flows which are known to couple large shear deformations with small normal loads; see for example; Tüzün (1987).

3. Newtonian Mechanics Numerical Integration

The model simulates the flow of discs in a two dimensional hopper and spheres in a three dimensional axial symmetric hopper using an explicit time stepping numerical integration of the forces acting on the discs. At every time step, Δt, the model calculates the resultant force and moment acting on each particle and hence generates its linear and angular accelerations. Numerical integration of the velocities produces the positions and orientations of the particles for the next time step. All contact and gravitational forces on a particle are summed at each time step to provide the linear acceleration on each particle. The tangential forces give rise to moments on the particles which are summed to form the angular acceleration. An appropriate modification of the Verlet Algorithm was used :

$$x_v (t+\Delta t/2) = x_v (t-\Delta t/2) + x_a(t) . \Delta t \qquad (9)$$

$$x_a(t) = - F(t)/m - \beta x_v(t) \qquad (10)$$

which describe the linear motion of particles, where β is the global damping constant which is set to zero in the current study. Instead, explicit contact damping is applied in order to allow for energy dissipation through mechanical work at the contact zone. The damping force at any instant is kept proportional to the normal and tangential components of the relative velocity. Setting the contact damping to a fraction of the critical damping ratio (Campbell (1993)) is found to achieve a rapid stabilization of the particle assembly during the hopper filling stage prior to discharge.

Similar expressions are used to calculate the resultant moments and the polar moment of inertia associated with angular motion. Particle positions are updated according to :-

$$x_p(t+\Delta t) = x_p(t) + x_v (t+\Delta t /2) . \Delta t \qquad (11)$$

The Verlet algorithm has been used extensively in the simulation literature and has been found to be extremely stable under a wide range of applications including non-equilibrium conditions analogous to the present study; see for example Fincham and Heyes (1985).

4. Hopper Geometry, Filling and Discharging Scenarios

It is considered that the *en masse* method of filling the model hopper produces an unrealistic arrangement of particles, unrepresentative of most hopper applications. For instance, it imparts much larger stresses on the walls on impact (as was shown in the previous simulations). It is also noted that the literature on hopper wall stresses divide the scenario into two regimes : (i) filling and static (which are treated as a single event); (ii) discharging. Hence the model filling process was modified to introduce a few particles at a time at the top of the hopper over a longer time period (Langston et al 1995). This process is closer to the way particles are in fact delivered into hoppers, for example by pneumatic conveying or using a conveyor belt. When all the particles have settled, the orifice is opened and the particles discharge under gravity.

Campbell et al (1993) have shown that mono-sized particles can give rise to special geometric effects. e.g. the packing arrangement can build up ordered "crystalline" stress patterns. In order to prevent such effects and mimic more closely real granular materials , a uniform distribution with a 10% standard deviation about the mean value is used in the simulation case studies.

5. Plane-Strain (2-D) Simulations

In *Fig.4*, the 2-D hopper behaviour is compared for three forms of the normal interaction. Internal distributions of the tangential displacement vectors (normalized to the maximum at gross sliding) are matched with the wall stress profiles in a tall bin with a flat bottom and a central slot orifice. In *Fig4 (a)* and *(b)*, the positions of the wall stress peaks are matched by the intersections of the "rupture zones" or localized "high shear bands" with the hopper walls when a long interaction range is introduced; i.e. $r = 1.2d$. In contrast, in *Figs..4(c) -(e)* the effect of an increasingly more abrupt cut-off of the normal force interaction curve is observed in the successive disappearance of the shear bands or rupture zones within the flow fields. This is also reflected by the disappearance of the localized stress peaks at the wall as the long-range connectivity in the flowing bulk is minimized by the introduction of a much more rigid Hertzian interaction with a force interaction cut-off set at $r = d$. The wall stress profiles in each case are compared with the hydrostatic stress (H) profile as well as the active (A) and passive (P) stress states resulting from soil mechanics analyses; see for example Nedderman (1992).

5.1 COMPARISONS WITH 2-D FLOW EXPERIMENTS

These involve photographing the discharge of large ($d = 10$ mm) discs from a vertical planar hopper and comparing these photographs with the graphical

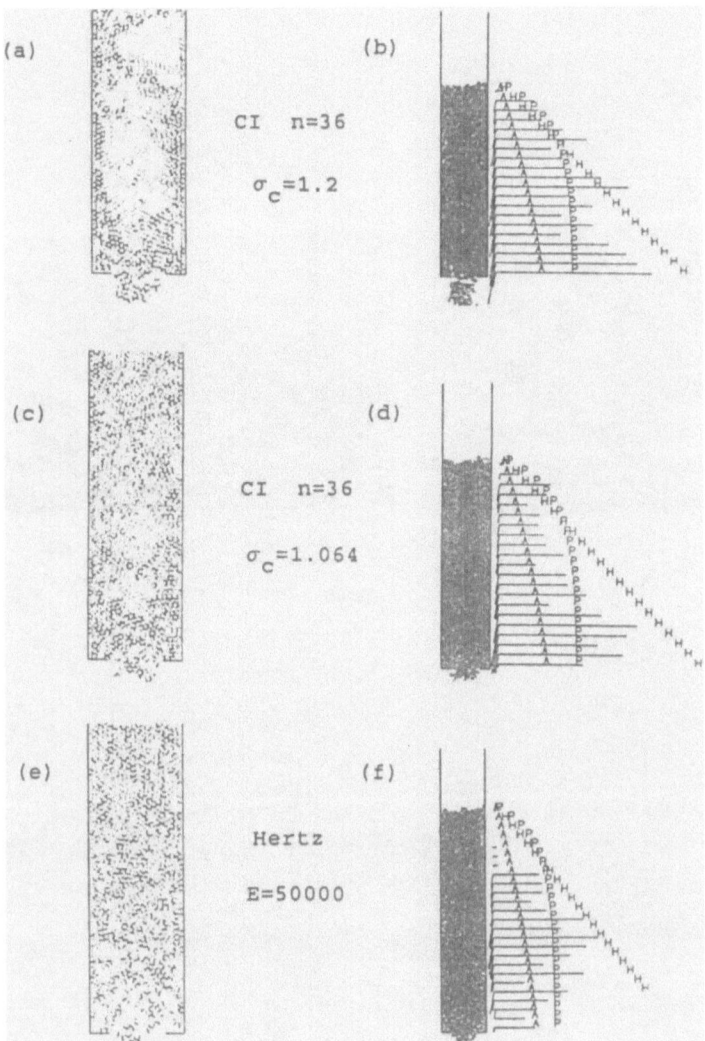

Figure 4. Internal tangential displacements and wall stresses predicted with different particle interactions (a), (b) CI with n=36 and r = 1.2d cut-off, (c), (d) CI with n=36 a cut-off at r = 1.064 d and (e) and (f) Hertzian interaction with E* = 50,000 and cut-off at r = d.

predictions from the simulation case studies. Visual as well as quantitative comparisons are made of the nature of flow, angle of the flow region boundary, particle velocities at the centre and near the walls, and the rotational motion of the particles during gravity flow. *Figure 5* shows the photographic measurements of the flow region bundary and the rotational movement of particles respectively in comparison with the simulations carried out using Hertzian Interaction Law with E* = 50,000 (reduced units). Both the stagnant regions in funnel-flow and the degree of rotational movement of the individual particles within the flow field appear to compare well with simulation results. For more details of quantitative comparisons, the reader is referred to Langston (1996). These comparisons are

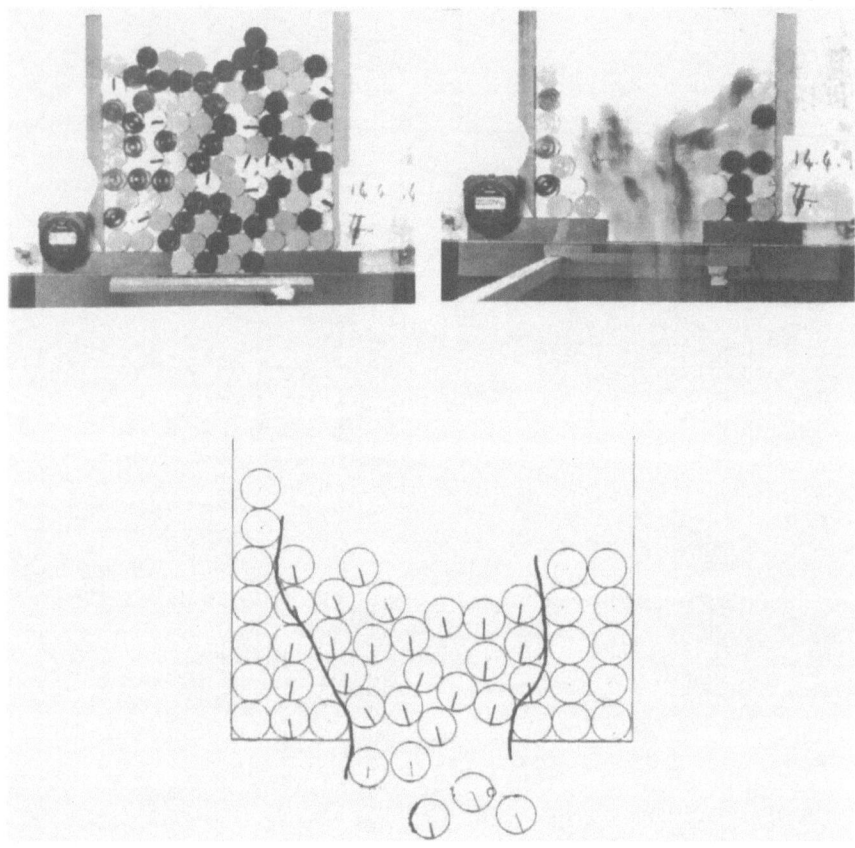

Figure 5. Flow region boundary and particle velocity vectors in 2-D funnel flow hopper compared with numerical simulations

believed to be the first of their kind in literature as they attempt to match precisely the dimensions of the particles and of the plane-strain hopper.

6. Axial -Symmetric (3-D) Simulations

Figure 6 compares the internal distributions of the particle velocity vectors (normalized to the maximum velocity during discharge) at the mid-planes of both funnel-flow and mass-flow containers. In the flat-bottom container, (*Fig.6(a))*, the position of the so-called stagnant zone boundary is predicted at a snapshot in time during discharge. It is most noticeable that the stagnant zones on either side of the converging flow zone are quite different in both size and shape thus confirming assymetry of flow about the centreline. A similar snapshot is shown in *Fig.6(b)* of the mass-flow flow field in a conical hopper. Again here, significant flow assymetry is observed coupled with noticeable cascading behaviour of the directions of the velocity vectors. As in 2-D results presented above. the effect of a smooth normal interaction potential (CI with n=36) with a long interaction cut-off (r = 1.2d) allows significant long-range connectivity to develop within the bulk. These effects are much less visible when Hertzian and Hooke's spring interactions are used.

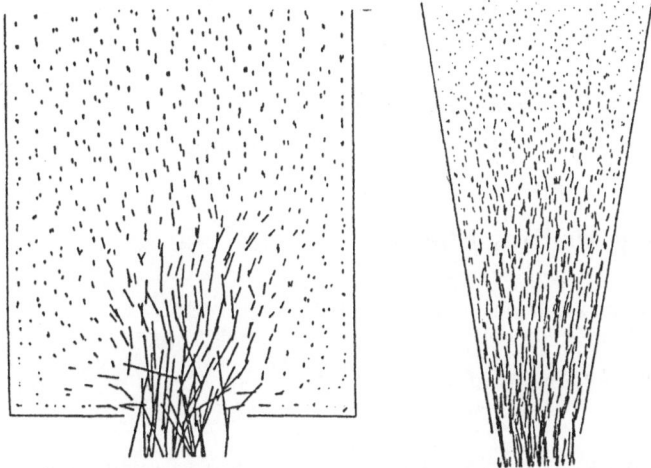

Figure 6. Comparisons of particle velocity vectors in a) funnel-flow and in b) mass flow hopper simulation snapshots during steady discharge at the mid-plane of axial-symmetric containers

6.1 COMPARISONS WITH 3-D FLOW EXPERIMENTS

The cross-sectional distributions of the solids and the interstitial voids are determined at different horizontal planes along the height of a model hopper using a CAT Scanner ring through the centre of which the model hopper can be moved to desired elevations as seen in *Fig.7* . Both the the static fill and time-averaged flowing values of the solids and void fractions are measured to facilitate direct comparisons with the simulation results based on the Hertzian Interaction. *Fig.8* compares the static and time-averaged flowing values of the solid fraction across a conical hopper at two different elevations. In each case, close agreement is found between experiment and simulation using almost identical particle size and shape and vessel geometry. It is particularly encouraging that the simulation results predict correctly the extent of assembly dilation accompanying bulk flow as borne out by the γ -ray tomographic measurements.

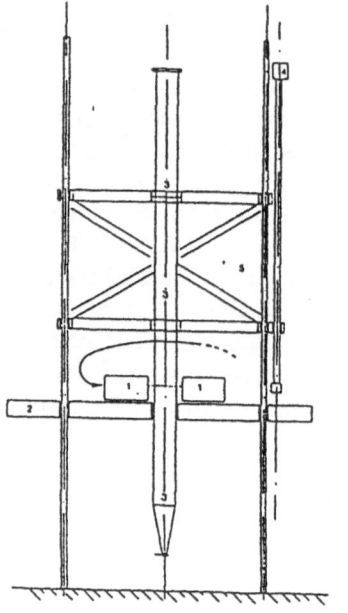

Schematic diagram of the experimental rig
with the parallel beam scanner: 1) tomographic scanner; 2) vibration-free table mount;
3) model hopper; 4) motorized vertical posi-
tioner; 5) model hopper support structure

Figure 7. Schematic of CAT scanner rig

7. Modelling Air-Drag Effects

The drag force on a spherical particle is given by :

$$F_D = (\Delta P / L) \; [\; V_p / (1 - \varepsilon)] \tag{12}$$

where V_p is the particle volume. This is a vector which acts in the direction of the
air-particle relative velocity. The interstitial air velocity is assumed to be radial,
the particle velocity , however, is not radial and hence F_D need not be radial . The
pore pressure drop term $(\Delta P / L)$ is calculated using a modified form of the
Ergun (1952) equation where the interstitial fluid velocity is replaced by the
modulus of the relative air particle velocity vector $|\, U_p - U_i \,|$. The value of the
local voidage, ε around a particle is calculated by averaging the void space
within the individual grid spaces of a fixed Eulerian Grid through which the
particles are allowed to move in Newtonian motion. The Ergun equation is then
used to calculate the local pressure drop which is then converted to a force from
the particle dimensions and the local voidage; for further details . see Langston et
al (1996).

In a series of simulations, both air-assisted (co-current) flow and air-impeded
(counter-current) flow cases are considered inside a hopper with a closed top .
The calculations are repeated with particles of density 100 and 1000 times the
surrounding interstitial fluid respectively to observe the effect of the magnitude of
the air drag on bulk flow. *Figure 9*compares the air-impeded flows of light and
heavy particles in a closed top hopper on the basis of the predicted particle velocity

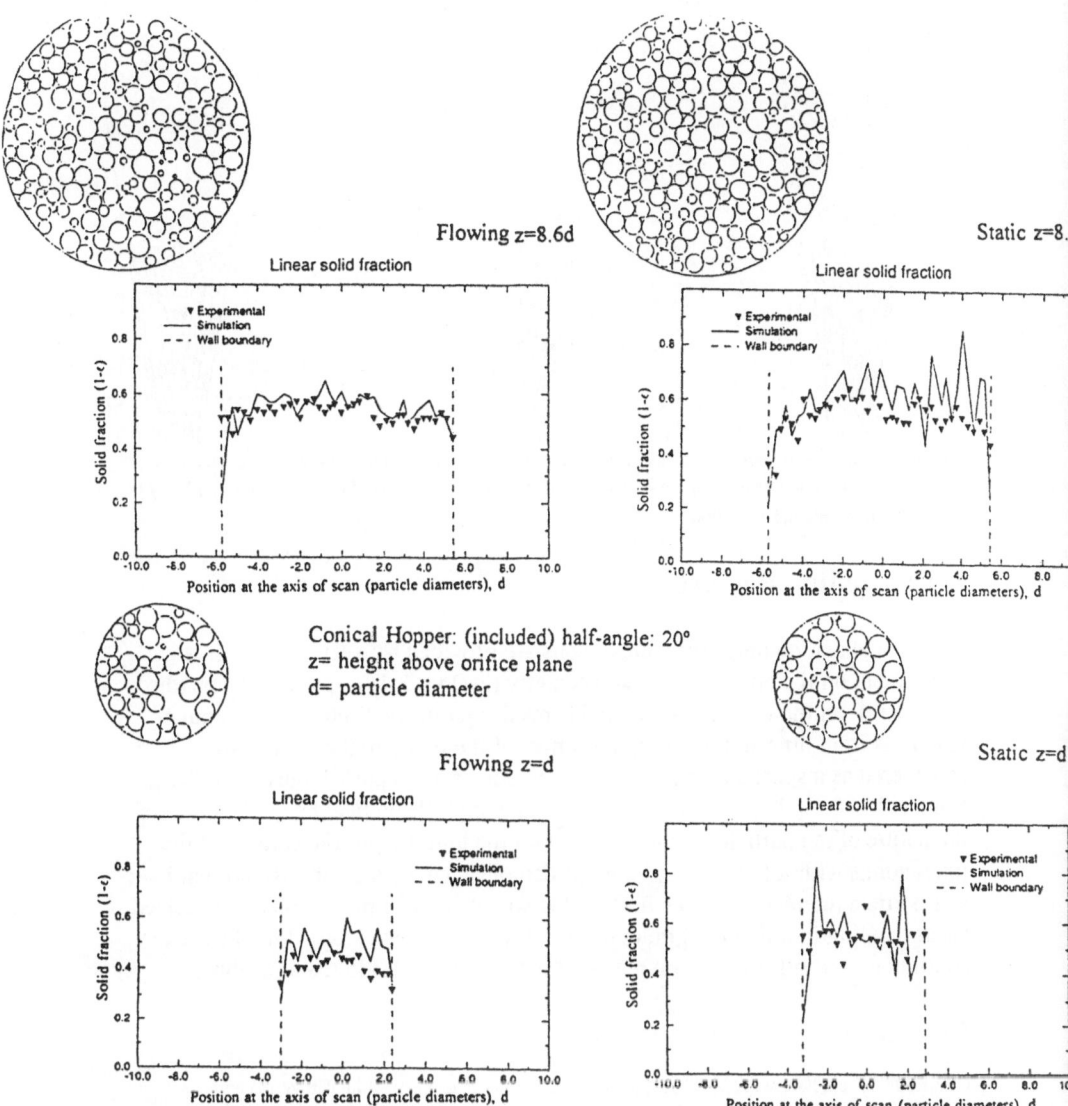

Figure 8. Comparisons of tomographic measurements with simulation predictions of static and
flowing solids fraction profiles in axially-symmetric flow

vectors normalized to the maximum velocity. Here. there is significant evidence of
the pulsing behaviour of the counter-current flow of interstitial air which is seen
clearly both from i) the synchronic shifts in the directions of the velocity vectors
and ii) the periodic arresting of particle flow at different heights . Such effects are
reported for the first time with Distinct Element calculations and are believed to
compare well with experimental observations; see Tüzün (1987).

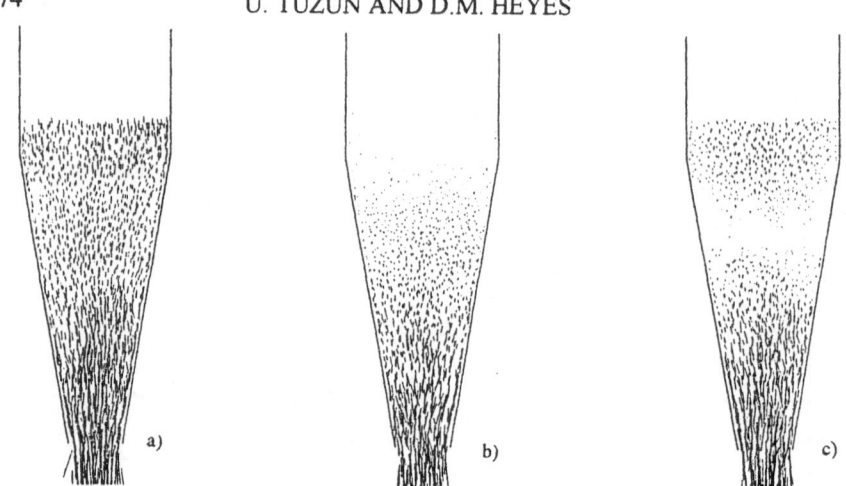

Figure 9. Snapshots of particle velocity vector fields during discharge at the mid-plane of a) an open-top conical hopper in comparison with a closed-top hopper; solid to fluid density ratio, ρ_s / ρ_f b) = 100 and c) = 1000.

8. Concluding Remarks

Both visual and quantitative comparisons are provided between simulation and experiment on the effects of vessel geometry (2-D vs. 3-D), particle shape (disc vs. sphere) on the flow and stress fields resulting during discharge of granular materials from bins and hoppers. The effect of the interstitial fluid is also considered as a special case and significant effects are reported during discharge from a closed-top hopper. At a more fundamental, theoretical level, the effect of the nature of the particle interaction laws on the long-range connectivity of the interactions within the bulk assembly is demonstrated by a comparison of hard and soft particle interactions. The first ever Distinct Element predictions of localized shear banding in a flowing granular assembly are presented based on the use of continuous and soft interaction potential between particles in the assembly.

References

Campbell C.S. and Potapov A.V. (1993), Recent Applications of Computer Simulations to Granular Systems, in *the Proceedings of the 1st Nisshin Engineering Particle Technology Int. Seminar*, January 18-20, Osaka, Japan.

Ergun , S. (1952), Fluid Flow through Packed Columns, *Chem. Eng. Progress*, **48**, 49.

Fincham D. and Heyes D.M. (1985), Recent Advances in Molecular Dynamics Computer Simulation, *Adv. Chem. Phys.* , **LXIII**, 475.

Johnson, K.L. (1985) , *Contact Mechanics*, Cambridge University Press, Cambridge.

Langston P.L. , Tüzün U. and Heyes D.M. (1995), Distinct Element Simulations of Granular Flow in 2D and 3D Hoppers, *Chem. Eng. Sci.*, **50**, 967.

Langston P.L. (1996) . *Distinct Element Modelling of Granular Flow in Hoppers*. Ph.D. Thesis, University of Surrey, Guildford.

Langston P.L. , Tüzün U. and Heyes D.M. (1996), Distinct Element Simulation of Interstitial Air Effects in Axially-symmetric Granular Flows in Hoppers, *Chem. Eng. Sci.*, **51**, 873.

Mindlin, R.D. and Deresiewicz, H. (1953), Elastic Spheres in Contact under Varying Oblique Forces, *J. Appl. Mech. (Trans. ASME)* , **20**, 327.

Nedderman R.M. (1992) , *Statics and Kinematics of Granular Materials*, Cambridge University Press, Cambridge. Tüzün U. (1987) , Critique of Recent Theoretical Advances in Particulate Solids Flow, in T. Ariman and T.N. Veziroglu (eds), *Particulate and Multiphase Processes*, Hemisphere Publ. , New York , **1**, 713.

GRANULAR AVALANCHES ON COMPLEX TOPOGRAPHY

J.M.N.T. GRAY
Institut für Mechanik
Technische Hochschule Darmstadt
64289 Darmstadt, Germany

1. Introduction

The Savage-Hutter theory for the one-dimensional gravity driven free-surface flow of cohesionless granular materials down inclines (Savage & Hutter, 1989) and curved beds (Savage & Hutter, 1991) was generalised to two-dimensions by Hutter *et al.* (1992) and Greve *et al.* (1994). Savage & Hutter (1991) introduced a simple curvilinear coordinate system Oxz with the x coordinate parallel to, and the z component normal to, the local one-dimensional chute geometry. In the two-dimensional theories a similar coordinate system $Oxyz$ was adopted with a lateral (or cross slope) coordinate y perpendicular to the x, z coordinates. Thus, the curvilinear surface $z = 0$ followed a quasi one-dimensional chute topography, with down slope but no cross slope variation. Agreement between experiments performed on such a chute and the predicted two-dimensional spreading of these unconfined granular avalanches was extremely good (Koch *et al.*, 1994). Recently the theory has been generalised (Gray *et al.*, 1996) to allow for *shallow* down and cross slope variation in the basal chute geometry. In this paper a comparison is made between the theoretical predictions and experimental results from two chutes with complex down and cross slope topography.

2. Governing equations

In the Savage-Hutter theory the avalanche is treated as an incompressible granular material satisfying a Mohr-Coulomb constitutive relation with internal angle of friction ϕ. The kinematic free surface is traction free and at the base the avalanche slides on a rigid impenetrable basal topography. A Coulomb-type rate-independent dry friction law, with basal friction angle δ, is used to model the *slip* at the base of the avalanche. The simple

275

N. A. Fleck and A. C. F. Cocks (eds.),
IUTAM Symposium on Mechanics of Granular and Porous Materials, 275–286.
© 1997 *Kluwer Academic Publishers.*

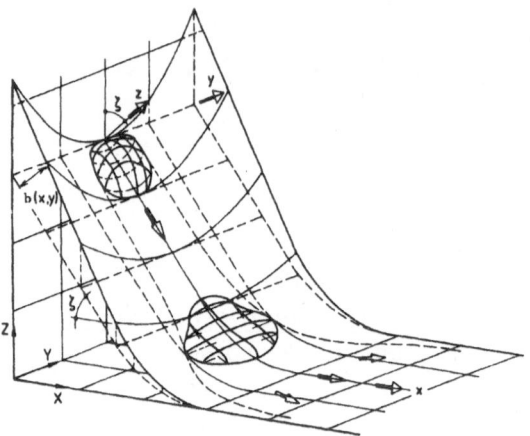

Figure 1.   The simple curvilinear coordinate system $Oxyz$ consists of an inclined plane connected to a horizontal plane by a smooth transition. The local inclination angle of the curvilinear surface $z = 0$ (dashed lines) is ζ. The actual basal topography $z = b(x, y)$ (solid lines) is superposed to top of the $z = 0$ curvilinear surface.

curvilinear coordinate system $Oxyz$ introduced in §1 is adopted here, however, following Gray *et al.* (1996) the $z = 0$ curvilinear surface no longer represents the basal topography. Instead the actual chute geometry is prescribed by defining its height $b(x, y)$ above the curvilinear surface $z = 0$. The curvilinear coordinate system (dashed lines) and the *superposed* basal topography (solid lines) is illustrated in Figure 1. For notational purposes we shall continue to refer to x as the down slope and y as the cross slope directions, respectively, even though the physical downslope direction may no longer be in the x direction. Extensive use is made of the fact that the avalanche and the basal geometry are shallow. That is the typical downslope length scale for order unity changes in the avalanche and the basal geometry are much greater than their heights. This together with the fact that the down and cross slope velocity components are nearly independent of depth allows the mass and momentum equations to be integrated through the avalanche depth to remove one space dimension.

The depth integrated mass balance reduces to

$$\frac{dh}{dt} + h\left(\frac{\partial u}{\partial x} + \frac{\partial v}{\partial y}\right) = 0, \tag{1}$$

where h is the avalanche thickness and u, v are the downslope and cross slope components of the velocity $\boldsymbol{v}$. The leading order depth integrated momentum balances reduce to

$$
\left.
\begin{array}{l}
\dfrac{du}{dt} = g\sin\zeta - \dfrac{u}{|\boldsymbol{v}|}\tan\delta(g\cos\zeta + \kappa u^2) - g\cos\zeta\left(K_x\dfrac{\partial h}{\partial x} + \dfrac{\partial b}{\partial x}\right), \\[3mm]
\dfrac{dv}{dt} = \qquad - \dfrac{v}{|\boldsymbol{v}|}\tan\delta(g\cos\zeta + \kappa u^2) - g\cos\zeta\left(K_y\dfrac{\partial h}{\partial y} + \dfrac{\partial b}{\partial y}\right),
\end{array}
\right\}
\tag{2}
$$

where g is the gravitational acceleration, ζ, $\kappa = -\partial\zeta/\partial x$ are the local inclination angle and the local curvature of the curvilinear surface $z = 0$ and K_x, K_y are earth pressure coefficients. The essential difference between these equations and those of Greve et $al.$ (1994) are the basal topography gradients $\partial b/\partial x$ and $\partial b/\partial y$ on the right-hand side of (2). The remaining terms on the right-hand side are the gravity acceleration, the basal friction and the earth pressure terms that arise from the stress divergence.

The earth pressures, which relate the $limiting$ stresses normal and parallel to the inclined plane, were introduced by Savage & Hutter (1989) and subsequently generalised to two-dimensions by Hutter et $al.$ (1992). The earth pressure coefficients are either $active$, or $passive$, depending on whether the motion is dilatational, or compressional

$$
K_x = \begin{cases} K_{x_{act}} & \partial u/\partial x > 0, \\ K_{x_{pas}} & \partial u/\partial x < 0, \end{cases}
\tag{3}
$$

$$
K_y = \begin{cases} K_{y_{act}}^{x_{act}} & \partial u/\partial x > 0, & \partial v/\partial y > 0, \\ K_{y_{act}}^{x_{pas}} & \partial u/\partial x < 0, & \partial v/\partial y > 0, \\ K_{y_{pas}}^{x_{act}} & \partial u/\partial x > 0, & \partial v/\partial y < 0, \\ K_{y_{pas}}^{x_{pas}} & \partial u/\partial x < 0, & \partial v/\partial y < 0. \end{cases}
\tag{4}
$$

Their values are given by

$$
K_{x_{act/pas}} = 2\sec^2\phi\left(1 \mp \{1 - \cos^2\phi\sec^2\delta\}^{1/2}\right) - 1,
\tag{5}
$$

$$
K_{y_{act/pas}}^x = \frac{1}{2}\left(K_x + 1 \mp \{(K_x - 1)^2 + 4\tan^2\delta\}^{1/2}\right),
\tag{6}
$$

and it is easy to show that $K_{x_{act}} \leq K_{x_{pas}}$ and $K_{y_{act}}^{x_{act}} \leq K_{y_{act}}^{x_{pas}} \leq K_{y_{pas}}^{x_{act}} \leq K_{y_{pas}}^{x_{pas}}$. That is the active earth pressure coefficients are always less than the passive earth pressure coefficients.

3. Comparison between experiment and theoretical predictions

Laboratory experiments have been performed on two chutes in order to validate the theory. The first geometry (V02) has a shallow concave parabolic cross slope profile on the inclined section of the chute and broadens out into a flat horizontal plane (Fig. 2a). The second topography (ISO3)

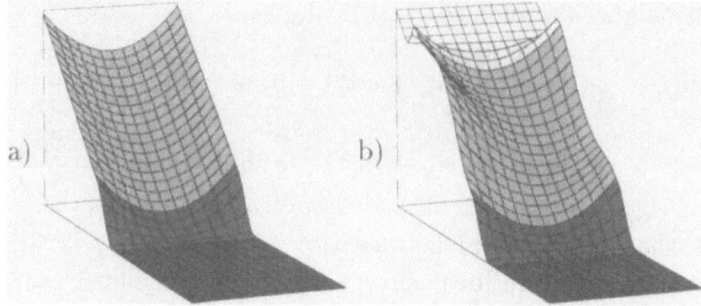

Figure 2. The basal chute geometry used in the experiments and numerical computations: a) symmetric parabolic chute (V02), b) the snaking valley chute (ISO3). The
shading highlights the different zones in the respective chutes.

consists of an inclined flat plane, which narrows into a shallow parabolic
channel that has an additional shallow *snaking valley* topography in the x-
direction, before once again broadening out into the horizontal plane (Fig.
2b). The snaking valley is achieved by a sinusoidal down slope shift of the
cross slope parabolic profile.

TABLE 1. Material parameters and chute geometry

Experiment	V02	ISO3
Bed friction angle, δ	30°	29°
Internal angle of friction, ϕ	40°	35°
Chute inclination, ζ	40°	40°
Channel radius of curvature, R	110 cm	85 cm
Amplitude of sinusoid	0 cm	20 cm
Cap radius, r_c	32 cm	18 cm
Cap height, h_c	22 cm	18 cm

Numerous laboratory experiments have been performed on both chutes.
In this paper we describe only one experiment, on each of the chutes, that is
representative of the general characteristics of the flow. The chute geometry
and the properties of the granular materials are summarised in Table 1.
Typical run times are of the order of 1.5-2 seconds for V02 and 2-2.5 seconds
for experiment IS03. The flowing avalanche is recorded on video and high
speed camera. This allows the avalanche edge to be determined at a series
of time steps. Two experimental parameters are required for the theoretical
predictions: the granular material's internal angle of friction ϕ and the bed
friction angle δ. The internal angle of friction is determined by measuring

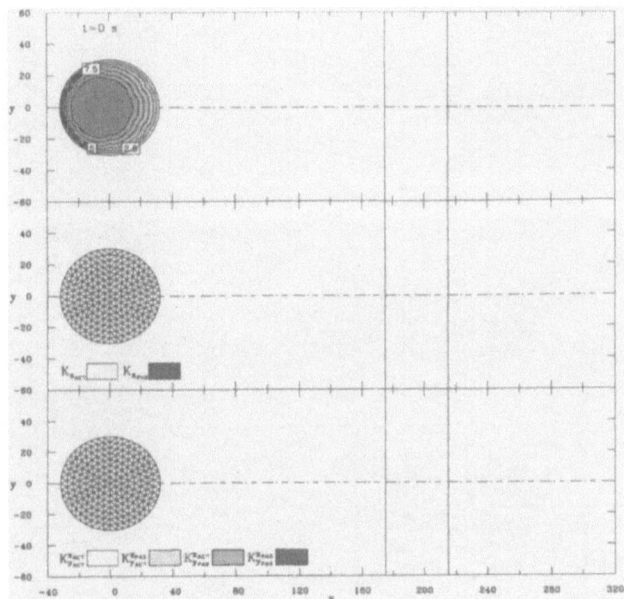

Figure 3. The initial configuration of experiment V02 is illustrated in three panels using projected curvilinear coordinates (x, y). The avalanche runs from left to right and all lengths are in cm. The solid lines at $x = 175$ and $x = 215$ cm indicate the position of the transition zone between the 40° inclined parabolic section to the left and the horizontal plane to the right of the chute. The dashed line along $y = 0$ indicates the Talweg. The top panel shows the avalanche thickness (in cm), the middle panel the downslope earth pressure and the bottom panel the cross slope earth pressure. The active/passive earth pressure states are indicated by shading the elements, a key is given on the plot. The time is indicated in the top left corner.

the steepest angle of inclination of a conical pile of the granular material. Whilst the basal angle of friction is measured by tilting a planar section of the bed topography and measuring the angle at which a static sample of the granular material begins to slide.

A Lagrangian explicit finite difference scheme is used to solve the system of equations (1)–(6). This has been developed by Wieland *et al.* (1996) and is based on the original one-dimensional method of Savage & Hutter (1989). The avalanche is initially discretised into a finite number of triangular elements. It follows from the thickness continuity equation (1) and Reynolds Transport theorem that the initial volume of granular material contained in each of the elements is preserved throughout the flow. At each time-step the algorithm then proceeds as follows, (1) the new positions of the element corners are computed by an explicit forward step in time, (2) the area of the new triangle is computed, (3) the thickness of the element is then determined by dividing the initial element volume by its current area, (4) the new velocity is then computed by an explicit step in time. The grid

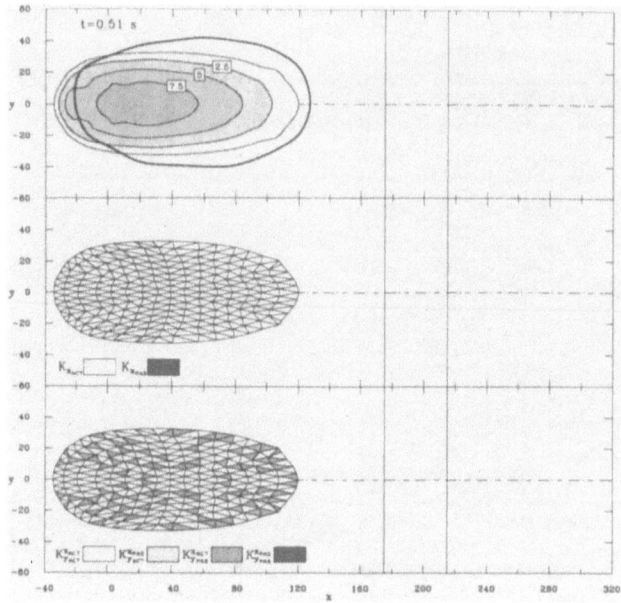

Figure 4. The same as Fig. 3. The thick line in the top panel indicates the experimentally determined position of the avalanche edge.

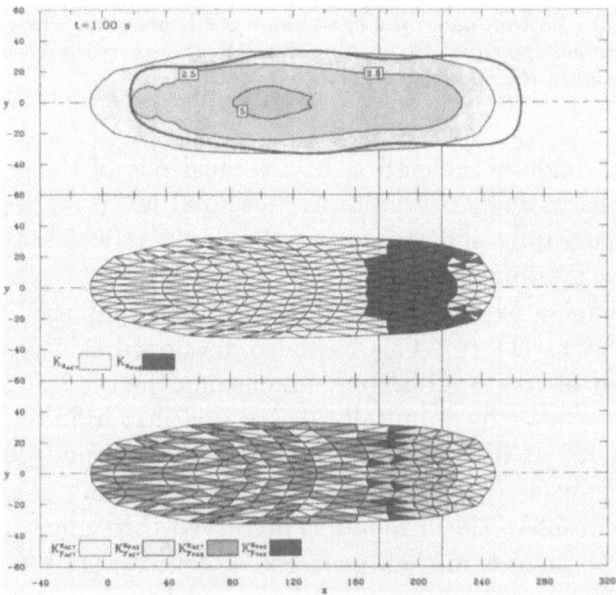

Figure 5. The same as Fig. 3. The thick line in the top panel indicates the experimentally determined position of the avalanche edge.

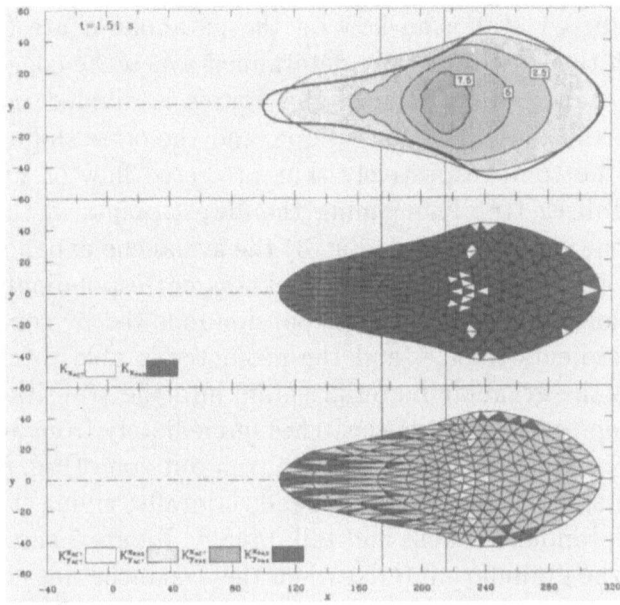

Figure 6. The same as Fig. 3. The thick line in the top panel indicates the experimentally determined position of the avalanche edge.

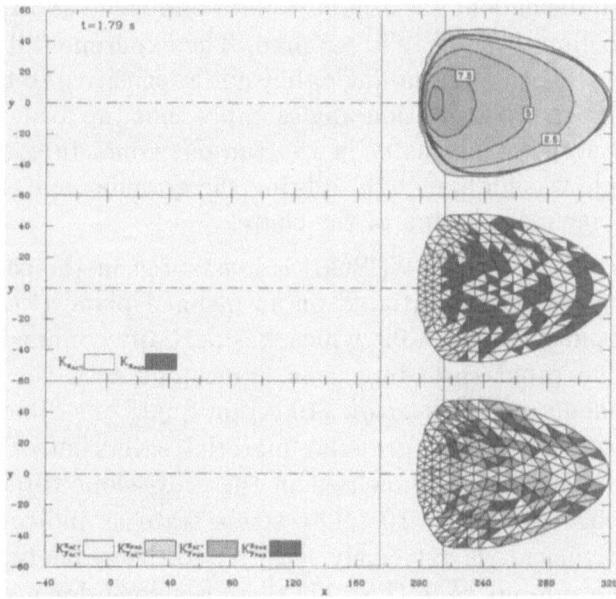

Figure 7. The same as Fig. 3. The thick line in the top panel indicates the experimentally determined position of the avalanche edge.

mesh therefore moves and deforms with the avalanche.

The predicted avalanche flow on the parabolic chute (V02) and comparison with the experimentally determined avalanche edge is shown in the sequence of figures 3–7. Each of the figures is divided into three panels showing the thickness, the down slope, and the cross slope earth pressures from top to bottom, respectively. The predicted flow of V02 is symmetric about the Talweg (the line joining the lowest points in the valley) as expected. As the cap is released (Fig. 3) the avalanche expands rapidly in the down-slope direction reducing in thickness correspondingly. In the cross-slope direction there is an initial expansion followed by convergence, as the thickness gradients reduce, and the geometry is able to channel the flow (Fig. 4). As the granular material enters into the transition zone (Fig. 5) the down slope earth pressure switches immediately from active to passive. And, as it proceeds into the horizontal run-out zone (Fig. 6) it compresses in the down-slope direction and expands laterally, giving it a *tadpole* structure with a pronounced nose and tail. Finally, Figure 7 shows the predicted position of the granular material when the avalanche has come to rest, this is in extremely good agreement with the experimental results. During the flowing phase of the predicted avalanche the tail section has a tendency to move somewhat slower than the experiments suggest. This can be compensated for by linearly reducing the bed friction angle along the avalanche length. A physical explanation for this phenomenon is that the bed friction is not rate-independent (as assumed here) and some additional velocity dependent Voellmy-type drag is required. The experimental results are quite sensitive to the bed-friction angle, but not as sensitive to the internal angle of friction. Larger bed friction angles imply that the avalanche moves more slowly and stops much faster in the run-out zone. In some cases the bed friction angle was high enough to bring the granular material to rest whilst still on the inclined section of the chute.

The second experiment (ISO3) is illustrated in the sequence of figures 8–13. The avalanche is initiated on an inclined plane (Fig. 8) and spreads far more rapidly than in V02, which was partially confined in the parabolic channel. The rapid spreading is in agreement with the initial phases of the experiments of Koch *et al.* (1994) on quasi one-dimensional surfaces. However, as soon as the granular material enters into the valley section (Fig. 9) it is strongly channelised in the cross-slope direction and *steered* by the topography (Fig. 10). The stress state as indicated by the earth pressure coefficients is extremely sensitive to the geometry of the chute. As it enters the run-out zone (Fig. 11) there is strong down-slope convergence and cross-slope spreading again giving it a nose and tail like tadpole structure. Figure 12 shows the position of the predicted avalanche at the time that the real avalanche comes to rest. Again we see that the tail moves

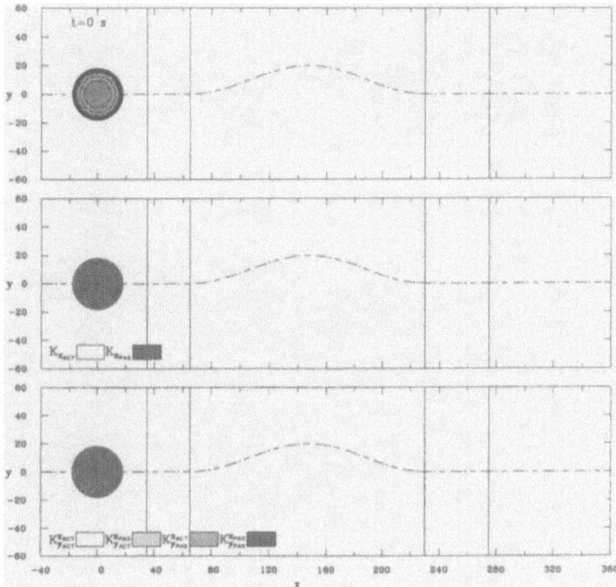

Figure 8. The same as Fig. 3 except for experiment ISO3. Initially the granular material is released on a 40° inclined plane. The lines at $x = 35, 65, 230, 275$ cm indicate the position of the transition zones between the inclined plane, the sinusoidal snaking valley and the horizontal plane. The dashed line shows the position of the Talweg.

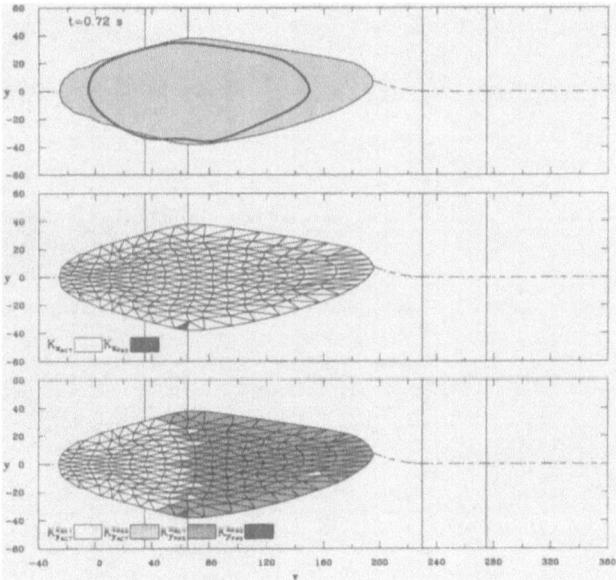

Figure 9. The same as Fig. 8. The thick line in the top panel indicates the experimentally determined position of the avalanche edge.

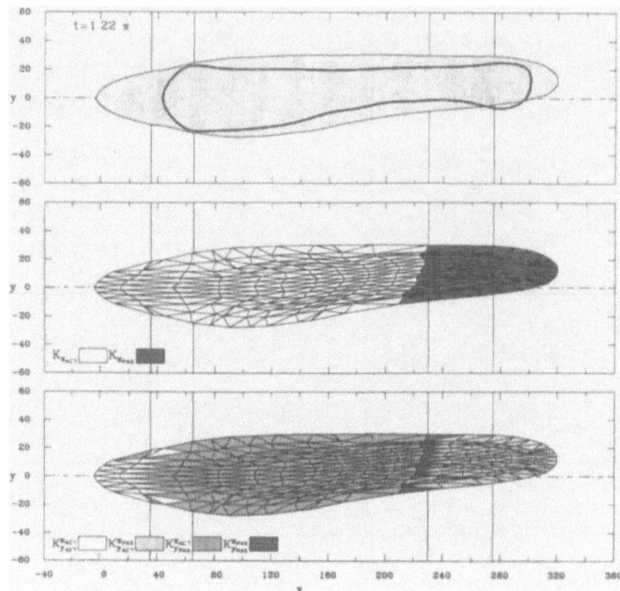

Figure 10. The same as Fig. 8. The thick line in the top panel indicates the experimentally determined position of the avalanche edge.

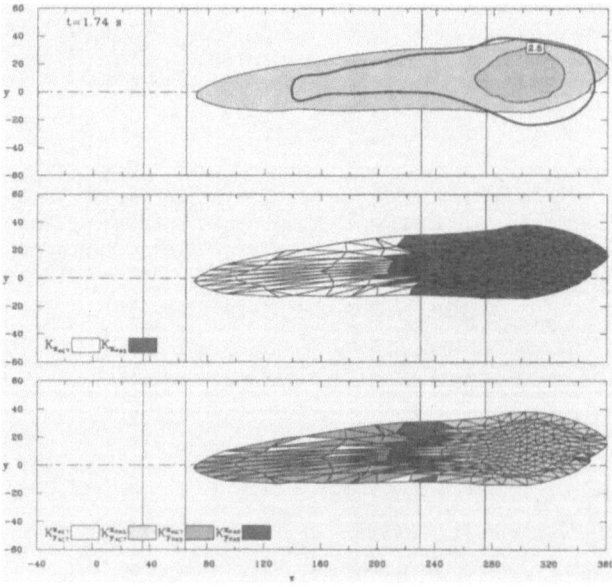

Figure 11. The same as Fig. 8. The thick line in the top panel indicates the experimentally determined position of the avalanche edge.

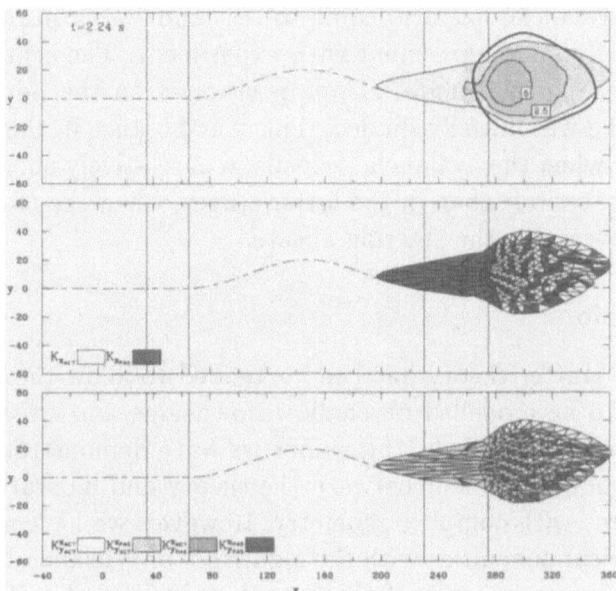

Figure 12. The same as Fig. 8. The thick line in the top panel indicates the experimentally determined position of the avalanche edge.

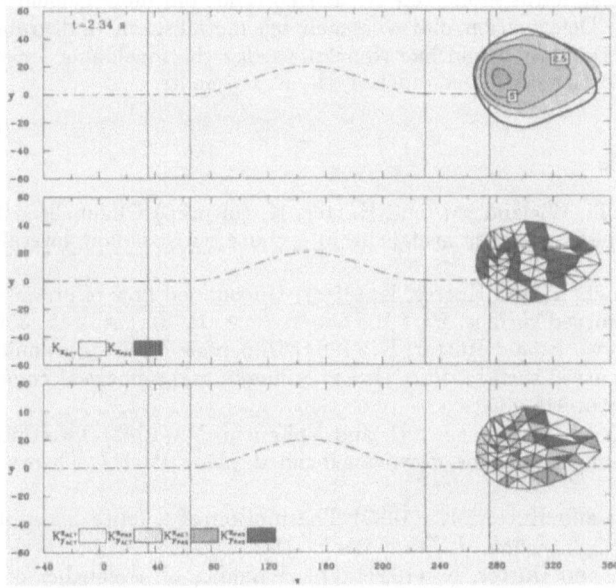

Figure 13. The same as Fig. 8. The thick line in the top panel indicates the experimentally determined position of the avalanche edge.

somewhat too slowly, however, a tenth of a second later in figure 13 the computed avalanche has also come to rest and the position of the deposit is in extremely good agreement with experiment. The centre of the resulting pile of avalanched material was positioned on the side towards which the avalanche was initially guided. This was because in the initial stages of the motion, when the avalanche velocity was relatively small, the geometry played a greater role than in the latter stages, when the momentum is high enough to overcome the steering effect.

4. Discussion

The Savage-Hutter theory has been extended to allow the flow of granular avalanches to be modelled on shallow downslope and cross slope topography (Gray et al., 1996). In this paper we have demonstrated that there is extremely good agreement between the theory and laboratory experiments on two chutes with complex geometry. However, we have also encountered some numerical difficulties with the algorithm of Wieland et al. (1996) when there is rapid convergence of the granular material. In particular the fine mesh computation illustrated in figure 12 becomes numerically unstable shortly after $t = 2.24$ s, and the picture of the final deposit (Fig. 13) was produced by using a courser mesh. Improvement of the numerical algorithm is a major thrust of our current research efforts.

This research was supported by the Deutsche Forschungsgemeinschaft SFB 298 project "Deformation und Versagen bei metallischen und granularen Strukturen". Dr Gray would like to acknowledge the invaluable assistance of Herr Wall and Herr Tai in conducting the experiments.

References

Gray, J.M.N.T., Wieland, M. and Hutter, K. (in prep) Channelised free surface flow of a cohesionless granular avalanche in a chute with shallow lateral curvature. Part I: theory.

Greve, R., Koch, T. and Hutter, K. (1994) Unconfined flow of granular avalanches along a partly curved surface. Part I: Theory. Proc. R. Soc. A, 445, 399-413.

Koch, T., Greve, R. and Hutter, K. (1994) Unconfined flow of granular avalanches along a partly curved surface, Part II: Experiments and numerical computations. Proc. R. Soc. A, 445, 415-435.

Hutter, K. Siegel, M. Savage, S.B. and Nohguchi, Y. (1992) Two-dimensional spreading of a granular avalanche down an inclined plane Part I. Theory. Acta Mech., 100, 37-68.

Savage, S. B. and Hutter, K. (1989) The motion of a finite mass of granular material down a rough incline. J. Fluid Mech., 199, 177-215.

Savage, S. B. and Hutter, K. (1991) The dynamics of avalanches of granular materials from initiation to runout. Part I: Analysis. Acta Mech., 86, 201-223.

Wieland, M. Gray, J.M.N.T. and Hutter, K. (in prep) Channelised free surface flow of a cohesionless granular avalanche in a chute with shallow lateral curvature. Part II: Numerical predictions and comparison with experiments.

TRACKING PARTICLES IN TUMBLING CONTAINERS

GUY METCALFE
CSIRO/DBCE Advanced Fluid Dynamics Laboratory
Box 56, Highett, Victoria 3190, Australia

> *Here are sands, ignoble things, ...*
> *–Francis Beaumont (1586-1616)*

1. Introduction

In answer to the question why is fluid dynamics more developed as a science than granular dynamics, one may cite that the Navier-Stokes equations of fluid motion are more developed and more extensively tested. A collection of grains—moving or static—inherently involves the intermediate or mesoscopic scales of motion: the strong separation of scales from the molecular to the macroscopic flow (Batchelor, 1967) that permits the continuum equations to well embody the physics of flowing fluids does not exist with granular materials, even for the smallest sized grains. Though this is undoubtedly true and represents a substantial theoretical challenge, perhaps in answering our question, we should not discount the ability of the æsthetic qualities of the flows themselves to motivate investigations. Fluid flows are beautiful. Typically opaque grains, on the other hand, mask what motivational beauty might lie in granular flow patterns. In this paper I describe some recent experiments visualising the granular transport in tumbling containers. In §2 I describe quasi-2-dimensional experiments in a rotating disk where the patterns can be observed by eye. These are experiments with the desirable qualities that the mixing behaviour can be explicitly viewed, quantified, and substantially modelled. In §3 I describe experiments in the 3-dimensional extension of the disk—a rotating tube. The 3-d flow is visualised with a non-invasive magnetic resonance imaging (MRI) technique which gives direct information on the radial and axial flow patterns. Non-invasive imaging (Nakagawa *et al.*, 1993; Broadbelt *et al.*, 1993; Nikitidis *et al.*, 1994; Jaeger *et al.*, 1996) seems on the verge of providing a wealth of new experimental information on 3-d granular flows.

N. A. Fleck and A. C. F. Cocks (eds.),
IUTAM Symposium on Mechanics of Granular and Porous Materials, 287–298.
© 1997 *Kluwer Academic Publishers.*

While aesthetics may pique our interest, practical concerns may also motivate us, and these experiments bear on the mixing and segregating properties of grains and powders, a subject of much engineering interest through the years (Fan *et al.*, 1990; Bridgwater, 1976; Lacey, 1954) and now coupled to a recent explosion of interest in granular mechanics in the physics community (Jaeger *et al.*, 1996; Metcalfe *et al.*, 1995; Mehta, 1994).

2. 2-Dimensional Tumblers

Tumblers are common solids processing devices, and for these experiments, are partially filled with coloured particles and rotating about an axis. A quasi-2-d tumbler can be a disk (figure 2) with an aspect ratio (diameter/depth) of 6/1, where the large-scale motions are confined to a plane. With increasing rotation speed, the flow in tumblers changes through a variety of regimes (Fan *et al.*, 1990). Here we confine ourselves to *slow flow*. Slow is defined so that one avalanche ceases completely before the next one begins. Material below the surface layer, which is only ≈ 6 particle diameters deep, rotates as a solid body with the container. For slow rotation the material's surface remains stationary until its angle with the horizontal exceeds a critical value θ_f where the surface fails and flows, creating an avalanche that returns the surface to its angle of repose θ_r; $\theta_r < \theta_f$. Iterated surface avalanches mix the material.

2.1. MIXING GEOMETRY AND DYNAMICS

The following observation motivates a very simple, yet powerful, model for tumbling mixers. The angle $|\theta_f - \theta_r|$ defines a wedge of material (see Metcalfe *et al.* (1995) or McCarthy *et al.* (1996) for wedge construction details). An avalanche is supposed to have two actions: the first moves the upper surface wedge to the corresponding position on the lower surface; the second rearranges the avalanching material. This divides avalanche transport neatly into two parts: geometric transport *of* wedges and dynamic transport *within* wedges. Surprisingly, quantitative agreement with experiment does not require the correct dynamics. We may represent the dynamics with a map of the wedge into itself, choosing the map from deep physical principles, or pulling one out of thin air and intuition. A random map is the simplest to choose. Mixing between wedges occurs because wedges overlap; the size of the overlap determines the mixing rate. Wedge overlap increases with decreasing fill level f and disappears at $f = 1/2$, above which a non-mixing "core" appears. The core area grow as f increases, but as avalanches never enter the core, core material does not move. Outside the core for $f > 1/2$ material mixes because wedge intersections re-emerge.

Tracking the positions of the coloured particles' oscillating (due to disk

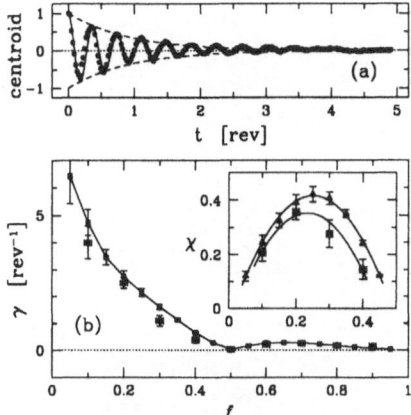

Figure 1. (a) Experimental centroid position normalised to that of the whole material. When perfectly mixed, the colour centroids coincide at the origin. Dots are salt data; solid lines a fit to $\exp(\gamma t)\cos(2\pi t/T)$. (b) Mixing rate γ versus fill level. Open symbols from model; filled symbols from experiments. For $f > 1/2$ the core is removed from the calculation. Inset shows $\chi \equiv \gamma V(f)$ the volume mixing rate versus f. Adapted from Metcalfe *et al.* (1995).

rotation) and exponentially decaying (due to iterative avalanches) centroids allows measurement of a mixing rate from the decay time γ. Figure 1a shows data as a function of time for one centroid's orbit. The experiments and simulations agree remarkably well, considering the correct dynamics is unknown. With a mixing rate, we may also find the optimum mixing efficiency from $\chi \equiv \gamma V(f)$ the volume mixed per characteristic time (inset of figure 1b). For less symmetric containers the arguments relating f to mixing are more complicated and depend on both f and the container's centre of mass, one consequence of which is χ increases for less symmetric containers, e.g. $\chi_{\max}$ for a triangle is almost 10% higher than for a circle (McCarthy *et al.*, 1996; Wolf, 1995). For all containers carrying similar particles, flow geometry determines the mixing behaviour.

2.2. ADDED COMPLEXITY: COHESION AND SEGREGATION

The experimental and modeling ease of 2-d tumbling devices make them useful testbeds for adding realistic complications in a controlled way. In this section I briefly outline some effects on patterns due to particle cohesion and size segregation. McCarthy *et al.* (1996) describe various baffling schemes.

Figure 2 shows initial conditions and one effect of mixing with a cohesive component. In the top row the particles are non-cohesive and differ only in colour. In the bottom row the darker particles on the left are non-cohesive, while the lighter particles on the right are cohesive; particles are identical otherwise. Results are preliminary, but one visually apparent difference is that after an avalanche the free surface is no longer flat during the initial stages of mixing. The cohesion supports slope discontinuities that influence where subsequent avalanches start. Also, material streaks appear in both

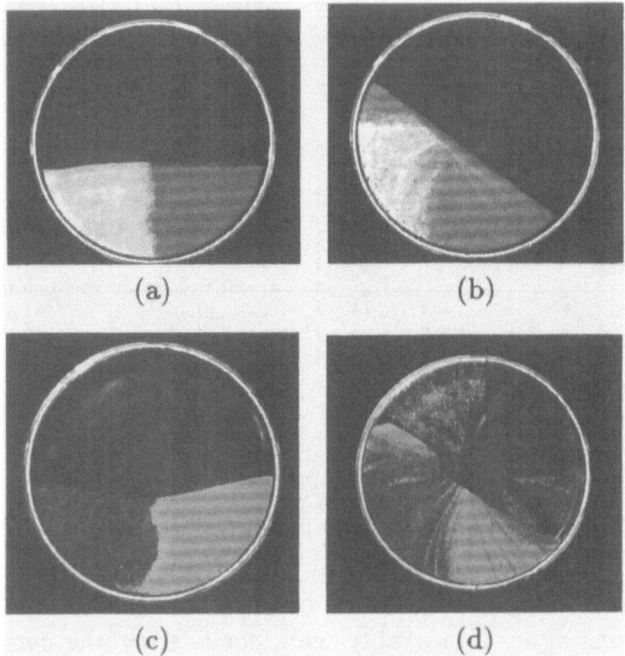

Figure 2. Mixing cohesive and non-cohesive materials. The containers are about 1/3 full and slowly rotated. (a) and (c) are initial conditions with the material on the right being cohesive and the material on the left being non-cohesive. (b) and (d) show the results after 2 container revolutions. Cohesion supports discontinuities in the free surface slope. Streaks of material appear in both cases but the streaking is more pronounced with the cohesive component. In the cohesive case these streaks may act as "lubricating" layers to boost the mixing rate. In (d) some grains stick to the glass cover.

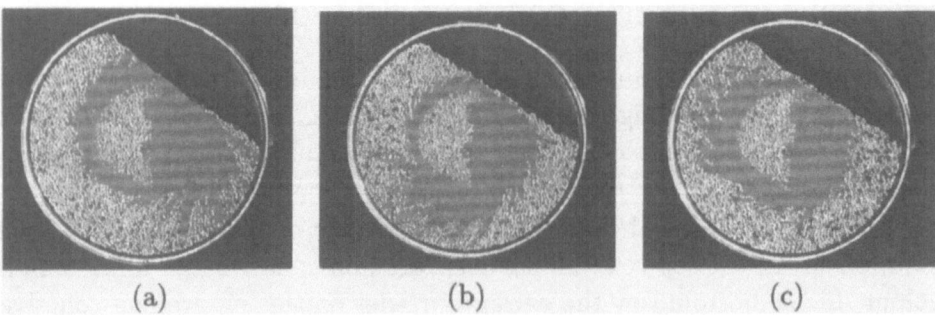

Figure 3. Pattern precession outside the core during tumbling with segregating materials: (a) after 3 container revolutions; (b) after 4; (c) after 5. Container rotation is clockwise. The layered pattern precesses ahead about 90°/container revolution. Aside from the precession the pattern is statistically stationary. Materials are 1.8 mm sugar balls and 0.6 mm salt cubes initially separated as in figure 2.

cases but the streaking is more pronounced with cohesion, where streaks appear when non-cohesive particles flow from beneath cohesive particles.

Figure 3 shows an example of radial segregation where larger particles move to the outside of the container and form a layered pattern that rotates about 90° ahead of the container rotation. This precessing pattern is essentially unchanging with time. While the geometrical decomposition has much to say about mixing granular materials—e.g. a core is prominent in figure 3—for the geometric model to capture segregation it must have the correct dynamics. However, hybrid models (McCarthy *et al.*, 1996) combining the geometric spirit with a particle dynamics simulation of the avalanche can capture several large-scale features of segregating materials.

2.3. CORE PRECESSION AND MODELLING CHALLENGES

While figure 3 shows a precessing segregation pattern, there is also a subtler precessing motion of the core illustrated in figure 4. In the model of §2.1, the core has a purely geometrical origin, implying the core should not move. However, figure 4 shows that the core does move. After every full revolution of the container, the line demarking the initial interface between the coloured grains should return to its original orientation. However, the line rotates past its original position. Precession depends on the container shape; it seems to happen only in containers of about equal width to height ratios. The precession rate, i.e. how many degrees the core precesses per container revolution $\dot{\theta}_p$ depends on f. $\dot{\theta}_p$ is highest just above half full and falls to zero for completely full. The precession rate also depends on the material. There is an order of magnitude difference in $\dot{\theta}_p$—the slopes of the points graphed in figure 4—between 0.6 mm salt cubes and 1.8 mm sugar balls. The physical mechanism driving core precession is at present obscure, but, since the precession rate depends on the material, it is an effect that does not seem explainable from geometry alone.

The detailed features of these patterns and their motions are most likely *not* in and of themselves important for industrial process design. However, they are important as validating tests for models of granular motion. To my knowledge no granular computation reports these very noticeable macroscopic motions. This may be a challenging problem.

3. 3-Dimensional Tumblers

When particle dissimilarities lead to segregation instead of mixing, some questions of interest are: How closely matched must particle properties be for segregation to dominate? Do all segregation mechanisms produce similar effects? How do mixing patterns change as segregation effects increase? Here I consider only size ratio s and density ratio d differences in a slowly rotating

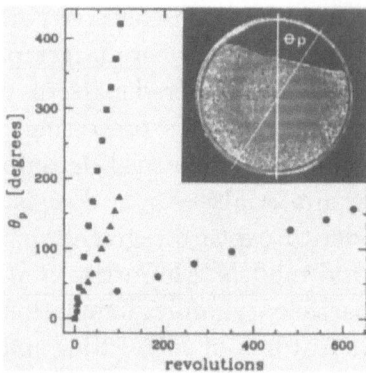

Figure 4. Core precession with different materials. The photograph shows an equal volume mix of sugar balls and salt cubes. The core precesses θ_p degrees after 20 container revolutions. The graph plots θ_p versus number of container revolutions for several materials: 1.8 mm sugar balls □, 0.6 mm salt cubes ○, and "art" sand △. The precession rate $\dot{\theta}_p$ is the slope of these lines.

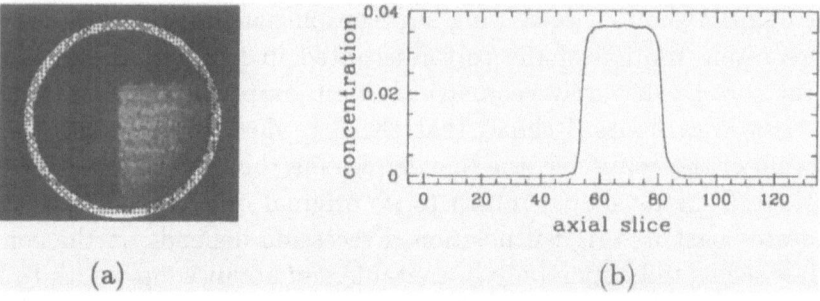

(a)   (b)

Figure 5. (a) Initial condition in the tube for radial mixing experiments. Seeds show as white. Other grains do not appear in the MR image. (b) Initial condition in the tube for axial mixing experiments is a "blob" of seeds approximately in the middle 20% of the tube. The signal is summed over each slice, normalised, and plotted versus the axial slice number.

tube. Granular flow patterns are tracked non-invasively by MRI techniques. MRI is usually blind to solids, but oil-bearing seeds (Nakagawa *et al.*, 1993) act analogously to blobs of dyed fluid to trace out their dispersion (Metcalfe & Shattuck, 1996). The "visible" grains are brown or yellow (white in the images) mustard seeds; invisible grains are glass beads or sugar balls. The glass and sugar are smooth spheres; the mustard seeds are approximately smooth and spherical. The size and density of brown mustard is 1.4 ± 0.2 mm, 1.7 ± 0.2 g/cm^3; Metcalfe & Shattuck (1996) list the other materials' properties. These materials allow four size and density ratios with which to observe different segregation effects.

The container tube is plexiglass, 5 cm in diameter and 15 cm long—or, based on a 1.5 mm particle, 33×100 particle diameters. Image resolution in planes perpendicular to the tube axis is 128^2 with 128 planes along the tube axis. Experiments begin with segregated volumes of seeds and other grains. Figure 5a shows the middle slice perpendicular to the long axis, the initial condition for radial mixing experiments, where equal volumes of seeds and balls fill 3/4 of the tube volume. Figure 5b shows a "blob" of seeds filling approximately the middle 20% of the tube, the initial condition in the tube for axial mixing experiments, where the MRI signal is summed over each slice, normalised, and plotted versus the axial slice number; and, balls fill out the rest of the tube volume to 3/4 full. During experiments, the tube slowly rotates about its long axis 1/4 revolution, stops during image acquisition, then resumes rotation. During image acquisition, grains are not moving.

3.1. RADIAL TRANSPORT

Figure 6 shows radial transport experiments with different combinations of s and d. The figure shows 5 slices, from the middle, near the ends, and in between, from the tube. Figure 6a shows mixing with no segregation of equal volumes of sugar/brown mustard: $s = 1.1 \pm 0.2, d = 0.8 \pm 0.2$. After 8 revolutions, grains are well mixed throughout the tube. Axial banding is not a factor in these measurements. Segregation arises from size and density differences (among others) of the individual grains (Bridgwater, 1976). For similar densities and different sizes, larger particles segregate to the outside of a tumbling container, e.g. figure 3. Figure 6b shows density segregation with similar sizes and different densities using smaller glass beads/brown mustard: $s = 1.1 \pm 0.2, d = 1.4 \pm 0.2$. The denser glass beads go to the centre of the tumbling container. Figure 6c uses larger glass beads/brown mustard: $s = 1.4 \pm 0.2, d = 1.4 \pm 0.2$. It shows a surprising outcome when both size *and* density segregation forces (if they may be thought of as forces) operate at the same time, which is the usual case both in nature and in industry. The glass is both denser and larger. Density segregation tends to move the glass to the inside of the tumbling container; size segregation tends to move the glass to the outside. The end result is that the segregation forces approximately cancel, giving good mixing where it would not normally be expected. In contrast to figure 6c, figure 6d shows the same density ratio but with the larger yellow mustard and smaller glass beads: $s = 0.8 \pm 0.1$. Now, size segregation moves the larger mustard to the outside and density segregation moves the denser glass to the inside, with the result that by manipulating the segregation mechanisms, we create a more strongly segregating mixture. Figure 7 plots results on the size ratio-density ratio (s-d) plane. The data suggests there is a line in the s-d plane where mixing is always good—perhaps in the region suggested by the dashed lines—and where segregation becomes ever stronger the farther removed materials are from this line. More data is needed to map out the mixing/segregating regions of the size-density plane.

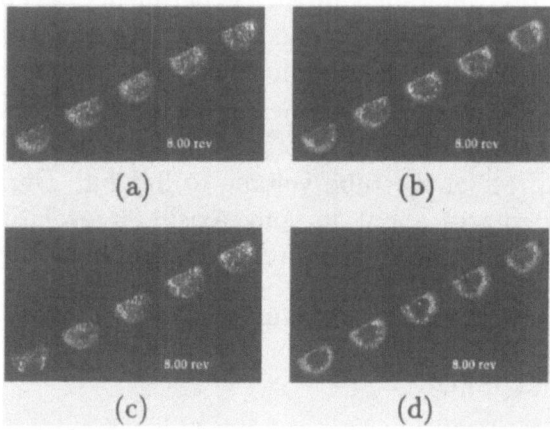

Figure 6. Mixing with and without segregation effects. Five slices from experiments initialised as in figure 5a, all after 8 revolutions of the tube. (a) Particles of similar size and density mix. (b) Particles of similar size and different density segregate. (c) Particles of different size and different density can, in some cases, mix due to cancellation of the segregation forces. (d) Reverse the sense of one of the segregation mechanisms in (c) to get the strongest segregation.

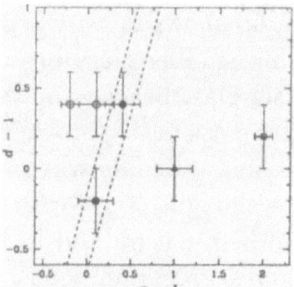

Figure 7. The size ratio-density ratio (s-d) mixing plane. Filled (open) symbols show good mixing (segregation). ○ tube; □ disk with sugar balls and small salt cubes; △ disk with sugar balls and sugar crystals. Dashed lines suggest region of suppressed segregation. Chute flow experiments (Bridgwater, 1994) suggest the rate of change of segregation decreases for $d > 3$.

In the quasi-2-d mixers, a stable non-mixing core forms whenever the container is more than half full, and segregation induces stationary rotating patterns, e.g. figure 3. In contrast in 3-d, videos of the motion seem to show a core forming but subsequently breaking up after ≈ 3 revolutions. The elimination of the 2-d segregation patterns and core from 3-d flows may be due to the axial motions described below.

3.2. AXIAL TRANSPORT

Figure 8a shows the progress of a non-segregating axial transport experiment initialised with a blob of seeds as in figure 5b. As the tube rotates, the

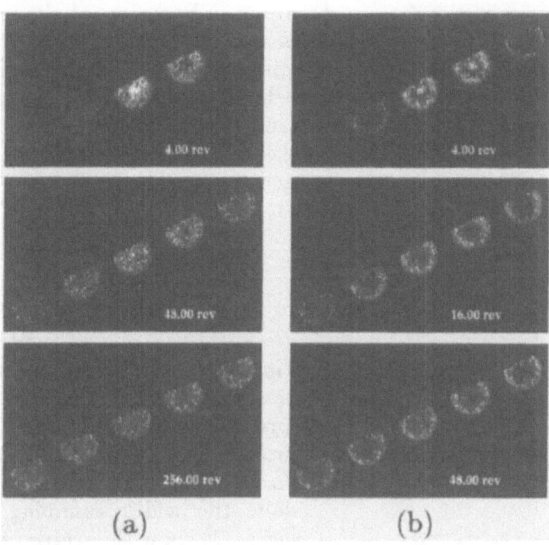

Figure 8. (a) Axial transport with no segregation; brown mustard and sugar balls. (b) Axial transport with segregation; brown mustard and smaller glass beads. Five slices from experiments initialised as in figure 5b. Frames are labelled by the number of tube revolutions.

seeds spread out. Figure 8b shows the same experimental set-up but with segregating particles. With segregation added, the central region in the initial slab is quickly depleted of seeds. The seeds disperse along the container boundary but cannot mix radially to the interior due to segregation forces. In both cases axial transport is much slower than radial transport, taking > 200 revolutions to disperse similar seeds, > 40 even when the seeds do not mix into the interior. Radial motions accomplish their work in under 10 revolutions. A closer look at the data reveals a radial diffusion rate gradient and allows an estimate of the segregation or *demixing* rate.

3.2.1. *Differential diffusion*
Axial motion in a rotating tube of particles is commonly treated (Bridgwater, 1976; Fan *et al.*, 1990, and the references of these reviews) as a 1-d diffusion process, i.e. the radial coordinate is averaged away. Experiments validating this averaging are mechanical equivalents of the radially averaged data of figure 5b. The spread of radially averaged concentration distributions from the MRI data also suggests 1-d diffusion. However, MRI data allows a closer look at the axial tansport and, while the results are preliminary, they suggest the real case is more involved than simple 1-d diffusion, that material diffuses axially some 40% faster through the region along the tube boundary than through the interior.

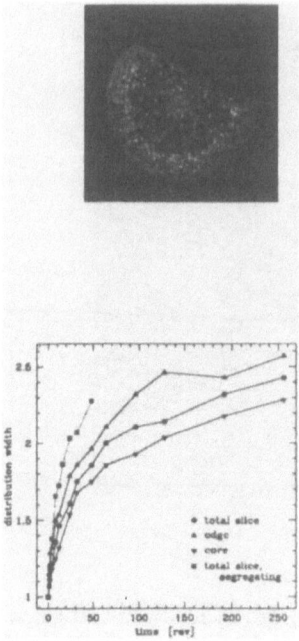

Figure 9. Partition of the image slices into two equal area zones, one covering the edge material near the tube boundary (lighter), the other covering the material in the core (darker). The bright spots are the imaged mustard seeds.

Figure 10. Distribution widths from slice partitions versus time. The grains spread more rapidly along the free surface and tube boundary ($\triangle$) than through the core of material ($\triangledown$). $\bigcirc$s show the width summing over whole slices. $\square$s show the width summing over whole slices for a segregating flow. Lines guide the eye.

Consider partitioning each tube slice into two zones of equal area (figure 9). One zone (lighter) covers the grain area from the tube edge inwards. The other zone (darker) covers the grain area in the tube core. Then ask, what is the axial flux through each of these zones? Measure distributions similar to those in figure 5b for each zone and at successive times, and plot the spread of the distributions' width versus time in figure 10. Edge diffusion (triangles) along the boundary is some 40% faster initially than that through the core region (inverted triangles). The total spreading rate (circles) lies in between. The axial spread saturates as the ends of the tube come into play. For comparision squares are for the segregating data of figure 8b, which spreads even faster. Clearly, there are significant departures from a 1-d diffusion description of axial grain transport. Seeds disperse axially along the tube boundary and then mix radially. Given that the only motions moving the material are the avalanches across the free surface, this is physically intuitive.

I would note two implications of segregation for any mixing measure based on volume averaging particle concentration (Harnby, 1985). (1) Such measurements would not distinguish any of the cases of figure 6. (2) Such measurements applied to figure 10 would show segregation enhances axial transport. It is important to identify the mixing patterns along with any coarse-grained measures of mixing.

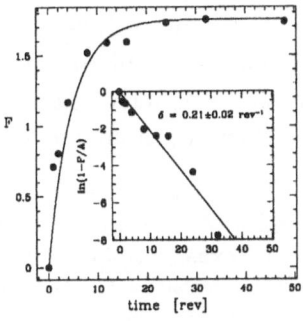

Figure 11. Measured demixing rate. F is the ratio of the signal from the tube's edge to the tube's core normalised to its initial value and 1 subtracted. The solid lines are fits of $F = A(1 - \exp(-\delta t))$ to the data. δ is the demixing rate; $A = 1.75$; and the errors on δ estimate the 1σ variance of the data around the fitted curve.

3.2.2. *De-mixing rate*

Figure 1b measures the mixing rate for no segregation γ. It is also of interest to ask about γ's "inverse" which I will call δ the segregation or de-mixing rate. While the results are preliminary, we can estimate an upper bound on δ from the data of figure 8b. With segregation, the central axial region in the initial blob quickly depletes of seeds. If we take the middle slice of the tube, divide it into equal-area inner and outer zones (figure 9), and measure the ratio of the signal from the edge zone to that of the core zone, this ratio will rise from 1 to some final value and the rise time of this ratio then will estimate δ. As this estimate does not separately account for particle flux from radial and axial motions, δ is an upper bound on the segregation rate. However, as radial motion is at least $5\times$ greater than axial, the bound is likely to be acceptably sharp.

Figure 11 shows the data. The demixing rate estimate is $\delta = 0.21 \pm 0.02$ for a demixing time estimate of $\tau_d = 4.76$ rev for a slowly rotating tube. Compare τ_d to the measured mixing time of the 3/4 full 2-d disk from figure 1 $\tau_m = 4.55$ rev. Keeping in mind the comparisons are between 2- and 3-d experiments with different materials of different sizes and densities, and that significant features of the flows—presence or absence of a core or of axial flow—are different, the mixing and demixing rates are, nonetheless, strikingly similar. This would seem to suggest that the 3-d radial flow and radial transport properties are also dominated by the effects of avalanche geometry just as they are in 2-d.

4. Future Directions

Recent studies of granular flow and mixing raise a host of questions with combined technological and scientific import, e.g.: What are the detailed segregation mechanisms in a device for given particle characteristics? What are the strategies for suppressing *or* enhancing segregation? Where, between the 10 particle diameter depths of the 2-d container and the 100 diameters of the 3-d, is the crossover between 2- and 3-d behaviour? How do 3-d motions (mostly axial) start? A goal of these and future experiments is to measure granular transport systems in enough detail to answer some of these questions, leading hopefully to greater predictive facility and a deeper level of physical understanding of the flow of granular materials.

With this work I have benefited greatly from collaboration with: K. Liffman of CSIRO; T. Shinbrot, J.J. McCarthy, E. Wolf and J.M. Ottino of Northwestern University; and M. Shattuck of the University of Texas. I thank G.A. Johnson and R.P. Behringer, directors of respectively Duke University's Center for *In Vivo* Microscopy and Center for Nonlinear and Complex Systems for the use of facilities and other support.

References

Batchelor, G.K. (1967) *An Introduction to Fluid Dynamics*, Cambridge University Press.

Bridgwater, J. (1976) Fundamental Powder Mixing Mechanisms, *Pow. Tech.* **15** 215-235.

Bridgwater, J. (1994) Mixing and Segregation Mechanisms in Particle Flow, in A. Mehta (ed.), *Granular Matter: An Interdisciplinary Approach*, Springer, 161–193.

Broadbelt, C.J., Bridgwater, J., Parker, D.J., Keningley, S.T., & Knight, P. (1993) Phenomenological Study of a Batch Mixer Using a Positron Camera, *Pow. Tech.* **76** 317-325.

Fan, L.T., Chen, Y., & Lai, F.S. (1990) Recent Developments in Solids Mixing, *Pow. Tech.* **61** 255-277.

Harnby, N. (1985) Characterization of Powder Mixtures, in Harnby, N., Nienow, A.W., & Edwards, M.F. (eds.), *Mixing in the Process Industries*, Butterworths.

Jaeger, H.M., Nagel, S.R., & Behringer, R.P. (1996) Physics of Granular Materials, *Rev. Mod. Physics* (in press); *Physics Today*, April 1996, 32–38.

Lacey, P.M. (1954) Developments in the Theory of Particle Mixing, *J. Appl. Chem.* **4** 257-268.

McCarthy, J., Wolf, E., Shinbrot, T., Metcalfe, G., & Ottino, J.M. (1996) Mixing of Granular Materials in Slowly Rotated Containers, *AIChEJ* (in press).

Mehta, A. (ed.) (1994) *Granular Matter: An Interdisciplinary Approach*, Springer.

Metcalfe, G., Shinbrot, T., McCarthy, J.J., & Ottino, J.M. (1995) Avalanche Mixing of Granular Solids, *Nature* **374** 39–41*.

Metcalfe, G., & Shattuck, M. (1996) Pattern Formation During Mixing and Segregation of Flowing Granular Materials, *Physica* **A** (in press)*.

Nakagawa, M., Altobelli, S.A., Caprihan, A., Fukushima, E., & Jeong, E.K. (1993) Non-invasive Measurements of Granular Flows by Magnetic Resonance Imaging, *Exp. Fluids* **16** 54-60.

Nikitidis, M., Tüzün, U., & Spyrou, N.M. (1994) Tomographic Measurements of Granular Flows in Gases and Liquids, *KONA Powder and Particle* **12** 43–56.

Wolf, E. (1995) Geometrical Aspects of Granular Solids Mixing, M.S. thesis, Northwestern University.

*Available online through http://www.mel.dbce.csiro.au/~guy/papers

ON THE REALITY OF ANTISYMMETRIC STRESSES IN FAST GRANULAR FLOWS.

HANS-B. MÜHLHAUS and PETER HORNBY
CSIRO, Division of Exploration & Mining
PO Box 437, Nedlands WA 6009, Australia

Abstract

Starting from the virtual power expression for fast flow of dense assemblies of rough, slightly inelastic spheres, we derive expressions for the average stress tensor, momentum and angular momentum balances and corresponding variational boundary conditions. We show that the stress tensor can be decomposed into two parts with distinct physical significance. The first part can be interpreted as an area averaged stress tensor, which is nonsymmetric in general. The second part is obtained as the divergence of a third order tensor which can be interpreted as the difference between the area and the volume averages of the stress tensor. Accordingly, the stress tensor is symmetric for quasistatic flow.

1. Introduction

It has long been recognised that internal degrees of freedom play an important role in the continuum mechanical modelling of material behaviours. What perhaps has been less widely appreciated is that these same internal degrees of freedom, and in particular the granular rotation or spin, may need to be recognised explicitly in the kinematical description of flow and deformation processes. The inclusion of non-standard degrees of freedom leads to a so called generalised continuum theory. In the case of independent rotational degrees of freedom (besides the spin of a material element of the continuum) one speaks of a Cosserat continuum.

From a microscopic point of view, coupling between the average granular velocity **v** and the spin field **w** is a manifestation of noncentral intergranular forces. A hint to a macroscopic description of the same effect emerges from the observation that a line element of flowing granular material will rotate with the vorticity 1/2 curl **v** but in doing so will meet the resistance or stress due to the internal spin if vorticity and spin are not synchronised. Such resistance is a function of 1/2 curl **v** - **w**. This observation was first made in a gas dynamics context by Max Born (1920) who suggested antisymmetric

299

N. A. Fleck and A. C. F. Cocks (eds.),
IUTAM Symposium on Mechanics of Granular and Porous Materials, 299–311.
© 1997 *Kluwer Academic Publishers.*

stress as the appropriate resistance. Born seemed unaware of the work of the brothers Cosserat (1909).

The recent renaissance of generalised continuum theories was mainly triggered by difficulties of conventional continuum models in dealing with strain non-uniformities and strain softening. Indeed, an adequate choice of the additional degrees of freedom and/or higher order gradient terms restores the well-posedness of the differential problem and makes it possible to describe spatial features of the deformation or flow behaviour (Walgraef and Aifantis, 1985; Triantafillides and Aifantis, 1986; Aifantis, 1987; Mühlhaus and Vardoulakis, 1987; Vardoulakis and Sulem, 1995; deBorst and Mühlhaus, 1993).

Mühlhaus and Vardoulakis assumed that slow granular flow can be described within the framework of an incremental Cosserat plasticity theory. In a linear instability analysis they predicted shear band thicknesses of the order of 10-20 average grain diameters, a result which is in good agreement with the range of shear band thicknesses observed in experiments. However this does not prove that granular materials are indeed Cosserat media. It merely shows that higher order gradients, and in this case rotation gradients, become significant once the characteristic wavelength of the macroscopic deformation pattern is of the order of 10-20 grain diameters. Indeed, Jenkins (1993) and Mühlhaus and Oka (1996) have derived continuum theories for slow deformations of granular materials where the higher order stress depends on gradients of the relative deformation gradient rather than the gradient of the granular spin alone, as it would in Cosserat-type theories. A similar situation is found in connection with Chapman-Enskog-type models for fast granular flow or the dynamics of dense gases of rough molecules (McCoy et al 1966; Lun, 1991). Collisional fluxes associated with collisional anti-invariants (such as angular momentum) vanish identically (McCoy et al, 1966) and consequently a collisional couple stress tensor in the conventional sense does not exist in such a theory. Yet the collisional stress is nonsymmetric in general and the balance of the angular momentum contains a higher order stress tensor of the order of the second gradient of the stress tensor (McCoy et al, 1966). Since the stress tensor depends on the relative velocity gradient, the higher order stress tensor in McCoy et al's theory is of the same type as the ones in Jenkins (1993) and Mühlhaus and Oka's (1996) theories. It should be mentioned that in their purely elastic theory McCoy et al only mentioned the existence of a higher order stress and do not elaborate on it. In Lun's (1991) paper higher order stresses are ignored.

In this paper we derive explicit expressions for the stress and the higher order stress tensor, the corresponding balance laws and variational boundary conditions. For simplicity, we concentrate on the flow of dense packings (ie free flight length of granules between collisions much smaller than the average granule diameter) of rough, inelastic, identical spheres.

2. Kinetic Considerations

We employ the usual notations of the kinetic theories (Chapman and Cowling, 1970) with the exception of the average particle diameter, which we designate as D instead of σ. The translational velocity of a granule's centre of mass will be denoted by c and its angular velocity about its centre by s. The mass of a granule is denoted by m, and its moment of inertia tensor by I. It follows in the usual way (Goldstein 1980) that the kinetic energy of a granule is

$$K(m, c, I, s) = \frac{m}{2} c^T c + \frac{1}{2} s^T I s \tag{2.1}$$

Next we define a probability function associated with the ensemble of possible grain configurations. The grain configurations are assumed compatible with the boundary conditions and the time evolution of the system up to some instant in time t. By considering such an ensemble, we can replace spatial and time averaging procedures, (which break down in non-stationary inhomogeneous systems) with averages over the ensemble. Let

$$P(dm, dc, dI, ds, dx; t) \tag{2.2}$$

denote the probability that a granule with mass in the range dm, velocity in the range dc and so on, has its centre of mass in a volume dx at time t. Then consider the definition of the kinetic energy density $\kappa(x,t)$ given by

$$\kappa(x, t)dx = \int_{m,c,I,s} K(m, c, I, s) \, P(dm, dc, dI, ds, dx; t) \tag{2.3}$$

It is shown in the Appendix that for spheres of diameter D this may be written

$$\kappa(x, t) = \frac{\rho}{2} \left[v^2 + <C^2> + \frac{D^2}{10} \left(\omega^2 + <\Omega^2> \right) \right]$$
$$= \frac{\rho}{2} \left[v^2 + \frac{D^2}{10} \omega^2 + 3T_T + 3\frac{D^2}{10} T_R \right] \tag{2.4}$$

where v and ω are the mean velocity and spin, and C and Ω are the velocity and spin fluctuations, and $\rho(x,t)$ is the mass density (see the Appendix for definitions). We have also introduced the translational temperature T_T and rotational temperature T_R associated with their respective fluctuations,

$$T_T = \frac{1}{3} <C^2>, \quad T_R = \frac{1}{3} <\Omega^2> . \tag{2.5}$$

In the spirit of the kinetic theories for very dense, non-uniform gases (Chapman and Cowling, 1970, p297; Jenkins and Savage, 1986) we assume that granule momentum is transferred exclusively through collisions. We also suppose that all energy loss to the molecular substructure is also mediated by collisions. These losses are assumed to be converted to molecular fluctuations (true thermal temperature) and stored potential energy through the generation of dislocations within granules. (We do not consider comminution processes). The inelasticity of the collisions is characterised in the simplest possible way (Lun and Savage, 1987) by means of two coefficients e and β, the coefficient of restitution in the normal direction and roughness coefficient in the tangential direction respectively.

It follows that the major determinants of such collisional energy loss are the frequency and magnitude of collisions. Firstly, we suppose that the *frequency* of collisions is a function of the translational temperature and density (through the solid fraction). This assumption is reasonable. Secondly, the *magnitude* of collisional energy loss is *estimated* by projecting the collisional impulse on the mean granular velocity and spin fields (neglecting the fluctuational part), and then averaging. This is harder to justify, but does find some vindication in the fact that we are deriving a continuum theory involving stresses generated in the context of the mean fields. In a sense, then, it is reasonable to deal only with that part of the energy loss that is directly attributable to those mean fields when defining the continuum stress and moment couple tensors.

It turns out that the fluctuational part of the collisional impulse is orthogonal to the mean fields anyway, and hence the "fluctuation/mean field" cross terms are zero after averaging. The reasoning is precisely analogous to that leading to the decomposition (2.4). Thus the unaccounted collisional energy loss terms are of the "fluctuation/fluctuation" type, and hence should appear in the granular temperature evolution equations (which we do not consider in this paper).

In addition to collisional losses, the granular kinetic energy density at a point can also change with time as a result of net mass flux, the action of body forces and moments, and the diffusion of granular velocity and spin fluctuations. Of these, we shall deal with mass flux, body forces and body moments in connection with continuum stresses.

3. Gradient Continuum Model

Consider two spherical granules 1 and 2 having translational velocities c_1 and c_2 and angular velocities s_1, and s_2 respectively. The total relative velocity at the point of contact just prior to collision is

$$g_{21} = c_{21} + \frac{D}{2}(k \times s_{21}) \qquad (3.1)$$

where $c_{21} = c_2 - c_1$, $s_{21} = s_{12} = s_1 + s_2$ and $\mathbf{k}$ is the unit vector along the centre line from granule 1 to granule 2. During a collision g_{21} is changed such that

$$\mathbf{k} \cdot \mathbf{g}'_{21} = -e(\mathbf{k} \cdot \mathbf{g}_{21}), \quad \mathbf{k} \times \mathbf{g}'_{21} = -\beta(\mathbf{k} \times \mathbf{g}_{21}), \tag{3.2}$$

where the primes denote post collisional quantities. The case $\beta = -1$ corresponds to collisions between perfectly smooth spheres. The case $\beta = 1$ represents collisions between perfectly rough, perfectly elastic spheres (McCoy et al, 1966). For further discussions and comparisons of (3.2) with more elaborate collision models for simple shear flows we refer to the paper by Lun and Bent (1994).

Combining (3.1, 3.2) and

$$\frac{mD^2}{10}(\mathbf{s}'_1 - \mathbf{s}_1) = -\frac{D}{2}(\mathbf{k} \times \mathbf{J}_1), \quad \mathbf{J}_1 = m(\mathbf{c}'_1 - \mathbf{c}_1) \tag{3.3}$$

yields

$$\mathbf{J}_1 = -\mathbf{J}_2 = m\left(\frac{1+e}{2}\mathbf{g}^n_{21} + \frac{2}{7}\frac{1+\beta}{2}\mathbf{g}^t_{21}\right) \tag{3.4}$$

and

$$\mathbf{s}'_1 - \mathbf{s}_1 = \mathbf{s}'_2 - \mathbf{s}_2 = \frac{10}{7D}\frac{1+e}{2}(\mathbf{k} \times \mathbf{g}_{21}), \tag{3.5}$$

where

$$\mathbf{g}^n_{21} = (\mathbf{k} \cdot \mathbf{g}_{21})\mathbf{k} \tag{3.6a}$$

and

$$\mathbf{g}^t_{21} = \mathbf{g}_{21} - \mathbf{g}^n_{21}. \tag{3.6b}$$

As in the previous section we decompose the granule velocities c_α and spins s_α, $\alpha = (1, 2)$ into average values v_α and ω_α and fluctuations (deviations from the average values) C_α and Ω_α respectively.

In view of the desired continuum theory we assume that

$$\mathbf{v}_{21} = \mathbf{v}_2 - \mathbf{v}_1 = D[\mathbf{k} \cdot \nabla]\mathbf{v} + \frac{D^2}{2}[\mathbf{k} \cdot \nabla][\mathbf{k} \cdot \nabla]\mathbf{v} + \cdots \tag{3.7}$$

$$\omega_{21} = \omega_2 - \omega_1 = D[\mathbf{k} \cdot \nabla]\omega + \frac{D^2}{2}[\mathbf{k} \cdot \nabla][\mathbf{k} \cdot \nabla]\omega + \cdots \tag{3.8}$$

With (3.7) and (3.8), the mean field part of g_{21} is

$$\bar{\mathbf{g}}_{21} = D\left(1 + \frac{D}{2}[\mathbf{k} \cdot \nabla]\right)\gamma\mathbf{k} \tag{3.9a}$$

where the matrix elements of γ are

$$\gamma_{ij} = \left[\nabla \mathbf{v} - \mathbf{W}^c\right]_{ij} = v_{i,j} - W_{ij}^c \tag{3.9b}$$

and $\mathbf{W}^c$ is the spin tensor corresponding to ω (viz $\omega \times \mathbf{k} = \mathbf{W}^c \mathbf{k}$).

We base the derivation of our continuum theory on the expression for the average specific power

$$\dot{w} = \frac{n}{2} < \mathbf{F}_{21} \cdot \overline{\mathbf{g}}_{21} > \quad , \quad \mathbf{F}_{21} = \mathbf{J}_1 / \tau_{21} \tag{3.10}$$

where $\rho = mn$ is the density, n is the number of granules per unit volume, the factor 1/2 considers the fact that one contact is shared by two granules, $<1> = 1$ is a suitable averaging operator and $1/\tau_{21}$ is the frequency of collisions between granules 1 and 2.

For convenience, in the following derivations we replace $1/\tau_{21}$ by the average collision frequency $1/\tau$. We will show later (eq. 3.18) how the directional dependency of the collision frequency may be considered in a simplified way. For $1/\tau$ we adopt the value derived in Chapman and Cowling (1970, p86, 298):

$$1/\tau = \frac{24v}{\pi D} h \sqrt{\pi T_T} = 4D^2 n h \sqrt{\pi T_T} , \tag{3.11}$$

where v is the solid volume fraction, T_T the translational temperature and h is the equilibrium radial distribution function at contact (see Lun and Bent, 1994 and Chapman and Cowling, p298 for possible forms). An example is the function proposed by Lun and Savage (1986),

$$h = \left(1 - v / v_m\right)^{-5v_m/2} \tag{3.12}$$

where v_m represents the maximum solids fraction of the system. For random packing of spheres $v_m \approx 0.64$, and for regular packing $v_m \approx 0.7407$.

Next we insert (3.6) into (3.4) and the resulting impulse, together with (3.9), into (3.10). The result is

$$\dot{w} = \sigma_{ij}\gamma_{ij} + m_{ijk}\gamma_{ij,k} , \tag{3.13}$$

where

$$\sigma_{ij} = \frac{\rho D^2}{2\tau}\left[\left(\frac{1+e}{2} - \frac{2}{7}\frac{(1+\beta)}{2}\right)\langle k_i k_j k_n k_m\rangle + \frac{2}{7}\frac{(1+\beta)}{2}\delta_{in}\delta_{rm}\langle k_r k_j\rangle\right]\gamma_{nm} , \tag{3.14}$$

$$m_{ijk} = \frac{\rho D^4}{8\tau}\left[\left(\frac{1+e}{2} - \frac{2}{7}\frac{(1+\beta)}{2}\right)\langle k_i k_j k_n k_m k_k k_s\rangle + \right.$$
$$\left. \frac{2}{7}\frac{(1+\beta)}{2}\delta_{in}\delta_{rm}\langle k_j k_k k_r k_s\rangle\right]\gamma_{nm,s} .$$
$$(3.15)$$

If higher order velocity gradients are neglected (3.13) can be written as

$$\dot{w} = \sigma_{ij}\gamma_{ij} + \mu_{ijk}W^c_{ij,k} , \qquad (3.16)$$

which corresponds to the specific power of a Cosserat Continuum. The couple stress tensor is obtained as

$$\mu_{ijk} = \frac{\rho D^4}{8\tau}\left[\frac{2}{7}\frac{(1+\beta)}{2}\delta_{in}\delta_{rm}\langle k_j k_k k_r k_s\rangle\right]W^c_{nm,s} . \qquad (3.17)$$

The averages in (3.14, 3.15) and (3.17) are defined as

$$\langle(\cdot)\rangle = \int_{4\pi} dk A(\mathbf{k},\mathbf{r},t)(\cdot) , \quad A(\mathbf{k}) = A(-\mathbf{k}) , \qquad (3.18)$$

where $\mathbf{r}$ is the current position vector of a material point. The function A() accounts for the orientation distribution of collisions so that $(1/\tau\ A\ d\mathbf{k})$ is the probable collision frequency in the element $d\mathbf{k}$ centred at $\mathbf{k}$. If the distribution of collisions is isotropic and independent of time and position then $A = 1/(4\pi)$. It should be mentioned that we have used the symmetry property $A(\mathbf{k}) = A(-\mathbf{k})$ in the derivations of (3.14), (3.15) and (3.17). We now turn to the derivation of the field equations. By integration and application of the divergence theorem, the right hand side of (3.13) is converted into

$$-\int_v \tau_{ij,j}v_i dV + \int_v e_{ijk}\tau_{ij}\omega_k dV + \int_A \tau_{ij}n_j v_i dA + \int_A m_{ijk}n_k\gamma_{ij}dA \qquad (3.19)$$

where we have introduced the notation

$$\tau_{ij} = \sigma_{ij} - m_{ijk,k} . \qquad (3.20)$$

and e_{ijk}, ($e_{123} = 1$) designates the permutation symbol. The stress τ_{ij} has a physical significance to which we return in the next section. Equation (3.19) suggests the form

$$\dot{W}_{ext} = -\int_V \rho \left(\dot{v}_i v_i + \frac{D^2}{10} \dot{\omega}_i \omega_i \right) dV + \int_A t_i v_i dA + \int_A t_{ij} \gamma_{ij} dA +$$

$$\int_V \rho g_i v_i dV + \int_V \rho \mu_i \omega_i dV , \tag{3.21}$$

for the power of the external forces. In the definition of $\dot{W}_{ext}$ we have included the inertial terms. The terms ρg_i and $\rho \mu_i$ are volume forces and couples respectively. From the invariance properties of $\int_V \dot{w} \, dV - \dot{W}_{ext} = 0$ it follows that

$$\tau_{ij,j} + \rho g_i = \rho \dot{v}_i \quad , \quad -e_{ijk} \tau_{ij} + \rho \mu_k = \rho \frac{D^2}{10} \dot{\omega}_k . \tag{3.22}$$

We note that $\tau_{ij} = \tau_{ji}$ if $\mu_k = 0$ and $(D^2 / 10)\dot{\omega}_k = 0$. Structurally, the present theory is very different from a Cosserat theory. A Cosserat theory is obtained if higher order velocity gradients are neglected (see eq. 3.15-17). In general however $D^2 v_{i,jk}$ is of the same order as $D^2 W^c_{ij,k}$.

4. Interpretation

Here we outline a heuristic interpretation of how stress tensors of the form

$$\tau_{ij} = \sigma_{ij} - m_{ijk,k} \tag{4.1}$$

as in (3.20) can come about. A similar interpretation was presented by Mühlhaus (1995) within the context of laminated materials (the higher order stresses (13.40) and (13.41) should have opposite sign in this reference). We consider a characteristic volume $V^c = $ (2a, 2b, 2c) which, for convenience we assume as rectangular with sides parallel to the coordinate axes (x_1, x_2, x_3). In the absence of volume forces, we have the identity

$$\int_{V^c} \sigma^m_{ij} dV = \int_{A^c} \sigma^m_{ik} n_k x_j dA = \int_{A^c} \sigma^m_{ik} x_j e_{mnk} dx_m dx_n , \tag{4.2}$$

where $\sigma^m_{ij}(r)$ designates the microstress distribution within V^c. In particular

$$\int_{V^c} \sigma^m_{12} dV = b \int \left(\sigma^m_{12}(x_2 = b) + \sigma^m_{12}(x_2 = -b) \right) dx_1 dx_3 +$$

$$\int \left(\left(\sigma^m_{11}(x_1 = a) - \sigma^m_{11}(x_1 = -a) \right) x_2 \right) dx_2 dx_3 + \tag{4.3}$$

$$\int \left(\left(\sigma^m_{13}(x_3 = c) - \sigma^m_{13}(x_3 = -c) \right) x_3 \right) dx_1 dx_2 .$$

This expression can be written as

$$\tau_{12} = \sigma_{12} - m_{12k,k} \,, \tag{4.4}$$

where

$$\tau_{12} = \frac{1}{8abc} \int \sigma_{12}^m dV \,, \tag{4.5}$$

$$\sigma_{12} = \frac{1}{4ac} \int \sigma_{12}^m (x_2 = 0) dx_1 dx_3 \,, \tag{4.6}$$

$$m_{122,2} = \frac{b^2}{8ac} \int \sigma_{12,22}^m (x_2 = 0) dx_1 dx_3 \,, \tag{4.7}$$

$$m_{121,1} = \frac{1}{8abc} \int \left(\left(\sigma_{11}^m (x_1 = a) - \sigma_{11}^m (x_1 = -a) \right) x_2 \right) dx_2 dx_3 \,, \tag{4.8}$$

$$= \frac{1}{4bc} \int \sigma_{11,12}^m (x_1 = 0, x_2 = 0) x_2^2 dx_2 dx_3 \,,$$

$$m_{123,3} = \frac{1}{8abc} \int \left(\left(\sigma_{13}^m (x_3 = c) - \sigma_{13}^m (x_3 = -c) \right) x_2 \right) dx_1 dx_2 \tag{4.9}$$

$$= \frac{1}{4ab} \int \sigma_{13,32}^m (x_2 = 0, x_3 = 0) x_2^2 dx_1 dx_2 \,,$$

Analogous expressions are obtained for the other components. The result suggests the interpretation of τ_{ij} as the volume average of the microstress σ_{ij}^m and σ_{ij} as the area average of σ_{ij}^m. The volume average is symmetric if σ_{ij}^m is symmetric, however σ_{ij} is nonsymmetric in general if σ_{ij}^m is inhomogeneous within V^c.

5. Boundary Conditions

From the surface integrals appearing in (3.19) we can deduce the boundary conditions necessary for a complete specification of the mechanical problem. The first of these terms already involves $\mathbf{v}$ directly, and hence will cause no special difficulty. The last term can be written as

$$\int_A m_{ijk} n_k \gamma_{ij} dA = \int_A m_{ijk} n_k v_{i,j} dA + \int_A m_{ijk} n_k e_{ijn} \omega_n dA \tag{5.1}$$

and again, the second term on the RHS of (5.1), involving ω directly, also causes no special difficulties. However, the remaining term is considerably less accommodating, due to the fact that tangent derivatives of $\mathbf{v}$ are determined by specifying $\mathbf{v}$ on the boundary, while the normal derivatives must be specified separately. Thus the first term on the RHS of (5.1) must be decomposed into two parts, and the part determined by $\mathbf{v}$

must be written so that the integrand contains **v** explicitly. For brevity, define $\phi_{ij} = m_{ijk} n_k$. Then the integrand of the term of interest is

$$
\begin{aligned}
\phi_{ij} v_{i,j} &= \phi_{ij}\left(v_{i,j} - v_{i,m} n_m n_j\right) + \phi_{ij} v_{i,m} n_m n_j \\
&= \left[\partial_j - n_j n_m \partial_m\right]\phi_{ij} v_i - \left(\phi_{ij,j} - \phi_{ij,m} n_m n_j\right)v_i + \phi_{ij} v_{i,m} n_m n_j
\end{aligned}
\tag{5.2}
$$

The surface divergence theorem

$$
\int_S \left[\partial_j - n_j n_m \partial_m\right] a_j \, dA = \int_S a_m n_m \left[\partial_j - n_j n_m \partial_m\right] n_j \, dA
\tag{5.3}
$$

follows from decomposing the vector **a** into tangent and normal parts, and noting that the integral of the surface divergence of the tangent part vanishes by virtue of the closedness of S. Applying the chain rule to the remaining normal part shows that one of the terms in the chain rule vanishes identically, leading to (5.3). Putting all this together leads to the three essential and corresponding natural boundary condition pairs

$$
\begin{aligned}
&\tau_{ij} n_j + m_{ijk,n} n_j n_k n_n + \\
&\left[\left(n_{n,n} - n_{n,m} n_n n_m\right) m_{ijk} + m_{ijm} n_{m,k}\right] n_j n_k + \\
&m_{ijk,j} n_k - m_{ijk} n_{k,j} \qquad\qquad\qquad\qquad \text{or} \quad v_i \qquad\qquad (5.4)\\
&m_{ijk} n_j n_k \qquad\qquad\qquad\qquad\qquad\qquad\quad \text{or} \quad v_{i,j} n_j \\
&m_{ijk} n_k e_{ijn} \qquad\qquad\qquad\qquad\qquad\qquad\quad \text{or} \quad \omega_n
\end{aligned}
$$

6. References

Aifantis, E.C. (1987) The physics of plastic deformation, *Int. J. Plasticity,* **3**, 211-247.

Born, M. (1920) The mobility of electrolytic ions, *Zeitschrift der Physik* **1**, 221-249.

deBorst, R. and Mühlhaus, H.-B. (1992) Gradient dependent plasticity: formulation and algorithmic aspects, *J. Num. Meth. in Engng.* **35**, 521-539.

Chapman, S. and Cowling, T.G. (1970) *The mathematical theory of non-uniform gases,* Third Edition, Cambridge, University Press.

Cosserat, E. and F. (1909) *Theory des corps deformable,* Herman et fils, Paris.

Goldstein, H. (1980) *Classical Mechanics,* Second Edition, Addison-Wesley Publishing Company.

Haff, P.K. (1983) Grain flow as a fluid-mechanical phenomenon, *J. Fluid Mech.* **134**, 401-430.

Jenkins, J.T. (1991) Anisotropic elasticity for random arrays of identical spheres. In: *Modern theory of anisotropic elasticity and applications,* J. Wu (ed) SIAM, Philadelphia.

Jenkins, J.T. and Richman, M.W. (1985) Kinetic theory for plane flows of a dense gas of identical, rough, inelastic, circular disks, *Phys. Fluids* **28**, 3485-3494.

Jenkins, J.T. and Savage, S.B. (1983) A theory for the rapid flow of identical, smooth, nearly elastic particles, *J. Fluid Mech.* **130**, 187-202.

Lun, C.K.K. (1991) Kinetic theory for granular flow of dense, slightly inelastic, slightly rough spheres, *J. Fluid Mech.* **233**, 539-559.

Lun, C.K.K. and Bent, A.A. (1994) Numerical simulation of inelastic frictional spheres in simple shear flow, **258**, 335-353.

Lun, C.K.K. and Savage, S.B. (1987) A simple kinetic theory for granular flow of rough, inelastic, spherical particles, Trans. ASME E: *J. Appl. Mech.* **54**, 47-53.

McCoy, B.J., Sandler, S.I. and Dahler, J.S. (1966) Transport properties of polyatomic fluids.iv. The kinetic theory of a dense gas of perfectly rough spheres, *J Chemical Physics* **45**, 3485-3512.

Mühlhaus, H.-B. (1995) A Relative Gradient Model for Laminated Materials. In: *Continuum Models for Materials with Microstructure,* Ch. 13, H.-B. Mühlhaus (ed.), John Wiley & Sons.

Mühlhaus, H.-B. and Aifantis, E.C. (1991) A variational principle for gradient plasticity, *Int. J. Solids and Structures* **28**, 217-231.

Mühlhaus, H.-B. and Oka, F. (1996) Dispersion and wave propagation in discrete and continuous models for granular materials, *Int. J. Solids and Structures* **33**, 2841-2858.

Mühlhaus, H.-B. and Vardoulakis, I. (1987) The thickness of shear bands in granular materials, *Geotechnique* **37**, 271-283.

Savage, S.B. (1992) Numerical simulation of couette flow of granular materials: spatio-temporal coherence of 1/f noise. In: *Physics of granular media*, Bideau and Dodds, J. (ed's), Nova Science Publishers Inc., NY., 343-362.

Triantafyllidis, N. and Aifantis, E.C. (1986) A gradient approach to the localisation of deformation-I. Hyperelastic materials, *J. Elasticity* **16**, 225-238.

Vardoulakis, I. and Sulem, J. (1995) Chapters 9 and 10 of: *Bifurcation analysis in geomechanics*. Chapman & Hall. 334-423.

Walgraef, D. and Aifantis, E.C. (1985) On the formation and stability of dislocation patters - 1: One-dimensional considerations; - II: Two-dimensional considerations; - III Three-dimensional considerations, *Int. J. Engng Sci.,* **23**, 1351-1372.

Appendix

The integral (2.3) can be decomposed in the obvious way by virtue of the additive decomposition of (2.1),

$$\kappa(x, t)dx = \frac{1}{2} \int_{m,c} mc^T c\, P(dm, dc, M_3^+, \Re^3, dx; t) +$$

$$\frac{1}{2} \int_{I,s} s^T Is\, P(\Re^+, \Re^3, dI, ds, dx; t) \tag{A.1}$$

where, for example

$$P(dm, dc, M_3^+, \mathfrak{R}^3, dx; t) = \int_{I,s} P(dm, dc, dI, ds, dx; t) \qquad (A.2)$$

denotes the probability measure after integrating over s and I ($\mathfrak{R}^3$ being the range of s, and M_3^+ is the range of I). Equation (A.2) has the interpretation that P is the probability of finding at time t, a granule of mass in the range dm, velocity in the range dc, any spin, and any moment of inertia, with its centre of mass in dx. Next, we define the average velocity v at the point x by

$$v(x, t) = \frac{1}{Z} \int_{m,c} mc\, P(dm, dc, M_3^+, \mathfrak{R}^3, dx; t) \qquad (A.3a)$$

where

$$Z = \int_m m\, P(dm, \mathfrak{R}^3, M_3^+, \mathfrak{R}^3, dx; t)$$
$$= \rho(x, t)\, dx \qquad (A.3b)$$

With this definition of v, we obtain the following alternative expression for the first integral of (A.1),

$$\frac{1}{2} \int_{m,c} mc^T c\, P(dm, dc, M_3^+, \mathfrak{R}^3, dx; t) = \frac{1}{2}\rho v^2\, dx +$$
$$\frac{1}{2} \int_{m,C} mC^T C\, P_C(dm, dC, M_3^+, \mathfrak{R}^3, dx; t) \qquad (A.4a)$$

where

$$C = c - v \qquad (A.4b)$$

are the velocity fluctuations at (x,t) and P_C is the probability measure after the change of variable (A.4b). This decomposition relies only upon the symmetry of the dot product, and consequently a similar decomposition applies to the rotational contributions by virtue of the symmetry and positivity of the inertial moment tensors. Specialising to grains of fixed mass m_0, we obtain

$$\frac{1}{2} \int\limits_{m,C} mC^TC \, P_C(dm, dC, M_3^+, \Re^3, dx; t)$$

$$= \frac{1}{2} m_0 \int\limits_C C^2 \, P_C(\Re^+, dC, M_3^+, \Re^3, dx; t)$$

$$= \frac{1}{2} \int\limits_m m \, P_C(dm, \Re^3, M_3^+, \Re^3, dx; t) \int\limits_C C^2 \, \frac{P_C(\Re^+, dC, M_3^+, \Re^3, dx; t)}{P_C(\Re^+, \Re^3, M_3^+, \Re^3, dx; t)}$$

$$= \frac{1}{2} \rho < C^2 > dx$$

(A.5)

Where the expectation $< C^2 >$ is conditional upon dx containing a centre of mass. Specialising further to spheres of diameter D, and equal mass, and following a similar line of reasoning for the rotational terms leads to equation (2.4).

INTERNAL FRICTION ANGLES: CHARACTERIZATION USING BIAXIAL TEST SIMULATIONS

D. CORRIVEAU AND S. B. SAVAGE

McGill University
Department of Civil Engineering and Applied Mechanics
817 Sherbrooke Street West, Montreal, H3A 2K6, Canada.

AND

L. OGER

Université de Rennes I
Groupe Matière Condensée et Matériaux, URA CNRS 804,
Avenue du Général Leclerc, Bât B11, 35042 Rennes, France.

The internal angles of friction for cohesionless assemblies of particles under compression were determined using molecular dynamics type simulations of biaxial tests. The objective was to determine appropriate internal friction angles to be used for continuum modelling of broken ice covers on waterways. The particles were modelled as random-sized circular disks. Constant confining pressures as well as a constant displacement rate were used for the biaxial tests. The stress-strain curves were obtained for a range of confining pressures. The yield envelopes follow a Mohr-Coulomb criterion of failure. The effect of the interparticle friction coefficient on the global angle of friction was examined. It was found that the internal angle of friction was essentially independent of the interparticle friction angle. Furthermore, assemblies with different particle size distributions were used in order to investigate the effects of this parameter on the internal angle of friction. It was found that as the particle size distribution is made wider, the magnitude of the internal angle of friction increases. Finally, the results obtained using two different contact models were compared in order to verify the effect of this parameter on the results.

1. Introduction: The Internal Angle of Friction

Yielding in a granular materials is influenced by the magnitude of the mean pressure. One description of this process is the Mohr-Coulomb yield crite-

N. A. Fleck and A. C. F. Cocks (eds.),
IUTAM Symposium on Mechanics of Granular and Porous Materials, 313–324.
© 1997 *Kluwer Academic Publishers.*

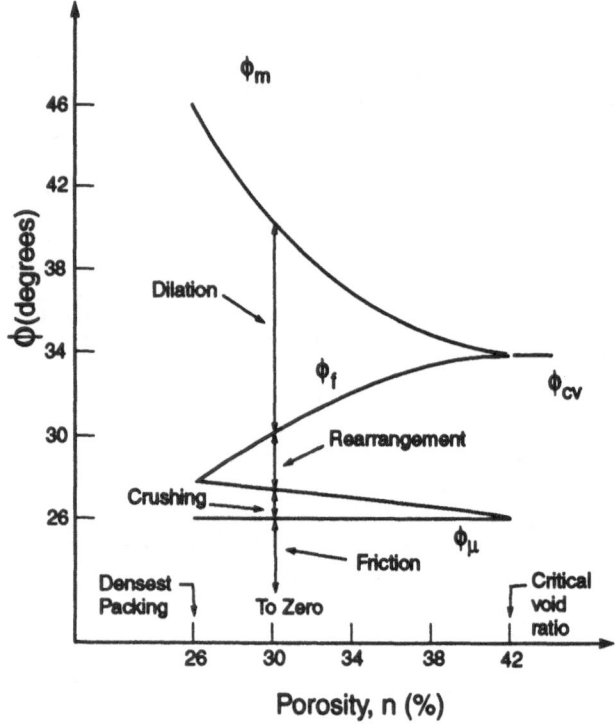

Figure 1. Contributions to shear strength of granular materials (Adapted from Mitchell, 1976).

rion which states that yielding will occur on a plane element when

$$\tau_f = c + \sigma_f \tan\phi , \qquad (1)$$

where c is the cohesion, σ_f and τ_f are the normal and shear stresses on the failure plane and ϕ is the internal angle of frcition. Thus, the internal angle of friction is a measure of the shear resistance of the granular material. It is a combined measure of four components (Mitchell, 1976) rather then solely the result of intergranular friction. First, there is the contact friction angle, ϕ_μ, between the particles which arises because of surface roughness. Then, there is the rearrangements contribution, ϕ_r, among adjacent particles under shear deformation which accounts for an important part of the internal friction as the porosity of the assembly becomes relatively high. This can be explained by the fact that at high porosity, the particles can roll and slide along planes inclined at various angles. Dilation contributes also to the internal angle of friction, and arises from the property that a granular materials has to alter its volume in accordance with a change in arrangement of its grains. Dilation is necessary in a granular material closely packed in order for deformation to be initiated. Finally, if the confining pressure is very high, there will be reduced dilation and particle crushing will become

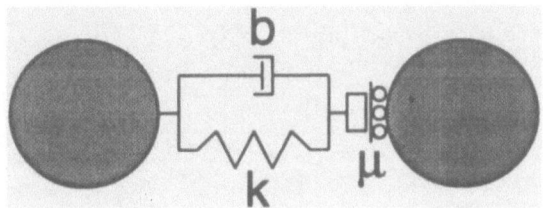

Figure 2. Schematic representation of the Coulombic friction force model.

an important contributor to the internal friction angle. However, particle crushing is not taken into account at this stage in the present study. Figure 1 shows the relative importance of the components of the internal friction angle (Mitchell, 1976). In this figure, ϕ_f is the friction angle corrected for the work of dilation. The global angle of friction can be defined as the constant volume friction angle, ϕ_{cv}. This value can be measured when an assembly of particles has reached the critical void ratio. At this stage no more dilation occurs. The peak friction angle ϕ_m is attained only for initially dense assemblies. Among the factors which influence the values of the internal friction angle are the type of confinement, the magnitude of the confining pressure, the strength of the granular material, as well as the size distributions of the material. Thus, the curves shown in figure 1 are only valid for a particular set of conditions.

2. Contact Models

The simulations were performed using two different contact models. The first contact model consists of a simple linear spring and dashpot combination in parallel to simulate the normal interaction between the particles. The normal force is given as

$$F_n = k\delta - bv_n , \qquad (2)$$

where k is the spring constant, δ is the relative normal displacement, v_n is the relative normal (departing) velocity and b is the dashpot constant. As no tensile strength is allowed, F_n is equal to zero in tension. The tangential forces are calculated using the Coulomb friction law

$$F_t = \mu F_n , \qquad (3)$$

where μ is the contact friction coefficient and F_t is the tangential force. A schematic of the contact model is shown in figure 2. However, with this model the particles are considered rotationally rigid. No shear deformation of the particles is considered. Thus, the particles in contact are continuously

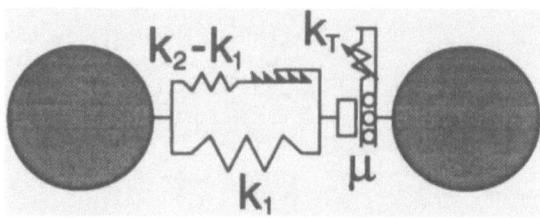

Figure 3. Schematic representation of the Walton & Braun model.

slipping. This model will be referred to as the *Coulombic friction force model* in the present paper. It was used for comparison purposes and only a very few results are shown in the present paper.

The second contact model used was developed by Walton & Braun (1986). This model allows for shear deformation and it is believed to better simulate quasi-static contacts. The normal forces are simulated by using a partially-latching-spring model. That is, the normal force is modeled using a linear spring with a stiffness constant which is smaller in the compression phase than in the recovery phase and in this way simulates energy dissipation. The tangential forces are modeled by using the incrementally slipping friction model of Walton & Braun that is based upon the theoretical work of Mindlin (1949) and Mindlin & Deresiewicz (1953) to model friction forces acting between elastic spheres. The incremental tangential force is calculated from the effective tangential stiffness K_T and the tangential displacement between two contacting particles Δs. The effective tangential stiffness K_T is given by

$$K_T = K_o \left(1 - \frac{T - T^*}{\mu N - T^*} \right)^\gamma \tag{4}$$

when the slip is in the direction of increasing tangential force and by

$$K_T = K_o \left(1 - \frac{T^* - T}{\mu N + T^*} \right)^\gamma \tag{5}$$

for slip in the other direction. In these equations, T is the total tangential force, T^* is the tangential force T when the relative tangential slip reverse direction, K_o is the initial tangential stiffness, μ is the friction coefficient and N is the total normal force. The exponent γ is set to 1/3 in Mindlin's theory. However, a value of γ between 1 and 2 better simulates frictional contacts according to Walton (1993) and thus a value of 1 was used for the present investigation. Figure 3 shows a sketch of the model.

3. Simulation Procedures

In order to determine the internal angle of friction, the yield envelope which is the locus of all stress states at failure had to be plotted using the results from strength tests performed on assemblies of particles. One of the most widely used laboratory strength tests is the triaxial test on granular material contained in a cylindrical rubber membrane. It basically consists in filling a cylindrical rubber membrane with a granular material. The two extremities of the cylinder are fitted with two frictionless, rigid end plates. The confining pressure on the membrane is applied by a surrounding fluid. The axial force can be provided by a piston acting on the end plates. Both the confining pressure and/or the axial stress are increased in order to reach a stress state which will cause the assembly to yield. If this test is performed for several different stress states which result in yielding of the assembly, then a failure envelope is obtained. If the yield envelope follows a Mohr-Coulomb yield criterion for a cohesionless material, the slope of the yield envelope is given by

$$\tan \phi = \tau_f / \sigma_f \,, \tag{6}$$

where τ_f and σ_f are the shear and normal stresses on the failure plane and ϕ is the internal angle of friction. Expressed in terms of the major and minor principal stresses, σ_1 and σ_2 respectively, the following is obtained for the slope of the yield envelope in the (σ_1, σ_2) space:

$$\text{slope} = \frac{(1 + \sin \phi)}{(1 - \sin \phi)} = \frac{\sigma_1}{\sigma_2} \,. \tag{7}$$

Using computer simulations, it is possible to perform the equivalent of the triaxial test in two dimensions. This *biaxial test* was simulated as follows. A test sample was created by filling a two dimensional box with circular disks randomly generated as shown in figure 4a. At the left and right boundaries of the packing, two flat frictionless walls moved inward at constant velocity during the simulation. These are the equivalent of the rigid end platens in the triaxial tests subjected to the piston force. The particles at the top and bottom edge of the assembly made up the flexible boundaries on which constant confining pressures were applied as shown in detail in figure 4b. These boundaries were the equivalent of the rubber membrane in a triaxial test. The magnitude of the force applied on the particles of the flexible boundaries was proportional to the projection of the particles in the horizontal and vertical direction. Following the model developed by Bardet & Proubet (1991), internal particles attempting to move across the flexible boundaries were immediately integrated into them. This prevents particles from escaping the assembly in the event the space between two

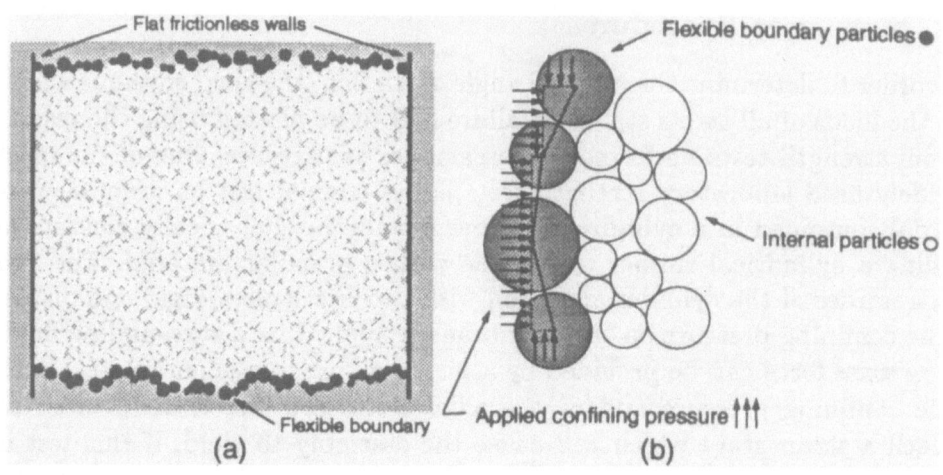

Figure 4.   (a) Initial configuration of the sample (b) Detail of the flexible boundary

particles in the flexible boundaries becomes too large. The same mechanical model was used for the particles making up the flexible boundaries as for those inside the sample. The advantage of flexible boundaries over periodic boundaries is that they do not bias the motion of the particles.

4. Results

Several biaxial tests were performed for assemblies of particles having different size distributions and numbers of particles. For each test, the axial stress versus axial strain curve was plotted. The axial stress for each time step is found by averaging the magnitude of all the contact forces on the end platens. The axial strain is obtained by dividing the instantaneous change in distance between the end plates by the initial distance. The stress-strain curves for a sample made of 1267 particles are shown in figure 5. The radii of the particles are expressed in dimensionless form by dividing the radius of the particles by the mean diameter D_{mean} of the assembly. The assembly used for the results presented in figure 5 has particles with dimensionless radii varying from 0.3 to 0.7. Since the distribution is uniform, this corresponds to an average particle radius of 0.5 for the assembly with a spread Δr of ± 0.2. Tests were performed for confining pressures in the range 10 - 150 KPa. The Walton & Braun contact model was used. The contact friction coefficient is 0.4. At low confining pressures, the axial stresses reach a plateau very rapidly. The fluctuations in the stress values are fairly small and of constant magnitude over the whole test. For the present study, the upper envelope of the stress-strain curves is chosen as the yield stress.

The 'volumetric strain' was also calculated. The results are shown in

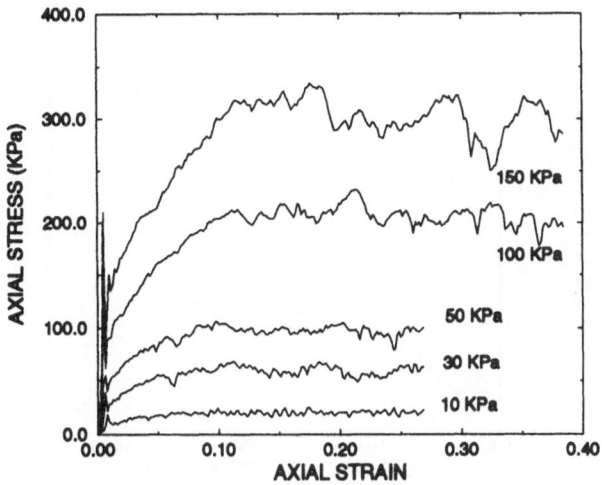

Figure 5. Axial stress versus axial strain, Walton & Braun contact model.

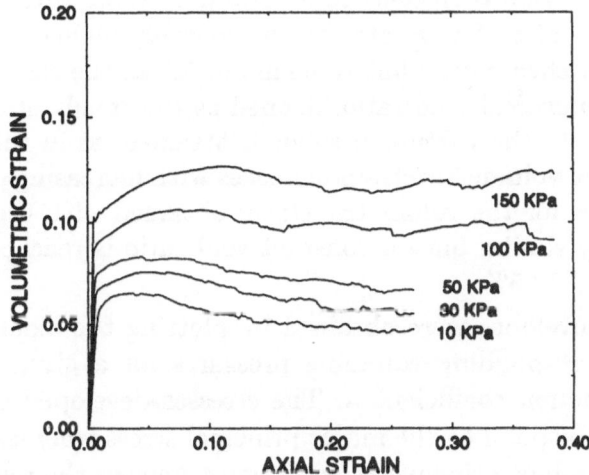

Figure 6. 'Volumetric strain' versus axial strain, Walton & Braun contact model.

figure 6 for the Walton & Braun model. The volumetric strain was taken to be the change in area of the assembly from its initial area divided by its initial area. Positive values of volumetric strain are taken as a reduction in volume. Initially, the volumetric strain increases sharply. The fact that no dilation is observed upon the initiation of compression is a good indication

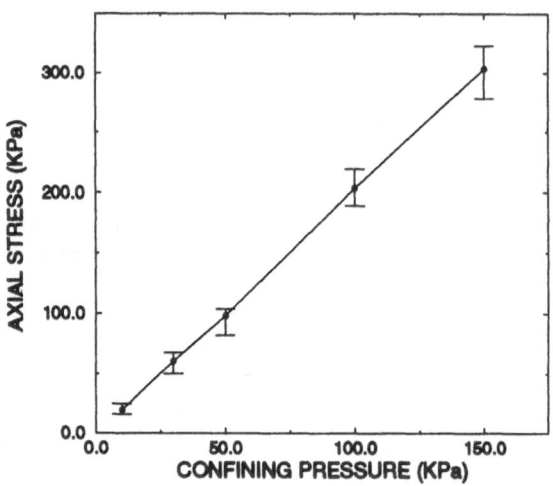

Figure 7.   Yield envelope; friction coefficient $\mu=0.4$;

that the sample was in a loose state at the beginning of the simulations. The volumetric strain then reached a maximum value with the volumetric strain increasing at a slower rate. At considerably higher strain, the volumetric strain reached a constant value in almost all the cases studied. This indicates that a critical void ratio, defined as the steady state ratio of the volume of voids to the volume of solids is attained, as in the case of soils. As expected, the volumetric strain increases with increasing confining pressure. The values for the volumetric strain obtained with the two different models are very similar once a constant void ratio is reached and they lie between 5% and 12.5%.

The yield envelopes were obtained by plotting the axial yield stresses against the corresponding confining pressures for a given assembly and interparticle friction coefficient μ. The stresses developed on the moving end platens correspond to the major principal stress whereas the confining pressure is the minor principal stress. Figure 7 shows the resulting plot for a sample of 1267 particles with the dimensionless radius varying between 0.3 and 0.7 and a coefficient of friction of $\mu=0.4$. These simulations were performed using the Walton & Braun model. A straight line is obtained with a y-intercept very close to the origin. This is consistent with a Mohr-Coulomb failure criterion for a cohesionless granular material. Error bars are included since uncertainties arise when determining the yield stresses on the stress-strain diagram.

The internal angle of friction ϕ_g was determined from the yield envelope

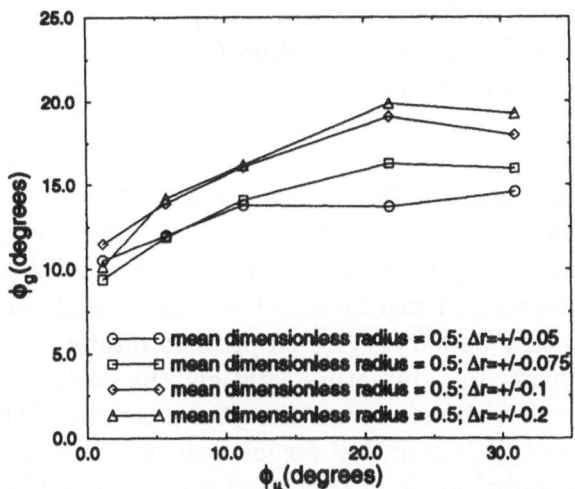

Figure 8.   Global angle of friction versus contact angle of friction for samples of about 1300 particles having different size distributions (uniform random with spread Δr). Walton & Braun model is used.

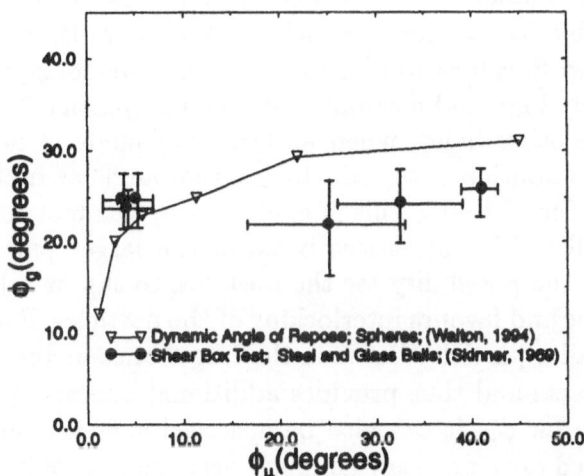

Figure 9.   Results from other investigations obtained using assemblies of identical particles.

and using equation (7). Figure 8 shows plots of the global angle of friction versus the contact friction angle for assemblies having different size distributions but approximately the same number of particles.   It can be seen that the internal angle of friction is practically independent of the contact angle of friction, except in the lower range of ϕ_μ values. This is in agreement with the results of Skinner (1969) who performed shear box tests on glass and steel balls. Furthermore, Walton (1994) observed a similar independence of the dynamic angle of repose of sphere on the particle-particle friction angle. The results obtained by Skinner and Walton are shown in figure 9. However, these results are in contradiction with the theories of Bishop (1954) and Caquot (1934) which predict that the internal angle of friction will continue to increase with increasing interparticle friction coefficient. This discrepancy between the theoretical studies and the results obtained in numerical and experimental studies is explained by the fact that both Bishop and Caquot neglected particle rotation as a possible deformation mechanism. In eliminating rotation, the particles of the system have one less degree of freedom and thus the strength of the assembly is increased. In figure 8, it is seen that for low values of the interparticles friction angle ϕ_μ the global angle of friction shows a dependence on ϕ_μ. This is due to the fact that when the value of ϕ_μ is very small, particle rotation is inhibited and thus ϕ_g increases with ϕ_μ as predicted by the theories of Bishop and Caquot which neglect particle rotation.

In figure 8 also, it can be seen that increasing the spread in the size distribution for a given assembly resulted in higher global friction angles for the range of contact friction angles tested. This conclusion holds both for the Coulombic friction force model and Walton & Braun model. As an example, in figure 8, values around 14.6^o are obtained for ϕ_g when the value of ϕ_μ is relatively high and a sample with a small particle size distribution is used. On the other hand, when an assembly made of particles having a broad size distribution is tested, the internal angle of friction can reach values of the order of 19.9^o. This is explained by the fact that the smaller particles can fill in the interstices between the larger particles. Thus, it greatly inhibits the possibility for the particles to slip or roll to new equilibrium positions and favours interlocking of the particles. Furthermore, the presence of smaller particles between the larger ones increases the number of particle contacts and thus provides additional lateral support. In figure 8, given a value for ϕ_μ, it can also be observed that the internal angle of friction measured does not vary as much between assemblies with particle size distributions of 20% and 40% as compared with the variation observed between the samples having particle size distributions in the range of 10% to 20%. Hence, it appears that the effect of the particle size distribution on the internal angle of friction of an assembly of particles is more significant for the lower range of size distributions.

The results were verified using two different contact models, that is the Walton & Braun model and the Coulombic friction force model. From the

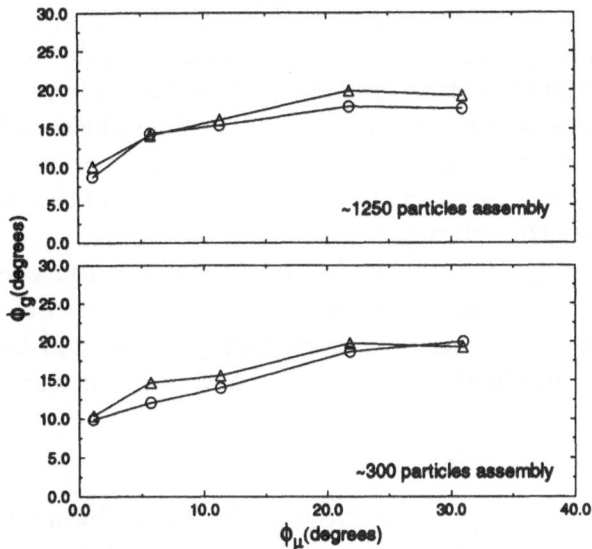

Figure 10. Global angle of friction versus contact angle of friction: Coulombic friction force model (◯) compared with Walton & Braun model (△) for assemblies with approximately 300 and 1250 particles. The mean dimensionless particle radius=0.5 ±0.2.

results presented in figure 10, it can seen that the two models yield results which are very similar for the global angle of friction even when the number of particles used is relatively small. However, in order to ensure a reasonable accuracy and obtained results free of finite size effects, it appears that assemblies made of at least 1000 particles should be used. Furthermore, since both the Walton & Braun model and the Coulombic friction force model yield very similar results for ϕ_g, this could be an indication that the overall averaged effects of the tangential particle contact forces do not make a significant contribution to the generation of the 'continuum' stress state.

5. Conclusion

The objective of the present work is to characterize the internal angle of friction of an assembly of disks using discrete element biaxial test simulations. The internal angles of friction measured show a weak dependence on the particle-particle friction angle except for the smaller values of ϕ_μ where particle rotation is negligible. Several particle assemblies having different particle size distributions were generated in order to evaluate the effect of this parameter on the internal friction angle. It was found that for assemblies having a wide range of particle sizes, the internal angle of friction is much greater than for samples composed of nearly uniform sized particles.

The results were obtained using both the Walton & Braun contact model and a simple spring-dashpot model. Both models yielded similar results for both small and large particle assemblies. However, it is recommended that a sufficiently large number of particles be used for the simulations in order to obtain a proper aggregate behaviour and a good accuracy for the yield stresses. Simulations performed with too few particles tend to overestimate or underestimate the internal angle of friction.

Further investigation are underway to include the effect of cohesive bonds between the particles in order to simulate the deformation of an intact ice cover. Furthermore, non-circular particles will be used in order to better represent the shape of most geomaterials encountered in nature.

6. Acknowledgements

This work was supported by an NSERC Strategic Grant titled "Mesoscale Ice Rheology for Regional Ice Forecasting". Acknowledgements are due to M. Sayed and R. Gutfraind for the numerous discussions and suggestions.

References

Bardet, J.P. and Proubet J. (1991) A Numerical Investigation of the Structure of Persistent Shear Bands in Granular Media, *Géotechnique* 41(4), 599–613.

Bishop, A.W. (1954) Correspondence on 'Shear Characteristics of a Saturated Silt, Measured in Triaxial Compression', *Géotechnique* 4(1), 43–45.

Caquot, A. (1934) *Equilibre des Massifs à Frottement Interne. Stabilité des Terres Pulvérulentes et Cohérentes.* Gauthier Villars, Paris.

Mindlin, R.D. (1949) Compliance of Elastic Bodies in Contact, *J. Appl. Mech. (Trans. ASME)* 16, 259–268.

Mindlin, R.D. and Deresiewicz, H. (1953) Elastic Spheres in Contact Under Varying Oblique Forces, *J. Appl. Mech. (Trans. ASME)* 20, 327–344.

Mitchell, J.K. (1976) *Fundamentals of Soil Behavior.* Wiley, New York.

Skinner, A.E. (1969) A Note on the Influence of Interparticle Friction on the Shearing Strength of a Random Assembly of Spherical Particles, *Géotechnique* 19(1), 150–157.

Walton, O.R. and Braun, R.L. (1986) Viscosity, Granular-Temperature, and Stress Calculations for Shearing Assemblies of Inelastic, Frictional Disks, *J. Rheology* 30(5), 949–980.

Walton, O.R. (1993) Numerical Simulation of Inclined Chute Flows of Monodisperse, Inelastic, Frictional Spheres, *Mechanics of Materials* 16, 239–247.

Walton, O.R. (1994) Effects of Interparticle Friction and Particle Shape on Dynamic Angles of Repose via Particle-Dynamics Simulation, in J. Jenkins and J. Goddard (eds.), *Proceeding of Workshop on Mechanics and Statistical Physics of Particulate Materials*, LaJolla, California, June 8-10, 1994.

A MECHANISTIC MODEL FOR ATTRITION OF PARTICLES IN FLOW SYSTEMS

D.C. CLUPPER, B.V. SANKAR, Z. CHEN, M. GUNDEPUDI, AND
J.J. MECHOLSKY, JR
University of Florida
Gainesville, Florida, USA

1. Introduction

Particle-particle contact affects many industrial systems such as powder processing, particle handling, and slurry flows. The control of fracture processes during contact is the key to the success of these systems. Many investigators have studied fracture of materials due to the contact between particles (Shipway and Hutchings, 1993). There are many correlations between the contact load and the amount of damage observed. However, until recently, the specific mechanism of damage was still unknown. This paper reports the experimental results and analytical models for the contact of brittle spheres in several contact configurations. Fracture analysis is used to determine the cause of failure and the stress at the failure site. The mechanism of failure is primarily Mode I (tensile) crack propagation. The location of failure is shown to be influenced by Young's modulus, the radius of curvature and the method of contact. The fracture origin for a single glass sphere compressed by two flats is at the equator and for a single glass sphere compressed by three spheres or by either a flat or sphere is at the contact points. Fracture originates from the contact point for alumina spheres for all testing configurations. The experimental stress is in agreement with theoretical predictions once the location of the primary fracture origin is known. Results for soda-lime-silica and borosilicate glasses and alumina balls are reported. The significance of the location of failure origin to the processing of particles are discussed.

The long-range objective of this investigation is to develop an analytical methodology for the prediction of attrition in flow systems. The immediate objective of this paper is to present a basis for the determination of a mechanistic model for particle-particle and particle-wall contacts.

2. Materials And Methods

Spheres of alumina, borosilicate glass, and soda-lime-silicate glass were loaded in various compressive loading configurations using an Instron model 1125 (0.005 or 0.2 in/min loading rate). Force vs. displacement was recorded on a strip chart. Samples ranged from 3 to 10 mm in diameter.

The various loading configurations investigated are shown in Figure 1. For tests B, C, D, and E, samples were constrained with a stainless steel mold. For all tests (A-F) a Si_3N_4 top indenter was used. However, Si_3N_4 bottom supports were used in only tests A, B,

N. A. Fleck and A. C. F. Cocks (eds.),
IUTAM Symposium on Mechanics of Granular and Porous Materials, 325–334.
© 1997 *Kluwer Academic Publishers.*

D, and F. Tests C and E used supports made of the sample material. For tests B, C, D, and E, three support spheres were used, whereas only one bottom support sphere was used for test F.

After fracture, the sample pieces were collected for fracture surface analysis using optical microscopy and SEM. The fracture stress was calculated for selected samples by measuring fracture surface features such as the size of the failure origin and the mirror radius. The fracture stress (σ_c) was computed as follows:

$$\sigma_c = K_{IC} / [1.24 \ c^{1/2}] \tag{1}$$

where K_{IC} is the mode I fracture toughness and c is the crack size at fracture and 1.24 is a geometric and load factor. The fracture toughness values used for alumina, soda-lime glass, and borosilicate glass were 3.5, 0.72, and 0.71 MPa•m$^{1/2}$, respectively (Mecholsky, 1993). Similarly, the fracture stress (σ_M) was also computed from the length of the mirror-mist boundary according to the following equation:

$$\sigma_M = M_1 / r^{1/2} \tag{2}$$

where M_1 is a material constant (1.8, and 1.9 MPa•m$^{1/2}$ for borosilicate glass and soda-lime glass, respectively.) and r is the distance between the origin and the mirror-mist boundary (Mecholsky, 1993).

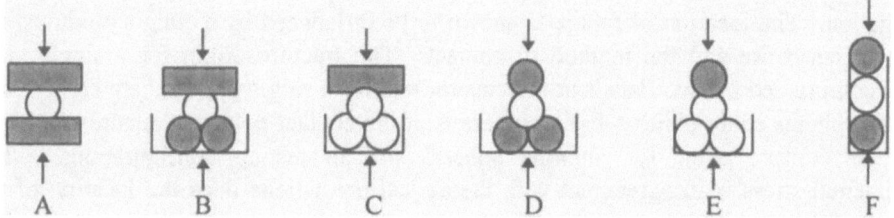

Figure 1. Loading Configurations (A-F) Shaded forms represent Si$_3$N$_4$, whereas the sample material is unshaded.

The silicon nitride platen was chosen for its intermediate elastic (Young's) modulus value relative to alumina and glass. As seen in Table I, the modulus of alumina is greater than silicon nitride, and glass has a modulus much less than that of silicon nitride. The magnitude of the contact stress is quite dependent on the Young's modulus of both the indenter and sample. Elastic mismatch between the sample and platen also leads to the development of shear stress and a reduction of Mode I loading (Warren and Hills, 1994).

Table I. Elastic Properties of Spheres and Platen Material

Material	Modulus (GPa)	Poisson's Ratio
Alumina	390	0.24
Silicon Nitride	300	0.27
Sodalime-Silica Glass	73	0.25
Borosilicate Glass	64	0.25

3. Theoretical Background

The Hertzian contact stress was calculated for each contact point for tests A-F for the various samples and sizes (Timoshenko and Goddier 1970). The theoretical contact stress (s) was calculated using the following equations. The legend is shown in Table II.

Plate on Sphere
$$r = [(0.75\ P\ R_s)/\ E^*]^{1/3} \tag{3}$$

$$E^* = 1\ /\ \{[(1 - v_s^2)/E_s] + [1 - v_i^2)/E_i]\} \tag{4}$$

$$\sigma = P\ /\ \pi\ r^2 \tag{5}$$

Sphere on Sphere
$$r = \{(0.375\ P\ /\ E^*)\ /\ [(1/(2R_r)) + (1/(2R_i))]\}^{1/3} \tag{6}$$

$$E^* = 1\ /\ \{[(1 - v_s^2)/E_s] + [1 - v_i^2)/E_i]\} \tag{4}$$

$$\cos \theta = (R_r - R_i)\ /\ (R_i + R_r) \tag{7}$$

$$F = P\ /\ (3 \sin \theta) \tag{8}$$

$$\sigma = F\ /\ \pi\ r^2 \tag{9}$$

Table II. Legend for Equations 3-9

Symbol	Represents	Symbol	Represents
r	contact radius	v_s	Poisson's (sample)
P	load	v_i	Poisson's (indenter)
E_s	sample modulus	R_r	ring radius
E_i	indenter modulus	R_i	indenter radius

4. Experimental Results and Discussion

4.1 SODALIME SILICA GLASS

The load at fracture was measured for soda-lime-silica glass spheres (3 mm diameter) loaded in five different configurations (Fig. 2). The fracture loads could be grouped into three categories: highest fracture loads for single particle test (A), intermediate fracture loads for multiple particle tests with a flat top platen (B and C), and lowest fracture loads for the multiple particle tests with a spherical top platen (D and E).

The composition of the bottom platen spheres (soda-lime-silica glass or Si_3N_4) had no significant effect on the load at fracture. The three bottom spheres in multiple particle tests B and C were soda-lime-silica glass and Si_3N_4 , respectively. There was no significant difference between the fracture force (870 N) for configurations B and C, and between the fracture forces for tests D and E.

For the 3 mm soda-lime glass samples, the platen geometry affected the fracture load. The fracture loads for tests D and E are significantly lower than those for test B and C. This difference is due to the increased stress caused by the spherical indenter.

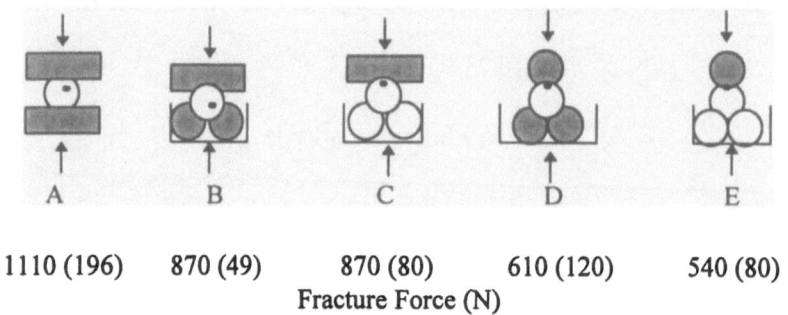

1110 (196)   870 (49)   870 (80)   610 (120)   540 (80)

Fracture Force (N)

Figure 2. Effect of Loading Configuration on Fracture Force and Location of Fracture Origin for 3 mm Soda-lime Silica Glass Spheres. Standard deviation is given in parentheses. The location of the fracture origin is given by the dark dot. The shaded objects are silicon nitride.

Variation in the fracture load with loading configuration for the 3 mm soda-lime glass samples suggests that different fracture modes exist for each. Fracture surface analysis (using optical microscopy and SEM) supports this hypothesis. The fracture origins are shown in Figure 2 for the various loading configurations. For single particle tests (A), the fracture originated from surface flaws on the equator. The fracture stress for these samples was determined to be 95-100 MPa. At loads slightly less than the fracture load, soda-lime-silica glass particles underwent local fracturing at the contact points.

Although the bottom support material did not affect fracture load, it did influence the location of the fracture origin. For test B (Si_3N_4 supports) fracture originated from the lower sphere-on-sphere contact point. For test C which used less stiff soda-lime silica glass supports, the fracture originated from the upper contact point with the top platen.

The particle fines created at high loads just below the fracture load essentially allow the spherical platen to "indent" the sample surface and create a flaw of critical size. The top flat platen, however, is constrained from indenting the sample due to the intact portion of the sample.

For the multiple particle tests using a spherical ball on the top loading point, the fracture originates from the top contact point. As was the case for the other multiple particle tests, fracture with spherical top platens is also caused by indentation of the sample surface by fines created by local fracturing. For the 3 mm soda-lime silica glass samples the force at the top contact area is approximately three times greater than at the lower contact points.

Comparison of the upper and lower contact stresses for tests B, C, D and E (equations 3-8) revealed that for each case the larger stress corresponded to fracture origin locations shown in Figure 2, i.e. for test B, the sphere-on-sphere (lower) contact stress was higher than that of the plate-on-sphere (upper) contact point. The theoretical contact stress for both the upper and lower contact points is plotted in Figure 3 for Test D with a 3 mm diameter soda-lime silica glass sphere. It is seen that the upper contact point stress is higher than the lower contact point stress over the entire loading range.

The SEM micrograph in Figure 4a, shows the fracture surface of soda-lime silica glass for Test D. The upper contact area is seen in the lower right hand corner. An enlargement of the area is shown in Figure 4b and the twist hackle marks radiating from the origin indicates the location of the origin.

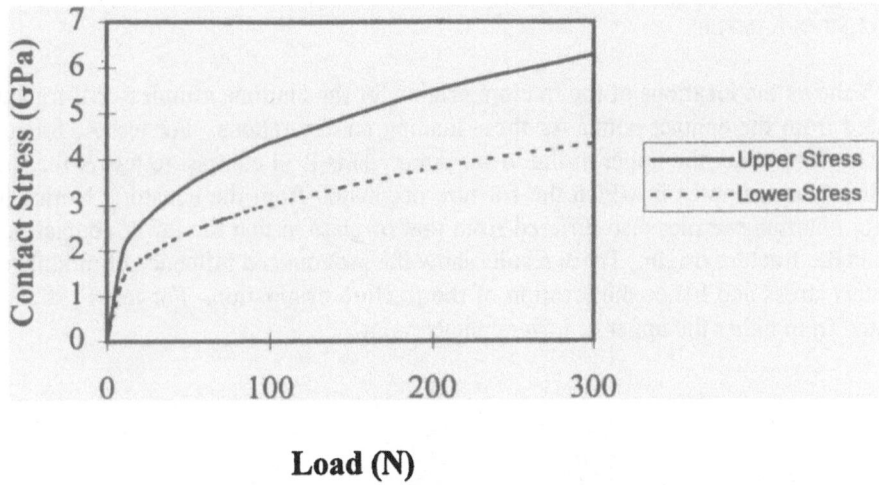

Figure 3. Contact Stress vs. Load for 3 mm Soda-lime Silica Glass (Test D)

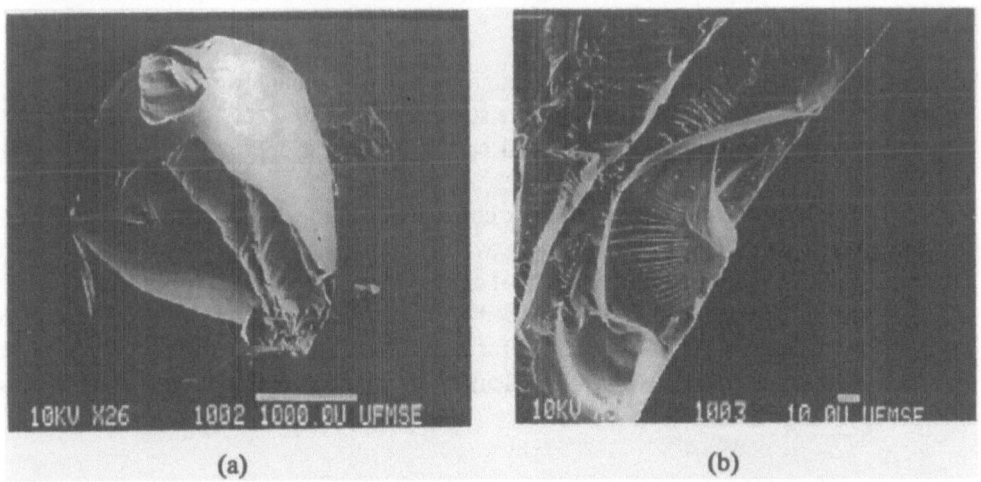

Figure 4. SEM micrographs of the fracture surface of soda-lime silica glass (Test D). The upper contact point is shown in the lower right hand corner (a) and magnified in (b)

It is thus shown that Hertzian contact theory can be used to predict the location of the fracture origin for the multiple particle tests (B, C, D, and E) for soda-lime silica glass. Although the contact stresses are up to two orders of magnitude greater that the fracture stresses, contact stresses are very localized, whereas fracture stresses are more global. A localized tensile stress can only cause a crack to propagate over relatively short distances, whereas global stresses can cause critical failure.

4.2 ALUMINA SAMPLES

Figure 5 shows the locations of the fracture origins for the alumina samples. All fractures originated from the contact points for these loading configurations. For test A, fractures originated from either the upper or the lower point. This is in contrast to test A for soda-lime silica glass samples in which the fracture originated from the equator. Notice that test C for alumina samples also differed from that of glass in that the lower contact point contained the fracture origin. These results show the pronounced influence of modulus on the contact stress and hence the location of the fracture origination. For test F, fracture originated from either the upper or lower contact point.

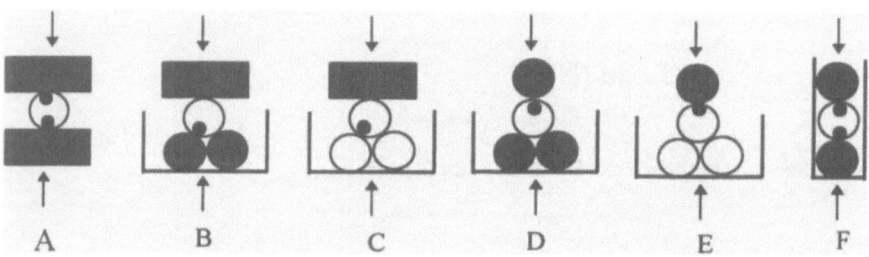

Figure 5. Effect of Loading Configuration on the Location of Fracture Origins for 3/8" Alumina Samples

The catastrophic failure of alumina particles is thought to initiate from radial cracks which form at the contact points at a stress level of 5-9 GPa. The radial cracks are more pronounced than the circumferential cracks formed at approximately 5 GPa (Fig. 6a).

The largest circumferential crack corresponds to the approximate contact area at the maximum applied load. In most cases two of the radial cracks were colinear. Figure 6b shows an SEM micrograph of two partial circumferential cracks and one radial crack.

Table III contains the fracture load with standard deviation (n=10) and fracture stress for selected tests. The fracture load for test F was significantly different than test A, B, and C. The fracture stress values, as calculated from the size of the origin using optical microscopy, range from 150 to 186 MPa.

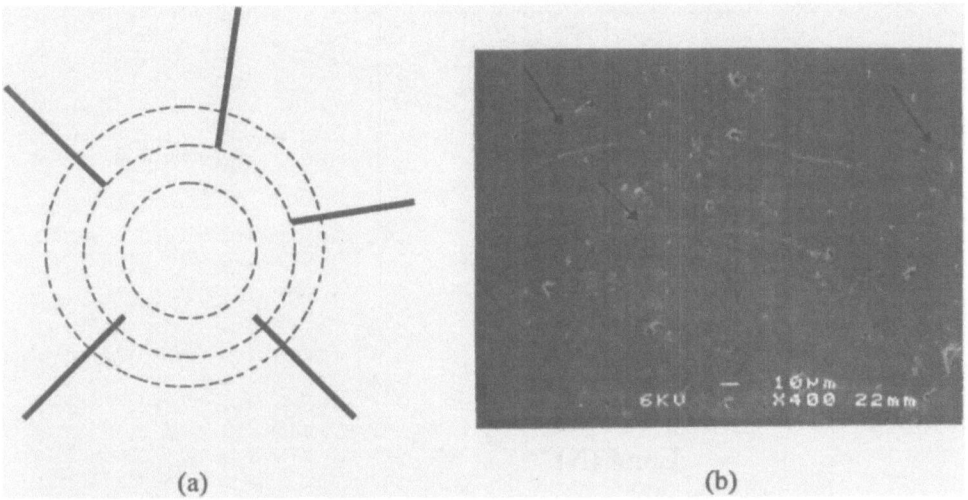

(a)   (b)

Figure 6a. Schematic of Initial Crack Formation at Contact Points in Alumina. Note that the radial cracks are more pronounced than the circumferential cracks. (b) SEM Micrograph of Initial Crack Formation at Contact Point (Alumina, load was 90% of catastrophic fracture load). Note that the left and middle arrows point to circumferential cracks whereas the radial crack is shown by the right arrow.

Table III. Fracture Load and Flaw Size Data for 5/16" Al_2O_3 Samples

Testing Configuration	Fracture Load (N)	Fracture Stress (MPa)
A	5603 (350)	165
B	6003 (525)	186
C	6018 (736)	150
F	4305 (182)	150

As shown for soda-lime silica glass, Hertzian contact theory can also be used to predict the location of the fracture origin for alumina samples. In Figure 7, the stresses for the upper and lower contact points are plotted as a function of load for test B for alumina. Here it is seen that the lower contact stress was higher than the upper contact point stress over the entire loading range. Therefore, it is expected that fracture should originate from the lower contact point. Indeed, the fracture originated from the lower contact point, (Figure 8.) The left SEM micrograph shows the fracture surface of alumina for Test B. The hackle markings which point back toward the origin are seen here. The right micrograph is an enlargement of the lower contact point. Here the "thumbnail" crack which caused failure is shown in the lower area. Hackle marks point back towards this fracture origin.

4.3 COMPARISON OF GLASS AND ALUMINA SAMPLES

Figure 9 shows the effects of several variables on the load at fracture: loading configuration, sample composition and sample diameter. For glass samples the maximum fracture load corresponded to the single particle loading configuration (test A). However, for crystalline alumina samples, the maximum fracture load was seen for a multiple particle loading configuration (test B). For both alumina and glass samples, test D produced the lowest fracture force.

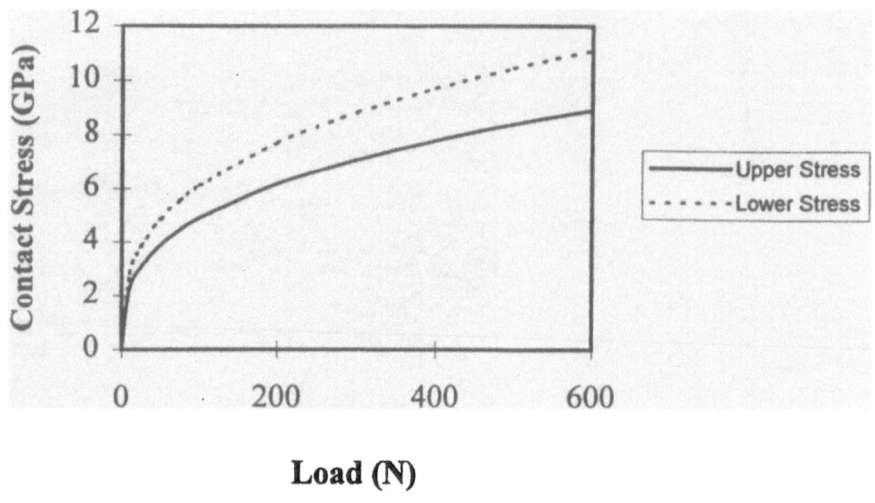

Figure 7. Contact Stress vs. Applied Load for Alumina Samples (Test B)

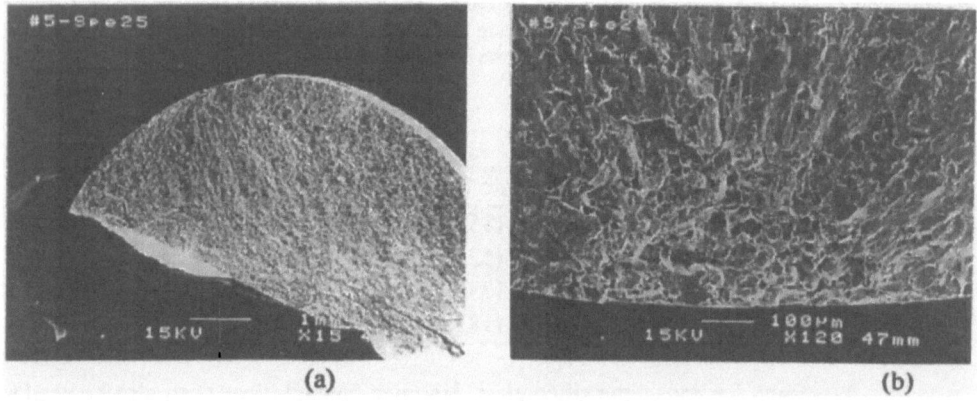

Figure 8. SEM micrographs of fracture surface for alumina (Test B). Micrograph (b) shows the fracture origin.

4.4 IMPACT TESTS

Impact testing was performed using pressurized gas to accelerate an approximately spherical sample into a steel plate. Impact velocity was recorded electronically. In Figure 10, the fracture force for test A for soda-lime silica glass (10 mm) is compared with the fracture forces for impact at various velocities. The impact force was calculated from equation 10:

$$F = (5/3 \; \pi \; r)^{0.6} \; (3/4 \; k)^{-0.4} \; V^{1.2} \, R^2 \tag{10}$$

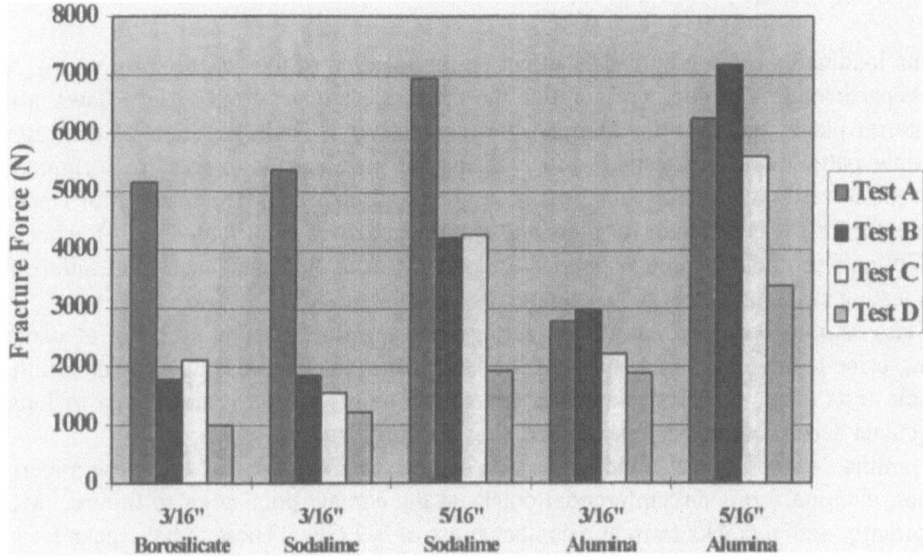

Figure 9. Comparison of fracture loads for alumina and glass samples.

where r is the sample density, k is a function of Poisson's ratio and Young's modulus, V is the impact velocity, and R is the sample radius (Knight et al., 1977). At a low impact force (270 N) there is no macroscopic surface damage. At a higher load of 1470 N the formation of a Hertzian cone crack was clearly evident and the contact stress was determined experimentally using the impact force and the measured contact area. Contact stresses which initiated "attrition" or loss of material without total particle failure were 1 to 4 GPa. The fracture force for impact was 9200 N. This is comparable to the fracture force for the static test A of 7700 N. One difference between the impact test and the static test A was that silicon nitride platens were used for Test A, whereas, the sample impinged on a steel plate during the impact test.

Less than 10% of alumina samples fractured upon impact with the steel plate. Significant plastic deformation took place in the steel plate. The impact fracture stress for the alumina sample was approximately 190 MPa, which is comparable to that measured for the static tests for alumina.

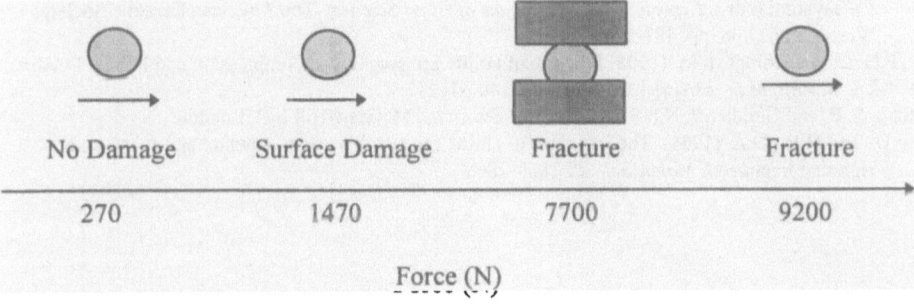

Figure 10. Comparison of Impact Forces with Static Test Failure Force.

5. CONCLUSIONS

1. The loading configuration has an effect on the location of the fracture origin. For the single particle glass testing configuration the fracture originates from surface flaws in the equatorial plane; however for alumina, the contact point is the location of the origin. Multiple particle fracture testing with a flat upper platen leads to fractures originating from the lower contact points for Test B in glass and alumina. The location of fracture origins for Test C is different for glass and alumina, and this is attributed to differences in the Young's modulus. Use of spherical indenters caused upper contact point failure for each testing case, regardless of the sample or support material.

2. Glass samples undergo significant surface damage due to cone cracking at contact points prior to fracture. It is thought that the lower fracture load values for multiple particle tests using spherical platens is due to the ability of the round platen to indent particulate debris into the sample surface, thus creating flaws.

3. Alumina samples do not undergo surface damage in the sense that they lose material. Rather, alumina forms circumferential cracks at the contact point prior to failure. More importantly, radial cracks form at a contact stress of 7-9 GPa. These radial cracks lead to total particle failure.

4. The fracture forces for impact testing soda-lime silica glass correlate with those for static compression test A.

5. In the development of a theoretical model for the transport and handling of particles, three major factors must be considered: (a) indenter geometry (b) Young's modulus and c) loading configuration.

6. ACKNOWLEDGMENTS

The authors acknowledge the financial support of the Engineering Research Center for Particle Science and Technology at the University of Florida, the National Science Foundation grant #EEC-94-02989, and the industrial partners of the ERC.

7. REFERENCES

Knight, C.G. Swain, M.V. and Chaudhri, M.M. (1977) Impact of small steel spheres on glass surfaces , J. Mater. Sci., 12 1573.
Mecholsky, Jr., J.J. (1993) Quantitative fracture surface analysis of glass materials, in C.J. Simmons and O. H. El-Bayoumi (eds.), Experimental Techniques of Glass Science, The American Ceramic Society, Westerville, Ohio, pp. 483-520.
Shipway, P.H. and Hutchings, I. M. (1993) Fracture of Brittle Spheres Under Compression and Impact Loading, I & II, Phil. Mag. A67[6] 1389-1404 and 1405-1421.
Timoshenko, S. P. and Goddier, J. N.(1970) Theory of Elasticity, McGraw-Hill Inc., London.
Warren, P.D. and Hills, D.A. (1994) The influence of elastic mismatch between indenter and substrate on Hertzian fracture, J. Mater. Sci., 29 2860-2866.

SIMPLE SHEAR OF POROUS MATERIALS AT LARGE STRAINS

D. DURBAN AND O. YAGEL
Faculty of Aerospace Engineering
Technion
Haifa 32000
Israel

1. Introduction

An early suggestion by Schleicher (1926) for the plastic yield condition of porous materials reads

$$\tau^2 + 3(Y_C - Y_T)p - Y_C Y_T = 0 \tag{1}$$

where $\tau = ((3/2)\mathbf{S} \cdot \cdot \mathbf{S})^{1/2}$ is the Mises effective stress, $\mathbf{S} = \sigma - p\mathbf{I}$ is the stress deviator, σ - the Cauchy stress tensor, $\mathbf{I}$ - the 2nd order unit tensor, $p = (1/3)\mathbf{I} \cdot \cdot \sigma$ is the hydrostatic stress and (Y_C, Y_T) are the uniaxial yield stresses in compression and in tension, respectively.

A later proposal by Drucker and Prager (1952) is given by

$$\tau + 3\left(\frac{Y_C - Y_T}{Y_C + Y_T}\right)p - \frac{2Y_C Y_T}{Y_C + Y_T} = 0 \tag{2}$$

where, in common with (1), the coefficient of p represents the "strength differential" of the solid.

The yield criteria (1)-(2) are particular cases of the family

$$\tau^m + 3\left(\frac{Y_C^m - Y_T^m}{Y_C + Y_T}\right)p - \frac{Y_C^m Y_T + Y_C Y_T^m}{Y_C + Y_T} = 0 \tag{3}$$

where, in agreement with existing experimental data, $1 \le m \le 2$. Conditions (1)-(2) are recovered from (3) with m=2 and m=1, respectively. Denoting by k the yield stress in pure shear $(\tau = \sqrt{3}k = \tau_0, \ p - 0)$, and by t the yield stress in hydrostatic tension ($\tau=0$, p=t) we can rewrite (3) in the form

$$\left(\frac{\tau}{\tau_0}\right)^m + \frac{p}{t} = 1 \qquad\qquad 1 \le m \le 2 \tag{4}$$

which is characteristic of pressure sensitive solids. Here

N. A. Fleck and A. C. F. Cocks (eds.),
IUTAM Symposium on Mechanics of Granular and Porous Materials, 335–342.
© 1997 *Kluwer Academic Publishers.*

$$\tau_0^m = \frac{Y_C^m Y_T + Y_C Y_T^m}{Y_C + Y_T} \qquad t = \frac{Y_C^m Y_T + Y_C Y_T^m}{3\left(Y_C^m - Y_T^m\right)} \qquad (5)$$

Notice that with $Y_C = Y_T \equiv Y$ we have from (5) that $\tau_0 = Y$, t goes to infinity, and the entire family (4) is reduced to the standard Mises condition $\tau = Y$.

Strain hardening (softening) and non-associativity can be accounted for in (3)-(4) by defining an effective stress σ_e and a plastic potential ϕ as

$$\tau^m + \mu\sigma_e^{m-1}p - \sigma_e^m = 0 \qquad (6)$$

$$\tau^m + \eta\phi^{m-1}p - \phi^m = 0 \qquad (7)$$

where μ and η are parameters that reflect the pressure sensitivity of the material; Associated behaviour is obtained with $\eta=\mu$, implying that $\phi=\sigma_e$.

It is now straightforward (Yagel 1955) to construct the flow and deformation theories, based on (6)-(7), by following the usual recipe which states that

(a) the total strain rate $\mathbf{D}$ is the sum of an elastic part $\mathbf{D}^E$ and a plastic part $\mathbf{D}^P$.
(b) the elastic part $\mathbf{D}^E$ is Hookean hypoelastic.
(c) the plastic part $\mathbf{D}^P$ is normal to the potential hypersurface ϕ, namely

$$\mathbf{D}^P = \Lambda \frac{\partial\phi}{\partial\sigma} \qquad (8)$$

Λ being a proportionality factor.
(d) the principle of plastic power equivalence is expressed as

$$\sigma \cdot\cdot \mathbf{D}^P = \sigma_e \dot{\varepsilon}_p \qquad (9)$$

where the total plastic strain ε_p is a known function of the effective stress and the superposed dot denotes differentiation with respect to a time like parameter.

Just to give an example, for the associated Drucker Prager model (m=1, $\eta=\mu$) we find the flow theory formulation (Durban and Fleck 1996)

$$\mathbf{D} = \frac{1}{2G}\left(\overset{\nabla}{\sigma} - \frac{\nu}{1+\nu}\mathbf{II}\cdot\cdot\overset{\nabla}{\sigma}\right) + \dot{\varepsilon}_p\left(\frac{3S}{2\tau} + \frac{1}{3}\mu\mathbf{I}\right) \qquad (10)$$

where $\overset{\nabla}{\sigma}$ is the Jaumann stress rate.

The corresponding deformation theory formulation is simply

$$\mathbf{E}_L = \frac{1}{2G}\left(\sigma - \frac{\nu}{1+\nu}\mathbf{II}\cdot\cdot\sigma\right) + \varepsilon_p\left(\frac{3S}{2\tau} + \frac{1}{3}\mu\mathbf{I}\right) \qquad (11)$$

where $\mathbf{E}_L = \ell n\left(\mathbf{F}\cdot\mathbf{F}^T\right)^{1/2}$ is the finite logarithmic strain tensor with $\mathbf{F}$ standing for the deformation gradient.

2. Simple Shear

Simple shear is a homogeneous deformation pattern (Figure 1) which has served as a thought experiment in evaluating large strain constitutive relations (Durban 1990, Durban et al. 1990, Durban 1994). The Eulerian strain rate tensor and the spin tensor are

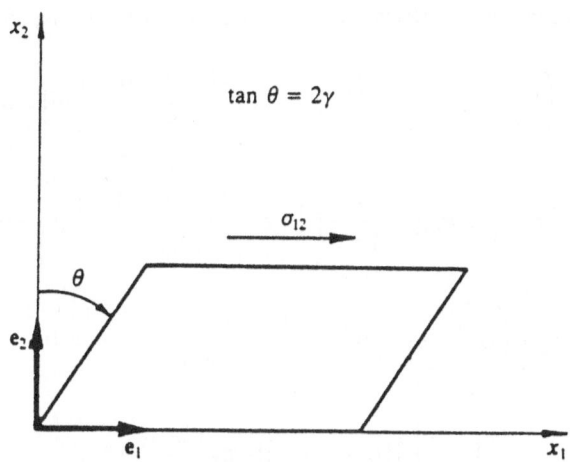

Figure 1. Notation for simple shear

$$\mathbf{D} = \dot{\gamma}(\mathbf{e}_1\mathbf{e}_2 + \mathbf{e}_2\mathbf{e}_1) \qquad\qquad \mathbf{W} = -\dot{\gamma}(\mathbf{e}_1\mathbf{e}_2 - \mathbf{e}_2\mathbf{e}_1) \qquad (12)$$

where $(\mathbf{e}_1, \mathbf{e}_2)$ is the unit Eulerian (spatial) orthonormal base.

Notice that θ is the shear angle while $\gamma = (1/2)\tan\theta$ is a convenient measure of shear strain. The logarithmic strain tensor is given by

$$\mathbf{E}_L = \frac{\Gamma}{\cosh\Gamma}\left[(\mathbf{e}_1\mathbf{e}_2 + \mathbf{e}_2\mathbf{e}_1) + \gamma(\mathbf{e}_1\mathbf{e}_1 - \mathbf{e}_2\mathbf{e}_2)\right] \qquad (13)$$

where

$$\gamma = \sinh\Gamma \qquad (14)$$

3. Flow Theory Solution

Assuming (as will be verified later) that the active stress components are $(\sigma_{11}, \sigma_{22}, \sigma_{33}, \sigma_{12})$ we have the stress tensor

$$\sigma = \sigma_{11}\mathbf{e}_1\mathbf{e}_1 + \sigma_{22}\mathbf{e}_2\mathbf{e}_2 + \sigma_{33}\mathbf{e}_3\mathbf{e}_3 + \sigma_{12}(\mathbf{e}_1\mathbf{e}_2 + \mathbf{e}_2\mathbf{e}_1) \qquad (15)$$

along with the Jaumann rate, with the aid of (12),

$$\overset{\triangledown}{\sigma} = \dot{\sigma} + \mathbf{W} \cdot \sigma - \sigma \cdot \mathbf{W} = \left(\dot{\sigma}_{11} - 2\dot{\gamma}\sigma_{12}\right)\mathbf{e}_1\mathbf{e}_1 + \left(\dot{\sigma}_{22} + 2\dot{\gamma}\sigma_{12}\right)\mathbf{e}_2\mathbf{e}_2$$

$$+ \dot{\sigma}_{33}\mathbf{e}_3\mathbf{e}_3 + \left[\dot{\sigma}_{12} + \dot{\gamma}\left(\sigma_{11} - \sigma_{22}\right)\right]\left(\mathbf{e}_1\mathbf{e}_2 + \mathbf{e}_2\mathbf{e}_1\right) \tag{16}$$

Inserting now the Eulerian strain rate **D**, from (12), the stress rate (16) and the stress (15) in the constitutive equation (10) we arrive at four scalar relations, namely

$$\frac{1}{2G}\left(\dot{\sigma}_{11} - 2\dot{\gamma}\sigma_{12} - \frac{3\nu}{1+\nu}\dot{p}\right) + \dot{\varepsilon}_P\left[\frac{3(\sigma_{11} - p)}{2\tau} + \frac{1}{3}\mu\right] = 0 \tag{17a}$$

$$\frac{1}{2G}\left(\dot{\sigma}_{22} + 2\dot{\gamma}\sigma_{12} - \frac{3\nu}{1+\nu}\dot{p}\right) + \dot{\varepsilon}_P\left[\frac{3(\sigma_{22} - p)}{2\tau} + \frac{1}{3}\mu\right] = 0 \tag{17b}$$

$$\frac{1}{2G}\left(\dot{\sigma}_{33} - \frac{3\nu}{1+\nu}\dot{p}\right) + \dot{\varepsilon}_P\left[\frac{3(\sigma_{33} - p)}{2\tau} + \frac{1}{3}\mu\right] = 0 \tag{17c}$$

$$\frac{1}{2G}\left[\dot{\sigma}_{12} + \dot{\gamma}\left(\sigma_{11} - \sigma_{22}\right)\right] + \dot{\varepsilon}_P\left(\frac{3\sigma_{12}}{2\tau}\right) = \dot{\gamma} \tag{17d}$$

Relations (17) form a system of four equations with four unknown stress components. We shall show that the solution can be obtained in quadrature expressions with the effective stress as the independent variable. To this end, upon adding (17a)-(17c), we find that

$$\frac{\dot{p}}{\kappa} + \mu\dot{\varepsilon}_P = 0 \tag{18}$$

where $p = (1/3)\left(\sigma_{11} + \sigma_{22} + \sigma_{33}\right)$ and $\kappa = E/3(1-2\nu)$ is the elastic bulk modulus. Integrating (18) along the loading path, with a stress free initial state, results in the hydrostatic stress as a function of the effective stress (strain), viz.

$$p = -\mu\kappa\varepsilon_P \tag{19}$$

Thus, the level of plastic deformation determines the value of p. In the pre-yield range, where $\varepsilon_P = 0$, there is no hydrostatic stress in agreement with the analysis by Durban (1990) for the Mises model. In fact, relation (19) implies that for the isochoric simple shear pattern positive plastic volume changes ($\mu\varepsilon_P$) are balanced by negative elastic dilation (p/κ). Inserting (19) in the Drucker-Prager relation, (6) with m=1, we obtain the Mises stress τ in terms of the effective stress σ_e:

$$\tau = \sigma_e + \mu^2\kappa\varepsilon_P \tag{20}$$

Also, upon inspecting the sum of (17a) and (17b) and comparing with (17c) we find, with the aid of (19), that

$$\sigma_{33} = \frac{1}{2}\left(\sigma_{11} + \sigma_{22}\right) = p = -\mu\kappa\varepsilon_P \tag{21}$$

Next, we generate, from (10) and the first of (12), the double scalar multiplication $\mathbf{S}\cdot\mathbf{D}$ to obtain the identity

$$2\sigma_{12}\dot{\gamma} = \frac{1}{3G}\tau\dot{\tau} + \tau\dot{\varepsilon}_P \tag{22}$$

Finally, we subtract (17b) from (17a) and use (22) to get an ordinary differential equation for the stress difference $\left(\sigma_{11} - \sigma_{22}\right)$ in the form

$$\left(\sigma_{11} - \sigma_{22}\right)' + \frac{3G\varepsilon_P'}{\tau}\left(\sigma_{11} - \sigma_{22}\right) = 2\tau\left(\frac{\tau}{3G} + \varepsilon_P\right)' \tag{23}$$

where the prime denotes differentiation with respect to the effective stress σ_e and we have used the transformation $(\dot{\ }) = (\ )'\dot{\sigma}_e$. Since both τ and ε_P are known functions of σ_e it is possible to solve (23) for $\left(\sigma_{11} - \sigma_{22}\right)$ as a function of σ_e. In fact, the solution of (23) is given by the quadratures

$$\sigma_{11} - \sigma_{22} = F\int_0^{\sigma_e} \frac{2\tau}{F}\left(\frac{\tau}{3G} + \varepsilon_P\right)' d\sigma_e \quad \text{with} \quad F = \exp\left(-3G\int_0^{\sigma_e} \frac{\varepsilon_P' d\sigma_e}{\tau}\right) \tag{24}$$

For certain hardening characteristics the integrals in (24) can be evaluated in closed form expressions. Once the solution (24) has been established we have, with the aid of (21), the normal stresses $\left(\sigma_{11}, \sigma_{22}\right)$ as a function of σ_e.

The dependence of the applied shear stress σ_{12} on the effective stress follows at once from the definition of the Mises effective stress, and (21),

$$\sigma_{12}^2 = \frac{1}{3}\tau^2 - \left(\frac{\sigma_{11} - \sigma_{22}}{2}\right)^2 \tag{25}$$

with $\left(\sigma_{11} - \sigma_{22}\right)$ given by (24). Combining that relation with (22) we find the shear strain γ in the form

$$\gamma = \int_0^{\sigma_e} \frac{\tau}{2\sigma_{12}}\left(\frac{\tau}{3G} + \varepsilon_P\right)' d\sigma_e \tag{26}$$

which completes the solution since all field quantities $\left(\sigma_{11}, \sigma_{22}, \sigma_{33}, \sigma_{12}, \gamma\right)$ have been determined in terms of σ_e. The solution is valid also for strain softening response only with the effective plastic strain being the independent variable.

4. Deformation Theory Solution

The constitutive relations follow from (11), (13)-(14) and (15), in the form of four algebraic equations

$$\frac{1}{2G}\left(\sigma_{11} - \frac{3\nu}{1+\nu}p\right) + \varepsilon_P\left[\frac{3(\sigma_{11}-p)}{2\tau} + \frac{1}{3}\mu\right] = \gamma S \qquad (27a)$$

$$\frac{1}{2G}\left(\sigma_{22} - \frac{3\nu}{1+\nu}p\right) + \varepsilon_P\left[\frac{3(\sigma_{22}-p)}{2\tau} + \frac{1}{3}\mu\right] = -\gamma S \qquad (27b)$$

$$\frac{1}{2G}\left(\sigma_{33} - \frac{3\nu}{1+\nu}p\right) + \varepsilon_P\left[\frac{3(\sigma_{33}-p)}{2\tau} + \frac{1}{3}\mu\right] = 0 \qquad (27c)$$

$$\frac{1}{2G}\sigma_{12} + \varepsilon_P\left(\frac{3\sigma_{12}}{2\tau}\right) = S \quad , \qquad S = \frac{\Gamma}{\cosh\Gamma} \qquad (27d)$$

which may be compared with the flow theory equations (17a)-(17d). It is a matter of ease to verify that relations (19)-(21) and (25) are valid also for the deformation theory and we proceed with deriving a simple relation between the shear strain γ and the effective stress σ_e. To this end rewrite the shear relation (27d) as

$$\sigma_{12} = 2G_sS \qquad\qquad \frac{1}{2G_s} = \frac{1}{2G} + \frac{3\varepsilon_P}{2\tau} \qquad (28)$$

where G_s is the secant shear modulus. Likewise, upon subtracting (27b) from (27a),

$$\frac{1}{2}(\sigma_{11} - \sigma_{22}) = 2G_s\gamma S \qquad (29)$$

Inserting (28)-(29) in (25) results in the desired relation

$$2\sqrt{3}\sqrt{1+\gamma^2}S = 2\sqrt{3}\Gamma = \frac{\tau}{G_s} \qquad (30)$$

with Γ defined in (14). This implies, by (28), that the shear stress σ_{12} is now a known function of the effective stress σ_e or, alternatively, of the shear strain γ. The in-plane normal stress components follow from (29) and (21) as

$$\sigma_{11} = -\mu\kappa\varepsilon_P + 2G_s\gamma S \qquad\qquad \sigma_{22} = -\mu\kappa\varepsilon_P - 2G_s\gamma S \qquad (31)$$

which completes the solution. On balance, the deformation theory analysis is much simpler since the mathematical model involves only algebraic equations.

5. Discussion

Simple shear of pressure sensitive solids is dominated by a strong elastoplastic coupling (19) which requires a hydrostatic pressure that increases linearly with the total plastic strain. Thus, $dp/d\varepsilon_p = -\mu\kappa$ for both flow and deformation theories. Sample calculations have shown that flow theory models a considerably stiffer shear response than the one obtained from deformation theory, particularly at low levels of porosity. The two theories yield maximum values for the shear stress (Figure 2), but flow theory predictions are not always at realistic stress levels. The analysis can be used to assess and compare flow versus deformation models for porous solids in the presence of large strains and finite rotations. The solutions derived here can also be used to model the primary path in shear bands of finite thickness and to examine possible bifurcations and loss of stability within the bands (Ogden and Connor 1995). It may be expected, in view of earlier results, that deformation theory will predict bifurcation loads at much lower strain levels than flow theory.

6. Acknowledgement

Part of this study was supported by the fund for the promotion of research at the Technion.

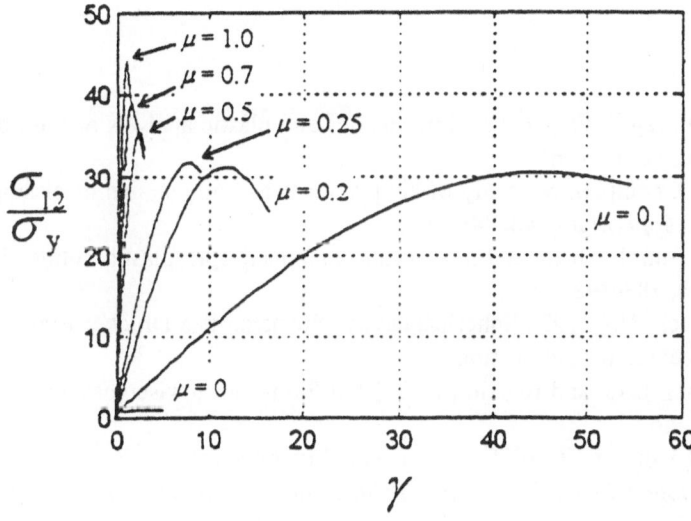

Figure 2a. Shear stress-shear strain curves for a power hardening material. Flow theory results. σ_y is a nominal yield stress. The hardening exponent is n=15.62.

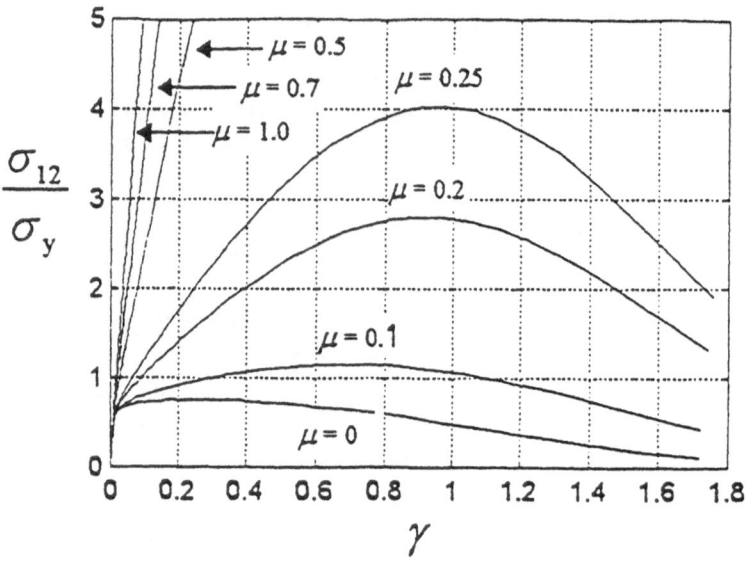

Figure 2b. Shear stress-shear strain curves for a power hardening material. Deformation theory results. σ_y is a nominal yield stress. The hardening exponent is n=15.62.

7. References

Drucker, D.C. and Prager, W. 1952 Soil mechanics and plastic analysis or limit design, *Quart. Appl. Math.*, **12**, 157-165.

Durban, D. 1990 A comparative study of simple shear at finite strains of elastoplastic solids, *Q. Jl Mech. Appl. Math.*, **43,** 449-465.

Durban, D. 1994 Simple shear at finite strains of anisotropic plastic solids, *Eur. J. Mech., A/Solids*, **13,** 783-792.

Durban, D. and Fleck, N.A. 1996 Spherical cavity expansion in a Drucker-Prager Solid, *J. Appl. Mech.*, accepted for publication.

Durban, D., Zeitoun, D.G. and Benaim, H.E. 1990 Finite linear viscoleasticity, *J. Eng. Mech.*, **116,** 2449-2462.

Ogden, R.W. and Connor, P. 1995 On the stability of shear bands, *IUTAM Symp. Anisotropy, Inhomogeneity and Nonlinearity in Solid Mechanics*, Kluwer Academic Publishers, 217-222.

Schleicher, F. 1926 Der Spannungszustand an der Fliessgrenze (Plastizitätsbedingung), *ZAMM*, **6**, 199-216.

Yagel, O. 1995 *Simple Shear of a Porous Material at Large Strain*, M.Sc. Research Thesis, Technion, Haifa.

FORMATION OF SHEAR BANDS IN MODELS OF GRANULAR MATERIALS

M. SHEARER, F. X. GARAIZAR AND M. K. GORDON
Department of Mathematics
North Carolina State University
Raleigh, NC 27695
USA

1. Introduction

In this paper we analyze the behavior of a granular material just prior to the appearance of a shear band. Shear band formation has long been associated with stress and strain localization – the deformation develops abruptly and locally near the site of the shear band (Rudnicki and Rice, 1975). It is also recognized that the phenomenon is connected with a change of type in the governing partial differential equations resulting from a continuum theory (Schaeffer, 1990).

Here, we analyze localization itself through an asymptotic analysis that incorporates nonlinear effects. Indeed, localization is inherently a nonlinear phenomenon; linear analysis cannot provide the form for stresses and velocities as a shear band appears, merely a criterion for the loss of stability (actually, a loss of well–posedness) that is presumed to signal the formation of a shear band.

Our analysis follows the evolution of a sample in simple shear. The deformation is taken to be nearly uniform until localization occurs. Nonuniformity is introduced by assuming the material is almost homogeneous, but with a small amount of inhomogeneity; the weakest places locate where shear bands will first appear. The strength of nonuniformity is measured by a small parameter $\epsilon > 0$. If $t_b(\epsilon)$ is the time at which a shear band forms, we find that

$$t_b(\epsilon) = t^* - k\sqrt{\epsilon} + O(\epsilon), \qquad (1.1)$$

where t^* is the time at which the uniform solution breaks down, $k > 0$, and $O(\epsilon)$ indicates terms of order ϵ and higher in the asymptotic expansion.

N. A. Fleck and A. C. F. Cocks (eds.),
IUTAM Symposium on Mechanics of Granular and Porous Materials, 343–352.
© 1997 *Kluwer Academic Publishers.*

The order $\sqrt{\epsilon}$ decrease in $t_b(\epsilon)$ from the uniform case indicates a substantial sensitivity of the strength of the material to material nonuniformity.

Even more striking is the behavior of the velocity $v = v_\epsilon$ as localization occurs: as time t approaches $t_b(\epsilon)$, the velocity suffers an $O(1)$ perturbation from the uniform case, no matter how small the choice of ϵ. The leading order term in the asymptotic expansion of v_ϵ depends on an improper integral. Depending on whether the integral converges, v_ϵ is either continuous or discontinuous at the location of the shear band in the limit $t = t_b(\epsilon)$.

To simplify the analysis, and to focus on the primary nonlinear mechanism responsible for localization, we consider a simplified model. Asymptotic results are stated in Section 2 in the case of the quasistatic approximation. (The results are proved in a forthcoming paper (Schaeffer and Shearer, 1996).) In Section 3, we illustrate the results with numerical simulations of both the dynamic and quasistatic versions of the simplified model. These provide a check on the asymptotic results, but also serve to justify the validity of the quasistatic approximation right up to shear band formation.

2. Equations and Asymptotic Results

Consider the following system of equations describing simple shear.

$$(a) \quad \mu \partial_t v = \partial_x \sigma$$

$$(b) \quad \eta \partial_t \begin{pmatrix} \sigma \\ \tau \end{pmatrix} + \begin{pmatrix} 1 \\ -\alpha \end{pmatrix} \partial_t \gamma = \begin{pmatrix} \partial_x v \\ 0 \end{pmatrix} \qquad (2.1)$$

$$(c) \quad \sigma + \alpha \tau = [1 + \epsilon m(x)] H(\gamma)$$

with boundary conditions

$$v = \pm 1 \quad \text{if} \quad x = \pm 1. \qquad (2.2)$$

Discussion of the equations.

(i) In (2.1), (σ, τ) are components of the stress tensor. (Note that $\sigma(\pm 1, t)$ specifies the surface traction at $x = \pm 1$, the observable stress in an experiment, while $\tau(\pm 1, t)$ specifies an internal stress in the sample.) The variable v is velocity; γ denotes the accumulated plastic strain.

(ii) The dependent variables in (2.1) are functions of the spatial variable x and time t.

(iii) To derive (2.1), the circles $\{|\tau| = \text{const}\}$ are approximated by tangent lines, in both the yield condition and the normal to the yield surface. Continuing plastic loading is assumed. $H(\gamma)$ is the hardening function; it is a smooth increasing and concave function, with $\lim_{\gamma \to \infty} H(\gamma) = 1$. The function $m(x)$ represents the material nonuniformity.

(iv) The parameter $\alpha > 0$ in (2.1) measures the amount of nonassociativity in the flow rule; the small angle approximation is made, and powers of α larger than the first power are neglected.

(v) The equations are given in nondimensional form. The parameter η is of order one; the parameter μ is very small: $\mu \sim 10^{-12}$ in laboratory conditions (Vardoulakis and Graf, 1985).

The asymptotic results of this section pertain to the *quasistatic approximation*, in which $\mu = 0$. Numerical results in the next section compare solutions with $\mu = 0$ and those with small $\mu > 0$.

Equations (2.1) with $\epsilon = 0$ admit *uniform solutions* satisfying the boundary conditions (2.2), in which $v = x$, and $\sigma = \sigma_o(t), \tau = \tau_o(t)$ satisfy a system of two ordinary differential equations; $\gamma = \gamma_o(t)$ is then given by the yield condition $\sigma_o + \alpha\tau_o = H(\gamma_o)$. The trajectory (σ_o, τ_o) is unique up to translation along $\sigma + \alpha\tau = 0$, and is invariant under translations of time. Accordingly, for initial conditions, we take

$$\sigma(x,0) = \sigma_o(0), \quad \tau(x,0) = \tau_o(0). \tag{2.3}$$

In the quasistatic approximation, $v(x,0)$ is determined from the equations and these initial conditions. In the dynamic case ($\mu > 0$), we take $v(x,0) = x$, consistent with the uniform solution. $\gamma(x,0) = \gamma_o(x)$ is given implicitly in terms of σ_o, τ_o by (2.1(c)).

Although the hardening function $H(\gamma)$ is monotonic, equations (2.1) nonetheless lose well-posedness due to nonassociativity in the flow rule (2.1(b)). Specifically, for $\mu > 0$, the equations are linearly well posed at time t if and only if

$$[1 + \epsilon m(x)]H'(\gamma(x,t)) > \alpha^2/\eta \quad \text{for all} \quad x. \tag{2.4}$$

Let $t = t_b(\epsilon)$ denote the time at which (2.4) first fails. Then we say a shear band forms at the corresponding locations x^* for which the inequality becomes equality. As for the uniform deformations with $\epsilon = 0$, it can be shown that the observable traction $\sigma(x^*, t)$ achieves a maximum at $t = t_b(\epsilon)$.

For $\mu = 0$, the equations remain well posed, but the failure of (2.4) signals a singularity in the equations, more specifically a singular perturbation in which a coefficient in a principal order term becomes infinite. The asymptotic analysis treats this singular perturbation, with the following results, proved in (Schaeffer and Shearer, 1996).

Let $\tilde{m}(x) = m(x) - m_{min}$ where $m_{min} = \min_{1 \le x \le 1} m(x)$, and let

$$I = \int_{-1}^{1} \frac{dx}{\sqrt{\tilde{m}(x)}}$$

Theorem 2.1 Asymptotically as $\epsilon \longrightarrow 0$,

(i) $t_b(\epsilon) = t_* - \sqrt{\epsilon}\sqrt{\dfrac{k}{C}}\langle\sqrt{\tilde{m}}\rangle + \mathcal{O}(\epsilon)$, where

$$k = H(\gamma_0(t_*)) - \frac{\alpha^2}{\eta}\frac{H(\gamma_0(0))}{H'(\gamma_0(0))}, \quad C = -\frac{1}{2}H''(\gamma_0(t_*))$$

are positive constants and $\langle\sqrt{\tilde{m}}\rangle = \dfrac{1}{2}\displaystyle\int_{-1}^{1}\sqrt{\tilde{m}(x)}dx$.

(ii) If I is finite, then

$$\lim_{t\to t_b(\epsilon)} v(x,t) = -1 + \frac{2}{I}\int_{-1}^{x}\frac{dx'}{\sqrt{\tilde{m}(x')}} + \mathcal{O}(\sqrt{\epsilon}).$$

(iii) If I is infinite and if $m(x)$ assumes its minimum m_{min} at exactly one point x_0, then

$$\lim_{t\to t_b(\epsilon)} v(x,t) = \text{sgn}(x - x_0) + \mathcal{O}(\sqrt{\epsilon}).$$

3. Numerical Results.

In this section, we show results of numerical experiments with two examples of nonuniformity $m(x)$. In the first tests, $m(x) = |x|$, and in the second series of tests, $m(x)$ is unity outside a small interval around $x = 0$, and zero inside the interval. The experiments are set up to treat various values of the parameters. In the results reported here, we take $\eta = 1, \epsilon = 0.01$, and we vary $\mu = a^{-2}$. In subsection 3.1, we describe a simple algorithm for the quasistatic approximation and in subsection 3.2 we describe a modified second order Godunov method, the modification treating the yield condition as a constraint. Finally, in subsection 3.3 we compare numerical results for the quasistatic and dynamic equations, showing velocity profiles in the two examples.

3.1. QUASISTATIC APPROXIMATION.

The quasistatic approximation to the dynamic equations (2.1) describing plastic deformation is system (2.1) with $\mu = 0$:

$$\begin{array}{llr} \partial_x\sigma & = 0 & (a) \\ \eta\partial_t\sigma + \partial_t\gamma & = \partial_x v & (b) \qquad (3.1) \\ \eta\partial_t\tau - \alpha\partial_t\gamma & = 0, & (c) \end{array}$$

with yield condition $[1 + \varepsilon m(x)]H(\gamma) = \sigma + \alpha\tau$. We compute the time t_b (at which shear bands form for the system (3.1)), and the solution of (3.1) at $t = t_b$. For convenience, we exploit the symmetry of the system, and perform calculations on the interval $x \in [0, 1]$, with the boundary condition $v(0, t) = 0$.

Following (Schaeffer and Shearer, 1996), we may use the yield condition and the t integral of (3.1(c)) to express σ as a function of γ and x:

$$\sigma = G(\gamma, x) = (1 + \varepsilon m(x))H(\gamma) - \tfrac{\alpha^2}{\eta}(\gamma - \gamma_o(x)) - \alpha\tau_o \qquad (3.2)$$

where γ_o, τ_o come from the initial conditions (2.3). This allows us to reduce (3.1) to a system for v and γ:

$$\begin{aligned} \partial_x G(\gamma, x) &= 0 & (a) \\ \eta\partial_t G(\gamma, x) + \partial_t \gamma &= \partial_x v. & (b) \end{aligned} \qquad (3.3)$$

Notice (3.3(a)) implies that

$$G(\gamma(x, t), x) = G(\gamma(0, t), 0). \qquad (3.4)$$

Integrating (3.3(b)) with respect to t and x gives

$$\eta \int_0^1 (\sigma - \sigma_o)dx + \int_0^1 (\gamma - \gamma_o)dx = t \qquad (3.5)$$

From (Schaeffer and Shearer, 1996), we have that $\sigma(x, t_b) = \sigma_*$ where $\sigma_* = G(\gamma_*, 0)$ and γ_* satisfies $H'(\gamma_*) = \tfrac{\alpha^2}{\eta}$. Then (3.4) implies that

$$G(\gamma(x, t_b), x) = \sigma_*, \qquad (3.6)$$

and from (3.5) we deduce that

$$\eta \int_0^1 (\sigma_* - \sigma_o)dx + \int_0^1 (\gamma(x, t_b) - \gamma_o(x))dx = t_b. \qquad (3.7)$$

Using Maple V, we solve (3.6) for $\gamma(x, t_b)$ and insert the result into (3.7) to obtain t_b.

All that is left is to compute $v(x, t_b)$. Let $\omega(t) = \gamma(0, t)$. By (3.4), γ may be written as a function of ω and x, and

$$\frac{\partial\gamma}{\partial\omega} = \frac{\frac{\partial G}{\partial\gamma}(\omega, 0)}{\frac{\partial G}{\partial\gamma}(\gamma, x)}. \qquad (3.8)$$

From (3.3), we have that

$$\eta\partial_t G(\omega, 0) + \partial_t\gamma = \left[\eta\frac{\partial G}{\partial\gamma}(\omega, 0) + \frac{\partial\gamma}{\partial\omega}(\omega, x)\right]\omega'(t) = \partial_x v, \qquad (3.9)$$

which we may integrate in x to obtain

$$\left[\eta\frac{\partial G}{\partial \gamma}(\omega,0) + \int_0^1 \frac{\partial \gamma}{\partial \omega}(\omega,x)dx\right]\omega'(t) = 1. \tag{3.10}$$

Solving (3.10) for $\omega'(t)$, integrating (3.9) in x, and using (3.8), we obtain an expression for v:

$$
\begin{aligned}
v(x,t) &= \frac{\eta x \frac{\partial G}{\partial \gamma}(\omega,0) + \int_0^x \frac{\partial \gamma}{\partial \omega}(\omega,\tilde{x})d\tilde{x}}{\eta \frac{\partial G}{\partial \gamma}(\omega,0) + \int_0^1 \frac{\partial \gamma}{\partial \omega}(\omega,x)dx} \\[2mm]
&= \frac{\eta x + \int_0^x \frac{\partial G}{\partial \gamma}(\gamma,\tilde{x})^{-1}d\tilde{x}}{\eta + \int_0^1 \frac{\partial G}{\partial \gamma}(\gamma,x)^{-1}dx}.
\end{aligned}
\tag{3.11}
$$

Evaluating the right-hand side of (3.11) at $t = t_b$ is troublesome because $\frac{\partial G}{\partial \gamma}(\gamma,\tilde{x})$ vanishes at $x = 0$. In the case where $m(x) = x$, $\frac{\partial G}{\partial \gamma}(\gamma,\tilde{x}) \sim$ $constant/\sqrt{x}$ when $t = t_b$, so we can compute the integrals in (3.11) using Maple V. In the case where $m(x)$ is the step function defined by

$$m(x) = H(x - \delta), \tag{3.12}$$

$\frac{\partial G}{\partial \gamma}(\gamma(x,t_b),x)$ vanishes on the interval $0 \le x \le \delta$. We then find $v(x,t_b)$ by a limiting process described as follows. It follows from (3.6) and (3.12) that $\gamma(x,t_b)$ is also a step function which jumps at $x = \delta$ and so (3.11) reduces to

$$v(x,t) = \frac{\eta x + x\frac{\partial G}{\partial \gamma}(\gamma(0,t),0)^{-1}}{\eta + \delta\frac{\partial G}{\partial \gamma}(\gamma(0,t),0)^{-1} + (1-\delta)\frac{\partial G}{\partial \gamma}(\gamma(1,t),1)^{-1}}$$

for $x < \delta$, and

$$v(x,t) = \frac{\eta x + \delta\frac{\partial G}{\partial \gamma}(\gamma(0,t),0)^{-1} + (x-\delta)\frac{\partial G}{\partial \gamma}(\gamma(1,t),1)^{-1}}{\eta + \delta\frac{\partial G}{\partial \gamma}(\gamma(0,t),0)^{-1} + (1-\delta)\frac{\partial G}{\partial \gamma}(\gamma(1,t),1)^{-1}}$$

for $x \ge \delta$. Since $\frac{\partial G}{\partial \gamma}(\gamma(0,t),0) \to 0$ as $t \to t_b$, we conclude that

$$v(x,t_b) = \begin{cases} x/\delta & \text{if } x < \delta \\ 1 & \text{if } x \ge \delta. \end{cases}$$

3.2. DYNAMIC EQUATIONS

The numerical method used to solve the dynamic equations (2.1) is based on a modification of a second order Godunov scheme for systems of hyperbolic conservation laws.

The state U at a given cell has the components v, σ and τ. We divide the spatial domain into N cells of size Δx_j and centered at the point x_j for $1 \leq j \leq N$. We denote by $x_{j+\frac{1}{2}}$ the location of the right boundary for the jth cell; $x_{j+\frac{1}{2}} = x_j + \Delta x_j/2$. We write U_j^n for the average of the state over the jth cell at time $t = t_n$. We assume that for a given n, the cell averages U_j^n are known for $1 \leq j \leq N$. We describe an algorithm to compute U_j^{n+1}, the cell averages at time $t = t_n + \Delta t$, where Δt is the time increment.

System (2.1) can be written in conservation form as

$$G(U)_t + F(U)_x = 0, \tag{3.13}$$

where $F(U) = \begin{pmatrix} -\sigma \\ -v \\ 0 \end{pmatrix}$ and $G(U) = \begin{pmatrix} \mu v \\ \eta\sigma + \gamma \\ \eta\tau - \alpha\gamma \end{pmatrix}$. (Since we are assuming continuing plastic loading, right up until the shear band forms, we ignore elastic unloading in this description of the algorithm.)

Discretization of the equations yields

$$G(U_j^{n+1}) = G(U_j^n) - \frac{\Delta t}{\Delta x_j}\left(F_{j+\frac{1}{2}}^{n+\frac{1}{2}} - F_{j-\frac{1}{2}}^{n+\frac{1}{2}}\right) \tag{3.14}$$

where $F_{m+\frac{1}{2}}^{n+\frac{1}{2}} = F(U_m^n, U_{m+1}^n)$ is the numerical flux evaluated at the state on the cell edge given by $x_{m+\frac{1}{2}}$ at the intermediate time $t = t_n + \Delta t/2$. This flux is calculated as the solution to an exact or linearized Riemann Problem (Garaizar, 1994) with initial states computed using a characteristic tracing algorithm which achieves second order accuracy (Trangenstein and Colella, 1991).

We write the two nonzero components of the numerical flux in (3.14) as

$$\begin{aligned} f_v &= (\sigma_{j+\frac{1}{2}}^{n+\frac{1}{2}} - \sigma_{j-\frac{1}{2}}^{n+\frac{1}{2}})\Delta t/\Delta x_j \\ f_\sigma &= (v_{j+\frac{1}{2}}^{n+\frac{1}{2}} - v_{j-\frac{1}{2}}^{n+\frac{1}{2}})\Delta t/\Delta x_j. \end{aligned} \tag{3.15}$$

The velocity equation is solved by $v_j^{n+1} = v_j^n + a^2 f_v$.

We now write the stress components of (3.14), together with the yield condition (2.1(c)) in the discrete form

$$\begin{aligned} \eta\sigma_j^{n+1} + \gamma_j^{n+1} &= \eta\sigma_j^n + \gamma_j^n + f_\sigma, \\ \eta\tau_j^{n+1} - \alpha\gamma_j^{n+1} &= \eta\tau_j^n - \alpha\gamma_j^n, \\ (1 + em(x_j+))H(\gamma_j^{n+1}) &= \sigma_j^{n+1} + \alpha\tau_j^{n+1}. \end{aligned} \tag{3.16}$$

We note that (3.16) can be reduced to a single implicit equation for γ_j^{n+1}

$$\Sigma(\gamma_j^{n+1}) = \Sigma(\gamma_j^n) + f_\sigma \tag{3.17}$$

where $\Sigma(\gamma) = \eta[1 + \epsilon m(x)]H(\gamma) + (1 - \alpha^2)\gamma$ is a monotone function ($\Sigma'(\gamma) = \eta[1 + \epsilon m(x)]H'(\gamma) + (1 - \alpha^2) > 0$). This observation makes equation (3.17) easy to solve.

We also notice that in the examples, the most relevant part of deformation will take place in a small region around $x = 0$. To concentrate the numerical effort in that region, we choose cells of different size. That is, if Δx_s is the mesh size on the subinterval $[-\frac{1}{3}, \frac{1}{3}]$, the mesh size away from this subinterval is $3\Delta x_s$. The time increment Δt is chosen to satisfy the usual Courant Friedrichs-Lewy (CFL) stability condition for hyperbolic equations, on the fine mesh.

	$\epsilon = 0.01$, $m(x) = x$		
a	t_b (quasistatic)	t_b (dynamic)	difference
10	1.070975267	1.1341150127	0.0631397456
10^2	1.070975267	1.0765539148	0.0055786478
10^3	1.070975267	1.0712692852	0.0002940182
10^4	1.070975267	1.0710133205	0.0000380535

TABLE 1. Time of shear band formation.

3.3. COMPARISON OF RESULTS.

In Table 1 we compare the time of shear band formation for the quasistatic and dynamic equations for a fixed ϵ and varying a. Of course, the time in the quasistatic case does not depend on a. Notice that, as a increases, the time of shear band formation for the dynamic equations approaches that for the quasistatic equations. Figures 1 and 2 show profiles of velocity at the time of shear band formation for both the quasistatic and dynamic equations for a fixed ϵ and a fixed large value of a. In each case, the larger values of velocity correspond to the quasistatic approximation. (The difference between the two figures is the choice of m.) Notice that the difference between the quasistatic and dynamic solutions is small, although in the second case, there is more of a balance between the acceleration (dynamic) terms and the nonuniformity, resulting in a deceleration in the dynamic case. To illustrate this point, we have taken a smaller value of a than is suggested by the estimates of Vardoulakis and Graf (1985).

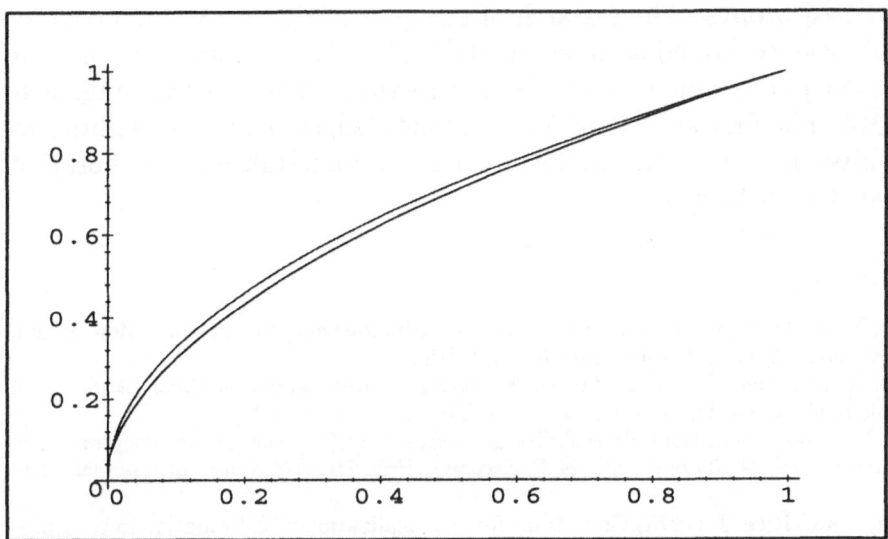

Figure 1. Velocity at $t = t_b$ for quasistatic and dynamic equations; $m(x) = x$, $\varepsilon = 0.01$, $a = 1000$.

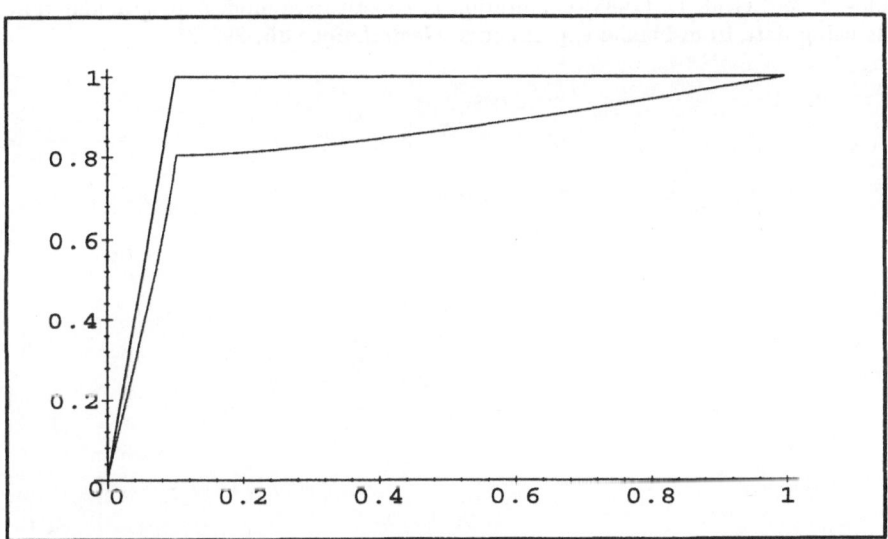

Figure 2. Velocity at $t = t_b$ for quasistatic and dynamic equations; $m(x) = H(x - \delta)$, $\varepsilon = 0.01$, $a = 1000$.

Acknowledgments. The research of this paper was partly supported by National Science Foundation grant DMS 9504583. Research of the first author was partly supported by Army Research Office grant DAAH04-94-G-0043. The first author gratefully acknowledges the use of facilities at Duke University, where part of the research was undertaken in collaboration with David G. Schaeffer.

References

Garaizar, X. (1994) Numerical computations for antiplane shear in a granular flow model, *Quarterly of Applied Mathematics* **52**, 289–309.

Garaizar, X. and Schaeffer, D.G. (1994) Numerical computations for shear bands in an antiplane shear model, *J. Mech. Phys. Solids* **42**, no. 1, 21–50.

Mandel, J. (1964) Conditions de stabilité et postulat de Drucker, in *Rheology and Soil Mechanics (ed. G. Kravtchenko & P. Sirieys), Proc IUTAM Symp. at Grenoble*, 58–68.

Rudnicki, J. and Rice, J. (1975) Conditions for the localization of deformation in pressure-sensitive solids, *J. Mech. Phys. Solids* **23**, 371–394.

Schaeffer, D.G. and Shearer, M. (1996) The influence of material nonuniformity preceding shear-band formation in a model for granular flow, *Euro. J. Appl. Math.*, to appear.

Schaeffer, D.G. (1990) Instability and ill-posedness in the deformation of granular materials, *Int'l J. Numer. Anal. Meth. Geomech.* **14**, 253–278.

Trangenstein, J.A. and Colella, P. (1991) A higher-order godunov method for modeling finite deformation in elastic-plastic solids, *Comm. Pure Appl. Math.* **44**, 41–100.

Vardoulakis, I. and Graf, B. (1985) Calibration of constitutive models for granular materials using data from biaxial experiments, *Géotechnique* **35**, 299–317.

MICRO-STRUCTURE DEVELOPED IN SHEAR BANDS OF DENSE GRANULAR SOILS AND ITS COMPUTER SIMULATION -- MECHANISM OF DILATANCY AND FAILURE--

M. Oda, K. Iwashita and H. Kazama
Department of Civil and Environmental Engineering
Saitama University, Urawa, Saitama 338, Japan

1. Introduction

Any soil consists of discrete particles and associated voids, which change their arrangement easily when subjected to overall shear distortion. A particle can move against neighboring particles, in general, by sliding and rolling at contacts. Sliding plays a dominant role in classical theories of strength and dilatancy of granular soils. For example, dilatancy was usually explained by using the sliding model proposed by Newland & Allely in 1957. It seems to the present writers, however, that the conventional model lays too much emphasis on sliding to interpret the dilatancy and failure of granular soils. In fact, some experimental observations has strongly suggested that rolling, rather than sliding, is a dominant micro-mechanism of deformation (Oda, 1972; Oda, Konishi & Nemat-Nasser, 1982).

Recently, numerical simulation methods, such as the distinct element method, have proved to be a powerful tool for the investigation of the micro-mechanism of dilatancy and failure of particle assemblies directly (e.g., Cundall, & Strack, 1979; Bardet & Proubet, 1991; Iwashita & Oda, 1996). However, a question must be answered: Does such a conventional numerical simulation method provide a sound basis for simulating the micro-mechanism of deformation, as well as the overall stress-strain behavior?

Our present objectives are to criticize the reliability of the sliding model on the basis

N. A. Fleck and A. C. F. Cocks (eds.),
IUTAM Symposium on Mechanics of Granular and Porous Materials, 353–364.
© 1997 *Kluwer Academic Publishers.*

of the micro-structure developed in shear bands, and to propose, instead, a realistic model for dilatancy and failure. Also a modified version of the distinct element method (called MDEM) will be proposed, by which even the micro-structure of shear band, as well as the overall stress-strain behavior, can be simulated with success.

2. Shear Band in Toyoura Sand

2.1 TESTING MATERIALS AND TESTING TECHNIQUE

Two sands, Toyoura and Ticino sands, were used to examine the micro-structure of a shear band. Only the result of Toyoura sand will be given here due to the limitation of space. It will be worth noting that the two sands yielded almost the same result so that any conclusion was extracted with making reference to the both tests.

Toyoura sand is a uniform dune sand with D_{50} of 0.206 mm. (Here D_{50} is the size obtained by a sieve analysis in which 50% of soil weight passes through the sieve, and is used as a synonym of the mean size.) The sand is composed mainly of sub-angular quartz and feldspar. Particles are so strong that little breakage occurs in a test at low stress. The maximum and minimum void ratios are 0.974 and 0.612, respectively (Yoshida, et al., 1994).

A drained plane strain test was carried out using a conventional testing machine with some minor modifications. Sample sizes were 180 mm high, 160 mm wide and 80 mm thick. A multiple sieve pluviation method was accepted to make a uniform layer with a void ratio of about 0.65. At every 25 mm deposition of the dry sand, dyed particles were poured through the same multiple sieves to make horizontal marker layers thinner than 1 mm.

The major principal stress σ_1 was increased by moving a top rigid platen downward at a constant rate while keeping the minimum principal stress σ_3 constant at 49.0 kN / m^2 under plane strain and drained conditions. The stress-strain relation is reported in Fig.1. The stress difference $(\sigma_1 - \sigma_3)$ dropped markedly after the peak, in parallel with progressive development of a shear band, and finally reached a steady state called the residual stress state. Then the axial stress was released down to the stress state marked by a in Fig.1. The sheared sample was then fixed in place by very carefully infiltrating a glue into the interstitial voids, so as not to give any change of stresses, through a line connected to the bottom of the specimen. The line was closed when the sample was saturated with the glue. About 24 hours later, the sample became so hard that one could handle it without introducing any further damage. The change of sample length during

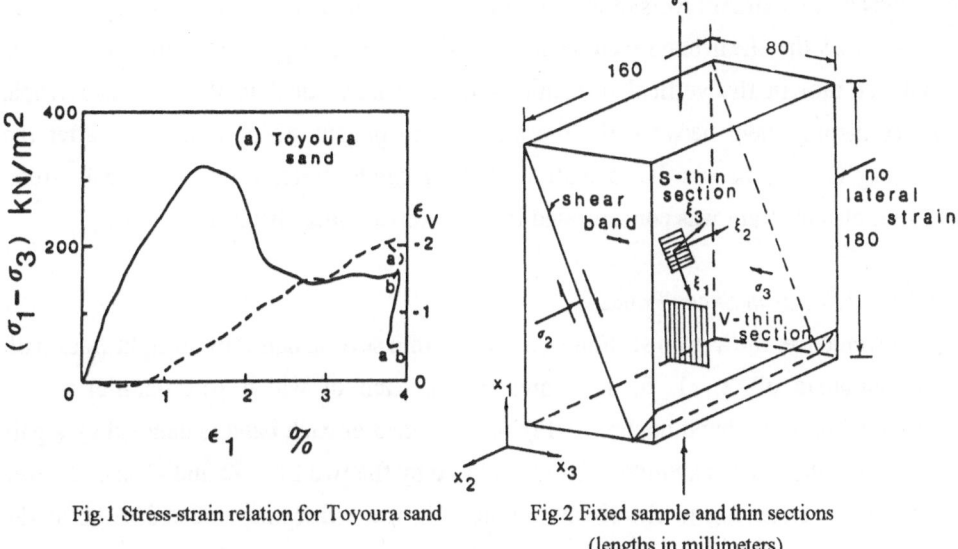

Fig.1 Stress-strain relation for Toyoura sand

Fig.2 Fixed sample and thin sections
(lengths in millimeters)

this procedure is represented by the path from a to b in Fig.1. The change of size was so small that the micro-structure in the shear bands was well preserved without leaving any significant damage.

Three thin vertical plates (5 mm to 10 mm thick) including the axes x_1 and x_3 were cut out of the central portion of the harden sample (Fig.2). X-ray was then applied to each of these thin plates, and X-ray photographs were taken to see the details of the shear band, such as location, direction, shape and so on.

Two thin sections (about 0.03 mm thick) were prepared for optical study using a microscope (Fig.2). A vertical thin section parallel to the axes x_1 and x_3, called the V-thin section, was made by grinding one of the vertical thin plates whose the X-ray photograph had already been taken. A thin section parallel to a shear band, called the S-thin section, was also made by grinding a thin plate sliced from a portion of a shear band.

2.2   RESULTS

2.2.1 *Shape and Orientation of Shear Band*

Fig.3 shows a sketch of an X-ray photograph, together with some measured values such as inclination angle of a shear band, shear band thickness and displacement jumps across the shear band. Two shear bands grow and join around the center. These two shear bands are inclined at $70.5°$ and $67.0°$ to the horizontal. Horizontal parallel lines on the X-ray photograph , about 25 mm apart, were sketched in Fig.3 to show the final positions of the marker layers of dyed particles. Using these, discontinuous displacement jumps across the shear band were estimated (Fig.3). The displacement discontinuity is about 4.1 mm in the vertical direction, and is exactly equal to the non-recoverable relative displacement between the top and bottom platens which took place after the peak (Fig.1). This means that evolution of shear bands started around the peak stress, and that plastic strain was concentrated into the narrow shear band (or bands).

2.2.2 Void Ratios in Shear Bands

For convenience, the V-thin section was divided into several bands by straight lines with serial numbers of -2, -1, 0, 1, 2 and so on, each of which was parallel to the corresponding shear band boundary (Fig.4(a)). Hereafter each band is denoted by a pair of serial numbers. For example, a band enclosed by the two lines -2 and -1 is called the band (-2,-1). Referring to the X-ray photograph, the 0-line was selected to be at the center of the shear band, and the -1 and 1 lines as the bottom and the top shear band boundaries, respectively. Later, however, it was found that these lines were not exactly the top and bottom boundaries, but were very close to them. It should also be noted that individual lines are 1 mm apart in the V-thin section.

A micro-photograph is shown in Fig.5, which was taken from the V-thin section by magnifying a hatched area of Fig.3 under a microscope. One of the most remarkable observations is that surprisingly large voids, like a, b, c and d in Fig.5, appear almost periodically along a narrow shear band (0,1). Since similar large voids were also observed in Ticino sand, it can be concluded that the appearance of such large voids is a common characteristic of the micro-structure developed in shear bands.

The void ratio was estimated using an optical method suggested by Oda & Kazama (1996). All measurements are summarized in Fig.3. The void ratio was estimated to be 0.63 in the right hand side of the shear band, and is considered to be the initial value. In the left hand side, on the other hand, the void ratio is 0.79 (much larger than 0.63). This is because the left hand side is just located on a prolongation of the upper shear band, and this effect can be seen on the X-ray photograph. In the shear band (-1,1), the void ratio ranges from 1.01 (V-thin section) to 1.13 (S-thin section). Both values are larger than the maximum void ratio e_{max} of 0.973 determined by the Japanese standard method. (In the case of Ticino sand, the void ratio is 0.66 in both sides of a shear band, and 1.09 to 1.14 in the shear band, which are also larger than the maximum void ratio e_{max} of 0.96.)

Oda & Kazama (1996) proposed a new dilatancy model for granular materials based not only on the development of large voids in   shear bands,   but on the experimental

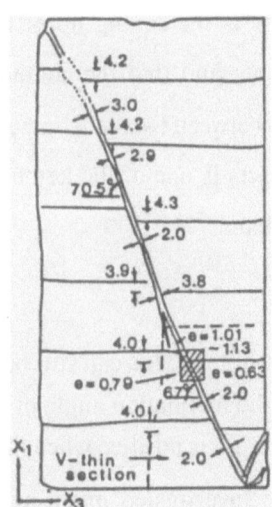

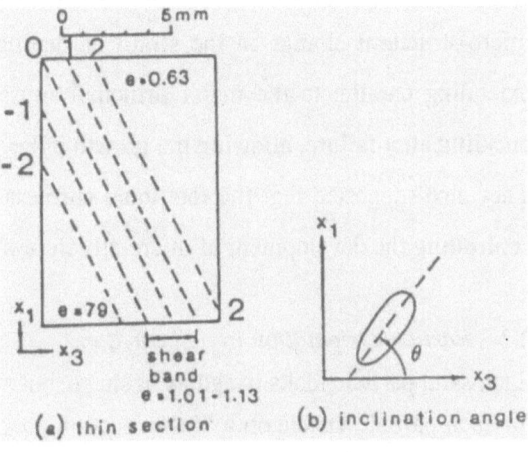

Fig.4 Zone division of thin section and particle orientation

Figure 3 Sketch of X-ray photograph

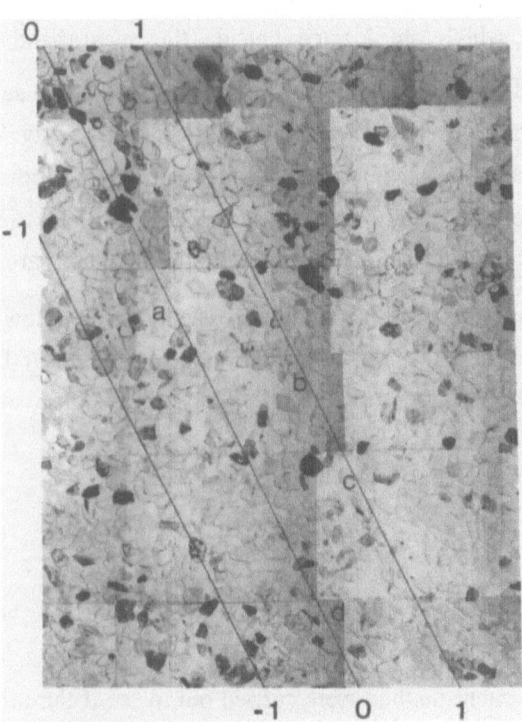

Figure 5 Micro-photograph showing micro-structure

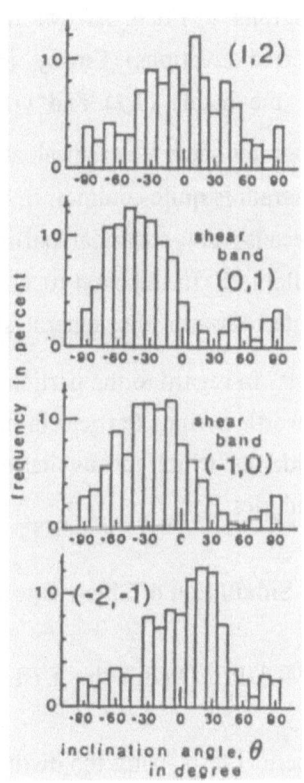

Fig.6 Particle orientation in shear band

observation of Oda, Konishi & Nemat-Nasser (1982). They concluded that the main micro-structural change in the strain hardening process is the setting up of columns extending parallel to the major principal stress direction, and that the columns start buckling after failure, allowing the growth of large voids between two buckling columns. They also suggested that the rotational stiffness at contacts is one of the key factors in controlling the development of micro-structure in shear bands.

2.2.3 *Particle Orientation in a Shear Band*

Each sand particle looks irregular in shape, but an apparent long axis can still be chosen for each particle visible on a V-thin section. Let θ be the inclination angle of such an apparent long axis to a reference axis x_3. (Fig.4(b)). Here θ is positive when measured counterclockwise. All particles in a band $(i, i+1)$ were investigated under an optical microscope to measure their inclination angles θ. The result is shown as frequency histograms in Fig.6. Arrows in the four figures denote the mean vector directions (or preferred directions) (Curray, 1956).

In the bands (1,2) and (-1,-2), which are located outside the shear band, the histograms show a clear peak around $\theta = 0°$. Oda (1972) has already shown that such a histogram is quite common in natural sands deposited under the action of gravity. This is because non-spherical particles lie on the horizontal plane with their long axes parallel to it. In the band (0,1), which is located inside the shear band, particles are re-oriented towards the general shear band direction to give a high frequency at $\theta = -44.3°$. In regard to the particle rotation taking place in the shear bands, the followings are worth noting: Particle orientation changes sharply when crossing the shear band boundaries, which means that a high gradient of the particle rotation takes place at the boundaries.

3.   Simulation of Shear Band by Modified Distinct Element Method

3.1 MODIFIED DISTINCT ELEMENT METHOD (MDEM)

Numerical tests using the distinct element method were carried out to see if the micro-structure developed in shear bands, as well as overall stress-strain behavior of granular soils, can be reproduced. To do this, the conventional distinct element method (DEM) was slightly modified so that moments can be transmitted through the contacts (see Iwashita & Oda, 1996).

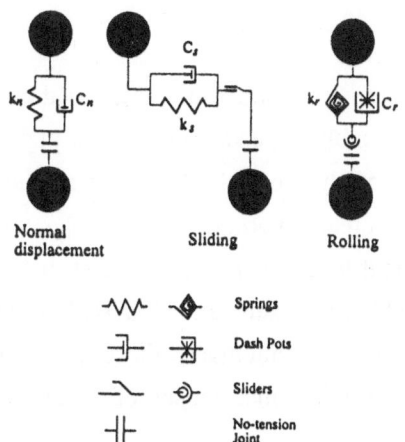

Number of Element	3960
Radii of the Particles	3, 4, 5 mm
increment of time step (dt)	1.00×10^{-5} sec
Particle Density	1800 Kg/m^3
Confining Pressure	1.32×10^5 Pa
Inter-particle Friction Coefficient	0.51
Inter-wall Friction Coefficient	0.00
Normal Spring Constant (kn)	6.00×10^7 N/m
Tangential Spring Constant (ks)	4.00×10^7 N/m
Damping Constant at the Contact	0.02
Rolling Spring Constant (kr)	7.00×10^2 Nm/rad
Coefficient of Rolling Damping	1.00×10^{-1} Nm sec/rad

Figure 7 Contact model   Table 1 Material constants used in MDEM

In the conventional DEM, each contact is replaced by a set of normal and shear springs, normal and shear dash pots, normal and shear no-tension joints and a shear slider which respond to a contact force F_i acting on it (Fig.7). The contact force, which is resolved into normal N_i and tangential T_i components, is in equilibrium with the resistance supplied by the springs and dash-pots. The shear slider starts working at a contact when T_i and N_i satisfy the inequality of $|T_i| > \mu N_i$ where μ is a coefficient of friction.

In the present MDEM, an additional set of elastic spring, dash pot, no-tension joint and slider were installed at each contact, which respond to the moment M_i of the couple force (Fig.7). Rotation resistance is supplied by the elastic spring and the dash pot. The elastic spring yields rotational resistance equal to $k_r \, \theta_r$. Here k_r is the rotational stiffness, and θ_r is the relative rotation (Iwashita & Oda, 1996). (The relative movement taking place at a contact during incremental deformation can be decomposed, in general, into two components; i.e., sliding and rolling (Oda et al., 1982; Bardet, 1994). The rolling component leads to the relative rotation between two particles with a common contact (Iwashita & Oda, 1996)). The dash-pot supplies rotational resistance equal to $C_r \, d\theta_r / dt$ where $d\theta_r / dt$ is the relative particle angular velocity and C_r is the viscosity. The sum of these resistance sources must be in equilibrium with the moment M_i. It is also assumed that the slider starts working if the moment M_i exceeds a threshold value M_{max}; i.e., $|M_i| \geq M_{max}$.

3.2   SIMULATION TESTS

Disc-like particles, with three different radii, 3 mm, 4 mm and 5 mm, were used in the simulations. The mechanical behavior of the assembly depends only on the physical constants dealing with contact properties, such as the normal, shear and rotational stiffness values. The choice of these constants is of great importance, in particular, when the simulation is compared with the results of real tests. In spite of this fact, most of these constants were chosen on a purely empirical basis as summarized in Table 1. This is admissible since the present objective is to show qualitatively the effect of the introduction of the rotational resistance on the micro-mechanism of shear band development.

Deformation of a particle assembly is controlled through the motion of four boundaries; i.e., top, bottom, left and right boundaries. The top and bottom boundaries move vertically, as loading platens, under a strain-controlled condition. The boundaries give no friction constraint on the particles, and have the same normal stiffness as the particles. The left and right boundaries are flexible, and can stretch or shrink like a membrane without giving a frictional effect on the particles (Iwashita & Oda, 1996).

An assembly of 3960 particles, with equal numbers of 3 mm, 4 mm and 5 mm particles, was generated inside the loading flame. The assembly was then consolidated by applying a uniform stress p_0 on the four boundaries until oscillation of the stress disappeared. The final assembly after consolidation will be referred as a reference state in later sections.

Biaxial compression tests were started from the reference assembly. The axial load was increased, while keeping the lateral stress $\sigma_{22} (= p_0)$ constant, by moving the upper and lower boundaries step by step. Using the same reference assembly, two simulation tests were carried out to examine the effect of the rotational resistance on the micro-structure in the shear bands: 1) Free rotation test (DEM): This is the case corresponding to the conventional DEM. No additional resistance against rolling was applied at any contact. 2) Moment transmission test (MDEM): Movement by rolling was restricted until the moment at a contact became larger than a threshold value M_{max}.

The variations of the stress ratio σ_{11}/σ_{22} and volumetric strain ε_v are shown, as a function of the axial strain ε_{11}, in Fig.8 for the two tests. The curves are similar in shape; the stress ratio increases up to the corresponding peak, and the volumetric strain is first compressive and is followed by dilatancy. However, the stress ratios mobilized at the peaks are quite different. No appreciable decrease of the stress ratio takes place after

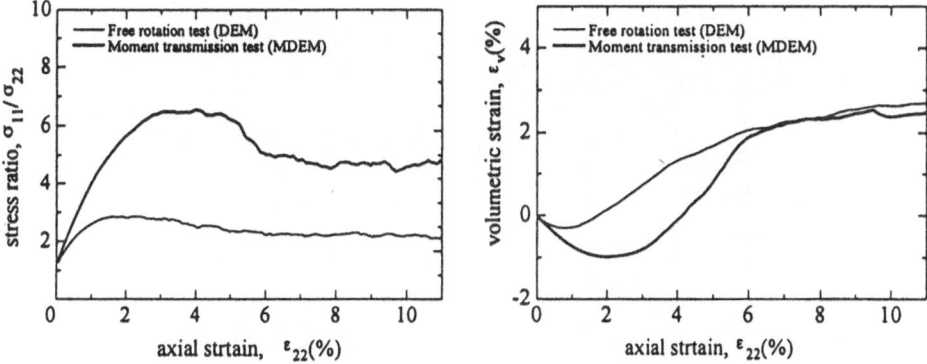

Figure 8 Stress-strain relations in numerical simulation tests

the peak in the free rotation test. On the other hand, the stress ratio clearly drops after the peak in the moment transmission test. This difference is important since strain localization is associated, in general, with strain softening behavior.

3.3   RESULTS OF SIMULATION OF SHEAR BAND FORMATION

Particles were chosen, which were occasionally aligned along some vertical and horizontal directions in the reference state. They were colored black such that the initial assembly was divided into many square regions. The movement of these colored particles was monitored during deformation step by step. Distortion of these square regions provides guidance for detecting the places in which strain was localized. The final positions of the colored particles at $\varepsilon_{11} = 10.94\%$ are shown in Fig.9.

In the free rotation test, clearly defined shear bands were not developed even after the axial strain was increased to over 10 % (Fig.9(a)). This does not necessarily mean that uniform deformation occurs at every instant, but rather means that the place of localization changes so quickly that the distribution of deformation looks uniform as a whole. In the moment transmission test, two shear bands clearly appear in the bottom half of the assembly (Fig.9(b)). Outside the shear bands, the strain is so small that the initial configuration of particles almost remains unchanged.

In Fig.10, black circles denote counterclockwise rotation by more than 9° (positive) taking place during the deformation from ε_{11} =0% to 10.94%., whilst white ones denote

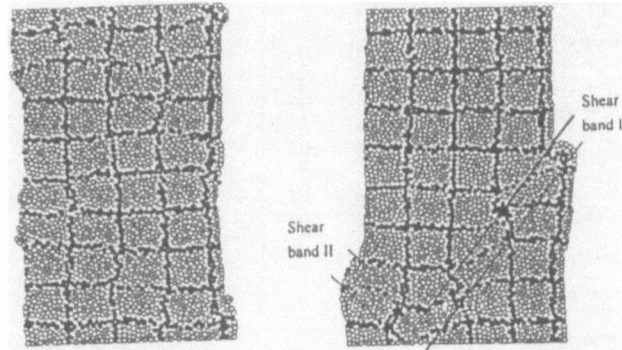

(a) Free rotation test (DEM)   (b) Moment transmission test (MDEM)
Figure 9 Development of shear band in MDEM

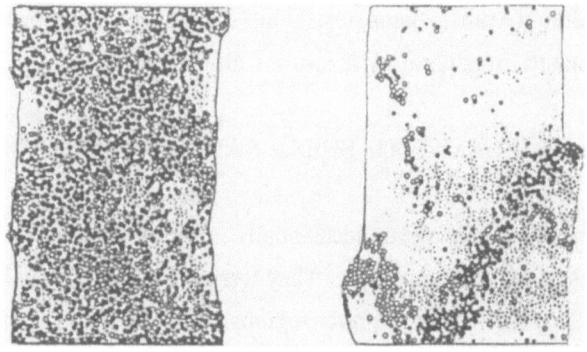

(a) Free rotation test (DEM)   (b) Moment transmission test (MDEM)
Figure 10 Particle rotation (● =counterclockwise; ○ =clockwise)

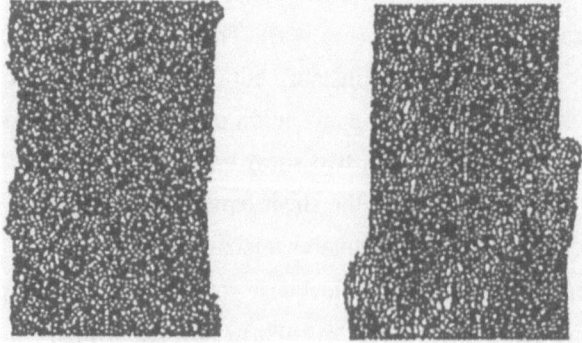

(a) Free rotation test (DEM)   (b) Moment transmission test (MDEM)
Figure 11 Distribution of voids (All particles are colored with black.)

clockwise rotation by more than 9° (negative). In both cases, particle rotation occurred extensively in some limited zones. Especially in the case of the moment transmission test (Fig.10(b)), the rotated particles only appear in two distinct zones in which shear strain is localized (Fig.9(b)). More importantly, there exists a marked contrast between (a) and (b). In the free rotation test, rotation takes place either clockwise or counterclockwise at the same rate. In the moment transmission test, on the other hand, counterclockwise rotation mainly appears in shear band 1 whilst clockwise rotation in shear band 2.

Why does such a difference arise? If rolling occurs between two neighboring particles, clockwise rotation of a particle causes counterclockwise rotation of another, as schematically shown in Fig.12(a), so that the clockwise and counterclockwise rotations have an equal chance of taking place during deformation. In fact, this happened in the free rotation test (DEM). In shear band 1 of the moment transmission test, however, particles belonging to a column were rotated by rolling as a whole, and upper particles roll over lower ones (Fig.12(b)). As a result, the column is bent counterclockwise as a whole, and counterclockwise rotation of particles may be more likely to occur than clockwise rotation. Columns were bent counterclockwise in a similar manner to Fig.12(b).

It should be noted that one side rotation in the shear band of Toyoura sand (Fig.6) is well simulated only by the moment transmission test. The conventional DEM fails to simulate the particle rotation.

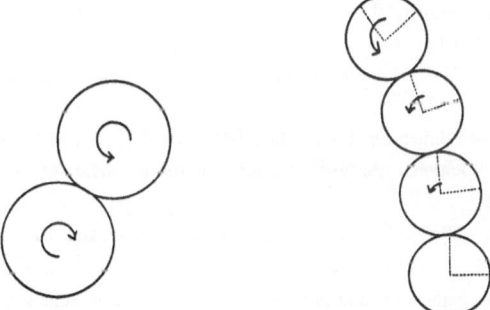

(a) Clockwise and counterclockwise   (b) Counterclockwise rotation
           rotation by rolling   by rolling
                 Figure 12 Mechanisms of particle rotation

The distribution of voids at $\varepsilon_{11} = 10.94\%$ is shown in Fig.11, in which particles are all colored black so that voids can be seen more clearly. Voids are uniformly distributed, without any alignment, in the free rotation test (Fig.11(a)). There is not any concentration of voids into specific zones. It seems quite natural to think that softening behavior may not take place in this case (Fig.8) since any evidence of weakening by concentration of voids is not observed along these shear zones. In the moment transmission test (Fig.11 (b)), however, extremely large voids are aligned along shear bands 1 and 2. This means that volumetric strain (dilatation) is concentrated, especially after failure, into these narrow bands, and that strain softening behavior must be a result of weakening along shear bands.

Such extremely large voids and void ratio also appeared in the shear band of Toyoura sand (see voids a, b, c and d in Fig.5). This result seems to prove that moment transmission through contacts takes place in natural soils, and that it is a key factor controlling the micro-structure of shear bands.

4. Conclusion

Not only the generation of large voids but also particle rotation in the shear bands were observed in tests using Toyoura sand Ticino sands. The micro-structure was well reproduced in numerical simulation tests by introducing rotational resistance at contacts into the conventional distinct element method.

5. References

Bardet, J.P. (1994). Observations on the effects of particle rotations on the failure of idealized granular materials. Mechanics of Materials 18, 159-182.

Bardet, J.P. & Proubet, J. (1991). A numerical investigation   of the structure of persistent shear bands in granular media. Geotechnique 41, No.4, 599-613.

Cundall, P.A. & Strack, O.D.L. (1979). A discrete numerical model for granular assemblies. Geotechnique 29 (1), 47-65.

Curray, J.A. (1956). Analysis of two-dimensional orientation data. J. Geology. 64, 117-131.

Iwashita, K. & Oda, M. (1996). Rotational resistance at contacts in the simulation of shear band development by DEM. (to be published in ASCE).

Newland, P.L. & Allely, B.H. (1957). Volume changes in drained triaxial tests on granular materials. Geotechnique 7, 17-34.

Oda, M. (1972). Deformation mechanism of sand in triaxial compression tests. Soils and Foundations 12, No.4, 45-63.

Oda, M & Kazama, H. (1996). Micro-structure of shear band and its relation to the mechanism of dilatancy and failure og granular soils. (To be published in Geotechnique).

Oda, M, Konishi, J. & Nemat-Nasser, S. (1982). Experimental micro-mechanical evaluation of strength of granular materials: Effect of particle rolling. Mechanics of Materials 1, 267-283.

Yoshida, T., Tatsuoka, F. ,Siddiquee, M.S.A., Kamegai, Y.& Park, C.-S. (1994). Shear banding in sands observed in plane strain compression. Symp. Localization and Bifurcation Theory for Soils and Rocks, (eds R. Cambou, J. Desrues and I. Vardoulakis), 165-179, Balkema.

THE EVOLUTION OF ANISOTROPY IN POROUS MATERIALS AND ITS IMPLICATIONS FOR SHEAR LOCALIZATION

M. KAILASAM AND P. PONTE CASTAÑEDA
Department of Mechanical Engineering and Applied Mechanics
University of Pennsylvania
Philadelphia, PA 19104
U. S. A

Abstract

A constitutive model is developed to characterize the effective response of nonlinearly viscous porous materials. The model is capable of accounting for the effects of changes in the microstructure that occur during finite-deformation processes. The model is composed of two parts: one, instantaneous constitutive equations for the porous material, which depend on appropriate state variables characterizing the state of the microstructure at any instant and two, evolution laws for these state variables. The model is used to estimate the overall behavior of initially isotropic, linear and nonlinear, viscous porous materials under different finite-deformation programs. In particular, attention is focused on the effects of changes in the aspect ratios and orientation of the voids which cause the material response to become anisotropic. Results are also presented for the limiting case of a perfectly plastic porous material and the implications of the evolution of the microstructure on the stability of the material are studied.

1. Introduction

Estimating the effective behavior of porous materials has been of interest to the solid mechanics community over the last several years. When a porous material is subjected to finite deformation, the microstructure of the material changes throughout the deformation process which in turn affects the overall response of the material. Thus, it becomes important to take into account any changes that occur in the microstructure and to be able to relate these changes to the effective behavior of the porous material.

   Among several works which have considered the behavior of nonlinear porous materials is that by Gurson (1977), who obtained estimates for the effective yield functions for perfectly plastic porous materials with dilute concentrations of spherical and cylindrical voids, incorporating the effects of void growth and nucleation. Budiansky et al. (1982) obtained solutions for the deformation of an isolated spherical void in a nonlinearly viscous matrix under axisymmetric loading conditions. Lee and Mear (1992) obtained analogous results for the case of spheroidal voids in a power law viscous matrix, also under axisymmetric loading. In addition to the volume and shape, Fleck and Hutchinson (1986) considered changes in the orientation of an ellipsoidal

N. A. Fleck and A. C. F. Cocks (eds.),
IUTAM Symposium on Mechanics of Granular and Porous Materials, 365–376.
© 1997 *Kluwer Academic Publishers*

void in a linearly viscous matrix and that of a 2-D void in a power-law viscous matrix, due to shear loading.

Using a different approach, Ponte Castañeda and Zaidman (1994) proposed a constitutive model for nonlinearly viscous porous materials which is capable of accounting for the evolution of the microstructure for moderate levels of porosity. This model is composed of instantaneous constitutive equations for the nonlinear porous material, obtained from the variational representation of Ponte Castañeda (1991), along with evolution equations for appropriate microstructural variables. However, in this work, it was assumed that the material is subjected to triaxial loads (with fixed loading axes), in which case, the porosity and the aspect ratios of the voids are the only relevant microstructural variables; the orientation of the voids was assumed to remain unchanged (on an average).

In the present work, we relax the restriction of triaxial loading with fixed loading axes and extend the model to predict the effective response of the porous material under more general loading conditions. In order to do so, we introduce the orientation of the voids as an independent microstructural variable. Evolution laws for the orientation of the voids are then developed from estimates for the average spin in the voids. Finally, the evolution equations for all the microstructural variables are integrated in conjunction with the instantaneous constitutive equations for the porous material in order to obtain the effective response of the material under any given finite deformation program.

Although the present model can be used to estimate the effective behavior of porous materials under arbitrary loading conditions, we present here only some preliminary results for the case of simple shear loading superposed with hydrostatic tension/compression. In the next section, we discuss the effective (instantaneous) constitutive relations for the porous material with a nonlinearly viscous matrix. Then, in the subsequent section we discuss the evolution laws for the microstructural variables and finally, we present some illustrative applications of the model.

2. Effective properties of porous materials

In this section, we will consider the effective properties of a porous solid which may be defined as a heterogeneous material composed of a distribution of voids in a matrix material. We assume that the size scale characterizing the typical void is small compared to the overall dimensions of the specimen and the scale of variation of the applied loads. This allows us to treat the porous material as a macroscopically homogeneous one, the effective properties of which can be expressed in terms of the relation between the average stress and the average strain rate in the material.

The constitutive behavior of the matrix material is assumed to be nonlinearly viscous and is defined in terms of a stress potential U by the relation

$$\mathbf{D} = \frac{\partial U}{\partial \boldsymbol{\sigma}}(\boldsymbol{\sigma}), \tag{1}$$

where $\boldsymbol{\sigma}$ is the Cauchy stress and $\mathbf{D}$ is the rate of deformation tensor. The effective constitutive relation for the porous material can then be written in terms of an effective potential $\tilde{U}$ as

$$\overline{\mathbf{D}} = \frac{\partial \tilde{U}}{\partial \overline{\boldsymbol{\sigma}}}(\overline{\boldsymbol{\sigma}}), \tag{2}$$

where $\overline{\mathbf{D}}$ is the average rate of deformation and $\overline{\sigma}$ is the average stress in the material. This effective potential, which is obtained from the principle of minimum complementary energy, can be written as

$$\tilde{U}(\overline{\sigma}) = \min_{\sigma \in S(\overline{\sigma})} \int_{\Omega} U(\mathbf{x}, \sigma) \, dv, \tag{3}$$

where $S(\overline{\sigma}) = \{\sigma | \nabla \cdot \sigma = 0 \text{ in } \Omega, \text{ and } \sigma \mathbf{n} = \overline{\sigma} \mathbf{n} \text{ on } \partial \Omega\}$ is the set of statically admissible stress fields corresponding to uniform traction on the boundary $\partial \Omega$ of the porous solid. As an example, the potential U, which characterizes the behavior of the matrix, may be taken to have the isotropic, pure power-law form

$$U(\sigma) = \phi(\sigma_e) = \frac{\sigma_y}{n+1} \left[\frac{\sigma_e}{\sigma_y} \right]^{n+1} = \frac{1}{3(n+1)\eta} \sigma_e^{n+1}, \tag{4}$$

which is widely used in high-temperature creep. In the above expression, n is the creep exponent, $\sigma_y = (3\eta)^{1/n}$ is the reference stress and $\sigma_e = [\frac{3}{2}\mathbf{s} \cdot \mathbf{s}]$ is the equivalent stress, where $\mathbf{s}$ is the deviatoric stress.

It is observed that when $n = 1$, the expression for U characterizes the behavior of an incompressible, linearly viscous fluid with viscosity $\eta = \mu$. In this case, the constitutive relation for the matrix can be written as $\sigma = \mathbf{L}^{(1)} \mathbf{D}$, where $\mathbf{L}^{(1)} = (\infty, 2\mu)$ is the viscosity tensor. The effective constitutive equation for the linearly viscous porous material may then be written as $\overline{\sigma} = \tilde{\mathbf{L}} \overline{\mathbf{D}}$, where $\tilde{\mathbf{L}}$ is the effective viscosity tensor for the porous material.

The effective viscosity tensor $\tilde{\mathbf{L}}$ (or the effective viscous compliance tensor $\tilde{\mathbf{M}} = \tilde{\mathbf{L}}^{-1}$) for the linearly viscous porous material may be characterized in terms of the Hashin-Shtrikman (HS) estimates for composites with microstructures exhibiting "ellipsoidal symmetry". These types of estimates, for composites with statistically isotropic microstructures, were introduced by Hashin and Shtrikman (1963) and generalized for the class of random microstructures exhibiting ellipsoidal symmetry by Willis (1977). Later, Willis (1978) showed that these results may be interpreted as estimates for the class of particulate composites with aligned ellipsoidal inclusions distributed in such a way that the inclusions and the function characterizing the distribution of their centers share the same aspect ratios. It is noted that the porous solid is a specific case of a more general particulate composite. This allows us to make use of the above results for the case of porous materials also. Although it is possible to consider more general microstructures where the aspect ratios of the voids may be different from those of the distribution function (Ponte Castañeda and Willis, 1995), we shall restrict our attention to the case where the two sets of aspect ratios are identical for reasons that will be discussed later.

The effective viscosity tensor can be expressed in a simple manner in terms of the "strain rate concentration" tensors $\mathbf{A}^{(r)}$ $(r = 1, 2)$ which provide us with estimates for the average rate of deformation in phase r by means of the relation $\mathbf{D}^{(r)} = \mathbf{A}^{(r)} \overline{\mathbf{D}}$. In this work, we will label the matrix material as phase 1 and the voids as phase 2. The effective viscosity tensor $\tilde{\mathbf{L}}$ of the linearly viscous porous material is then given by the simple expression $\tilde{\mathbf{L}} = (1 - f) \mathbf{L}^{(1)} \mathbf{A}^{(1)}$, where f is the porosity and $\mathbf{A}^{(1)}$ is obtained from the relation $(1 - f) \mathbf{A}^{(1)} + f \mathbf{A}^{(2)} = \mathbf{I}$. $\mathbf{A}^{(2)}$ for the porous material is derived from the Hashin-Shtrikman (HS) procedure and is given by the relation

$$\mathbf{A}^{(2)} = \left[\mathbf{I} - (1 - f)\mathbf{S}^{(1)}\right]^{-1}. \tag{5}$$

In the above expressions, $\mathbf{I}$ is the fourth order identity tensor and $\mathbf{S}^{(1)}$ is the Eshelby (1957) tensor corresponding to an isolated ellipsoidal void in a matrix of phase 1. Note that $\mathbf{S}^{(1)}$ depends only on the aspect ratios and the orientation of the void. The corresponding expression for the effective compliance tensor is then given by

$$\tilde{\mathbf{M}} = \left\{\mathbf{I} + \frac{f}{1 - f}\left(\mathbf{I} - \mathbf{S}^{(1)}\right)^{-1}\right\}\mathbf{M}^{(1)}. \tag{6}$$

Along with the expression for the strain rate concentration tensors, it is possible to make use of the HS procedure to obtain estimates for the average continuum spin in the phases. This is done in terms of "spin concentration" tensors $\mathbf{C}^{(r)}(r = 1, 2)$ by means of the relation $\mathbf{W}^{(r)} = -\mathbf{C}^{(r)}\overline{\mathbf{D}} + \overline{\mathbf{W}}$, where $\overline{\mathbf{W}}$ is the average continuum spin in the material. For the case of porous materials, the spin concentration tensor $\mathbf{C}^{(2)}$ is given by the expression

$$\mathbf{C}^{(2)} = -(1 - f)\mathbf{\Pi}^{(1)}\mathbf{A}^{(2)}, \tag{7}$$

where $\mathbf{\Pi}^{(1)}$ is a transformation tensor (similar to $\mathbf{S}^{(1)}$) introduced by Eshelby (1957) and also depends only on the aspect ratios and the orientation of the voids.

These results for linear materials can now be extended to porous materials with an incompressible, nonlinearly viscous matrix. This is accomplished by means of the variational statement of Ponte Castañeda (1991). This variational statement recasts the principle of minimum complementary energy in a form that allows the use of the effective potential of linear composites to obtain estimates for the effective potential of nonlinear composites which have the same microstructure as the linear composite. Thus, the effective potential of the nonlinearly viscous porous material may be written in the form

$$\tilde{U}(\overline{\sigma}) \geq \max_{\mu \geq 0}\left\{\frac{1}{2}\overline{\sigma} \cdot \left(\tilde{\mathbf{M}}\overline{\sigma}\right) - (1 - f)V(\mu)\right\}, \tag{8}$$

where $\tilde{\mathbf{M}}$ represents any estimate for the effective compliance tensor of the linearly viscous porous material. In the above expression, V is defined as

$$V(\mu) = \max_{\sigma}\left\{\frac{1}{6\mu}\sigma_e^2 - U(\sigma)\right\}, \tag{9}$$

where $U(\sigma)$ is the stress potential of the nonlinearly viscous matrix. We have seen that the HS procedure provides an estimate for $\tilde{\mathbf{M}}$, which may then be used to obtain the following expression for the effective potential of the nonlinearly viscous porous material:

$$\tilde{U}(\overline{\sigma}) \geq \frac{1}{3\eta(n + 1)}\left(\frac{1}{1 - f}\right)^{\frac{n-1}{2}}\left[\overline{\sigma} \cdot (\tilde{\mathbf{m}}\overline{\sigma})\right]^{\frac{n+1}{2}}, \tag{10}$$

where $\tilde{\mathbf{m}} = (3\mu)\tilde{\mathbf{M}}$ is independent of μ. Treating the right-hand side of (10) as an estimate for the effective potential, expression (2) then provides estimates for the instantaneous relations between the average stress and the average rate of deformation.

Finally, we consider the effective properties of a porous material with a perfectly plastic matrix. It is observed that when $n \to \infty$, expression (4) for the potential U

characterizes the behavior of a rigid/perfectly plastic solid of the Mises type with tensile yield strength σ_y. The function $\phi(\sigma_e)$ then takes the form

$$\phi(\sigma_e) = \begin{cases} 0 & \text{if } \sigma_e \le \sigma_y, \\ \infty & \text{otherwise,} \end{cases} \tag{11}$$

and relation (1) for the local behavior of the composite is replaced by the expression

$$\mathbf{D} = \dot{\lambda}\mathbf{s}, \tag{12}$$

where $\dot{\lambda}$ is the (non-negative) plastic loading parameter. In order to describe the effective behavior of such materials, it is convenient to introduce an effective yield function $\tilde{\Phi}(\overline{\sigma})$, such that

$$\tilde{U}(\overline{\sigma}) = \begin{cases} 0 & \text{if } \tilde{\Phi}(\overline{\sigma}) \le 0, \\ \infty & \text{otherwise.} \end{cases} \tag{13}$$

Then, it follows that

$$\overline{\mathbf{D}} = \dot{\Lambda}\frac{\partial\tilde{\Phi}}{\partial\overline{\sigma}}(\overline{\sigma}), \tag{14}$$

where $\dot{\Lambda}$ is a non-negative parameter to be determined from the consistency condition $\tilde{\Phi}(\overline{\sigma}) = 0$ and depends on the evolution of the microstructure. For the case of porous materials, the estimate (10) for $\tilde{U}$ leads to corresponding estimates for $\tilde{\Phi}$ given by

$$\tilde{\Phi}(\overline{\sigma}) \le \overline{\sigma} \cdot (\tilde{m}\overline{\sigma}) - (1-f)\sigma_y^2. \tag{15}$$

3. Evolution of the microstructure

In the previous section, we obtained expressions for the effective behavior of viscous porous materials. It is noted that these expressions depend on the microstructure at any given instant. Thus, it is clear that when the porous material undergoes finite deformation and the microstructure, as characterized by the porosity, the aspect ratios and the orientation of the voids changes, the effective response of the porous material is also altered. In particular, changes in the aspect ratios of the voids and their orientation modifies the anisotropic response of the porous material.

The evolution of the microstructure depends on the complex interaction between the various microstructural variables, the loading conditions and instantaneous constitutive response of the porous material, and therefore it constitutes an extremely difficult problem. In order to make some progress, we must make certain simplifying assumptions. Thus, in this work, we restrict the microstructure of the porous material to one with ellipsoidal symmetry (Willis, 1978) so that the voids and their distribution function have the same aspect ratios throughout the deformation process. We note that it is possible to consider more general microstructures where the two sets of aspect ratios are allowed to evolve independently. However, it has been observed that for the case of porous materials, the effects of changes in the distribution of the voids are relatively small compared to the effects of changes in the shape of the voids (Kailasam et. al., 1996).

It is known that under uniform boundary conditions, an isolated ellipsoidal inclusion in an infinite, linear matrix experiences uniform strain fields so that it deforms through a sequence of ellipsoidal shapes (Eshelby, 1957), with varying size, aspect ratios and orientation. In this work, it will be assumed that even for moderate values of porosity the initially spherical voids deform into ellipsoidal voids with different size,

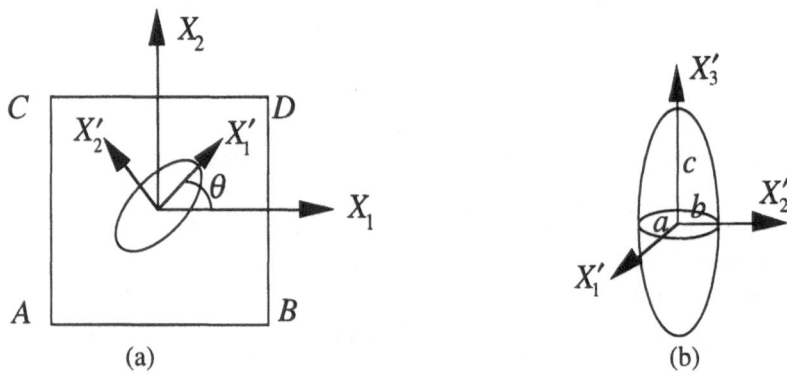

Figure 1. (a) $X_1 X_2$ cross-section showing the orientation of a void and (b) shape of a typical void.

shape and orientation. However, the average rate of deformation and spin in the voids which are used to determine the evolution of the size, shape and orientation of the voids are obtained from the "non-dilute" expressions (5) and (7) for the strain rate concentration tensor and the spin concentration tensor, respectively, instead of the corresponding dilute expressions. This provides a generalization of the work of Ponte Castañeda and Zaidman (1994) for triaxial loading conditions. For finite porosity, it is known that the rate of deformation and spin in the voids will not be uniform and as a consequence the voids will not remain ellipsoidal in shape. However, to the level of accuracy of the HS estimates for the effective properties of porous materials, the rate of deformation fields in the voids are uniform, which is consistent with the approximation that the spherical voids will deform into ellipsoidal voids. Besides, the fact that the HS procedure makes use of the averages of the state variables makes it sufficient to know how the average size, shape and orientation of the voids evolve; there is no need to know precisely how each void evolves.

Although the model can handle general loading conditions, in this work we present preliminary results for the response of porous materials to simple shear along with superposed hydrostatic tension or compression. Under these loading programs, we identify the porosity (f), the aspect ratios of the voids ($w_1 = b/a$ and $w_2 = c/b$) and their orientation (θ) (see Fig. 1) as the relevant microstructural variables. We note that the voids remain aligned along the X_3 axis throughout the deformation process as a result of which only one angle is required to define the orientation of the voids.

The evolution of the porosity is given by the well-known relation

$$\dot{f} = (1 - f)\overline{D}_{kk}, \qquad (16)$$

and the changes in the aspect ratios of the voids are governed by the relations

$$\dot{w}_1 = w_1\left(D_{22}'^{(2)} - D_{11}'^{(2)}\right) \text{ and } \dot{w}_2 = w_2\left(D_{33}'^{(2)} - D_{22}'^{(2)}\right) \qquad (17)$$

where $D_{11}'^{(2)}$, $D_{22}'^{(2)}$ and $D_{33}'^{(2)}$ are components of $\mathbf{D}^{(2)}$ relative to a coordinate system which is parallel to the principal axes of the voids (see Fig. 1).

In order to characterize the orientation of the voids, we must determine the spin of the eigenvectors of the right-stretch tensor corresponding to the deformation in the voids (Ogden, 1984). Then, for the case where the voids are initially spherical in shape, the evolution law for the orientation angle θ is given by

$$\dot{\theta} = \frac{1+(w_1)^2}{1-(w_1)^2} D_{12}'^{(2)} - W_{12}'^{(2)}. \tag{18}$$

In the above expressions, the rate of deformation in the voids can be obtained by making use of estimate (5) for the strain rate concentration tensor, recalling that $\mathbf{D}^{(2)} = \mathbf{A}^{(2)}\overline{\mathbf{D}}$. Similarly, the average spin in the voids is obtained by making use of expression (7) for the spin concentration tensor and the fact that $\mathbf{W}^{(2)} = -\mathbf{C}^{(2)}\overline{\mathbf{D}} + \overline{\mathbf{W}}$.

For a two-phase composite, the evolution of the microstructural variables is, in general, expected to depend on the relative properties of the constituent phases. However, for the special case of porous materials, it is found that the evolution equations for these variables are independent of the properties of the matrix (the viscosity and the nonlinearity exponent n). The evolution equations can now be used along with the corresponding instantaneous constitutive equations for the porous material in order to study the evolution of f, w_1, w_2 and θ, and to obtain the effective response of the porous material for the given loading program.

Finally, we note that these evolution equations can be used to obtain estimates for the effective hardening rate H for porous materials with a perfectly plastic matrix. First, we introduce the symbol $\circ$ to define the rate co-rotational with the spin of the voids; for example, $\overset{\circ}{\overline{\sigma}}$ is the average stress rate co-rotational with the spin of the voids. The consistency condition then yields

$$\dot{\tilde{\Phi}} = \overset{\circ}{\tilde{\Phi}} = \frac{\partial\tilde{\Phi}}{\partial\overline{\sigma}_{ij}}\overset{\circ}{\overline{\sigma}}_{ij} + \frac{\partial\tilde{\Phi}}{\partial f}\dot{f} + \frac{\partial\tilde{\Phi}}{\partial w_1}\dot{w}_1 + \frac{\partial\tilde{\Phi}}{\partial w_2}\dot{w}_2 = 0, \tag{19}$$

which along with the evolution equations for the porosity and the aspect ratios allows us to rewrite (14) in the form

$$\overline{D}_{ij} = \frac{1}{H}\frac{\partial\tilde{\Phi}}{\partial\overline{\sigma}_{ij}}\frac{\partial\tilde{\Phi}}{\partial\overline{\sigma}_{kl}}\overset{\circ}{\overline{\sigma}}_{kl}, \tag{20}$$

where H is the effective hardening rate and is given by

$$H = \left[(1-f)\frac{\partial\tilde{\Phi}}{\partial f}\frac{\partial\tilde{\Phi}}{\partial\overline{\sigma}_{kk}} + w_1\frac{\partial\tilde{\Phi}}{\partial w_1}\left(A_{22ij}^{(2)} - A_{11ij}^{(2)}\right)\frac{\partial\tilde{\Phi}}{\partial\overline{\sigma}_{ij}} + w_2\frac{\partial\tilde{\Phi}}{\partial w_2}\left(A_{33ij}^{(2)} - A_{22ij}^{(2)}\right)\frac{\partial\tilde{\Phi}}{\partial\overline{\sigma}_{ij}}\right].$$

4. Applications

In this section, some illustrative results are presented for the evolution of the identified microstructural variables and for the effective behavior of a porous material under different loading conditions. First, we present some results for the case where the porous material is subjected to velocity boundary conditions and later, those for traction boundary conditions. We recall that the porous material initially consists of a statistically isotropic distribution of spherical voids, with initial porosity f_0, in a matrix with potential given by (4). As seen earlier, $\mathbf{A}^{(2)}$ and $\mathbf{C}^{(2)}$ are independent of material properties in this case, so that the evolution of the microstructure is identical for different values of the nonlinearity exponent n. However, it is emphasized that the effective stress-strain relations for the porous material does depend on the nonlinearity of the matrix material.

When the porous material is subjected to velocity boundary conditions of the type

$$\overline{\mathbf{v}} = (\beta x_1 + 2x_2)\hat{\mathbf{e}}_1 + \beta x_2\hat{\mathbf{e}}_2 + \beta x_3\hat{\mathbf{e}}_3, \tag{21}$$

we have $\overline{D}_{11} = \overline{D}_{22} = \overline{D}_{33} = \beta \overline{D}_{12}$ and $\overline{D}_{23} = \overline{D}_{31} = 0$, along with $\overline{W}_{12} = \overline{D}_{12}$ and $\overline{W}_{23} = \overline{W}_{31} = 0$. For these boundary conditions, we obtain the evolution of the porosity, the aspect ratios of the voids, the orientation of the voids and that of the average stress in the porous material with respect to a strain measure $\overline{\gamma}$ ($\dot{\overline{\gamma}} = 2\overline{D}_{12}$).

In Fig. 2(a), we plot the evolution of porosity as a function of average shear strain $\overline{\gamma}$ in a porous material with an initial porosity of $f_0 = 0.15$ for different values of β. It is clear that for $\beta > 1$, the average dilation rate is positive, for $\beta = 0$ the dilation rate is zero, while for $\beta < 1$ it is negative. Thus, as expected, the porosity increases for $\beta = 0.1$, stays at its initial value for $\beta = 0$ and for $\beta = -0.1$, it decreases until it becomes zero at a strain of 1.08.

In Fig. 2(b), we plot the corresponding evolution curves for the aspect ratio of the voids. For $\beta = 0$, it is seen that the aspect ratio w_1 decreases while the aspect ratio w_2 increases. For $\beta = 0.1$, the same trend is observed, but the superposed hydrostatic tension slows the rate of change of the aspect ratios. On the other hand, for $\beta = -0.1$, the opposite effect is observed; the superposed hydrostatic compression increases the rate at which the aspect ratios change with deformation. It is seen that w_1 goes to zero, while w_2 becomes large as the porosity goes to zero.

In Fig. 2(c), we plot the evolution of the orientation of the voids. It is seen that in all cases the voids start evolving at $\theta = 45°$ and the value of θ decreases with increasing deformation. Just as in Fig. 2(b), here also it is seen that superposed hydrostatic tension slows down the rate at which the voids change orientation, while superposed hydrostatic compression increases the rate. For $\beta = -0.1$, it is seen that the voids have an average orientation of around $29°$ when the porosity becomes zero.

In Fig. 2(d), we show the evolution of the normalized stress components $\hat{\sigma}_{11}$, $\hat{\sigma}_{22}$ and $\hat{\sigma}_{33}$ relative to $\overline{\gamma}$, where $\hat{\sigma}_{11} = \overline{\sigma}_{11} / (\Sigma)^{(n-1/n)}$, with $\Sigma = 3\mu(1-f)\overline{\mathbf{D}} \cdot (\tilde{\mathbf{L}}\overline{\mathbf{D}})$ and $\hat{\sigma}_{22}$ and $\hat{\sigma}_{33}$ are defined similarly. Due to this normalization, the stress-strain curves appear to be identical for all values of n, while in fact the physical components of the stresses depend on n. For $\beta = 0$, which corresponds to the case of simple shear, the normal components of the stress start at zero but become non-zero as the deformation increases. This is a consequence of the developing anisotropy of the material: initially, the material is isotropic and its response to simple shear produces no normal stresses, but as the deformation progresses, the material develops anisotropy leading to a "normal stress effect". For $\beta = 0.1$, all the normal components of the stress begin from an identical non-zero value (due to the superposed hydrostatic tension), but take on different values as the material becomes anisotropic. In this case, it is seen that the value of the stress components decrease steadily. This can be understood by noting that as the porosity increases, the material becomes more compliant. For $\beta = -0.1$, the stress state is compressive and the magnitude of the three normal components of the stresses increases. This can again be explained by noting that as the porosity of the material decreases, the material becomes less compliant. In this case also, the three components differ from each other even after a small amount of deformation due to the anisotropic response of the material.

In Fig. 2(e), we plot the evolution of the $\overline{\sigma}_{12}$ component of the average stress in the porous material normalized in a way similar to the normal stresses. For $\beta = 0$, the shear stress increases gradually with deformation, but begins to decrease after a certain level of deformation has been reached. For $\beta = 0.1$, the shear stress decreases with

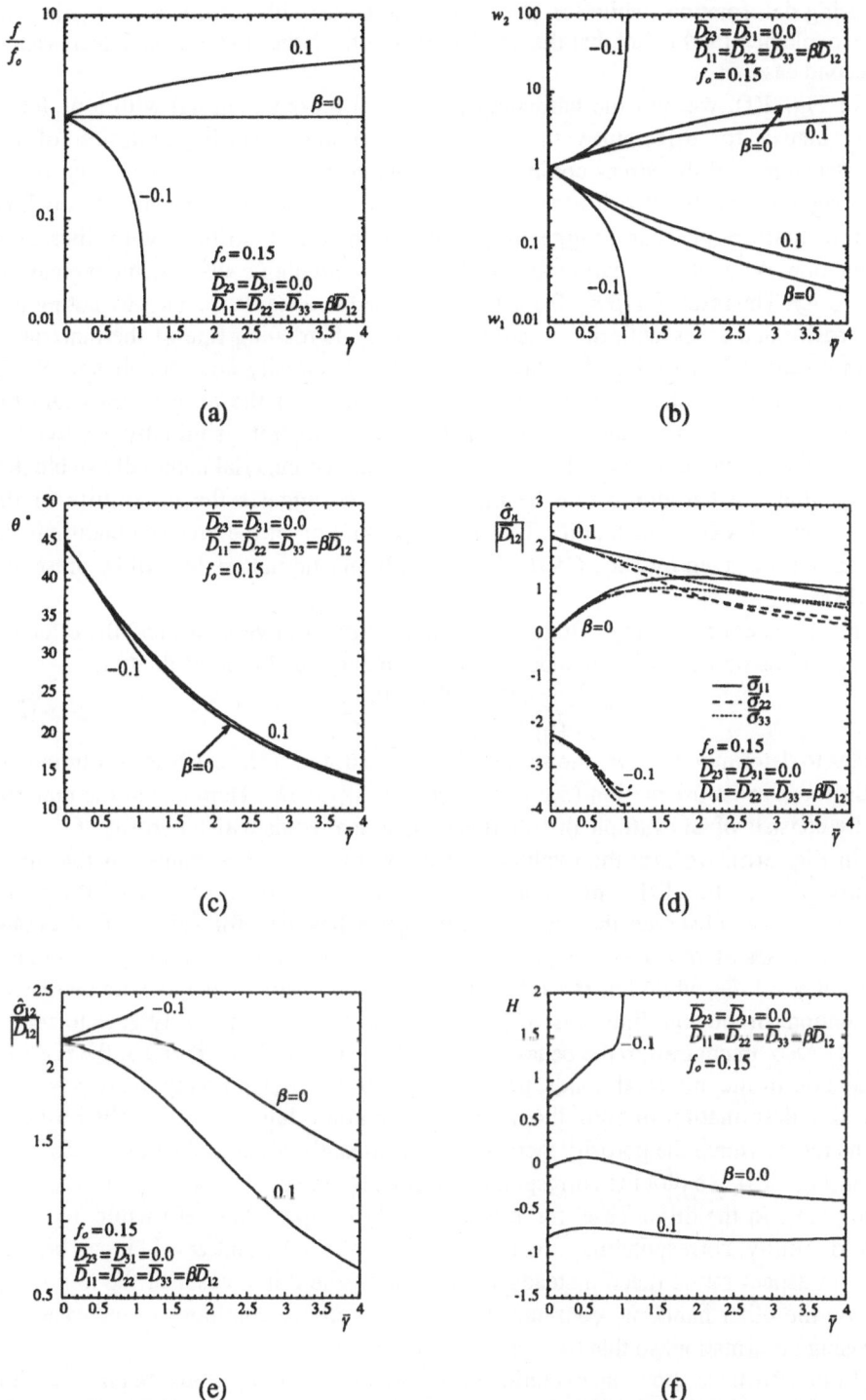

(a)

(b)

(c)

(d)

(e)

(f)

Figures 2. Evolution of porosity, aspect ratios and orientation of voids; normal stress components, shear stress component and the hardening rates.

increasing deformation, while for $\beta = -0.1$ it increases with deformation. Again, this can be explained by the fact that the porosity increases in the first case and decreases in the second case.

In Fig. 2(f), we plot the hardening rates for a porous material with a perfectly plastic matrix for different values of β. It is noted that the evolution of the microstructure and the stress components in this case are the same as shown in the earlier figures. For $\beta = 0.1$, it is seen that the hardening rate stays negative throughout the deformation process, suggesting that the material softens with increased deformation. As explained earlier, this is due to the softening caused by the increase in the porosity. The result for $\beta = -0.1$ can be explained in a similar manner by noting that the porosity decreases with deformation so that the hardening rate of the material is always positive. As mentioned earlier, for $\beta = 0$, the porosity does not change. So all the effects that are observed are due to the evolution of the aspect ratios and the orientation of the voids. Thus, it is seen that the hardening rate is initially positive and later becomes negative at $\bar{\gamma} = 1.11$. This suggests that the material is initially stable, but the fact that the hardening rate becomes negative suggests the possibility of the development of a shear instability. Exact implementation of the precise conditions for localization, as given by Rice (1976) for perfectly plastic materials, will be presented elsewhere.

Next, we consider the evolution of the microstructural variables and the effective response of the porous material under traction boundary conditions of the type

$$\bar{\sigma}_{23} = \bar{\sigma}_{31} = 0,$$
$$\bar{\sigma}_{11} = \bar{\sigma}_{22} = \bar{\sigma}_{33} = \alpha\bar{\sigma}_{12}. \tag{22}$$

In order to determine the average spin of the material, the surface AB is constrained to remain fixed in its orientation (see Fig. 1) so that $\bar{W}_{12} = \bar{D}_{12}$. Here again, the material initially consists of an isotropic distribution of spherical voids with a porosity f_o.

In Fig. 3(a), we plot the evolution of the porosity for two values of the initial porosity f_0 and for different values of α. For dilute concentration of the voids ($f_0 \to 0$), it was observed that the porosity approaches zero for values of $\alpha < 0.44$, while for values of $\alpha > 0.44$, the porosity increases. Thus, for $\alpha = 0$, the porosity goes to zero at $\bar{\gamma} = 1.79$, while for $\alpha = 0.5$ and $\alpha = 1$, the porosity increases with increasing deformation. In the non-dilute case ($f_0 = 0.15$), for $\alpha = 0$, the porosity goes to zero at $\bar{\gamma} = 2.1$. Also, in contrast to the behavior of the dilute material, for $\alpha = 0.5$, the porosity of the non-dilute material tends to zero. For $\alpha = 1$, the porosity increases with increasing deformation in both, the dilute and the non-dilute cases, but it is observed that the rate at which the porosity increases is significantly faster in the dilute case.

In Fig. 3(b), we plot the corresponding results for the evolution of the aspect ratios of the voids. In the dilute case, for $\alpha = 0$, $w_1 = b/a$ approaches zero while $w_2 = c/b$ tends to infinity, corresponding to void collapse. For $\alpha = 0.5$ and $\alpha = 1$, it can be seen that both aspect ratios reach a steady final value in the dilute case. In the non-dilute case, on the other hand, w_1 continues to decrease and w_2 continues to increase with increasing deformation, so that the voids collapse in this case.

In Fig. 3(c), we show the evolution of the orientation of the voids. In all cases, it is observed that the voids initially evolve into ellipsoidal shapes with $\theta = 45°$. In the dilute case, for $\alpha = 0$ it is seen that $\theta = 23°$ when the porosity becomes zero. For the same value of α, in the non-dilute case, $\theta = 21.6°$ when the porosity becomes zero. For

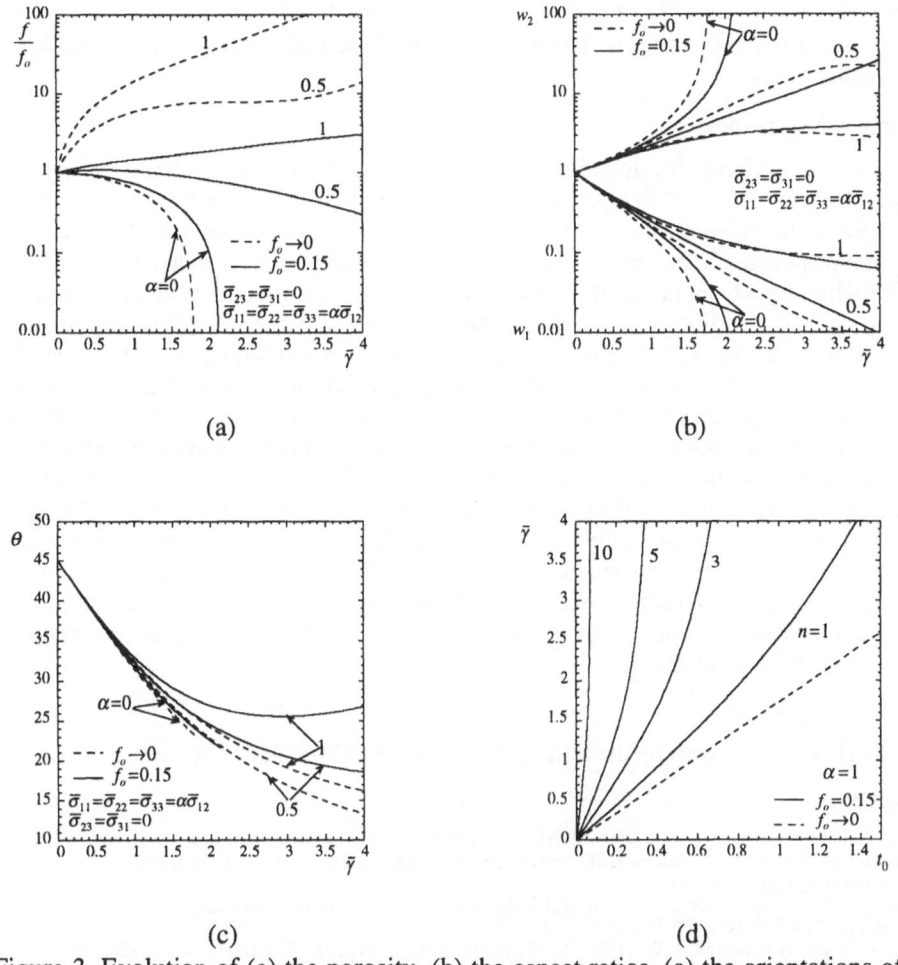

(a)   (b)

(c)   (d)

Figure 3. Evolution of (a) the porosity, (b) the aspect ratios, (c) the orientations of the voids and (d) the average shear strain in the porous material.

other values of α, in the dilute case, θ continues to decrease with increasing deformation. As mentioned earlier, in the dilute case, the porosity goes to zero when $\alpha \leq 0.44$. Thus, the final orientation of the voids depends on the amount of deformation required for the porosity to become zero.

In Fig. 3(d), we plot the evolution of the shear strain $\bar{\gamma}$ relative to the normalized time $t_0 = t\left(\sqrt{3}\bar{\sigma}_{12}\right)^n / \mu$ for $\alpha = 1$ and for different values of n. In the dilute case, we observe that the evolution curves for $\bar{\gamma}$ are linear and are identical for all values of n. This is as expected, since for dilute volume fractions, $\bar{\gamma}$ corresponds to the behavior of the matrix material. For $f_0 = 0.15$, the evolution curves for $\bar{\gamma}$ are nonlinear and a strong dependence on n is seen.

We note that these results for the evolution of the porosity, the aspect ratios and the orientation of the voids in the dilute case agree with the results of Fleck and Hutchinson (1986) for an isolated void in an infinite, linearly viscous matrix. Also, the present results for porous materials with 3D voids are qualitatively similar to the results obtained by Fleck and Hutchinson (1986) and Kailasam and Ponte Castañeda (1996) for porous materials with 2D voids. Finally, we note that this model is meant to be used only for low triaxialities, where the changes in the shape and orientation of the voids are significant effects.

5. Concluding remarks

In this work, we have developed a constitutive model for porous materials which can account for the effects of changes in the orientation of the voids, in the porosity and in the shape of the voids on the overall properties of the material. The model was used to study the response of a porous material under velocity and traction boundary conditions. Unlike a homogeneous material, which would be expected to behave identically under the two loading conditions, it was found that the porous material behaves significantly differently under the two loading conditions. This is a consequence of the fact that microstructure of the porous material evolves differently in the two cases. For simple shear velocity boundary conditions, significant normal stresses develop in the porous material and these components are all different from each other. This is a result of the developing anisotropy in the material caused by changes in the aspect ratios and orientation of the voids. For the case of a perfectly plastic porous material, it was found that the evolving anisotropy could cause the material response to become unstable. It is thus essential to take into consideration changes in the microstructure of heterogeneous materials in order to predict their effective responses accurately. In particular, it is important to account for the evolution of anisotropy in a heterogeneous material undergoing finite deformations.

Acknowledgement

This work was supported by the NSF under Grant No. CMS-9622714.

References

Budiansky, B., Hutchinson, J. W. and Slutsky, S. (1982) Void growth and collapse in viscous solids. In *Mechanics of Solids, The Rodney Hill Anniversary Volume*, edited by H. G. Hopkins and M. J. Sewell, Pergamon Press, Oxford, 13-45.
Eshelby, J. D. (1957) The determination of the elastic field of an ellipsoidal inclusion and related problems. *Proc. R. Soc. Lond.* A **241**, 376-396.
Fleck, N. A. and Hutchinson, J. W. (1986) Void growth in shear. *Proc. R. Soc. Lond.* A **407**, 435-458.
Gurson, A. L. (1977) Continuum theory of ductile rupture by void nucleation and growth. *J. Engng. Mater. Technol.* **99**, 2-15.
Hashin, Z. and Shtrikman, S. (1963) A variational approach to the theory of the elastic behavior of multiphase materials. *J. Mech. Phys. Solids* **11**, 127-140.
Kailasam, M. and Ponte Castañeda, P. (1996) Constitutive relations for porous composites: The effect of changing void shape and orientation. In IUTAM symposium on *Micromechanics of Plasticity and Damage of Multiphase Materials*, edited by A. Pineau and A. Zaoui, Kluwer, New York.
Kailasam, M., Ponte Castañeda, P. and Willis, J. R. (1996) The effect of particle size, shape, distribution and their evolution on the constitutive response of nonlinear composites — II. *Phil Trans. R. Soc. Lond. A, to appear*.
Lee, B. J. and Mear, M. E. (1992) Axisymmetric deformation of power-law solids containing a dilute concentration of aligned spheroidal voids. *J. Mech. Phys. Solids* **40**,1805-1836.
Ogden, R. W. (1984) In *Non-linear elastic deformations*, Halsted Press, New York, 128-130.
Ponte Castañeda, P. (1991) The effective mechanical properties of nonlinear isotropic composites. *J. Mech. Phys. Solids*, **39**, 45-71.
Ponte Castañeda and Willis, J. R. (1995) The effect of spatial distribution on the effective behavior of composite materials and cracked media. *J. Mech. Phys. Solids* **43**, 1919-1951
Ponte Castañeda, P. and Zaidman, M. (1994) Constitutive models for porous materials with evolving microstructure. *J. Mech. Phys. Solids* **42**, 1459-1497.
Rice, J. R. (1976) The localization of plastic deformation. In *Proceedings of the 14th International Congress of Theoretical and Applied Mechanics*, edited by W. T. Koiter, North-Holland, 207-220.
Willis, J. R. (1977) Bounds and self-consistent estimates for the overall moduli of anisotropic composites. *J. Mech. Phys. Solids* **25**, 185-202.
Willis, J. R. (1978) Variational principles and bounds for the overall properties of composites. In *Continuum Models for Discrete Systems*, edited by J. W. Provan, University of Waterloo Press, 185-215.

SHEAR BAND LOCALIZATION IN FLUID-SATURATED GRANULAR ELASTO-PLASTIC POROUS MEDIA

W. EHLERS

Institut für Mechanik (Bauwesen), Lehrstuhl II,
Universität Stuttgart, Pfaffenwaldring 7, D-70 569 Stuttgart

1. Introduction

Shear band localization phenomena as appearing, e. g., in the classical base failure problem of geotechnical engineering, occur in non-saturated and saturated soils as a result of local concentrations of plastic solid strains (de Borst, 1991 a; Schrefler *et al.*, 1995, 1996). The present contribution concentrates on the saturated case, where the elasto-plastic deformations of the granular soil can be described by use of a macroscopic continuum mechanical approach within the well-founded framework of the Theory of Porous Media (TPM), compare, e. g., the work by Bowen (1982), de Boer and Ehlers (1986), Ehlers (1989, 1993) and de Boer (1996).

In the present article, the TPM formulation of the skeleton material is extended by micropolar degrees of freedom in the sense of the *Cosserat* brothers (de Borst, 1991 b; Steinmann, 1994). Proceeding from two basic assumptions, material incompressibility of both constituents (skeleton material and pore-fluid) and geometrically linear solid deformations, the non-symmetric effective skeleton stress and the couple stress tensor are determined by linear elasticity laws. In the framework of the ideal plasticity concept applied to cohesive-frictional soils, the plastic yield limit is governed by a smooth and closed single-surface yield function together with non-associated flow rules for both the plastic strain rate and the plastic rate of curvature tensor. Fluid viscosity is taken into account by the drag force.

The inclusion of micropolar degrees of freedom, in contrast to the usual continuum mechanical approach to the TPM, allows, on the one hand, for the determination of the local average grain rotations and, on the other hand, additionally yields a regularization effect on the solution of the strongly coupled system of governing equations when shear banding occurs. The numerical example exhibits the base failure problem induced by elasto-plastic consolidation. The computations are carried out by use of finite element discretization techniques.

N. A. Fleck and A. C. F. Cocks (eds.),
IUTAM Symposium on Mechanics of Granular and Porous Materials, 377–388.
© *1997 Kluwer Academic Publishers.*

2. Governing Equations

In the framework of the Theory of Porous Media, fluid-saturated porous solid skeleton materials are considered as a mixture of immiscible constituents φ^α with particles X^α ($\alpha = S$: solid skeleton; $\alpha = F$: pore-fluid), where at any time t, each spatial point $\mathbf{x}$ of the current configuration is simultaneously occupied by material points X^α of all constituents φ^α (superimposed continua). These particles proceed from different reference positions $\mathbf{X}_\alpha$ at time t_0. Thus, each constituent is assigned an individual motion function

$$\mathbf{x} = \chi_\alpha(\mathbf{X}_\alpha, t). \tag{1}$$

The volume fractions

$$n^\alpha = n^\alpha(\mathbf{x}, t) \tag{2}$$

are defined as the local ratios of the constituent volumes v^α with respect to the bulk volume v. Thus, the saturation condition yields

$$n^S + n^F = 1. \tag{3}$$

Associated with each constituent φ^α is an effective (realistic or material) density $\rho^{\alpha R}$ and a partial (global or bulk) density ρ^α. The effective density $\rho^{\alpha R}$ is defined as the local mass of φ^α per unit of v^α, whereas the bulk density ρ^α exhibits the same mass per unit of v. The density functions are related by

$$\rho^\alpha = n^\alpha \rho^{\alpha R}. \tag{4}$$

Proceeding from (4), it is obvious that the property of material incompressibility of any constituent φ^α (defined by $\rho^{\alpha R} = \text{const.}$) is not equivalent to global incompressibility of this constituent, since the partial density functions can still change through changes in the volume fractions n^α.

It follows from (1) that each constituent is assigned its own velocity field. Thus, by use of either the *Lagrangean* or the *Eulerian* description,

$$\overset{'}{\mathbf{x}}_\alpha = \frac{\partial \chi_\alpha(\mathbf{X}_\alpha, t)}{\partial t}, \qquad \overset{'}{\mathbf{x}}_\alpha = \overset{'}{\mathbf{x}}_\alpha(\mathbf{x}, t). \tag{5}$$

Assume that Γ is any arbitrary, continuous and sufficiently often continuously differentiable function of $(\mathbf{x}, t)$. Then, the material time derivatives of Γ corresponding to the individual motion functions of φ^α are given by

$$(\Gamma)'_\alpha = \frac{\partial \Gamma}{\partial t} + \text{grad}\,\Gamma \cdot \overset{'}{\mathbf{x}}_\alpha. \tag{6}$$

Therein, the operator "grad($\cdot$)" characterizes the partial derivative of ($\cdot$) with respect to the position vector $\mathbf{x}$ of the actual configuration.

Concerning the problem under study, it is convenient to consider the solid motion in the frame of the *Lagrangean* description by the introduction of the displacement vector

$$\mathbf{u}_S = \mathbf{x} - \mathbf{X}_S, \tag{7}$$

whereas the pore-fluid is better described by a modified *Eulerian* description proceeding from the introduction of the seepage velocity $\mathbf{w}_F$:

$$\mathbf{w}_F = \overset{\prime}{\mathbf{x}}_F - \overset{\prime}{\mathbf{x}}_S . \tag{8}$$

From (1) and (7), the material deformation gradient and the displacement gradient of the solid skeleton are

$$\mathbf{F}_S = \text{Grad}_S \, \mathbf{x} , \qquad \mathbf{H}_S = \text{Grad}_S \, \mathbf{u}_S , \tag{9}$$

where the operator "$\text{Grad}_S \, (\, \cdot \,)$" defines the partial derivative of $(\, \cdot \,)$ with respect to the reference position $\mathbf{X}_S$ of φ^S. From the displacement gradient, one easily derives the classical measures of the geometrically linearized theory

$$
\begin{aligned}
\mathbf{H}_{S \text{ sym}} &= \tfrac{1}{2} \, (\mathbf{H}_S + \mathbf{H}_S^T) =: \boldsymbol{\varepsilon}_S , \\
\mathbf{H}_{S \text{ skw}} &= \tfrac{1}{2} \, (\mathbf{H}_S - \mathbf{H}_S^T) =: \boldsymbol{\omega}_S \times \mathbf{I} ,
\end{aligned}
\tag{10}
$$

where $\boldsymbol{\varepsilon}_S$ is the linear *Lagrangean* solid strain tensor, $\boldsymbol{\omega}_S$ is the so-called continuum rotation vector, $\mathbf{I}$ is the second-order identity (fundamental tensor of second order), and $(\, \cdot \,)^T$ is the transpose of $(\, \cdot \,)$. The external tensor product between vectors and tensors (de Boer, 1982) is defined by

$$\boldsymbol{\omega}_S \times \mathbf{I} = - \overset{3}{\mathbf{E}} \boldsymbol{\omega}_S ; \tag{11}$$

$\overset{3}{\mathbf{E}}$ is the *Ricci* permutation or the fundamental tensor of third order, respectively. Given $(10)_2$, it is straight forward to obtain $\boldsymbol{\omega}_S$ as a function of $\mathbf{H}_S$:

$$\boldsymbol{\omega}_S = - \tfrac{1}{2} \overset{3}{\mathbf{E}} \, (\mathbf{H}_{S \text{ skw}}) . \tag{12}$$

In extension of the continuum mechanical description of the skeleton material, additional micropolar degrees of freedom are introduced via the independent rotation $\overset{*}{\boldsymbol{\omega}}_S$, thus defining the total average grain rotation $\bar{\boldsymbol{\omega}}_S$ as the sum of the continuum rotation and the additional micropolar rotation, viz.:

$$\bar{\boldsymbol{\omega}}_S = \boldsymbol{\omega}_S + \overset{*}{\boldsymbol{\omega}}_S . \tag{13}$$

By use of the additional micropolar degrees of freedom, the linear *Cosserat* strain tensor $\boldsymbol{\varepsilon}_{Sc}$ and the curvature tensor $\bar{\boldsymbol{\kappa}}_S$ yield:

$$\boldsymbol{\varepsilon}_{Sc} = \mathbf{H}_S + \overset{3}{\mathbf{E}} \bar{\boldsymbol{\omega}}_S , \qquad \bar{\boldsymbol{\kappa}}_S = \text{Grad}_S \, \bar{\boldsymbol{\omega}}_S . \tag{14}$$

The symmetric and skew symmetric parts of the *Cosserat* strain are

$$
\begin{aligned}
\boldsymbol{\varepsilon}_{Sc \text{ sym}} &= \tfrac{1}{2} \, (\mathbf{H}_S + \mathbf{H}_S^T) , \\
\boldsymbol{\varepsilon}_{Sc \text{ skw}} &= \tfrac{1}{2} \, (\mathbf{H}_S - \mathbf{H}_S^T) + \overset{3}{\mathbf{E}} \bar{\boldsymbol{\omega}}_S .
\end{aligned}
\tag{15}
$$

In this representation, $\varepsilon_{Sc\ sym}$ equals the linearized *Lagrangean* strain ε_S of non-polar materials, whereas

$$\varepsilon_{Sc\ skw} = \overset{3}{\mathbf{E}}\,(\bar{\omega}_S - \omega_S) = \overset{3}{\mathbf{E}}\,\overset{*}{\omega}_S \tag{16}$$

exhibits a tensorial measure for the additional micropolar rotation $\overset{*}{\omega}_S$. Finally, with the aid of $(15)_1$ and (16), the *Cosserat* strain ε_{Sc} yields

$$\begin{aligned}\varepsilon_{Sc} &= \tfrac{1}{2}(\mathbf{H}_S + \mathbf{H}_S^T) + \overset{3}{\mathbf{E}}\,\overset{*}{\omega}_S \\ &= \varepsilon_S + \overset{3}{\mathbf{E}}\,\overset{*}{\omega}_S.\end{aligned} \tag{17}$$

Neglecting inertia effects and excluding mass exchanges between the solid and the fluid materials, the governing balance equations read (Ehlers, 1993; Diebels and Ehlers, 1996):

Balance of mass:

$$0 = (\rho^\alpha)'_\alpha + \rho^\alpha \mathrm{div}\,\overset{\prime}{\mathbf{x}}_\alpha. \tag{18}$$

Balance of momentum:

$$\begin{aligned}0 &= \mathrm{div}\,\mathbf{T}^\alpha + \rho^\alpha \mathbf{b} + \hat{\mathbf{p}}^\alpha, \\ \hat{\mathbf{p}}^S + \hat{\mathbf{p}}^F &= \mathbf{0}.\end{aligned} \tag{19}$$

Balance of moment of momentum:

$$\begin{aligned}0 &= \mathbf{I} \times \mathbf{T}^\alpha + \mathrm{div}\,\bar{\mathbf{M}}^\alpha + \hat{\mathbf{m}}^\alpha, \\ \hat{\mathbf{m}}^S + \hat{\mathbf{m}}^F &= \mathbf{0}.\end{aligned} \tag{20}$$

In these relations, $\mathbf{T}^\alpha$ is the non-symmetric *Cauchy* stress tensor of micropolar materials, $\mathbf{b}$ is the volume force per unit of φ^α-th mass, $\hat{\mathbf{p}}^\alpha$ is the momentum production of φ^α which, in the case of a binary material, can be interpreted as the local interaction force (per unit of bulk volume) between the solid and the fluid phases, $\bar{\mathbf{M}}^\alpha$ is the couple stress tensor, $\hat{\mathbf{m}}^\alpha$ is the moment of momentum production, and "$\mathrm{div}\,(\,\cdot\,)$" is the divergence operator corresponding to "$\mathrm{grad}\,(\,\cdot\,)$". The external vector product between the tensors $\mathbf{I}$ and $\mathbf{T}^\alpha$ (de Boer, 1982) indicates twice the axial vector of $\mathbf{T}^\alpha$ via

$$\overset{A}{\mathbf{t}}{}^\alpha = \tfrac{1}{2}(\mathbf{I} \times \mathbf{T}^\alpha) = \tfrac{1}{2}\overset{3}{\mathbf{E}}\,(\mathbf{T}^\alpha)^T. \tag{21}$$

Proceeding from the assumption that only the skeleton material exhibits micropolar properties (and not the pore-fluid), one obtains by use of $\hat{\mathbf{m}}^F = \mathbf{0}$ (de Boer *et al.*, 1991) that

$$\begin{aligned}0 &= \mathbf{I} \times \mathbf{T}^S + \mathrm{div}\,\bar{\mathbf{M}}^S, \\ \mathbf{T}^F &= (\mathbf{T}^F)^T,\end{aligned} \tag{22}$$

(22) thus substituting (20).

Based on the property of material incompressibility of both constituents ($\rho^{\alpha R} =$ const.), solid skeleton and pore-fluid, the sum of the individual mass balance equations (18), the saturation condition (3) and the density relation (4) combine to yield:

$$\operatorname{div}\left(n^F \mathbf{w}_F + (\mathbf{u}_S)'_S\right) = 0. \tag{23}$$

This equation represents the volume balance relation or the incompressibility constraint of the binary material under study.

3. Constitutive Equations

To close the set of governing equations for the binary model under study, constitutive equations must be formulated for the stress tensors $\mathbf{T}^\alpha$ of the solid and fluid materials, the solid couple stress tensor $\bar{\mathbf{M}}^S$ and for the interaction force (momentum production) $\hat{\mathbf{p}}^F = -\hat{\mathbf{p}}^S$. As a result of the incompressibility constraint of both materials, the expressions for the stress tensors and for the interaction force consist of two terms each:

$$
\begin{aligned}
\mathbf{T}^\alpha &= -n^\alpha p\,\mathbf{I} + \mathbf{T}_E^\alpha, \\
\hat{\mathbf{p}}^F &= p\,\operatorname{grad} n^F + \hat{\mathbf{p}}_E^F.
\end{aligned}
\tag{24}
$$

The first term in these relations includes the *Lagrangean* multiplier p, which can be identified as the effective fluid pressure, while the second term, in the sense of an extra quantity, index $(\cdot)_E$, is governed by the solid deformation state and the fluid viscosity (Ehlers, 1989). As is usual in hydraulics, the friction force $\mathbf{z}^F = \operatorname{div}\mathbf{T}_E^F$ is neglected. Thus, one proceeds from the *a priori* assumption $\mathbf{T}_E^F = \mathbf{0}$. The internal friction between the skeleton and the viscous pore-fluid is taken into account by the effective drag force or the momentum production term, respectively, viz.:

$$
\hat{\mathbf{p}}_E^F = -\frac{(n^F)^2\,\gamma^{FR}}{k^F}\,\mathbf{w}_F.
\tag{25}
$$

Therein, $\gamma^{FR} = \rho^{FR}|\mathbf{b}|$ is the effective (true) specific weight of the pore-fluid, and k^F is the *Darcy* permeability coefficient providing information on both the local size and isotropic structure of the pore-space and the fluid viscosity.

Proceeding from the assumptions of geometrically linear solid deformations, both the non-symmetric *Cosserat* strain ε_{Sc} and the curvature tensor $\bar{\kappa}_S$ are additively decomposed into elastic and plastic parts:

$$
\begin{aligned}
\varepsilon_{Sc} &= \varepsilon_{Sce} + \varepsilon_{Scp}, \\
\bar{\kappa}_S &= \bar{\kappa}_{Se} + \bar{\kappa}_{Sp}.
\end{aligned}
\tag{26}
$$

In extension of the *Hookean* material model, the non-symmetric solid extra stress or the effective stress, respectively, is given by

$$\mathbf{T}_E^S = \overset{4}{\mathbf{C}}_e \, \varepsilon_{Sce} ,$$

$$\overset{4}{\mathbf{C}}_e = 2\,\mu^S \overset{4}{\mathbf{I}}_{\text{sym}} + 2\,\mu_c^S \overset{4}{\mathbf{I}}_{\text{skw}} + \lambda^S \, (\mathbf{I} \otimes \mathbf{I}) ,$$

(27)

wherein

$$\overset{4}{\mathbf{I}}_{\text{sym}} = \tfrac{1}{2} [(\mathbf{I} \otimes \mathbf{I})^{\overset{23}{T}} + (\mathbf{I} \otimes \mathbf{I})^{\overset{13}{T}}] ,$$

$$\overset{4}{\mathbf{I}}_{\text{skw}} = \tfrac{1}{2} [(\mathbf{I} \otimes \mathbf{I})^{\overset{23}{T}} - (\mathbf{I} \otimes \mathbf{I})^{\overset{13}{T}}] ;$$

(28)

the transposition $(\,\cdot\,)^{\overset{ik}{T}}$ indicates an exchange of the i-th and k-th basis systems included into the tensor basis of higher order tensors.

The application of the fundamental tensors of fourth order, $\overset{4}{\mathbf{I}}_{\text{sym}}$ and $\overset{4}{\mathbf{I}}_{\text{skw}}$, to an arbitrary tensor of second order yields its symmetric and its skew-symmetric parts, respectively. μ^S and λ^S are the *Lamé* constants of the porous skeleton material, whereas μ_c^S is an additional material parameter governing the influence of the skew-symmetric part of the *Cosserat* strain on the effective stress of the micropolar skeleton material. In comparison to the strain relation (17), the symmetric part of $\mathbf{T}_E^S$ equals the classical stress-strain relation of non-polar linear elastic materials, while the skew-symmetric part is directly related from the independent micropolar rotation.

Concerning the couple stress tensor, the constitutive relation by de Borst (1991 b) is considered. Thus,

$$\bar{\mathbf{M}}^S = \mu^S \, (l_c^S)^2 \, \bar{\kappa}_{Se} .$$

(29)

Therein, $\mu^S \, (l_c^S)^2$ is a proportional factor consisting of both the *Lamé* constant μ^S and an additional independent material parameter l_c^S, which is usually interpreted as the internal length scale, thus implicitly delivering the possibility to include the shear band width into the model.

Concerning the plastic material properties of the porous solid skeleton, use is made of the ideal plasticity assumption. Thus, one has to introduce a convenient yield function together with evolution equations for the non-symmetric plastic strain rate $(\varepsilon_{Scp})_S'$ and the plastic rate of curvature tensor $(\bar{\kappa}_{Sp})_S'$.

In the present article, use is made of an extension of the single-surface yield criterion by Ehlers (1993), compare Figure 1, which is extended towards micropolar cohesive-frictional materials, viz.:

$$
\begin{aligned}
F^c &= \Phi^{1/2} + \beta\, \mathrm{I} + \epsilon\, \mathrm{I}^2 - \bar{\kappa} = 0 , \\[4pt]
\Phi &= \mathrm{II}_{D\,\text{sym}} \, (1 + \gamma\,\vartheta)^m + \tfrac{1}{2}\alpha\, \mathrm{I}^2 + \delta^2 \, \mathrm{I}^4 , \\[4pt]
\vartheta &= \mathrm{III}_{D\,\text{sym}} / (\mathrm{II}_{D\,\text{sym}})^{3/2} , \\[4pt]
\bar{\kappa} &= \kappa - \bar{k}_1 \, (\tfrac{1}{2}\bar{\mathbf{M}}^S \cdot \bar{\mathbf{M}}^S)^{1/2} - \bar{k}_2 \, (\mathrm{II}_{D\,\text{skw}})^{1/2} .
\end{aligned}
$$

(30)

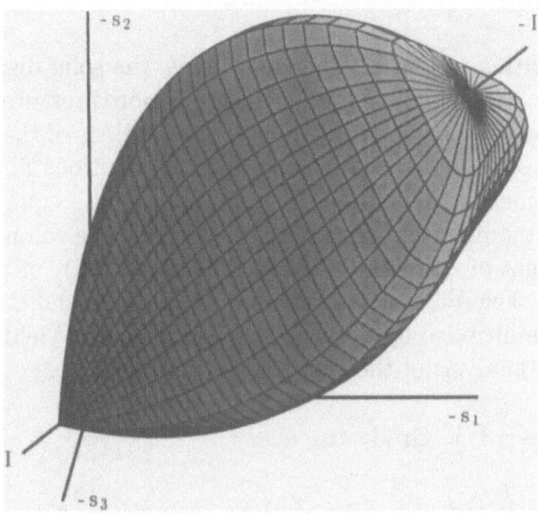

Figure 1. Single-surface yield criterion for non-polar cohesive-frictional materials; s_1, s_2, s_3: principal stresses of $\mathbf{T}^S_{E\,\text{sym}}$ (tension positive).

In (30), I, $\mathrm{II}_{D\,\text{sym}}$, $\mathrm{III}_{D\,\text{sym}}$ and $\mathrm{II}_{D\,\text{skw}}$ are the principal invariants of $\mathbf{T}^S_E$ and its deviatoric symmetric and deviatoric skew-symmetric parts. The material constants α, β, γ, δ, ϵ, κ and m govern the yield curve in the non-polar case, whereas $\bar{k}_1$ and $\bar{k}_2$ describe the softening influence of $\bar{\mathbf{M}}^S$ and $\mathbf{T}^{SD}_{E\,\text{skw}}$. The index $(\,\cdot\,)^D$ characterizes the deviatoric part of tensor quantities.

The description of the plastic range must be completed by evolution equations (flow rules) for both the plastic strain rate $(\varepsilon_{Scp})'_S$ and the plastic rate of curvature tensor $(\bar{\kappa}_{Sp})'_S$. Considering an evolution equation of $(\varepsilon_{Scp})'_S$ first, it is known from various experiments (Lade and Duncan, 1973) that the usual associated flow rule concept cannot be used as far as frictional materials are concerned. Following this, use is made of an extension of the non-associated flow rule for non-polar materials by Ehlers (1993):

$$(\varepsilon_{Scp})'_S = \Lambda \left[\tfrac{1}{2} \left(\mathrm{II}_{D\,\text{sym}}\right)^{-1/2} \mathbf{T}^{SD}_{E\,\text{sym}} + \tfrac{1}{2}\,\bar{k}_2 \left(\mathrm{II}_{D\,\text{skw}}\right)^{-1/2} \mathbf{T}^{SD}_{E\,\text{skw}} + \sqrt{\tfrac{1}{6}}\,\tan\nu_p\,\mathbf{I} \right]. \tag{31}$$

In good accordance with experimental data, the flow rule (31) exhibits co-axial behaviour of the deviatoric part of $(\varepsilon_{Scp})'_S$, whereas the hydrostatic part is governed by an additional material parameter function, the dilatation angle ν_p, governing the deviation of the total strain rate direction from the purely deviatoric one.

In analogy to the second term of (31), the following evolution equation for the plastic rate of curvature tensor is used,

$$(\bar{\kappa}_{Sp})'_S = \Lambda\,\tfrac{1}{2}\,\bar{k}_1 \left(\tfrac{1}{2}\bar{\mathbf{M}}^S \cdot \bar{\mathbf{M}}^S\right)^{-1/2} \bar{\mathbf{M}}^S, \tag{32}$$

thus representing both a co-axial and an associated relation for $(\bar{\kappa}_{Sp})'_S$.

4. Weak Formulation of the Governing Field Equations

Based on consideration of four independent fields, the solid displacement $\mathbf{u}_S$, the seepage velocity $\mathbf{w}_F$, the effective fluid pressure p (pore pressure) and the total average grain rotation $\bar{\omega}_S$, the corresponding four equations of the weak formulation can be obtained from the kinematics, the balance relations and the constitutive equations of the model under study. In particular, these equations are given by the solid and fluid momentum balance equations (19), the volume balance relation (23) and the moment of momentum balance equation $(22)_1$ of the solid material.

Concerning the skeleton material, the sum of the solid and the fluid momentum balance equations multiplied by the weighting function $\delta \mathbf{u}_S$ yields in the framework of the small-strain analysis of the skeleton displacements:

$$
\int_V \left(\mathbf{T}_E^S - p\,\mathbf{I} \right) \cdot \mathrm{Grad}_S\, \delta \mathbf{u}_S \; dv - \\
- \int_V \left(n^F \rho^{FR} + n^S \rho^{SR} \right) \mathbf{b} \cdot \delta \mathbf{u}_S \; dv = \int_{\partial V} \mathbf{t} \cdot \delta \mathbf{u}_S \; da .
\tag{33}
$$

Therein, $\mathbf{t}$ is the external load vector acting upon both constituents. The weak formulation of the pore-fluid momentum balance results from a multiplication of $(19)_1$ by the weighting function $\delta \mathbf{w}_F$. Thus, one obtains

$$
\int_V \left(\frac{\gamma^{FR}}{k^F}\, n^F \mathbf{w}_F - \rho^{FR} \mathbf{b} \right) \cdot \delta \mathbf{w}_F \; dv - \\
- \int_V p \,\mathrm{div}\, \delta \mathbf{w}_F \; dv = - \int_{\partial V} p\,(\delta \mathbf{w}_F \cdot \mathbf{n})\, da ;
\tag{34}
$$

$\mathbf{n}$ is the outward oriented unit surface normal. The third equation

$$
\int_V \left(n^F \mathbf{w}_F + (\mathbf{u}_S)_S' \right) \cdot \mathrm{grad}\, \delta p \; dv = \int_{\partial V} \left(n^F \mathbf{w}_F + (\mathbf{u}_S)_S' \right) \cdot \mathbf{n}\, \delta p \; da
\tag{35}
$$

represents the incompressibility constraint of both materials multiplied by the weighting function δp. Excluding external loading by couple stress vectors, multiplication of $(22)_1$ by the weighting function $\delta \bar{\omega}_S$ finally yields

$$
\int_V (\mathbf{I} \times \mathbf{T}^S) \cdot \delta \bar{\omega}_S \; dv + \int_V \bar{\mathbf{M}}^S \cdot \mathrm{Grad}_S\, \delta \bar{\omega} \; dv = 0 .
\tag{36}
$$

5. Example

The following numerical example carried out by the finite element method concerns the base failure problem as a result of plastic strain localization phenomena.

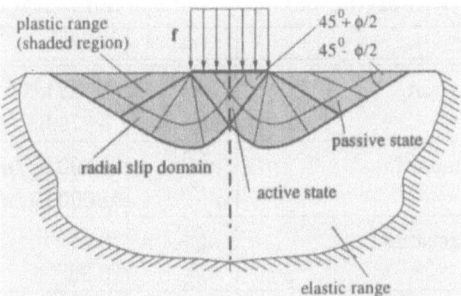

Figure 2. The base failure problem.

Figure 2 shows the well-known classical base failure problem (Terzaghi and Jelinek, 1954), where a critical external load **f** exerted onto the surface of the elasto-plastic half-space yields a failure mechanism represented by one of the possible principal slip lines separating the plastic range from the remainder of the elastic half-space. In the plastic domain, a slip line field occurs due to *Rankine*'s theory (Rankine, 1857), ϕ is the angle of internal friction.

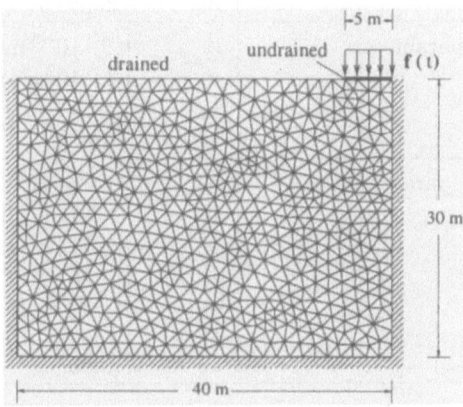

Figure 3. Initial mesh of the finite element discretization loaded by $f(t) = 0.5\,[\mathrm{kN}/(\mathrm{m}^2\mathrm{s})]$, $t\,[\mathrm{s}]$ $(0 < t \le 290\,\mathrm{s})$.

By consideration of the symmetry of the base failure problem, the two-dimensional (plane strain) finite element discretization proceeds from a liquid-saturated soil block of 30 m height and 40 m depth, initially represented by an irregular start mesh of triangular elements, compare Figure 3. The external load $f(t)$ linearly increases with time. The loaded part of the surface is undrained, while the free part is fully drained. The remainder of the external boundary is rigid and impermeable. The material parameters can be taken from Table 1, where the plastic material parameters, in the non-polar case, are fitted in comparison with *Schad*'s cap model (Schad, 1979) including an angle of internal friction of $\phi = 21°$. In addition,

$$\nu_p = \nu_1 \exp[a_1(I - I_{01})] + \nu_2 \exp[a_2(I - I_m)] + \nu_3 \exp[a_3(I - I_{02})] + \nu_4 , \quad (37)$$

where the included parameters are also included into Table 1.

TABLE 1. Material parameters of the numerical example

parameter	symbol	value
Lamé constants	μ^S	$5\,583\,\text{kN/m}^2$
	λ^S	$8\,375\,\text{kN/m}^2$
effective densities	ρ^{SR}	$2\,600\,\text{kg/m}^3$
	ρ^{FR}	$1\,000\,\text{kg/m}^3$
volume fractions	n_{0S}^S	0.67
	n_{0S}^F	0.33
effective specific fluid weight	γ^{FR}	$10 \cdot 10^3\,\text{kN/m}^3$
permeability coefficient	k^F	$2 \cdot 10^{-4}\,\text{m/s}$
parameter of the	α	$-1.653 \cdot 10^{-4}$
single-surface	β	0.119 6
yield criterion	γ	1.555
	δ	$1.1 \cdot 10^{-4}\,\text{m}^2/\text{kN}$
	ε	$6.21 \cdot 10^{-5}\,\text{m}^2/\text{kN}$
	κ	$10.27\,\text{kN/m}^2$
	m	0.593 5
Cosserat parameters	l_c^S	$2 \cdot 10^{-3}\,\text{m}$
	μ_c^S	$2 \cdot 10^2\,\text{kN/m}^2$
	$\bar{k}_1$	$5\,\text{m}^{-1}$
	$\bar{k}_2$	10^{-4}
dilatation parameters	I_{01}	$58\,\text{kN/m}^2$
	I_m	$-516\,\text{kN/m}^2$
	I_{02}	$-774\,\text{kN/m}^2$
	a_1	$1.25 \cdot 10^{-1}\,\text{m}^2/\text{kN}$
	a_2	$-8.50 \cdot 10^{-3}\,\text{m}^2/\text{kN}$
	a_3	$-7.30 \cdot 10^{-2}\,\text{m}^2/\text{kN}$
	ν_4	0.08

The numerical results shown in Figures 4 and 5 exhibit the accumulated plastic strains 290 s after having applied the external load. These strains implicitly indicate the widths and the directions of the different bands. The individual results are given both for the standard TPM formulation based on a non-polar skeleton material and for the extended TPM formulation accounting for additional micropolar solid rotations.

In comparison of the results given by Figures 4, both computed on the initial mesh, the fundamental tendency of shear banding can only be observed insofar as the principal slopes of the slip line fields can be found. On the other hand, both results exhibit a certain mesh dependence, where the extended formulation, in contrast to the standard TPM computation, shows two distinct bands together with a minor diffusive slip line domain.

In a second step, the same computations have been carried out on modified meshes. After two refinements, compare Figure 5, the standard TPM formulation only represents the dominant slip line, whereas the extended TPM formulation again exhibits two slip lines, namely the dominant one and a secondary one included into the passive slip line domain.

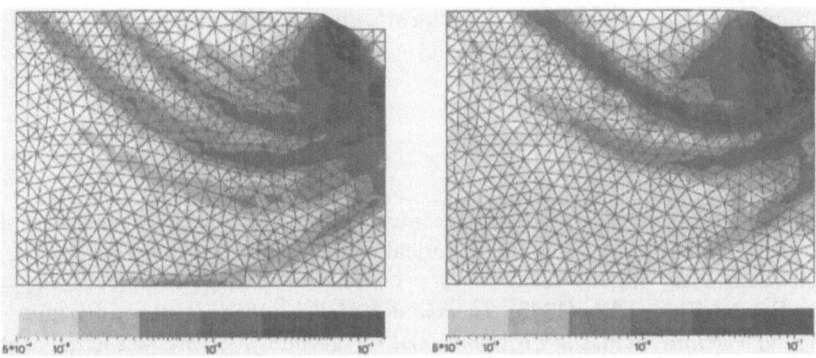

Figure 4. Accumulated plastic strains on the initial mesh
by use of the standard computation (left) and the extended computation (right).

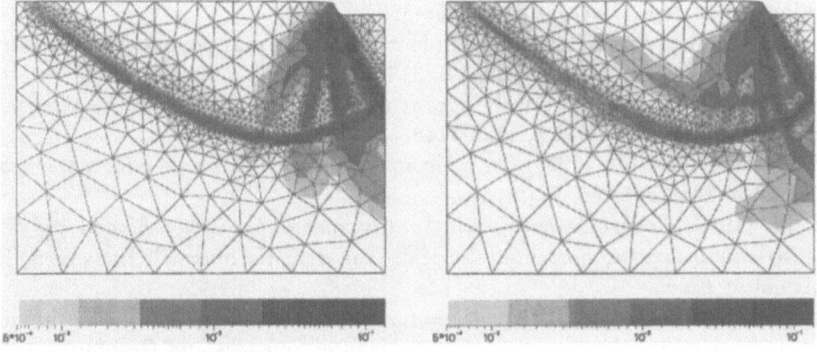

Figure 5. Accumulated plastic strains on a modified mesh (2nd refinement)
by use of the standard computation (left) and the extended computation (right).

6. Conclusions

In the present article, the Theory of Porous Media (TPM) has been applied to fluid-saturated elasto-plastic porous solid materials. In particular, the description of the porous skeleton material has been extended by the inclusion of additional micropolar degrees of freedom in the sense of the *Cosserat* brothers. Both TPM formulations, the standard and the extended ones, have been applied to the well-known base failure problem of geotechnical engineering. The numerical results reveal that both formulations are, in principal, appropriate for the solution of shear band phenomena. However, the Figures 4 and 5 clearly show the strong mesh dependence of the standard TPM computations, whereas the extended formulation produces, on each mesh, two bands corresponding to the solution of the initial boundary-value problem under consideration. Finally, it should be noted that to obtain a certain regularization within the standard TPM formulation, the viscosity effect of the pore-fluid must be included into the problem.

7. References

de Boer, R. (1982) *Vektor- und Tensorrechnung für Ingenieure*, Springer-Verlag, Berlin.

de Boer, R. (1996) Highlights in the historical development of the porous media theory, *Appl. Mech. Rev.* **49**, 201-262.

de Boer, R. and Ehlers, W. (1986) *Theorie der Mehrkomponentenkontinua mit Anwendung auf bodenmechanische Probleme, Teil I*, Forschungsberichte aus dem Fachbereich Bauwesen, Heft 40, Universität Essen.

de Boer, R., Ehlers, W., Kowalski, S. and Plischka, J. (1991) *Porous Media – A Survey of Different Approaches*, Forschungsberichte aus dem Fachbereich Bauwesen, Heft 54, Universität Essen.

de Borst, R. (1991 a) Numerical modelling of bifurcation and localisation in cohesive-frictional materials, *Pageoph.* **137**, 368-390.

de Borst, R. (1991 b) Simulation of strain localization: A reappraisal of the Cosserat continuum, *Engineering Computations* **8**, 317-332.

Bowen, R. M. (1980) Incompressible porous media models by use of the theory of mixtures, *Int. J. Engng. Sci.* **18**, 1129-1148.

Diebels, S. and Ehlers, W. (1996) On basic equations of multiphase micropolar materials, *Technische Mechanik* **16**, 77-88.

Ehlers, W. (1989) *Poröse Medien – ein kontinuumsmechanisches Modell auf der Basis der Mischungstheorie*, Forschungsberichte aus dem Fachbereich Bauwesen, Heft 47, Universität Essen.

Ehlers, W. (1993) Constitutive equations for granular materials in geomechanical context, in *Continuum Mechanics in Environmental Sciences and Geophysics* (edited by K. Hutter), pp. 313-402. CISM Courses and Lecture Notes No. 337, Springer-Verlag, Wien.

Rankine, W. J. M. (1857) On the stability of loose earth, *Phil. Trans. Roy. Soc. Lond.* **147**, 9-27.

Schad, H. (1979) *Nichtlineare Stoffgleichungen für Böden und ihre Verwendung bei der numerischen Analyse von Grundbauaufgaben*, Mitteilungen des Baugrundinstituts Stuttgart 10, Universität Stuttgart.

Schrefler, B. A., Majorna, C. E. and Sanavia, L. (1995) Shear band localization in saturated porous media, *Arch. Appl. Mech.* **47**, 577-599.

Schrefler, B. A., Sanavia, L. and Majorna, C. (1996) A multiphase media model for localisation and post-localisation simulation in geomaterials, *Mechanics of Cohesive-frictional Materials* **1**, 95-114.

Steinmann, P. (1994) A micropolar theory of finite deformation and finite rotation multiplicative elastoplasticity, *Int. J. Solids Structures* **31**, 1063-1084.

Terzaghi, K. and Jelinek, R. (1954) *Theoretische Bodenmechanik*, Springer-Verlag, Berlin.

CONSOLIDATION OF METAL COATED FIBERS

H.N.G. Wadley and J. M. Kunze
University of Virginia
Thornton Hall
Charlottesville, VA 22903, USA

1. INTRODUCTION

The emergence of large (100-140 µm) diameter silicon carbide (SiC) monofilaments with very high elastic stiffness and fracture strength has led to the development of a new generation of metal matrix composites with axial Young's moduli above 200 GPa, ultimate tensile strengths of around 1.8 GPa and a 600°C creep rupture life at 1 GPa of several thousand hours or more. These exceptional properties are only obtained when the matrix is fully dense, the fibers remain straight/unbroken and only minimal reaction occurs between the matrix and fiber during processing. One promising approach for accomplishing this involves a two-step vapor deposition/hot consolidation process [Deve, 1995].

Metal coated fibers form a dense random packing (see Fig. 1) which can be consolidated by either hot isostatic pressing (HIP), vacuum hot pressing (VHP) or roll bonding. Densification of the layup must be accomplished while simultaneously minimizing both the bending of fibers (a result of the usually small number of fiber-fiber crossovers present in a random packing [Deve, 1995) and the thickness of reaction product layers at the fiber-matrix interface [Cantonwine, 1994].

Models have played a key role for optimizing the consolidation processes used to densify alloy powders [Arzt, 1983] and metal matrix monotapes [Elzey, 1993]. These studies have sought to identify the dominant mechanisms/mechanics of densification and to then construct models whose output is the relative density as a function of the process conditions (pressure/temperature/time). By adopting a micromechanics approach in which material parameters such as yield strength, creep exponent and boundary diffusivity are used to define the material response, the approach can be made generic and applied to any material whose properties are known or can be estimated. The development of an in situ eddy current sensor to monitor densification has subsequently provided a sensitive means for evaluating these models [Wadley, 1991].

To develop a densification model for the metal coated fiber case, we have conducted eddy current instrumented HIP consolidation experiments to measure the evolution of density during a pressure/temperature/time cycle. Interrupted test samples have also been metallographically analyzed to identify the dominant geometrical features of the densification process, and the mechanisms of matrix deformation at the test temperatures and pressures have been determined [Warren, 1995]. Together, these have been used to develop a model for densification that can be compared to sen-

N. A. Fleck and A. C. F. Cocks (eds.), IUTAM Symposium on Mechanics of Granular and Porous Materials, 389–402.
© 1997 Kluwer Academic Publishers.

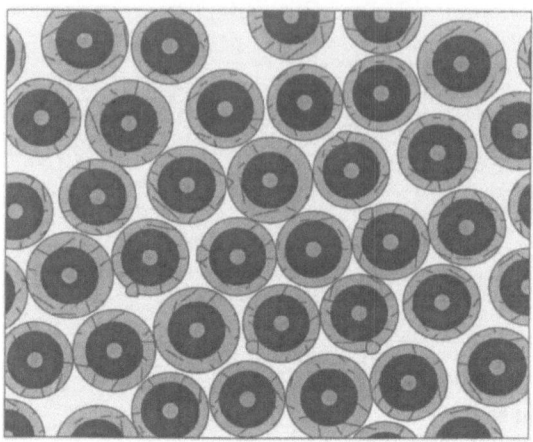

Figure 1   A schematic diagram of sputter coated silicon carbide monofilaments (mapped from a micrograph of a
                randomly packed array). The fibers are 100 μm in diameter.

sor data. The model addresses the important new effects of composite fiber-fiber contact deformation, the severe cusp-shape of the pores and the matrix microstructural evolution accompanying densification.

2.   CONSOLIDATION EXPERIMENTS

2.1   Materials

Tungsten cored Sigma 1240 SiC monofilaments were used for the consolidation study. The Sigma 1240 fiber has a nominal diameter of 100 μm and a carbon-TiB_2 duplex coating [Li, 1994]. The Ti-6Al-4V matrix was deposited on the fibers via a sputtering process at the 3M Company's Metal Matrix Composite Center (Mendota Heights, MN) using similar conditions to those reported by Warren et al [Warren, 1995]. The near line-of-sight sputtering process resulted in a slightly elliptical matrix coating. During deposition the temperature was kept low (~400°C) resulting in a ~30-100 nm matrix grain size. Upon cooling, the combination of an asymmetric coating and the thermal expansion mismatch between the fiber and matrix caused a large number of the coated fibers to bend. The resulting curvature made it difficult to produce a close packed fiber array and resulted in the random packing whose cross section is shown in Fig. 1.

2.2   Sample Geometry

Bundles of approximately 4000 of the matrix coated fibers were packed in cylindrical canisters, Fig. 2a. The canisters consisted of a tube and two endcaps of commercially pure (CP-2) titanium. The design of the canister allowed for a 51 mm long composite specimen. The effect of canister constraint was reduced by the use of two endcaps that had conically tapered ends as shown in Fig. 2a. With the taper, the greatest amount of the applied load was directed onto the coated fibers since the tube wall was unsupported near the composite specimen. The circular cross-section of the endcaps also helped to reduce canister buckling and to maintain a circular cross-section in the composite specimen region. Four longitudinal groves on the surface of each endcap were used to aid in evacuation of the canister prior to e-beam welding.

2.3 Density Sensor

An eddy current sensor, Fig. 2a, installed within the HIP chamber was used to measure the canister diameter during each consolidation experiment. Details concerning the operating principle of the sensor have been described elsewhere [Wadley, 1991, Wadley, 1988, Choi, 1993]. Briefly, the sensor used a two coil technique in which a primary coil was excited by a variable frequency (AC) current. This induced an electromagnetic field of variable frequency that linked the electrically conductive specimen resulting in eddy current induction in the sample and a perturbation to the electromagnetic field that would have existed in the sample's absence. A secondary coil, located near the sensor surface detected the field's perturbation. Due to the skin effect, the field perturbations at high frequencies were functions of only the geometry of the component (and not its electrical conductivity) and enabled deduction of the canister's diameter and thus its density, as the consolidation process evolved.

2.4 Experiments

An Asea Brown Boveri QIH-15 hot isostatic press equipped with a two-zone molybdenum furnace and four tungsten/rhenium thermocouples was used in this study. A total of five experiments were conducted. Here, we analyze in detail the representative case depicted in Fig. 2b.

2.5 Geometry Characterization

Metallographic analysis of partially consolidated samples indicated they contained no spherical voids. Rather, a distribution of cusped pores existed with a range of sizes and shapes. This observation has considerable significance since the densification rates of cusp shaped voids has been shown to be several times faster than their spherical void equivalents [Liu, 1994, Qian, 1995].

Early models for the densification of powders [Arzt, 1983] and more recent versions for plasma sprayed composite monotapes [Elzey, 1993] have all assumed that the voids present during final stage consolidation were spherical (the equilibrium shape). This assumption implies that surface diffusion along the void surface is fast enough to redistribute material being deposited from the contact by boundary diffusion. Recently, sintering models have been developed for quasi- and non-equilibrium shaped pores [Svoboda, 1995]. Using the results of Svoboda and Riedel [Svoboda, 1995] for the sintering of an hexagonal array of wires together with diffusion data for β-Ti, we estimate that it would take on the order of 10^6 hours for cusped pores to reach a quasi-equilibrium shape at 900°C because of both the low diffusivity of Ti-6Al-4V and the large radius of the coated fiber (~73 μm). Even a hexagonal array of wires of one-tenth the radius (7.3 μm) would still take on the order of 10^2 hours to reach a quasi-equilibrium shape at 900°C.

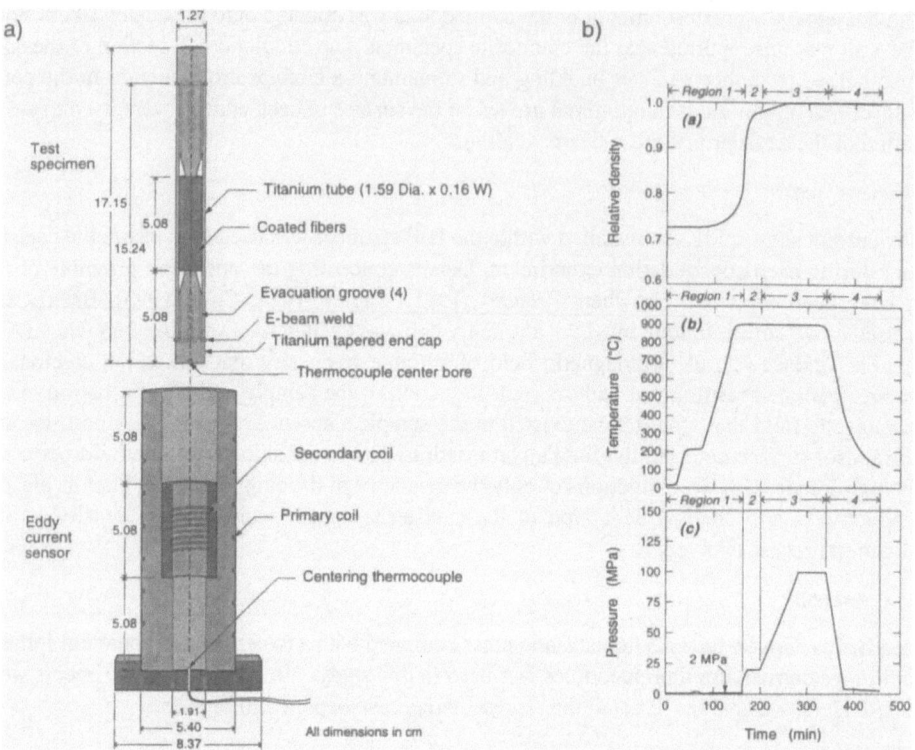

Figure 2   a) The cylindrical HIP canister assembly containing metal coated fibers and a schematic diagram of the
                eddy current sensor into which it was inserted. b) Time dependent data for a) density, b) temperature and
                c) pressure histories.

2.6   Microstructure Evolution

The extremely small (50-100nm) grain size of the PVD deposited Ti-6Al-4V matrix rapidly coarsens at the elevated temperature used for consolidation. In their study of the deformation of nomi-
nally identical PVD Ti-6Al-4V, Warren *et al* [Warren, 1995] determined the grain growth kinetics
to be well represented by an empirical expression of the form,

$$d = d_o + kt^a \tag{1}$$

where d is the grain size in microns after annealing for time, t, in seconds, and d_o, k, and a are temperature dependent parameters. Using Hilliard's single circle intercept method [ASTM, 1988], the matrix grain sizes were measured for the consolidated specimens. The results indicated that equation (1) again fitted the data with $k = 0.23$ and $a = 0.20$ at 900°C.

The final matrix grain size was less than 2 μm whereas conventionally processed Ti-6Al-4V typically has a grain size of 5 to 7 μm [Hamilton, 1982]. The result of this unusually small grain size is an enhanced superplastic contribution to deformation. Warren et al [Warren, 1995] have shown that within this temperature range, the behavior of the matrix is well represented by the Ashby-Verral model [Ashby, 1973] for superplastic flow. Since the Ashby-Verral strain rates are proportional to the inverse of the grain size squared, a grain size change from 100 to 2000 nm during consolidation leads to more than two orders of magnitude change in matrix creep rate (at a fixed stress) over the course of a consolidation cycle.

3. DENSIFICATION MODEL

3.1 Modelling Methodology

Here, we develop a model for the densification of metal coated fibers by hot isostatic pressing based upon the experimental observations above. To predict early stage (Stage I) densification, the model uses recently completed finite element analyses of coated fiber contacts [Wadley, 1996, Gampala, 1996] together with the measured coated fiber-fiber contact evolution. The densification at high relative density (Stage II) is modeled using a potential method which incorporates the important cusp-shape of the voids [Qian, 1995], Fig. 3. The enhanced superplastic properties of the nanocrystalline PVD matrix [Warren, 1995] and their evolution during the densification process are captured within the model by using the technique of Warren et al [Warren, 1996].

It has been shown that in principle, plasticity, diffusion accommodated grain sliding (DAGS) and power law creep can all contribute to the deformation of the matrix [Warren, 1995].

3.2 Stage I

3.2.1 Micromechanical Problem

If the Stage I representative volume element, Fig. 3, undergoes a displacement, h, in the vertical direction, then conservation of matrix volume gives an expression for the relative density;

$$\frac{h}{r_o} = \left(\frac{D}{D_o}\right)^{1/2} - 1 \qquad (2)$$

where r_o is the initial coated fiber radius, D_o is the initial relative density of the cell, and D is the density after suffering a densifying displacement, h.

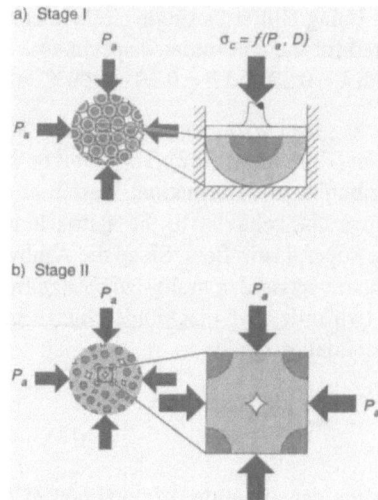

Figure 3   Representative volume elements for a) Stage 1 and b) Stage II.

By differentiating equation (2) with respect to time, the rate of densification under an imposed displacement rate, $\dot{h}$, is just;

$$\dot{D} = 2(DD_o)^{\frac{1}{2}} \frac{\dot{h}}{r_o} \tag{3}$$

In HIP consolidation, h and $\dot{h}$ are determined by the applied pressure (which must be supported by contacts) and by the inelastic deformation mechanism of the contacts. It is possible to capture the deformation resistance of a contact with a flow coefficient, F, representing the material's resistance to flow [Gampala, 1994, Wadley, 1996, Warren, 1996]. For a steady state creeping material, the relationship between the contact stress σ_c, and the strain rate at the contact may thus be written

$$\frac{\sigma_c}{\sigma_o} = F \left(\frac{\dot{\varepsilon}}{\dot{\varepsilon}_o} \right)^{1/n} \tag{4}$$

where F is the flow coefficient, $\dot{\varepsilon}$ is the overall strain rate perpendicular to the contact, while σ_o and $\dot{\varepsilon}_o$ are material dependent reference stress and reference strain rates respectively. F has been calculated for metal coated fibers as a function of strain for a wide range of n values and fiber volume fractions [Gampala, 1996].

The contact stress, σ_c, is just the ratio of the force supported by the contact, f_c, and the contact area, a_c;

$$\sigma_c = \frac{f_c}{a_c} \tag{5}$$

To relate this to density and applied pressure, we note that the average force acting on a contact for a random packing of coated fibers is determined by a force balance on the array.

$$\frac{f_c}{r_o L} = \frac{2\pi}{ZD} P_a \tag{6}$$

where P_a is the externally applied pressure, Z is the number of contacts for the given density, D, and L is the length of coated fiber and is similar to a result obtained by Molerus [Molerus, 1975] for powder consolidation. The relationship between the number of contacts, Z, and relative density, D, was experimentally determined to be

$$Z = 10.8D - 5.3 \tag{7}$$

Using equation (5) to eliminate f_c in [Duva, 1994], and substituting equation (4) for σ_c will give a relationship between the density, densification rate, applied pressure and the contact area, a_c, that can be used to predict density provided a_c and F are known.

Finite element analyses [Wadley, 1996, Gampala, 1996], have investigated the manner in which the area of a contact, a_c, increases with contact deformation. This is conveniently captured with an area coefficient, c, defined by

$$h = \frac{1}{2c^2} \frac{a^2}{r_o} \tag{8}$$

or

$$\frac{a}{r_o} = c\sqrt{2\frac{h}{r_o}} \tag{9}$$

where $a = a_c/2L$, h is the displacement normal to the contact, r_o is the initial radius of the matrix coating, and c is the contact area coefficient. An area coefficient of unity corresponds to a contact width equal to the length of a chord created by truncating a circle a distance, h, in from the radius. The area coefficient therefore represents the manner in which material is distributed after deformation at the contact. Previous research has shown that the area coefficient has a value of about 1.3 and is only a very weak function of normal displacement and so may be taken to be independent of the amount of deformation [Gampala, 1996].

3.2.2 Deformation Mechanisms

Finite element analyses have shown that the value of the flow coefficient, F, depends upon the operative mechanism and extent of contact deformation [Gampala, 1994, Wadley, 1996, Gampala, 1996]. A combination of DAGS and power law creep (PLC) well represents the superplastic behavior of PVD Ti-6Al-4V [Warren, 1995]. Ashby and Verrall [Ashby, 1973] have developed an expression for the strain rate of a material due to DAGS.

$$\dot{\varepsilon}_{DAGS} = \frac{100\Omega D_v}{kTd^2}\left(\sigma_c - \frac{0.72\Gamma}{d}\right)\left(1 + \frac{3.38\delta D_{gb}}{dD_v}\right) \tag{10}$$

where σ_c is the contact stress, T is the temperature, d is the grain size, Γ is the grain boundary energy, δ is the grain boundary width, Ω is atomic volume, D_v and D_{gb} are vacancy and grain boundary diffusion respectively, and k is Bolztmann's constant.

In the current notation of equation (4), it can be seen that $n = 1$ and the ratio of reference parameters, $\dot{\varepsilon}_o/\sigma_o$, may be defined as

$$\frac{\dot{\varepsilon}_o}{\sigma_o} = \frac{100\Omega D_v}{kTd^2}\left(1 + \frac{3.38\delta D_{gb}}{dD_v}\right) \tag{11}$$

Power Law Creep

The steady state strain rate of a power law creeping material under uniaxial loading may be described using the well known [Frost, 1982] power law creep relation,

$$\dot{\varepsilon}_{PLC} = A\frac{\mu b}{kT}\left(\frac{\sigma_c}{\mu}\right)^n D_v \tag{12}$$

where μ is the temperature dependent shear modulus, b is the burger's vector, A is a material property, and n is the creep exponent, also a material property.

Again, a comparison of equations (12) and to (4) reveals that

$$\frac{\dot{\varepsilon}_o}{\sigma_o^n} = A\frac{\mu b D_v}{kT\mu^n} \tag{13}$$

Plasticity

Plasticity is taken to be a time independent mechanism which yields an instantaneous strain when the load is increased. It is possible to analyze the perfectly plastic behavior by letting the power law exponent, n, approach infinity. Taking the limit of equation (4) as n approaches infinity results in

$$\sigma_c = F_p \sigma_y \tag{14}$$

where σ_o becomes the uniaxial yield strength, σ_y, of the matrix material, and F_p is the plastic flow coefficient.

Plasticity thus occurs when the stress at the contact, σ_c, equals or exceeds the matrix yield strength multiplied by a plastic flow coefficient, F_p, which represents the resistance of the matrix coating to plastic flow. F_p is about 2 for coated fibers deforming by plasticity. It is a weak function of both the extent of contact deformation and the volume fraction of fiber [Wadley, 1996, Gampala, 1996]. Using FEM results for F_p and c from [Wadley, 1996, Gampala, 1996], it is possible to determine the contact area as a function of the average normal displacement (or equivalent relative density) and thus obtain the density evolution.

3.3 Stage II

3.3.1 Micromechanical Problem

In Stage II, the representative volume elements are composed of individual cusp-shaped voids as shown in Fig. 3(b). Qian *et al* [Qian, 1995] have developed a strain rate potential for various types of multi-cornered cusp-shaped voids and have shown that they densify at much higher rates than the spherical voids of equivalent volume fraction used in powder densification models [Duva, 1994].

The Qian et al potential has been developed for cylindrical voids in a monolithic matrix and therefore does not account for the effect of the fiber on matrix flow. While this is likely to be small as the relative density approaches unity, at lower relative densities, the strain rate field can interact strongly with the fiber (especially as $n \to 1$), and finite element analyses [Gampala, 1996] indicate a significantly increased flow resistance. To approximately account for the presence of the fiber, we introduce a fiber restraint parameter, R, that enforces flow resistance continuity between the finite element (Stage I) contact analyses at the end of Stage I ($D = 0.9$) and the Qian et al potential at the same density. From the work of Qian et al, a mean stress-strain relation for a power law creeping monolithic body containing cusped voids can be obtained:

$$\frac{\sigma_m}{\sigma_o} = \left(\frac{3}{2}\right)^{1/n} \beta^{-(n+1)/2n} \left(\frac{\dot{\varepsilon}}{\dot{\varepsilon_o}}\right)^{1/n} \tag{15}$$

where for isostatic loading, $\sigma_m = P_a$ and

$$\beta = \frac{3\lambda}{1 - s\lambda} D^{(1-n)/(1+n)}$$

$$\lambda = \frac{1 - D^r}{s + D^r}$$

(16)

The coefficients r and s are dependent upon the pore shape and have been determined to be 5.375 and 1.6875 [Qian, 1995] for the type of voids observed above.

3.3.2 Deformation Mechanisms

DAGS and PLC

The uniaxial strain rate equations used in determining the densification rates due to both diffusion accommodated grain sliding and power law creep are the same as equations (10) and (12) respectively. From Qian *et al* [Qian, 1995], it is possible to relate the applied isostatic pressure, P_a, to the macroscopic strain rate, $\dot{\varepsilon}$.

$$\frac{P_a}{\sigma_o} = R\left(\frac{3}{2}\right)^{1/n} \beta^{-(n+1)/2n} \left(\frac{\dot{\varepsilon}}{\dot{\varepsilon}_o}\right)^{1/n}$$

(17)

where R is a density and fiber volume fraction dependent fiber restraint factor developed by [Kunze, 1996] and β is defined using equation (16).

Once the macroscopic strain rate, $\dot{\varepsilon}_{kk}$, has been determined using the strain rate potential, the densification rate is simply

$$\dot{D} = \dot{\varepsilon}_{kk} D$$

(18)

where compressive strains are taken to be positive. Under plane strain (in-plane isostatic stress) conditions $\dot{\varepsilon}_{kk} = 2\dot{\varepsilon}$.

Plasticity

The Stage II densification mechanism of plasticity has also been derived from the work of Qian *et al* [Qian, 1995]. We again determine the relationship for perfect plasticity by taking the limit of equation (17) as n approaches infinity.

$$P = R\frac{1}{\beta_p^{1/2}}\sigma_y$$

(19)

where $\beta_p = \left[\dfrac{3\lambda}{1-s\lambda}\right]\dfrac{1}{D}$ and λ, r, and s are as defined above. Inspection of equation (19) reveals that β_p is the only density dependent variable. Thus the densification due to plasticity may be computed in a manner similar to Stage I; by directly solving equation (19) for the density that results as a function of the applied pressure.

3.4 Matrix Constitutive Response

In order to effectively simulate the densification process, two factors related to the matrix deformation behavior must be addressed. First, Ti-6Al-4V is a two phase material within the temperature range of interest. Although each phase will deform with a different constitutive behavior, the effective response of the two phases can be estimated by partitioning the contact stress amongst both phases and invoking an isostrain-rate criterion [Warren, 1995]. Second, the PVD deposited Ti-6Al-4V matrix initially has an extremely small grain size that rapidly coarsens during consolidation. This is important because in equation (10), the DAGS strain rate is a function of $1/d^2$ making the densification rate a very sensitive function of the grain size. The coarsening rate (equation (1)) has been used to represent the grain growth kinetics of the matrix and other parameters of PVD Ti-6Al-4V alloy are found in [Warren, 1995].

4. MODEL RESULTS AND DISCUSSION

Simulations of the HIP experiments have been conducted using the model of Section 3. The strain rates for both of the time dependent mechanisms are computed at each density and summed resulting in the overall densification rate. The contribution of each mechanism is determined by the stress/temperature and is assumed to act independently of the others. As an input into the model, the volume fractions of fiber determined by an image analysis were used. In computing the flow and area coefficients and fiber restraint factor for volume fractions of fiber and creep exponents other than those for which values were explicitly determined, a linear interpolation method was used.

The simulation was initiated after thermal equilibrium had been achieved using, as the starting point, the calculated densities at which the pressure ramp began. The pressure in the model was ramped to the appropriate pressure hold at the same pressurization rate as in the experiments. During the process, the time dependent grain growth kinetics determined by equation (1) were used to calculate the average grain size in order to determine the contribution of the grain size dependent mechanisms using the models of [Warren, 1995]. The simulation ended when a density of 0.9995 was obtained or the elapsed time of the actual process cycle was reached.

The results of the process simulations are compared with the experimentally measured data in Fig. 4. The start of the simulation corresponded to a density of nearly 80% due to the low pressure densification which occurred as thermal equilibrium was attained. The model accurately predicts the

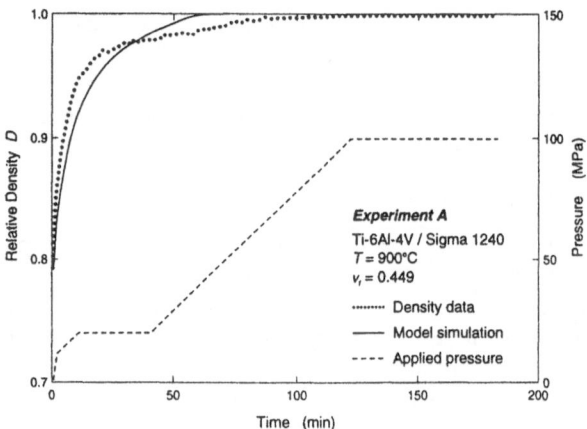

Figure 4·   Experimental densification profile and model simulation result.

early densification during Stage I. Experimentally, the densification rate remained quite high until about 96% and then decreases rapidly. The densification rate of the model was initially less than that of the data, but failed to "harden" in Stage II to the same degree as in the experiments. Experiments conducted at lower temperatures have yielded a similar level of agreement between sensed data and model prediction [Kunze, 1996].

Following the approach devised for powder consolidation [Arzt, 1983], the model can be used to investigate the effects of processing conditions by constructing HIP maps. In constructing the maps, two additional assumptions have been made. First, the pressure is raised to the setpoint at an infinite rate. That is to say, the external pressure is applied as a step function and plastic densification is computed first. The time dependent mechanisms then begin to contribute after this initial densification and the time contours are computed for these mechanisms. Secondly, it is assumed is that the grain size of the material does not change. In order to demonstrate the effect of varying the grain size, maps have been constructed for grain sizes of both 1 and 10 μm.

The HIP maps shown in Fig. 5 are for a 45% fiber volume fraction coated fiber array at 900°C and for grain sizes of either 1 or 10 μm. A 1 μm grain size is typical of PVD Ti-6Al-4V after briefly elevating it to a temperature of 900°C. The dominant time dependent densification mechanism is diffusion accommodated grain sliding. The DAGS mechanism remains dominant even with a grain size of 10 μm, however, with the larger grain size, the overall densification rate is greatly decreased as indicated by the longer time contours needed to reach a given density. The maps show that the anomalously high densification rates observed in experiments is in part a consequence of the fine grain size of the PVD alloy. The model predicts that the slower grain growth kinetics at lower temperatures leads to surprisingly high densification rates that can be exploited to reduce fiber-matrix reactions [Kunze, 1996].

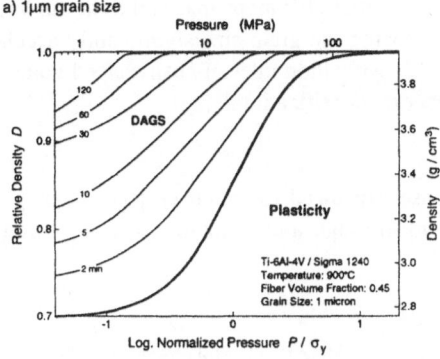

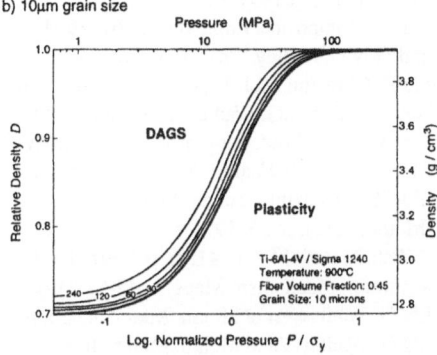

Figure 5   Hip maps of Ti-6Al-4V matrix and 45% fiber volume fraction at 900°C and grain sizes of a) 1 and b) 10 μm. Contours of constant time are shown

5. CONCLUSIONS

1. An eddy current sensor has been used to measure relative density during the HIP consolidation of SiC monofilaments PVD coated with Ti-6Al-4V. Very high densification rates are observed during early stages of consolidation.

2. A two stage model was developed for the HIP consolidation of these metal coated fibers. It incorporated the mechanics of contact deformation, the evolution of particle contacts, the cusp-shape of the final stage pores and the microstructural coarsening accompanying consolidation. The model successfully simulated HIP experiments.

3. The model was used to construct HIP maps that highlight the large effect of grain size upon the densification process. Reducing the grain size significantly accelerates the densification rate and, together with the higher densification rates of cusp-shaped voids, more than compensates for the retarding effect of the fiber on densification.

Acknowledgments

We are grateful to the Defense Advanced Research Projects Agency (Anna Tsao, Program Manager) for support of this work through a collaborative research program with ISI (Santa Clara, CA).

References

Annual Book of ASTM Standards, E112-88, ASTM (Philadelphia), p. 309, 1988.

E. Arzt, M.G. Ashby and K.E. Easterling, Met Trans. A, 14A, p. 211, 1983.

M.F. Ashby and R.A. Verall, Acta Metall., 21, 149, 1973.

P.E. Cantonwine and H.N.G. Wadley, Composites Eng., 4(1), p. 67, 1994.

B.W. Choi, K.P. Dharmasena, and H.N.G.Wadley, *Proc. Rev. Prog. in Quantitative Nondestructive Evaluation*, Ed. D.O. Thompson and D.E. Chimenti, Vol. 12, p. 1047, Plenum Press, New York, 1993.

H.E. Deve, D.M. Elzey, J.M. Warren and H.N.G. Wadley, Proc. 8th CIMTEC World Ceramic Congress and Forum on New Materials, "Consolidation Processing of PVD Ti-6Al-4V Coated SiC Fiber Composites," H.E. Deve, D. M. Elzey, J. Warren, H.N.G. Wadley, Advances in Science and Technology, 7, <u>Advanced Structural Fiber Composites</u>, Ed. P. Vincenzini, Techna (Florence, Italy), p. 313, 1995.

J.M. Duva and P. Crow, Mech. Mater., 17(1), p. 25, 1994.

D.M. Elzey and H.N.G. Wadley, Acta Metall. Mater., 41(8), p. 2297, 1993.

H.J. Frost and M.F. Ashby, Deformation Mechanism Maps, p. 43-52, Pergamon Press, Oxford, 1982.

R. Gampala, D.M. Elzey and H.N.G. Wadley, Acta Metall. Mater., 42(9), p. 3209, 1994.

R. Gampala, D.M. Elzey and H.N.G. Wadley, Acta Metall. Mater., In Preparation, 1996.

J.F. Groves, H.N.G. Wadley, "Functionally Graded Materials Synthesis Via Low Vacuum Directed Vapor Deposition," Composites Engineering, Submitted, 1996.

C.H. Hamilton, A.K. Gosh and M.W. Mahoney, in *Advanced Processing Methods for Titanium Alloys*, Ed. D.F. Hasson and C.H. Hamilton, TMS-AIME (Warrendale), p. 129, 1982.

J.M. Kunze, Mechanisms and Models of Metal Matrix Coated Fiber Densification, PhD Thesis (University of Virginia), 1996.

J.M. Kunze and H.N.G. Wadley, Acta Metall. Mater., Submitted, 1996.

J.H. Li, Z.X. Guo, P.S. Grant, M.L. Jenkins, B. Derby, and B. Cantor, *Composites*, 25 (9), p. 887, 1994.

Y. Liu, H.N.G. Wadley, and J.M. Duva, Acta Metall. Mater., 42 (7), p. 2247, 1994.

O. Molerus, Powder Technology, 12, p. 259, 1975.

Z. Qian, J.M. Duva, and H.N.G. Wadley, Acta Metall. Mater., Submitted October, 1995.

J. Svoboda and H. Riedel, Acta Metall. Mater., 43 (2), p. 499, 1995.

H.N.G. Wadley, A.H. Kahn, Y. Gefen, and M. Mester, *Proc. Rev. Prog. in Quantitative Nondestructive Evaluation*, Ed. D.O. Thompson and D.E. Chimenti, Vol 7B, p. 1589-1598, Plenum, New York, 1988.

H.N.G. Wadley et al, Acta Metall. Mater., 39, p. 979, 1991.

H.N.G. Wadley, T.S. Davison, and J.M. Kunze, Composites Engineering, In Press, 1996.

J. Warren, L.M. Hsiung and H.N.G. Wadley, Acta Metall. Mater., 43, p. 2773, 1995.

J. Warren, D.M. Elzey and H.N.G. Wadley, Acta Metall. Mater., In Press, 1996.

ANALYSIS FOR DIE COMPACTION OF METAL POWDERS

K.T. KIM, H.T. LEE AND J.S. KIM
Department of Mechanical Engineering
Pohang University of Science and Technology
Pohang, 790-784, Korea

AND

Y.S. KWON
Agency for Defense Development, Taejon, 305-600. Korea

Abstract. The effect of die wall friction on the densification of metal powders under die pressing is studied. The elastoplastic constitutive equations based on the yield functions by Fleck and Gurson and by Shima and Oyane were implemented into a finite element program to simulate die compaction processes. The friction coefficient between the powder and the die wall was determined from the relationship between the compaction pressure and the ejection pressure. Finite element calculations were compared with experimental data for densification and deformation of 316L stainless steel powder and D7 tool steel powder under single and double action die pressing.

1. INTRODUCTION

The main merit of powder metallurgy (P/M) is near net shape forming of mechanical parts (Lenel, 1980; German, 1984; Kuhn and Ferguson, 1990). Generally P/M forming process produces parts by cold compaction, sintering and finishing. The main manufacturing route in P/M products is via die compaction. P/M parts formed by die compaction have inhomogeneous density distribution because of the friction between the powder compact and the die wall and punches. The inhomogeneity in density may cause unwanted effects such as cracks and localized deformation in the P/M part (Lewis *et al.*, 1993; Gethin *et al.*, 1994). Process simulations by using a finite element analysis may be useful to control shape and density distribution of final P/M parts (Kuhn and Ferguson, 1990; Lewis *et al.*, 1993;

403

N. A. Fleck and A. C. F. Cocks (eds.),
IUTAM Symposium on Mechanics of Granular and Porous Materials, 403–413.
© 1997 *Kluwer Academic Publishers.*

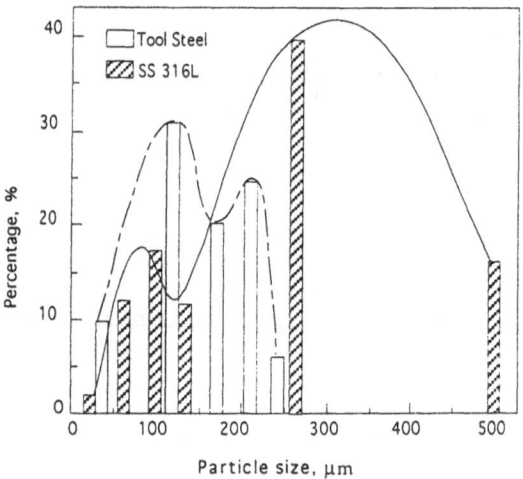

Figure 1. Comparison between the sieve analysis for 316L stainless steel powder and tool steel powder

Gethin *et al.*, 1994). The numerical modeling of the powder compaction process requires the appropriate constitutive models for densification of powder materials.

During the last two decades, several plastic yield functions for porous materials have been developed (e.g. Kuhn and Downey, 1971; Shima and Oyane, 1976; Gurson, 1977; Doraivelu *et al.*, 1984; Kim and Suh, 1989). Most of these yield functions, however, were developed for deformation of porous materials with high relative density. Recently, Fleck *et al.* (1992) proposed a constitutive model for densification of powder materials with low relative density.

In this study, we investigated densification and the effect of friction between powder and the die wall for stainless steel powder and tool steel powder under single and double action die pressing. Finite element calculations by using the yield function by Shima and Oyane (1976) were compared with experimental data for stainless steel powder and tool steel powder under cold die pressing. We also considered the yield functions by Fleck *et al.* (1992) for low density range and by Gurson (1977) for high density range to compare with the experimental data.

2. EXPERIMENTS

For theoretical analysis of powder compaction, Gas atomized spherical 316L stainless steel powder (Anval Co., Sweden) and D7 tool steel powder (Anval Co., Sweden) were used in this work. Fig. 1 shows that the sieve analysis results for stainless steel powder and tool steel powder from the manufacturer. Tool steel powder has a narrower particle size distribution than stainless steel powder.

For theoretical analysis of powder compaction, it is necessary to know the mechanical properties of the matrix material of a powder material, e.g., elastic modulus, yield strength and flow stress. We obtained these properties by uniaxial compression of matrix materials produced by hot isostatic pressing (HIP).

The compaction responses of powders at room temperature were investigated in a closed die under single and double action pressings. The die was made of cemented carbide having 20mm in inner diameter, 80mm in outer diameter and 70mm in height. The compacts were sintered without change in density to provide a strength to the compact. The relative density of the compact was measured by Archimedes' method.

3. ANALYSIS

Fleck *et al.* (1992) proposed a yield function Φ to predict plastic response of a porous material with low relative density ($D < 0.9$). Thus,

$$\Phi(\sigma, \bar{\varepsilon}_m^p, D) = \left(\frac{\sqrt{5}p}{3P_y}\right) + \left(\frac{5q}{18P_y} + \frac{2}{3}\right)^2 - 1 = 0 \tag{1}$$

$$P_y = 2.97 D^2 \frac{(D - D_0)}{(1 - D_0)}\sigma_m$$

where $p(= -\sigma_{kk}/3)$ is the hydrostatic stress, $q(= \sqrt{3\sigma'_{ij}\sigma'_{ij}/2})$ is the effective stress, and P_y is the yield strength under hydrostatic compression.

From plastic response of a hollow sphere model, Gurson (1977) proposed the yield function for a porous material with high relative density ($D > 0.9$). Thus,

$$\Phi(\sigma, \bar{\varepsilon}_m^p, D) = \left(\frac{q}{\sigma_m}\right)^2 + 2q_1(1 - D)\cosh\left(-\frac{3q_2 p}{2\sigma_m}\right)^2 \tag{2}$$

$$-1 - \{q_1(1 - D)\}^2 = 0$$

where the parameters q_1 and q_2 were introduced by Tvergaard (1981, 1982). When $q_1 = q_2 = 1$, (2) reduces to the yield function originally proposed by Gurson (1977). In this work, $q_1 = 1.25$ and $q_2 = 0.95$ were used as suggested by Becker *et al.* (1988).

To compare with experimental data, we used the yield function (1) by Fleck *et al.* (1992) for low density range ($D < 0.9$) and the yield function

(2) by Gurson (1977) and Tvergaard (1981, 1982) for high density range
($D > 0.9$) during powder compaction.

We also used the interpolation equation suggested by Fleck *et al.* (1992)
for the continuity of constitutive models in the density range where the
applied constitutive model differs. Thus,

$$\Phi = \left(\frac{D_2 - D}{D_2 - D_1}\right)\Phi_1 + \left(\frac{D - D_1}{D_2 - D_1}\right)\Phi_2 = 0 \tag{3}$$

where Φ_1 and Φ_2 are the yield functions of Fleck *et al.* (1992) and by
Gurson (1977), respectively. In (3), D_1 and D_2 are the relative densities at
which the transition begins and ends, respectively. We used $D_1 = 0.75$ and
$D_2 = 0.9$ (Fleck *et al.*, 1992).

Shima and Oyane (1976) proposed a plastic yield function for a porous
solid by determining the parameters in their yield function empirically from
uniaxial compression tests of sintered porous copper and iron. Thus,

$$\Phi(\sigma, \bar{\varepsilon}_m^p, D) = \left(\frac{q}{\sigma_m}\right)^2 + 2.49^2(1 - D)^{1.028}\left(\frac{p}{\sigma_m}\right)^2 - D^5 = 0 \tag{4}$$

The constitutive equations of porous material (Govindarajan, 1992;
Kwon *et al.*, 1996) were implemented into the user subroutine UMAT of
ABAQUS (1995) to analyze die compaction behaviors of metal powders.

4. RESULTS AND DISCUSSION

Mechanical properties of the matrix material of powder materials from
uniaxial compression test are as follows.

For 316L stainless steel powder :

$$\sigma_m = \begin{cases} 215 + 1322(\bar{\varepsilon}_m^p)^{0.678} \text{ MPa} & \text{for } \bar{\varepsilon}_m^p \leq 0.4 \\ 1306.5(\bar{\varepsilon}_m^p)^{0.353} \text{ MPa} & \text{for } \bar{\varepsilon}_m^p \geq 0.4 \end{cases}$$

$$\text{E=220GPa, G=86GPa}$$

For D7 tool steel powder :

$$\sigma_m = 574 + 1093(\bar{\varepsilon}_m^p)^{0.258} \text{ MPa}$$

$$\text{E=274GPa, G=103GPa}$$

Fig. 2 shows finite element meshes and boundary conditions for (a) double action pressing and (b) single action pressing of metal powders. One

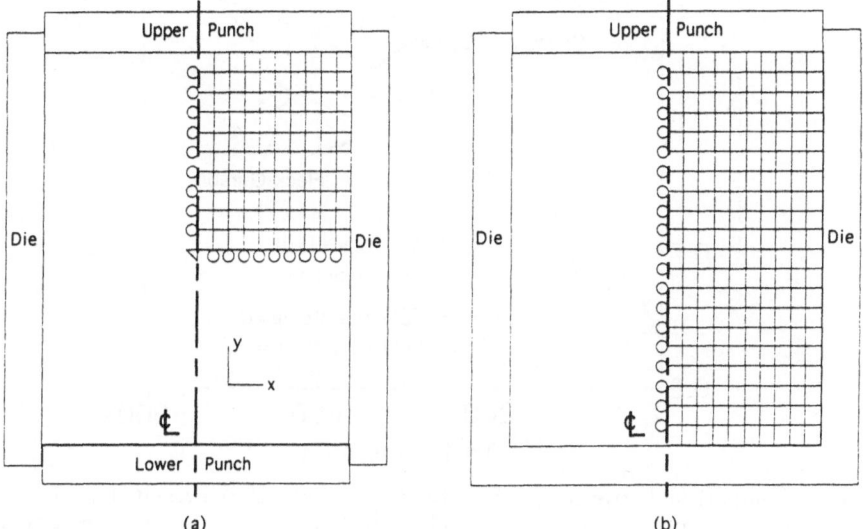

Figure 2.   Finite element meshes and boundary conditions for (a) double action pressing and (b) single action pressing of metal powders. The initial aspect ratio (height/diameter) of the powder compact is 1.24.

hundred four-node axisymmetric elements(CAX4) were used.

Fig. 3 and Fig. 4 show comparisons between experimental data and calculated results for the variation of relative density with axial stress during die compaction for stainless steel powder (Fig. 3) and tool steel powder (Fig. 4), respectively. The circular data points were obtained from single action die pressing of powders. The solid curves were obtained from finite element calculations without friction between the powder compact, the die wall and the punch. The dashed curves were obtained from finite element calculations by considering the friction effect with the friction coefficient $\mu = 0.2$.

In Fig. 3, the calculated curves by using the yield functions of Fleck *et al.* (1992) and of Gurson (1977) underestimate the experimental data for stainless steel powder. The calculated curves by using the yield function of Shima and Oyane (1976) show good agreements with the experimental data in the overall density range.

In Fig. 4, the calculated curves by using the yield function of Shima and Oyane (1976) overestimate the experimental data for tool steel powder. The calculated curves by using the yield function of Fleck *et al.* (1992) shows good agreements with the experimental data in low density range.

From the comparison between Fig. 3 and Fig. 4, the relative density of tool steel powder is lower than that of stainless steel powder under the same compaction pressure. It is also observed that the calculated curve with the

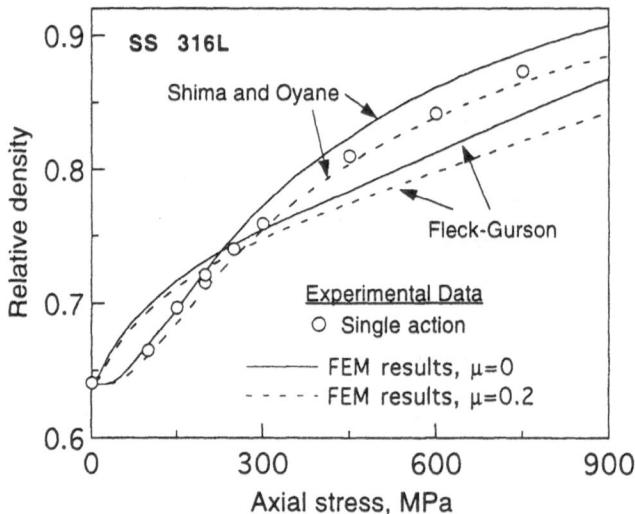

Figure 3. Comparison between experimental data and calculated results for the variation of relative density with axial stress of stainless steel powder during die compaction

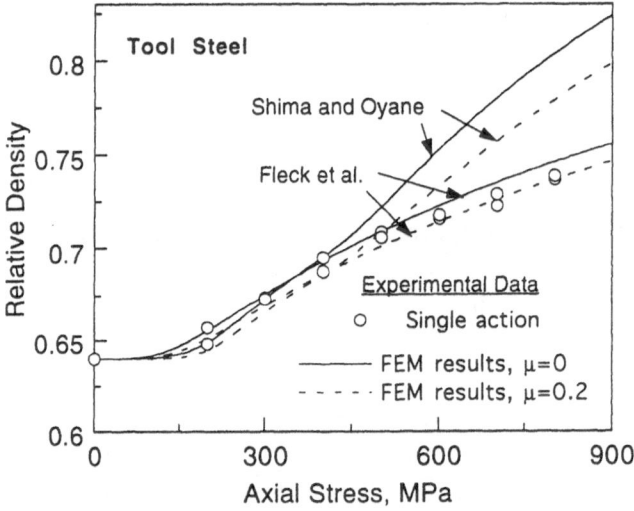

Figure 4. Comparison between experimental data and calculated results for the variation of relative density with axial stress of tool steel powder during die compaction

friction effect agrees better with the experimental data than that without the friction effect.

Fig. 5 shows experimental data and calculated results for the variation of ejection stress with axial stress of stainless steel powder during die compaction. The ejection force was measured by pushing the pressed compact out of the die. The circular data points were obtained for stainless steel powder under single action die pressing. The solid and dashed curves were obtained from finite element calculations by using the yield function of Shima and Oyane (1976) with various friction coefficients. Here, the solid

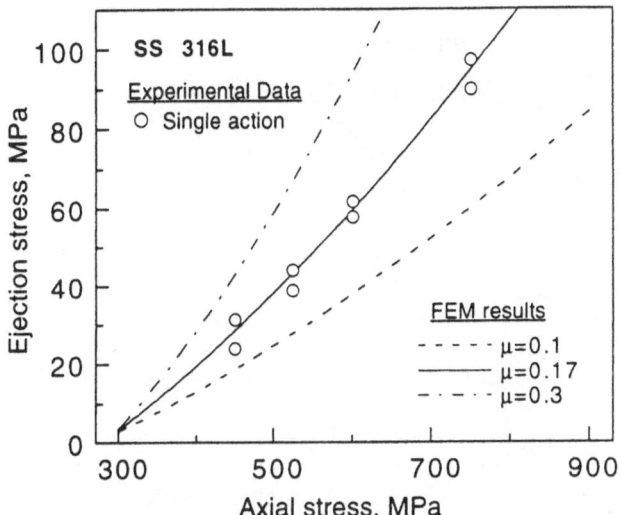

Figure 5.   Experimental data and calculated results for the variation of ejection stress with axial stress of stainless steel powder during die compaction

curve with the friction coefficient $\mu = 0.17$ shows very good agreement with experimental data. The friction coefficient $\mu = 0.17$ was found by a trial and error method.

Fig. 6 and Fig. 7 show comparisons between experimental data and theoretical predictions for the variation of relative density during die compaction with axial stress for stainless steel powder (Fig. 6) and tool steel powder (Fig. 7), respectively. The circular data points were obtained from double action pressing and the square data points from single action pressing. The solid curve was calculated by using the yield function of Shima and Oyane (1976) for stainless steel powder and Fleck *et al.* (1992) for tool steel powder without friction.

In Fig. 6, the dashed curve was obtained by using the yield function of Shima and Oyane (1976) under double action pressing with $\mu = 0.17$ and the dash-dotted curve under single action pressing with $\mu = 0.17$.

In Fig. 7, the dashed curve was obtained by using the yield function of Fleck *et al.* (1992) under double action pressing with $\mu = 0.16$ and the dash-dotted curve under single action pressing with $\mu = 0.16$. The friction coefficient $\mu = 0.16$ was found by a trial and error method with the relationship between the compacting pressure and relative density.

From Fig. 6 and Fig. 7, the calculated curves for both single and double action pressings with the friction effect agree very well with experimental data. The difference in relative density under double and single action pressings can be attributed to the effect of friction between the powder and the die wall.

Fig. 8 and Fig. 9 show comparisons between experimental data and

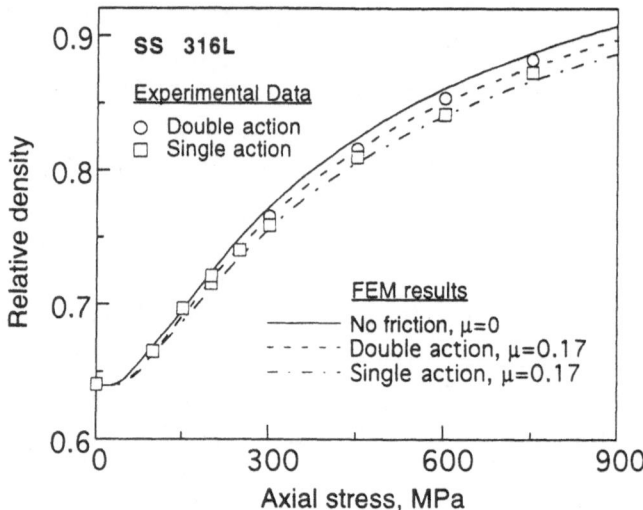

Figure 6. Comparison between experimental data and theoretical predictions for the variation of relative density with axial stress of stainless steel powder during die compaction

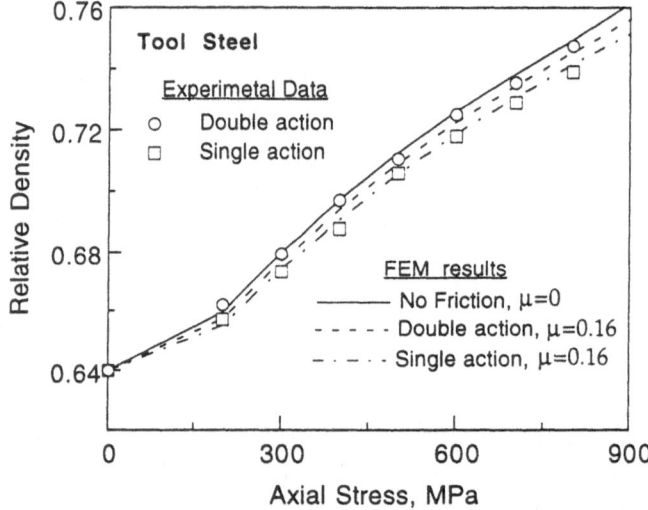

Figure 7. Comparison between experimental data and theoretical predictions for the variation of relative density with axial stress of tool steel powder during die compaction

finite element calculations for relative density contour plots of a stainless steel powder compact (Fig. 8) and a tool steel powder compact (Fig. 9) by single action pressing under axial stress of 600 MPa, respectively. It is understood that only right halves of the compacts are shown in Fig. 8 and Fig. 9. Fig. 8(a) and Fig. 9(a) were obtained from Rockwell hardness which can be converted into relative density by using the following relationships (Fleck *et al.*, 1992; Cho *et al.*, 1994).

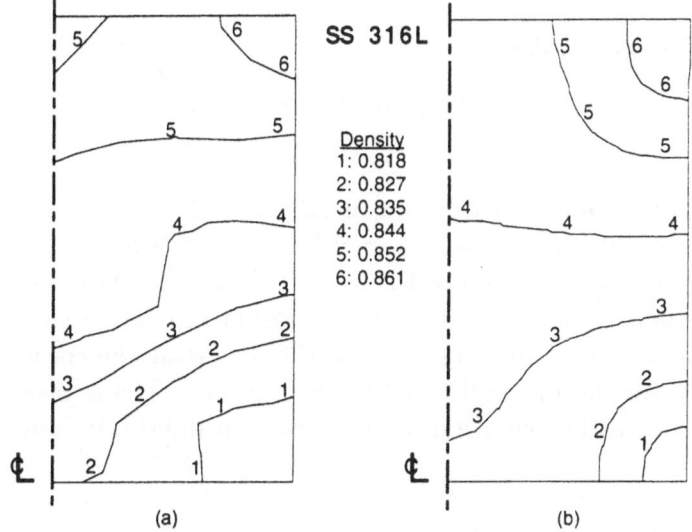

Figure 8. Comparison between (a) experimental data and (b) finite element calculations for relative density contour plots of a stainless steel powder compact by single action pressing under axial stress of 600MPa

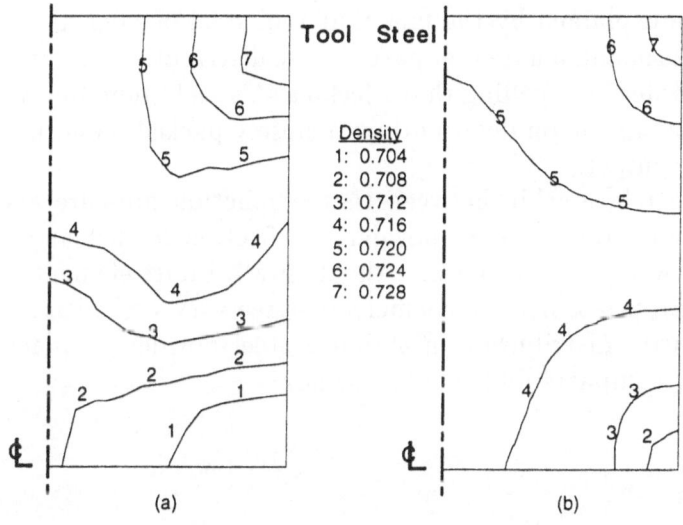

Figure 9. Comparison between (a) experimental data and (b) finite element calculations for relative density contour plots of a tool steel powder compact by single action pressing under axial stress of 600MPa.

For stainless steel powder :

$$D = 3.812 \times 10^{-3} HRB + 0.760 \qquad (5)$$

For tool steel powder :

$$D = 2.178 \times 10^{-3} HRB + 0.667 \qquad (6)$$

Fig. 8(b) was obtained from finite element calculations by using Shima and Oyane's yield function with $\mu = 0.17$ and Fig. 9(b) was obtained from Fleck et al.'s yield function with $\mu = 0.16$. In Fig. 8(a) and Fig. 9(a), the relative density is the highest at the corner of contact surface between the die wall and the upper punch, and the lowest at the corner of contact surface between the die wall and the lower punch. The agreement is good between theoretical calculations and experimental data, for both materials.

5. CONCLUSION

For stainless steel powder, the yield function by Shima and Oyane (1975) agrees better with experimental data than that by Fleck et al. (1992). For tool steel powder, however, the yield function by Fleck et al. (1992) agrees better with experimental data than that by Shima and Oyane (1975). This may be explained by the fact that tool steel powder has higher yield strength deform and a narrower particle size distribution compared to stainless steel powder by recalling that Fleck et al.'s yield function was obtained from the assumption on deformation of closely packed monosized spherical particles in contact.

From the relationship between the compaction pressure and the ejection pressure of stainless steel powder, the friction coefficient $\mu = 0.17$ was obtained between the powder and the die wall. Finite element calculations by using appropriate friction coefficients agree very well with experimental data for density distribution in stainless steel powder compacts and tool steel powder compacts under die pressing.

References

ABAQUS (1995) User's I and II Manual Version 5.3. Hibbitt, Karlsson and Sorensen.
Becker, R., Needleman, A., Richmond, O. and Tvergaard, V. (1988) Void Growth and Failure in Notched Bars, J. Mech. Phys. Solids, Vol. 36, pp. 317–351
Cho, H.K., Suh, J. and Kim, K.T. (1994) Densification of Porous Alloy Steel Preforms at High Temperature, Int. J. Mech. Sci., Vol. 36, pp. 317–328
Doraivelu, S.M., Gegel, H.L., Gunasekera, J.S., Malas, J.C. and Morgan, J.T. (1984) A New Yield Function for Compressible P/M Materials, Int. J. Mech. Sci., Vol. 26, pp. 527–535
Fleck, N.A., Kuhn, L.T. and McMeeking, R.M. (1992) Yielding of Metal Powder Bonded by Isolated Contacts, J. Mech. Phys. Solids, Vol. 40, pp. 1139–1162

Fleck, N.A., Otoyo, H. and Needleman, A. (1992) Indentation of Porous Solids, *Int. J. Solids Structures*, **Vol. 29**, pp. 1613–1636

German, R.M. (1984) *Powder Metallurgy Science*. Metal Powder Industries Federation, Princeton, New Jersey.

Gethin, D.T., Tran, V.D., Lewis, R.W. and Ariffin, A.K. (1994) An Investigation of Powder Compaction Processes, *Int. J. Powder Metall.*, **Vol. 30**, pp. 385–398

Govindarajan, R.M. (1992) *Deformation Processing of Porous Metals*. Ph. D. Thesis, University of Pennsylvania.

Gurson, A.L. (1977) Continuum Theory of Ductile Rupture by Void Nucleation and Growth - Part 1. Yield Criteria and Flow Rules for Porous Ductile Media, *J. Eng. Mat. Tech.*, **Vol. 99**, pp. 2–15

Kim, K.T. and Suh, J. (1989) Elastic-Plastic Strain Hardening Response of Porous Metals, *Int. J. Eng. Sci.*, **Vol. 27**, pp. 767–778

Kuhn, H.A. and Downey, C.L. (1971) Deformation Characteristics and Plasticity theory of Sintered Powder Metal Materials, *Int. J. Powder Metall.*, **Vol. 7**, pp. 15–25

Kuhn, H.A. and Ferguson, B.L. (1990) *Powder Forging*. Metal Powder Industries Federation, Princeton, New Jersey.

Kwon, Y.S., Lee, H.T. and Kim, K.T. (1996) Analysis for Cold Die Compaction of Stainless Steel Powder, submitted for publication.

Lenel, F.V. (1980) *Powder Metallurgy - Principles and Applications*. Metal Powder Industries Federation, Princeton, New Jersey.

Lewis, R.W., Jinka, A.G.K. and Gethin, D.T. (1993) Computer-Aided Simulation of Metal Powder Die Compaction Processes, *Powder Metall. Int.*, **Vol. 25**, pp. 287–293

Shima, S. and Oyane, M. (1976) Plasticity Theory for Porous Metals, *Int. J. Mech. Sci*, **Vol. 18**, pp. 285–291

Tvergaard, V. (1981) Influence of Voids on Shear Band Instabilities under Plane Strain Conditions, *Int. J. Fracture*, **Vol. 17**, pp. 389–407

Tvergaard, V. (1982) On Localization in Ductile Materials Containing Spherical Voids, *Int. J. Fracture*, **Vol. 18**, pp. 237–252

THE MODELLING OF THE INFLUENCE OF WALL FRICTION ON THE CHARACTERISTICS OF PRESSED CERAMIC PARTS

S.A. WATSON
M.J. ADAMS
Unilever Research
Port Sunlight Laboratory
Wirral L63 3JW

S.L. ROUGH
B.J. BRISCOE
T. PAPATHANASIOU
Chemical Engineering Department
Imperial College
London SW7 2AZ

1. Introduction

The manufacture of ceramic components will often involve the die compaction of a powder to form a durable porous body, called a green body, and subsequent densification of the green body by sintering. Invariably, due to a non-uniform green density, the sintered component is not a geometrically scaled replica of the die geometry. Where small dimensional tolerances are required, it may be necessary to finish the part by machining, e.g. in the manufacture of uranium dioxide pellets. The non-uniform green density is a result of high stress gradients that are primarily a consequence of the friction at the powder-die interface. However, additional important factors are the bulk constitutive behaviour of the powder, which is determined by the mechanical properties of the particles and the particle-particle interactions, and the geometry of the die.

There is a requirement to develop procedures that are capable of predicting the final geometry of sintered parts. It may then be possible to design a process where a near-net shape is achieved, thus eliminating the need for further finishing operations. A key step is the development of a capability for predicting the influence of the aforementioned variables on the density distribution of the green body. The scope for an analytical approach is limited, for example, the classical solution for the uniaxial compaction of powders as proposed by Janssen, and modified by Walker (Briscoe et al. (1987)), enables a first order estimate of the axial stress distribution in a cylindrical column of powder, from which the average axial slice density can be predicted. However, even for this relatively simple geometry, experimental measurements have shown that the density distribution is complex, varying both in the axial and lateral directions (Aydin et

415

N. A. Fleck and A. C. F. Cocks (eds.),
IUTAM Symposium on Mechanics of Granular and Porous Materials, 415–426.
© 1997 *Kluwer Academic Publishers.*

al. (1994)). Recently, attention has focused on the application of numerical modelling techniques. A potential approach is to model the powder as an assembly of individual agglomerates, or particles, using either granular dynamics (Thornton et al. (1987)), or discrete elements (Munjiza et al. (1995)). However, this approach is computationally expensive, and therefore, not practicable for modelling realistic compaction geometries at this time. An alternative is to use the finite element method, modelling the powder as a continuum using constitutive equations such as those originally developed in the field of soil mechanics. The compaction of a number of ceramic powders has been modelled using this technique (Aydin et al. (1994,1996),Gethin et al. (1994), Nagao et al. (1987), Trasorras et al. (1989)), and although further work is required in order to establish appropriate friction models and constitutive laws, the results suggest that there is considerable potential with this approach.

In the current paper, an experimental and computational investigation of the influence of die wall friction on the uniaxial compaction of alumina is reported. The friction coefficient at the die wall was derived from experimental uniaxial stress transmission data as a basis for modelling the compaction of the alumina using the finite element method. The bulk constitutive behaviour of the alumina was modelled using both a nonlinear elasticity model (Mroz et al. (1980)) and a cap model (DiMaggio et al. (1971)). The influence of friction on the computed stress transmission and green density was investigated, the resulting data being compared with those obtained from experiment.

2. Experimental Procedure

The experimental configuration is shown in figure 1. Using an Instron universal testing machine, a commercial spray dried alumina powder (Martinswerk GmbH, 99.7% Al_2O_3, 90% α-Al_2O_3), with a typical agglomerate size of 150μm, was compacted uniaxially in a 13mm diameter hardened steel die by a single side-acting punch travelling at a constant velocity of 5mm/min. The positioning of force transducers in line with the upper and lower punches enabled the force applied by the moving upper punch, and the transmitted force sensed by the stationary

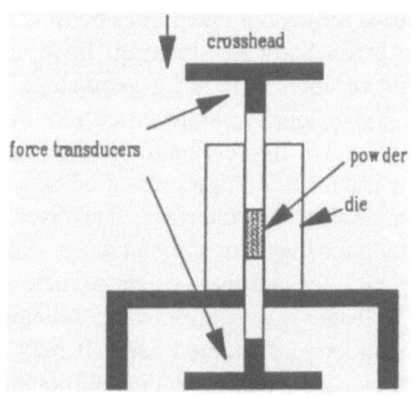

Figure 1.   The experimental arrangement.

lower punch, to be recorded continuously during compaction. The displacement of the upper punch was monitored using a displacement transducer.

The effect of the final compact aspect ratio (final compact height/die diameter) and the state of wall lubrication on the stress transmission ratio (transmitted pressure/applied pressure) was investigated. In those cases where the die wall was lubricated, zinc stearate (FSA Laboratory supplies) with a mean particle size of 3μm was used.

3. Experimental Results

The stress transmission ratio and mean green density for compacts of various aspect ratios, pressed to an ultimate punch pressure of 70MPa, under lubricated and unlubricated die wall conditions, are shown in table 1. It may be seen that for a particular aspect ratio, the stress transmission is greater in the case of a lubricated die. In this case, as a result of a smaller die wall friction, less of the applied force is transmitted to the die wall. Consequently, the overall green density is higher, and as a result of the smaller stress gradients, the internal

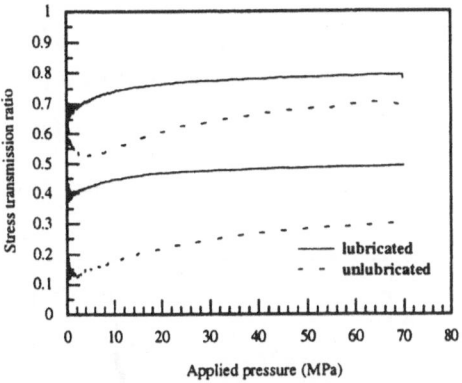

Figure 2. Measured stress transmission ratios as a function of applied pressure.

density distribution will be more homogeneous. In addition, the effect of decreasing the aspect ratio is to reduce the frictional effects. Consequently, as the aspect ratio is decreased, the stress transmission and overall density increases, and therefore the density distribution is more homogeneous.

Typical plots of the stress transmission ratio as a function of the applied pressure are shown in figure 2. It may be observed that this ratio is a function of the applied pressure. This is a result of the complex interactions between the die geometry, frictional properties of the interface, the bulk constitutive behaviour, and the way in which they determine the resultant shear stress at the die wall. This is discussed further in the following sections.

TABLE 1. Values of stress transmission ratio and mean green density at an applied pressure of 70MPa.

State of lubrication	Final aspect ratio (H/D)	Stress transmission ratio	Mean density (kg/m³)
Lubricated	0.5	0.79	2358
	1.0	0.66	2333
	1.5	0.49	2326
Unlubricated	0.5	0.70	2334
	1.0	0.48	2312
	1.5	0.30	2277

The nonlinear elasticity model used in the finite element analysis requires that the interrelationship between the axial stress, σ_{zz}, and axial natural strain, ε_{zz}, be determined. Figure 3 shows experimental data for the compaction of a green body with an aspect ratio of 0.1. A small aspect ratio was used in order to minimise wall friction effects, the resulting stress transmission ratio being close to unity. The data were adequately described by the following empirical relationships:

$$\sigma_{zz} = 3.6374e-2 + 5.0365\varepsilon_{zz} \qquad\qquad \varepsilon_{zz} < 0.0932 \qquad\qquad (1a)$$

$$\sigma_{zz} = 0.24945 \times 10^{3.0821\varepsilon_{zz}} \qquad\qquad 0.0932 \le \varepsilon_{zz} < 0.5415 \qquad\qquad (1b)$$

$$\sigma_{zz} = 2.1738e-3 \times 10^{7.1307\varepsilon_{zz}} \qquad\qquad \varepsilon_{zz} \ge 0.5415 \qquad\qquad (1c)$$

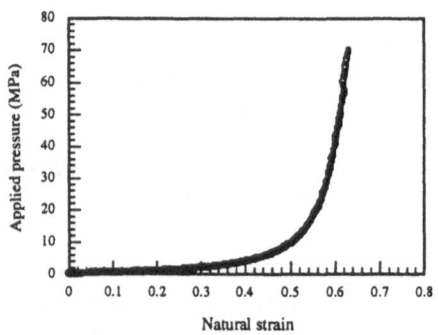

Figure 3. Fit to experimental applied pressure-natural strain data, aspect ratio = 0.1.

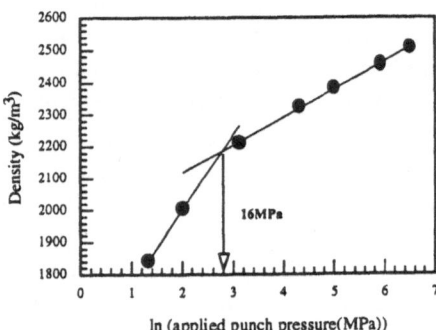

Figure 4. Compact density as a function of applied pressure, aspect ratio = 0.1.

In order to determine analytically the mean density distribution along the axis of a compact and to obtain the compact density distribution from the stress field computed by the nonlinear elastic finite element model, a density-axial stress relationship was determined. Over a range of applied pressures, compacts of a small aspect ratio of 0.1 were pressed in order to minimise friction effects, thus enabling the production of a homogeneous density distribution. Compact density is plotted as a function of the natural logarithm of the applied pressure in figure 4. This figure exhibits two distinct regions; after a joining pressure of approximately 16MPa, which represents the elimination of pores between primary particles (Lukasiewicz et al (1978)), the compaction behaviour can be described by the following empirical equation,

$$\rho = A + B\ln\sigma_{zz} \qquad\qquad (2)$$

where ρ is the compact density in kg/m^3, σ_{zz} is the axial stress(=applied punch pressure) in MPa, and A and B are empirical parameters determined from the experimental curve as 1943.9kg/m^3 and 87.1kg/m^3 respectively.

4.   Analysis of Results

The compaction process can be described by the Janssen-Walker method of differential "slices" (Ozkan and Briscoe (1987)). Considering the equilibrium of a "slice", and integrating along the axial coordinate, the ratio of the applied pressure, Q_a, and the mean axial stress, $\bar{\sigma}_{zz}$, at the axial coordinate z is given by

$$\frac{\bar{\sigma}_{zz}}{Q_a} = \exp^{-Cz} \tag{3}$$

The parameter C is given by,

$$C = \frac{4\mu_w K_w \Psi}{D} \tag{4}$$

where D is the diameter of the die, μ_w is the wall friction coefficient, Ψ is a stress distribution factor relating the axial stress at the wall to the mean axial stress, and K_w is the ratio of the radial and axial wall stresses such that

$$\sigma_{rr}\big|_w = K_w \sigma_{zz}\big|_w = K_w \Psi \bar{\sigma}_{zz} \tag{5}$$

The stress transmission ratio for a compact of height, H, is obtained by letting $z=H$, and thus $\bar{\sigma}_{zz} = Q_t$, the transmitted axial stress. Using the Janssen-Walker solution, and substituting for σ_{zz} in equation 2, it is possible to determine analytically the axial distribution of the compact density, or the density of a slice.

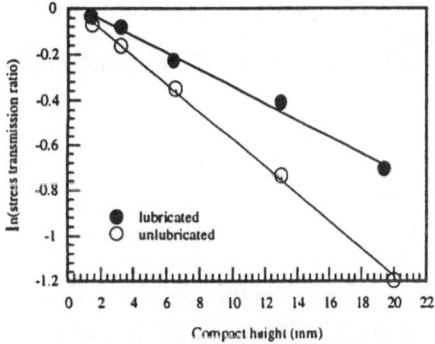

Figure 5. ln(stress transmission ratio)-compact height for compacts pressed to 70MPa.

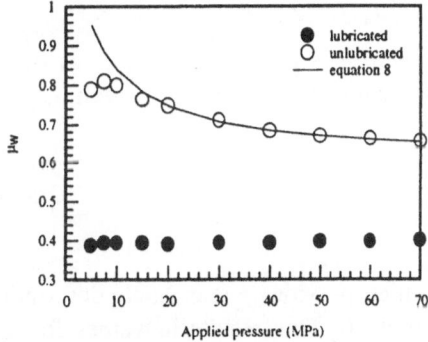

Figure 6.   Friction coefficient as a function of applied pressure determined using Janssen-Walker.

Assuming C to be a constant, the stress transmission ratio will be a linear function of the compact height. Such an interrelationship is shown in figure 5 for a number of compacts, of various aspect ratios, compacted by the same applied pressure of 70MPa; the values of C were obtained from the gradients of these data. The value of

Ψ for many granular materials is assumed to be unity, whilst a reasonable estimate of K_w is considered to be approximately 0.3 (Briscoe et al. (1991)). On this basis, the values of the wall friction coefficients for the lubricated and unlubricated die wall were calculated to be 0.399 and 0.654 respectively using equation 4.

The above procedure was repeated for compacts pressed to a range of applied pressures. For the lubricated case, the value of C was found to be not significantly sensitive to the applied pressure. Therefore, again assuming $K_w=0.3$, a relatively constant value of 0.394 was obtained for the friction coefficient, see figure 6. However, in the case of the unlubricated die, C was found to be very sensitive to the applied compaction pressure. A plot of the calculated friction coefficient as a function of the applied pressure indicates that the friction coefficient reduces as the applied pressure increases, see figure 6.

The observed pressure dependency of the friction coefficient is consistent with the adhesion model of friction (Bowden et al. (1955)). Using this approach, the coefficient of wall friction may be written as (Briscoe et al. (1987)),

$$\mu_w = \frac{\xi \tau_o}{\sigma_{rr}|_w} + \alpha \tag{6}$$

where ξ is the ratio of the true to the apparent contact areas, and τ_o and α are characteristic constants for a given interface. If it is assumed that ξ is a function of the radial wall stress,

$$\xi = \frac{\sigma_{rr}|_w}{b + \sigma_{rr}|_w} \tag{7}$$

where b is a constant, so that as $\sigma_{rr}|_w \rightarrow \infty$, $\xi \rightarrow 1$; and if $K_w = 0.3$, then substituting equations 5 and 7 into equation 6,

$$\mu_w = \frac{\tau_o}{b + 0.3\sigma_{zz}} + \alpha \tag{8}$$

which provides a reasonable description of the data for the unlubricated die shown in figure 6. This results in values for τ_o, b, and α of 1.047MPa, 1.522MPa and 0.607 respectively. Therefore, as the applied pressure increases, μ_w will decrease and reach an asymptotic value equal to α. The constant value of μ_w in the lubricated case can be explained by assuming that $\xi \tau_o / \sigma_{rr} \ll \alpha$ and hence $\mu_w \approx \alpha = 0.394$.

5.   Numerical Simulation of Compaction

The compaction process was modelled using the finite element method. Two commercial finite element codes were used, LUSAS (FEA Ltd, UK) and Elfen (Rockfield Software, UK). The nonlinear elastic model was implemented in LUSAS, and the cap model in Elfen. The alumina powder was modelled as a continuum using four-noded axisymmetric elements. The die wall and the upper and lower punches were modelled as rigid bodies. Contact at the powder/die, powder/punch, and punch/die interfaces was modelled using slidelines. For the lubricated case, a constant wall friction coefficient of 0.394 was used, see figure 6. For the unlubricated case, the wall friction coefficient was described by equation 8. The upper punch displacement was applied in increments until the applied displacement was equal to the maximum displacement (at the maximum applied pressure of 70MPa) obtained during the corresponding experiment.

5.1 . CONSTITUTIVE MODELS

5.1.1 . *The Nonlinear Elasticity Model*
This model, based on the work of Mroz (1980), was adopted by Aydin et al. (1994) to predict density distributions in alumina compacts. Although no computed applied pressure or stress transmission data was presented, computed density distributions showed an encouraging similarity with those obtained from experiment.

In this model, the incremental stress-strain relationship is given by,

$$d\sigma = D_t \, d\varepsilon \qquad (9)$$

where $d\sigma$ is the increment in the stress components, $d\varepsilon$ is the increment in the strain components, and D_t is the tangential stiffness matrix. The elements of the matrix, D_{ij} are functions of the tangent modulus, E, and Poisson's ratio, ν. In general, E and ν are functions of the invariants of the strain tensor, I_1', I_2', and I_3' (Chen et al. (1990)). The dependence of E on these invariants can be determined using data obtained from uniaxial compaction experiments. In such a configuration, assuming negligible friction effects, as in the case of small aspect ratios, the principal strains, ε_1, ε_2, and ε_3, coincide with the axial, radial, and tangential strains. Therefore, $\varepsilon_1 = \varepsilon_{rr} = \varepsilon_2 = \varepsilon_\theta = 0$, and ε_3 is equal to the compressive axial strain ε_{zz}, thus, $I_2' = I_3' = 0$, and $I_1' = \varepsilon_{zz}$. Assuming that the constitutive behaviour of the alumina is described by a unique relationship between the axial stress and axial strain, E as a function of ε_{zz} is obtained by differentiation of equations 1a-c

$$E = \frac{d\sigma_{zz}}{d\varepsilon_{zz}} = 5.0365 \qquad\qquad \varepsilon_{zz} < 0.0932 \qquad (10a)$$

$$\frac{d\sigma_{zz}}{d\varepsilon_{zz}} = 3.0821\ln(10)\times 0.24945\times 10^{3.0821\varepsilon_{zz}} \quad 0.0932 \le \varepsilon_{zz} < 0.5415 \qquad (10b)$$

$$\frac{d\sigma_{zz}}{d\varepsilon_{zz}} = 7.1307\ln(10)\times 2.1738e{-}3\times 10^{7.1307\varepsilon_{zz}} \quad \varepsilon_{zz} \ge 0.5415 \qquad (10c)$$

This model was implemented assuming v to be constant with a value of 0.3 (Aydin et al. (1994)).

5.1.2. The Cap Model

This model is derived from classical elasto-plasticity theory and was originally proposed in the form presented here by DiMaggio and Sandler (1971). Although originally intended as a model for geomaterials, it has been applied, with some success, to the prediction of the density distributions in a number of ceramic compacts, these including tungsten carbide (Trasorras et al. (1989)), an iron based powder (Gethin et al. (1994)), and alumina (Aydin et al (1996) compacts. Whilst Gethin et al. (1994)

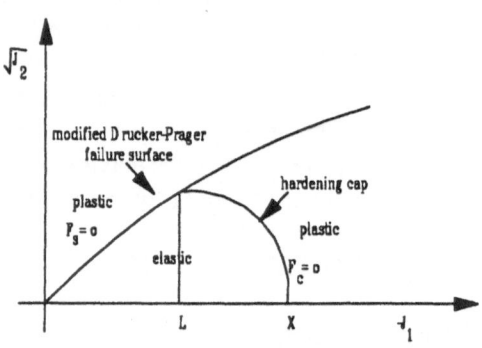

Figure 7. The cap model failure surface.

compared normalised computed and applied transmitted pressures with experimental data, these showing similar behaviour, there appears to be few instances in the literature where computed and experimentally determined tooling pressures/forces are compared.

TABLE 2. Parameters for the cap model

Elastic constants				
E = 303MPa	v = 0.07			
Cap parameters				
A = 0.0386MPa	B = 0.0MPa	C = 0.0MPa	Θ = - 0.1194	R = 1
Hardening function				
X = -0.561x10$^{-4.6695\varepsilon_v^p}$				

The cap model, shown in figure 7, is assumed to be isotropic and consists of two parts. The first part is a non-hardening modified Drucker-Prager failure surface, described by,

$$F_s = \sqrt{J_2} - \left(A - C \exp^{BJ_1} + \Theta J_1 \right) = 0 \qquad (11)$$

where J_1 is the first stress invariant, J_2 is the second invariant of the deviatoric stress tensor, and A, B, C, and Θ are material parameters. The second part of the model is a strain-hardening elliptic cap, described by,

$$F_c = \sqrt{J_2} - \frac{1}{R} \sqrt{\left[(X - L)^2 - (J_1 - L)^2 \right]} = 0 \qquad (12)$$

where L is the value of J_1 at the intersection of the cap with the failure surface, and X is the point of intersection of the cap with the J_1 axis. The position of the cap, X, is considered to be a function of the volumetric plastic strain, ε_v. R is a material parameter defining the shape of the cap (for R=1, the cap is circular).

If the state of stress is within the envelope described by the functions $F_s=0$ and $F_c=0$, then the material is assumed to be linear elastic, if not, then the material undergoes plastic deformation. Finally, in the present formulation, an associated flow rule was assumed such that the increment of the plastic strain is given by,

$$d\varepsilon_p = d\lambda \frac{dF}{d\sigma} \qquad (13)$$

where $d\lambda$ is the proportionality constant.

The material parameters, and the cap hardening function, were obtained using a triaxial test method (Chen et al. (1990)) and are given in table 3.

5.2. NUMERICAL RESULTS

Computed applied pressures and transmission ratios for green compacts of aspect ratio 0.5 and 1.5 are given in table 3. Considering those computations performed using the nonlinear elasticity model, it can be seen that for the lower aspect ratio of 0.5, both the applied pressure and the transmission ratio compare well with the experimental values. In the case of the higher aspect ratio, the difference between the computed and experimental applied pressure becomes more significant. This is especially so in the case of an unlubricated die, where the computed applied pressure, depending on the definition of the friction coefficient, is 50-60% greater than the experimental value of 70MPa. As expected from the experimental data, because of the increasing influence of the die-wall friction, the computed transmission ratios for this higher aspect ratio are smaller and also compare favourably with the experimental values. Using the cap model, in the case of a lubricated die, it can be seen that the applied pressure is significantly underestimated for both aspect ratios. Similarly, the computed stress transmission ratios compare poorly with the experimental data. However, the significant reduction in measured transmission ratio resulting from increasing the aspect ratio is reflected in the computational results.

TABLE 3 Computed applied pressures and stress transmission ratios using the nonlinear elasticity model and the cap model (experimental applied pressure = 70MPa).

State of lubrication	Aspect ratio	Applied stress (MPa)	Computed stress transmission ratio	Experimental stress transmission ratio
Lubricated	0.5	65.29	0.838	0.79
		40.40*	0.707	
	1.5	82.83	0.496	0.49
		56.50*	0.270	
Unlubricated	0.5	66.86	0.740	0.70
	1.5	104.92	0.276	0.30
		111.35+	0.227	

* Cap model
+ Constant μ_w = 0.7727

Figure 8 shows the measured density distribution inside an alumina green of aspect ratio 1.5, obtained by (Aydin et al. (1994)) using the "lead-shot" tracer method. Taking advantage of axial symmetry, half of the compact is shown. The salient features are that moving from the top to the bottom of the green body, there is a decrease in the mean density of a slice. Additionally, at the top corner and at the bottom of the compact, on the centre-line, regions of relatively high density exist. Also, at the bottom corner of the compact, and at the top, on the centre-line, there are regions of relatively low density. Figure 9 shows density distributions computed using both the nonlinear elasticity and cap models, the aspect ratio of the compact being 1.5, and both the lubricated and unlubricated cases being considered. As observed experimentally, for all the

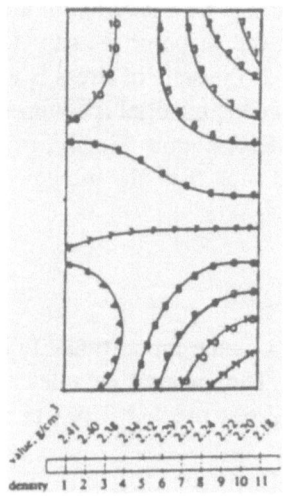

Figure 8. Measured density distribution in an alumina compact, aspect ratio = 1.5 (Aydin et al. (1994)).

computational models, it can be seen that the density of the green body decreases with depth, and that high and low density regions are predicted at the top and bottom corners of the die, respectively. The effect of increasing the die wall friction coefficient is to increase the inhomogeneity of the density distribution. Comparing the data obtained using the cap model with those obtained using the nonlinear elasticity model, it may be seen that the cap model predicts a high density region at the top of the compact, on the centre-line, that is not predicted by the nonlinear elasticity model. Neither models predict the high and low density spots located on the centre-line observed from experiment. It is also noted that the variation in density between the top and the bottom of the compact is considerably greater in the case of the cap model.

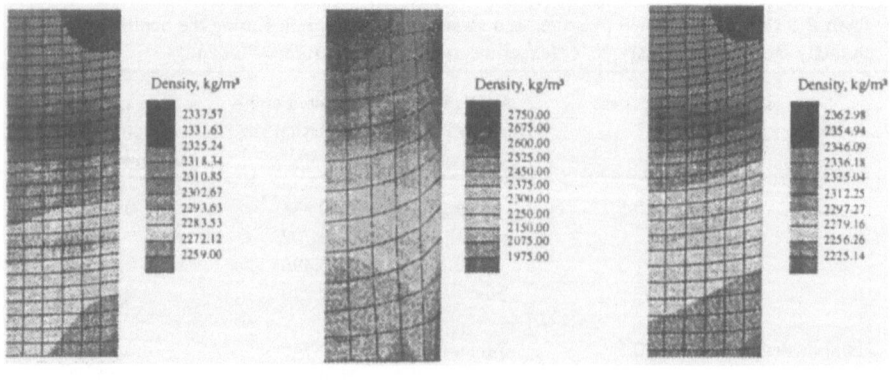

(a) Lubricated nonlinear elasticity

(b) Lubricated cap model

(c) Unlubricated nonlinear elasticity

Figure 9. Computed density distributions for compacts of aspect ratio = 1.5.

6. Discussion and Conclusions

The experimental and computational investigations carried out here exemplify the importance of die-wall friction in determining the stress gradients and, consequently, the density distribution in a green body. In addition to potentially large axial density gradients, significant gradients can be seen to exist in the lateral direction, particularly at the top and bottom of the green body. This observation underlines the importance of the approximations introduced in the Janssen-Walker model employed here to derive the coefficients of wall friction. However, the comparison between the experimental and computational results is encouraging. In order to improve the computational models, development work on the friction models, the constitutive laws, and the experimental procedures for determining the material parameters is required.

The over estimation of the applied pressure for the 1.5 aspect ratio green body suggests that the friction coefficients used in the analyses are too high. This is likely to be a result of the use of the Janssen-Walker analysis in determining the friction coefficients. This can be illustrated by examining the assumption that the ratio of the lateral to the axial stress at the die wall is constant and approximately equal to 0.3. The parameter is plotted as a function of the applied pressure at the top and bottom corner of the die in figure 10 using data originating from the finite element analysis involving the cap model. It can be seen that K_w is dependent on the applied pressure and position, and, with the exception of low pressures, is significantly greater than 0.3. Using equation 4, substituting a higher value of K_w would result in a smaller friction coefficient. Determining the exact dependence of the friction coefficient on the normal wall pressure will require improved experimental procedures and analysis.

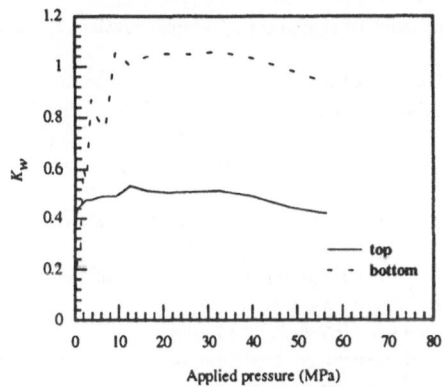

Figure 10. Computed K_w against applied pressure using the cap model.

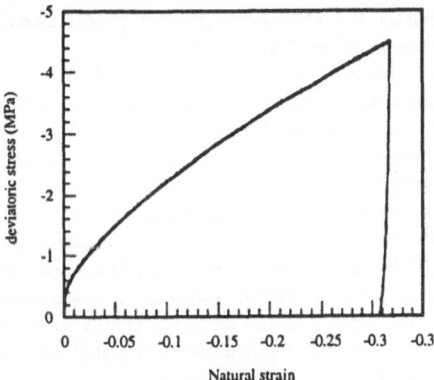

Figure 11. Deviatoric stress against axial natural strain for a cell pressure of 1.7MPa.

Describing the bulk powder behaviour using the nonlinear elasticity model outlined in this paper has the advantage that the required material constants are easily determined from uniaxial compaction tests. However, the assumption that the tangent modulus for the powder is described by a unique relationship between the axial stress and strain can be considered an oversimplification. Generally, the failure of a granular material will be dependent on all current components of the stress, this would suggest that the cap model may be more appropriate. However, this model is complex, and determining the required material parameters is significantly more difficult. For example, the relatively poor comparison between the computed applied pressures and the experimental data may be a result of the relatively small stresses that can be applied in the employed triaxial cell compared with those pervading during the uniaxial compaction of alumina. Therefore, the material parameters were obtained by fitting to data that were acquired at relatively low stress levels. In the context of the cap model, a possible improvement would be to include a dependence of the elastic properties on the stress invariants, and the inclusion of a hardening Drucker-Prager failure surface. The results of the triaxial experiments show that the deviatoric stress does not reach a steady value over the applied axial strain range, see figure 11, as would be expected in the case of a granular material with a perfectly plastic failure surface. This is further demonstrated when the alumina powder is subjected to a hydrostatic pressure that is subsequently removed. A coherent cylinder of alumina is obtained instead of a loose powder, thus indicating an increase in shear strength with hydrostatic pressure.

7. References

Aydin, I., Briscoe, B.J., and Sanliturk, K.Y. (1994) Density distributions during the compaction of alumina powders: A comparison of a computational prediction with experiment, *Computational Materials Science* 3, 55-68.

Aydin, I., Briscoe, B.J., and Sanliturk, K.Y. (1996) The internal form of compacted ceramic components: A comparison of finite element modelling with experiment, *Submitted to Powder Technology*.

Bowden, F.B., and Tabor, D. (1955) *Friction and Lubrication of Solids*, Cambridge University Press, Cambridge.

Briscoe, B.J., and Evans, P.D. (1991) Wall friction in the compaction of agglomerated ceramic powders, *Powder Technology* 65, 7-20.

Briscoe, B.J., Fernando, M.S.D., and Smith, A.C. (1987) The role of interface frciction in the compaction of maize, in B.J. Briscoe and M.J. Adams (eds.), *Tribology in Particulate Technology*, Adam Hilger Publishers, Bristol, pp. 220-233.

Chen,W.F., and Mizuno, E. (1990) *Nonlinear Analysis in Soil Mechanics*, Elsevier Science Publishers, Netherlands.

DiMaggio, F.L., and Sandler, I.S. (1971) Material model for granular soils, *Journal of the Engineering Mechanics Division, Proceedings of the American Society of Civil Engineers* 97, 935-950

Gethin, D.T., Lewis, R.W., Tran, D.V., and Ariffin, A.K. (1994) Finite element modelling of multilevel compaction of powders, *Advances in Powder Metallurgy and Particulate Materials* 7, 13-22.

Lukasiewicz, S.J., and Reed, J.S. (1978) in *American Ceramics Society Bulletin* 57, 798

Mroz, Z. (1980) On hypoelasticity and plasticity approaches to constitutive modelling of inelastic behaviour of soils, *Int. J. for Numerical and Analytical Methods in Geomechanics* 4, 45-55.

Munjiza, A., Owen, D.R.J., and Bicanic, N. (1995) A combined finite-discrete element method in transient dynamics of fracturing solids, *Engineering Computations* 12, 145-174.

Nagao, T., Hatamura, Y., Matsui, H., and Mikami, T. (1987) Stress analysis on the compaction of granular materials by FEM, in M. Satake and J.T. Jenkins (eds.), *Micromechanics of Granular Materials*, Elsevier Science Publishers B.V., Amsterdam, pp.215-224.

Thornton, C. (1987) Computer simulated experiments on particulate materials, in B.J. Briscoe and M.J. Adams (eds.), *Tribology in Particulate Technology*, Adam Hilger Publishers, Bristol, pp. 292-302.

Trasorras, J., Krauss, T.M., and Ferguson, B.L. (1989) Modeling of powder compaction using the finite element method, in *Advances in Powder Metallurgy - Volume 1*, Metal Powder Industries Federation, Princeton, pp. 85-104.

DRAINAGE AND DRYING OF DEFORMABLE POROUS MATERIALS: ONE DIMENSIONAL CASE STUDY.

P. DANGLA, O. COUSSY
Laboratoire Central des Ponts et Chaussées
58 bd Lefèbvre 75732 Paris cedex France

1. Introduction

The modeling of partially saturated porous materials is becoming a subject of main concern in various engineering fields, for example the safety of nuclear plants, the durability of building materials and the stability of unsaturated soils. The field equations governing the physics of porous materials are well established, whether obtained by mixture theories (Bedford and Drumheller, 1983, Truesdell, 1984) or by macroscopic approaches. For partially saturated deformable materials, the major problem lies in the formulation of constitutive equations. For soils they are often heuristically introduced, using an effective stress assumption by referring to the saturated situation for which the concept is firmly founded (Terzaghi, 1925). The thermodynamics can provide a consistent framework in constitutive modeling. In the first part of this paper this approach is applied to partially saturated porous deformable materials, using elements of thermodynamics of open continua (Biot, 1977, Coussy, 1995). It is shown how this approach receives experimental support.

In solving problems involving partially saturated materials, mathematical and numerical difficulties arise due to the strong non-linear character of diffusion processes. They are associated with the strong discontinuities which are generated by the vanishing diffusivity of the fluid phase which disappears at the interface between partially and totally saturated zones. In the second part of this paper, these discontinuities are systematically explored through the analysis of the drainage of a soil column.

2. Constitutive equations

An element of porous medium, whose deformation is identified by that of the skeleton, can gain or lose fluid masses due to external actions. In this sense any element of porous medium can be considered as an open thermodynamic system. In many situations three fluids can be distinguished: the liquid water (w), dry air (d) and water vapor (v). The laws of thermodynamics applied to this open system entail

$$\sigma_{ij} d\varepsilon_{ij} + \mu_w dm_w + \mu_v dm_v + \mu_d dm_d - SdT - d\Psi = 0 \tag{1}$$

Equation (1) is very similar to the Gibbs-Duhem equation encountered in physical

427

N. A. Fleck and A. C. F. Cocks (eds.),
IUTAM Symposium on Mechanics of Granular and Porous Materials, 427–438.
© *1997 Kluwer Academic Publishers.*

chemistry (Atkins, 1990): σ_{ij} and ε_{ij} are, respectively, the stress and the strain tensor. μ_α is the chemical potential of liquid phase α per mass unit, T the temperature, while m_α, S and Ψ are, respectively, the fluid masses, the total entropy and the total free energy per unit of initial volume of the bulk material. The left hand side of equation (1) represents the intrinsic dissipation and equals zero since only reversible transformations are considered in the following.

For ideal mixtures, the chemical potentials μ_α can be identified with the Gibbs-potentials g_α, i.e. the specific free enthalpies (Atkins, 1990). Standard thermodynamics of fluids yields

$$\mu_\alpha = g_\alpha = \psi_\alpha + (p/\rho)_\alpha \qquad\qquad dg_\alpha = (dp/\rho)_\alpha - s_\alpha dT \qquad (2)$$

where p_α, ρ_α, s_α and ψ_α are the pressure, mass density, specific entropy and specific free energy of the fluid α respectively. Combining (1) and (2) yields

$$\sigma_{ij} d\varepsilon_{ij} + p_w d\phi_w + p_v d\phi_v + p_d d\phi_d - S_s dT - d\Psi_s = 0 \qquad (3)$$

where ϕ_α is the volumetric fluid content of liquid phase α while S_s and Ψ_s are the entropy and free energy of the "skeleton" per unit initial volume of the bulk material:

$$m_\alpha = (\rho\phi)_\alpha \qquad\qquad S = S_s + m_\alpha s_\alpha \qquad\qquad \Psi = \Psi_s + m_\alpha \psi_\alpha \qquad (4)_{a,b,c}$$

The "skeleton" considered here is the system composed of both the solid grains and the phase interfaces having their own thermomechanical properties. Owing to the idealisation of the saturated mixture, the separate knowledge (2) of the thermodynamics of saturated fluids and that of the porous element considered as a whole (1) allows the separate identification (3) of the thermodynamics of the skeleton.

The dry air and the water vapor occupy the same space and form the wet air (subscript a). Consequently,

$$\phi_v = \phi_d (= \phi_a) \qquad\qquad p_a = p_v + p_d \qquad (5)_{a,b}$$

Let $\phi = \phi_w + \phi_a$ be the current total porosity referred to the initial volume of the skeleton. Assuming incompressible solid grains, the volumetric strain ε is only due to the variation of porosity, yielding

$$\phi = \phi_0 + \varepsilon \qquad (6)$$

Due to $(5)_{a,b}$ and (6) equation (3) can be written as

$$(\sigma_{ij} + p_a \delta_{ij})d\varepsilon_{ij} - p_c d\phi_w - S_s dT - d\Psi_s = 0 \qquad (7)$$

where $p_c = p_a - p_w$ is the capillary pressure. This relation implies that, in case of reversible transformations and incompressible grains, the skeleton free energy Ψ_s depends only on strain ε_{ij}, liquid water porosity ϕ_w and temperature T. Their conjugate state variables are respectively $\sigma_{ij} + p_a\delta_{ij}$, p_c and S_s according to

$$\sigma_{ij} + p_a\delta_{ij} = \frac{\partial \Psi_s}{\partial \varepsilon_{ij}} \qquad p_c = -\frac{\partial \Psi_s}{\partial \phi_w} \qquad S_s = -\frac{\partial \Psi_s}{\partial T} \qquad (8)$$

From now let us consider only isothermal transformations. Let us assume that the deviatoric behaviour is uncoupled from the volume change behaviour ($\Psi_s(\varepsilon_{ij}, \phi_w) = \Psi_s'(\varepsilon, \phi_w) + \Psi_s''(e_{ij})$). Then differentiating (8) yields the incremental relations

$$d\sigma + dp_a = K(\varepsilon, p_c)\, d\varepsilon + b(\varepsilon, p_c)\, dp_c \qquad d\phi_w = b(\varepsilon, p_c)\, d\varepsilon - N(\varepsilon, p_c)\, dp_c \qquad (9)_{a,b}$$

where $\sigma = \frac{1}{3}\sigma_{ii}$. Finally, assuming an isotropic linear deviatoric behaviour ($\Psi_s''(e_{ij}) = 2\mu e_{ij}e_{ij}$) yields

$$s_{ij} = 2\mu\, e_{ij} \qquad (10)$$

where s_{ij} and e_{ij} are the stress and strain deviatoric tensors respectively.

In equations (9)-(10) K and μ are the bulk and shear modulus, while b and N are generalized Biot's coefficients.

3. Experimental validation of the model for concrete

In order to validate this model, the theoretical predicted strain is compared with the experimentally observed strain in isothermal sorption experiments. Equation $(9)_b$ can be rewritten in the form

$$\phi\, dS_w = (b - S_w)\, d\varepsilon - N\, dp_c \qquad (11)$$

where $S_w = \phi_w/\phi$ is the water degree of saturation. For non deformable porous materials the degree of saturation is a function of the capillary pressure only, reading $S_w = S_w(p_c)$. Extending this assumption to deformable materials, (11) entails $b = S_w$. Thus the Biot-like coefficient b can be experimentally determined during an isothermal sorption experiment. During such an experiment the wet air pressure p_a and the total mean stress are constant and equal to the atmospheric pressure. According to $(9)_a$, this implies

$$d\varepsilon = -\frac{S_w}{K}\, dp_c \qquad (12)$$

Equation (12) and Kelvin's law, expressing the liquid-vapor equilibrium, yields

$$d\varepsilon = \frac{RT\rho_w}{M_v K} \frac{S_w}{h} dh \tag{13}$$

where h is the relative humidity, R the gas constant and M_v the water vapor molecular weight. In the isothermal sorption experiment, the relative humidity is controlled while the moisture content and the shrinkage are measured. The water degree of saturation is then evaluated from the moisture content. Figure 1 shows the measured sorption isotherm in ordinary concrete (Baroghel-Bouny, 1994). The predicted and measured shrinkage are compared in the figure 2, with $K = 25.4$ Gpa (Lassabatère T., 1994). Above 50% of humidity, a close agreement between theoretical and experimental results has been obtained. For values of humidity lower than 50% a discrepancy between the theoretical predictions and the experimental observations is recorded. This discrepancy could be attributed to phenomena involving liquid water contained in the nano-porosity the behaviour of which is not described by equation (2).

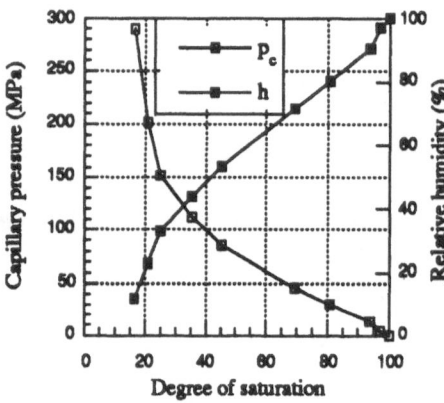

Figure 1 Sorption isotherm.

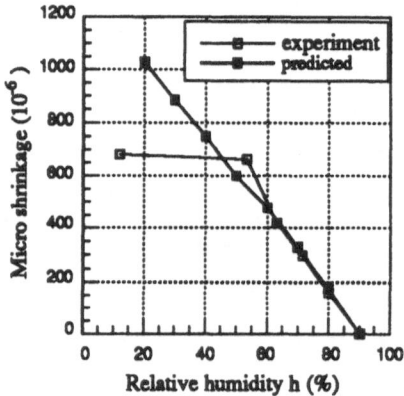

Figure 2 Comparaison between the predicted and observed shrinkage.

4 Governing equations

In the following sections only the air phase and the liquid water phase are considered. The governing equations are obtained in expressing the overall momentum balance equation and the mass balances for air and water. Neglecting inertial effects, the local momentum balance equation reads

$$\sigma_{ij,j} + \rho g_i = 0 \tag{14}$$

where g_i is the gravitational acceleration and ρ the bulk density. The balance of mass for water reads

$$\frac{\partial(\rho_w \phi S_w)}{\partial t} = -(\rho_w q_w)_{i,i} \qquad (15)$$

The water is assumed to be incompressible, its mass density ρ_w being constant (10^3 kg/m^3). In (15) q_{wi} is the flux density of water (relative to the solid grains) defined as the rate of fluid volume crossing a unit area of porous solid whose normal is in the i direction. Adopting Darcy's q_{wi} is given by

$$q_{wi} = k_w(-p_{w,i} + \rho_w g_i) \qquad (16)$$

In equation (16) k_w is the permeability coefficient relative to the water depending strongly upon the degree of saturation S_w (or upon the capillary pressure according to relation $S_w = S_w(p_c)$). This variation is commonly taken into account through the relative permeability to water k_{rw}, according to

$$k_w = \frac{k}{\mu_w} k_{rw} \qquad (17)$$

where k is the intrinsic permeability and μ_w the dynamic viscosity of water.

Similarly the mass balance of air, assumed to be an ideal gas, and Darcy's law with respect to the air phase lead to the set of equations

$$\frac{\partial(\rho_a \phi(1-S_w))}{\partial t} = -(\rho_a q_a)_{i,i} \qquad (18)$$

$$\rho_a = \frac{M_a}{RT} p_a \qquad (19)$$

$$q_{ai} = k_a(-p_{a,i} + \rho_a g_i) \qquad (20)$$

$$k_a = \frac{k}{\mu_a} k_{ra} \qquad (21)$$

where M_a the molecular weight of air. The constant M_a/RT is about $1.2 \ 10^{-5}$ s^2/m^2 for a temperature of 293 °K. The expressions of k_{rw} and k_{ra} remain to be specified. Typical expressions are proposed by Brooks and Corey (1964), Gardner (1958), Matyas (1967), Yoshimi and Osterberg (1963).

5 Drainage from one-dimensional soil column

5.1 DESCRIPTION OF THE PROBLEM

Liakopoulos (1965) conducted several experiments on the drainage of water from vertical columns of Del Monte sand. One of these experiments was chosen for the application of the present model. A 1m high column ($H=1$m) was packed with Del Monte sand and was instrumented with a sufficient number of tensiometers to measure continuously moisture tension at various points within the column. The physical properties of Del Monte sand were measured by Liakopoulos by an independent set of experiments. The sand had a porosity of about 30%. The dependencies of the degree of water saturation and water permeability are fitted by Schrefler (1993) on experimental data obtained by Liakopoulos. Since the problem is solved as a two-phase flow problem, the relative air permeability is also needed. This was represented by the Brooks and Corey (1966) relationship. Table 1 shows the constitutive functions and parameters used in this example.

Properties	Functions and parameters
Bulk density (saturated) ρ	$1700 \, \text{kg/m}^3$
Young's modulus E	$1.3 \, \text{MPa}$
Poisson's ratio ν	0.4
Porosity ϕ	0.2975
Water degree of saturation S_w	$1 - 0.09686 \, (p_c/\rho_w g H)^{2.428}$ $(H=1\text{m})$
Intrinsic permeability k	$4.4 \, 10^{-13} \, \text{m}^2$
Relative permeability to water k_{rw}	$1 - 2.207 \, (1 - S_w)^{0.953}$
Relative permeability to air k_{ra}	$(1-S_e)^2 (1-S_e^{5/3})$ $S_e=(S_w-0.2)/0.8$
Dynamic viscosity of water μ_w	$10^{-3} \, \text{Pa s}$
Dynamic viscosity of air μ_a	$1.8 \, 10^{-5} \, \text{Pa s}$
Gravity g	$9.81 \, \text{m}^2/\text{s}$
Air density ρ_a	$1 \, \text{kg/m}^3$
Water density ρ_w	$1000 \, \text{kg/m}^3$

TABLE 1 Soil properties

The initial conditions correspond to the saturated state and hydrostatical equilibrium of the water. These are defined by $p_a = p_{atm}$ and $p_w = p_{atm} + \rho_w g(H-z)$ where p_{atm} is the atmospheric pressure (0.101 MPa). At $t=0$ the lower boundary is rendered permeable to water by keeping the water pressure equal to the atmospheric pressure thereby allowing the water to drain freely from the column. The boundary conditions are the following. The lateral surfaces are made impermeable to water and air while the horizontal displacement is imposed to be equal to zero. For the bottom surface the two components of the displacement are kept equal to zero while the water pressure and the air pressure are maintained at p_{atm}. For the top surface the air pressure is maintained at p_{atm} and it is made impermeable to water. Except gravity forces, no other external forces are applied to the column. The boundary conditions relative to water and air continuity equations are summarized in figure 3.

A numerical method was used to work out this problem. In order to approximate the solution of equations (9)-(10) and (14)-(21), different methods can be used. Our attention was focused on exhibiting the local behaviour of unknown variables around the limit between the saturated and unsaturated areas. To this aim we selected a finite volume method for treating equations (15), (16), (18), (20), associated with a finite element method for treating equations (9) and (14). The spatial discretization of the column involved 1000 equally sized elements.

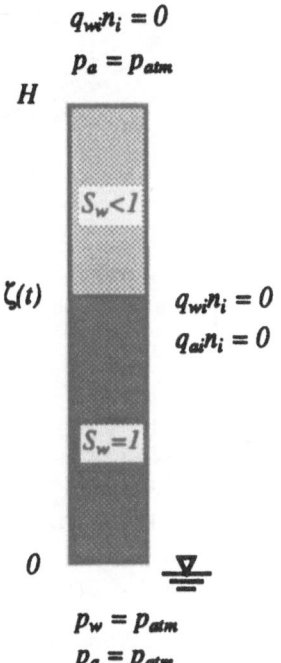

Figure 3 Boundary conditions and the two expected regions.

The resulting profiles of water pressure, air pressure, vertical displacement and water degree of saturation are shown in figure 4. The profiles of degree of saturation clearly exhibit the propagation of a water table separating a saturated region from a partially saturated region as pictured in figure 3. Contrary to standard assumptions, the air pressure varies considerably in the region of the column just above the water and strongly influences its velocity. Indeed, further calculations have shown that the dewatering process is by far the quickest when assuming erroneously that the air remains under atmospheric pressure.

The profiles of degree of saturation suggest that a discontinuity of the gradient of S_w propagates downwards. In fact this question can be directly solved from a theoretical point of view.

5.2 ASYMPTOTIC ANALYSIS AROUND THE WATER TABLE

Let us consider a continuous function $g(z,t)$ whose first derivatives (with respect to space and time variables) undergo a finite jump across the surface $z=\zeta(t)$, where $\zeta(t)$ is the vertical coordinate of water table at time t (figure 3). Here the jump $[a]$ across surface $z=\zeta(t)$ of any quantity a can be computed through the limiting procedure (figure 5)

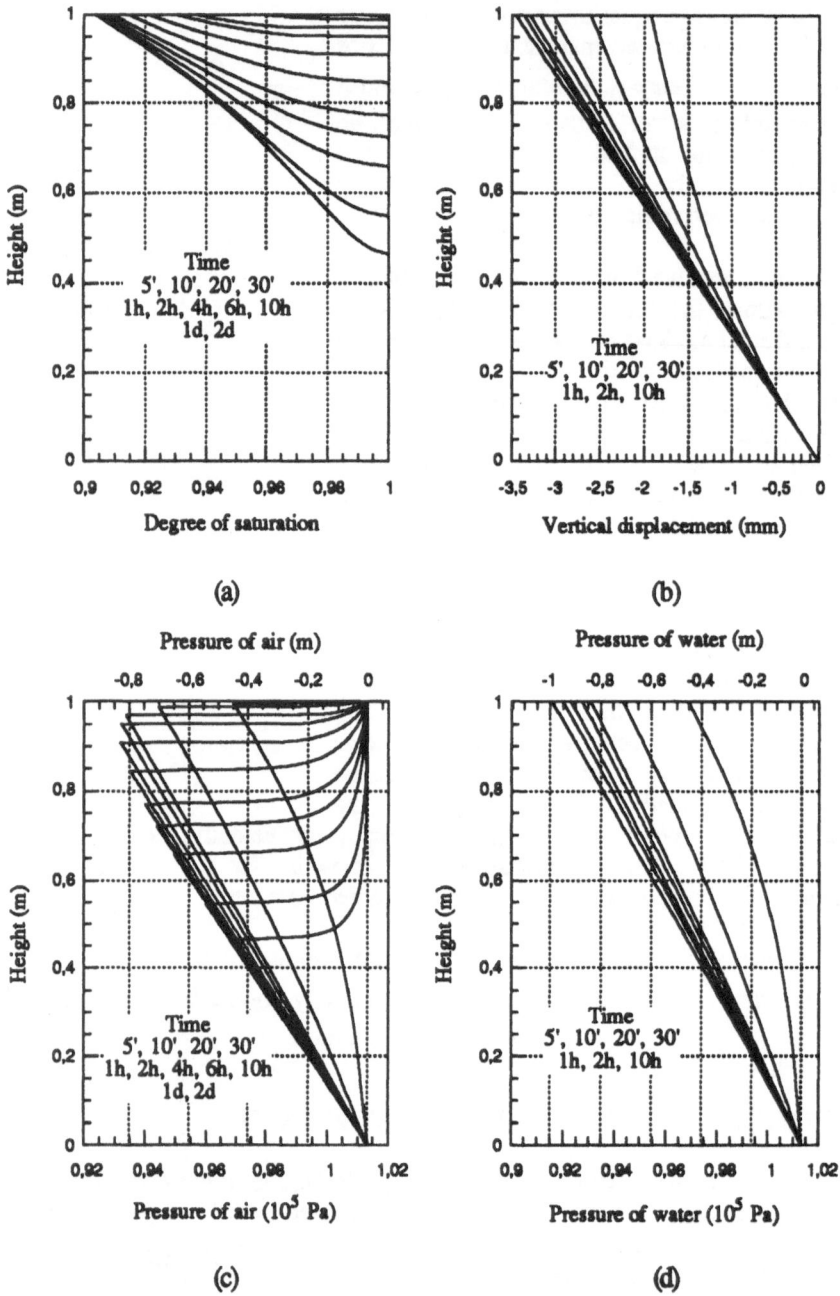

Figure 4 Resulting profiles: a) degree of saturation; b) vertical displacement;
c) pressure of air; d) pressure of water.

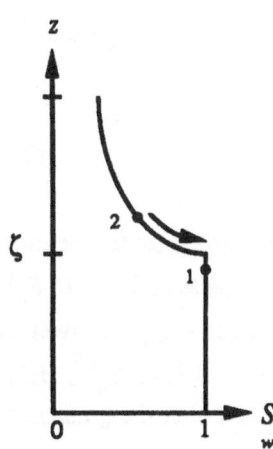

Figure 5 The jump of a across ζ is computed as

$$[a] = \lim_{s_w \to 1} (a_2 - a_1)$$

$$[a] = \lim_{S_w \to 1} (a_2 - a_1) \qquad (22)$$

Furthermore the theory of singular surfaces (Hadamard, 1903) imposes

$$\left[\frac{\partial g}{\partial t}\right] = -V\left[\frac{\partial g}{\partial z}\right] \qquad (23)$$

where $V = \dfrac{\partial \zeta}{\partial t}$. Let us apply these relations to the differential equations ruling the water table movement. A differential equation governing the degree of saturation S_w is obtained from the continuity equations (15) and (18). For the sake of simplicity our analysis is restricted here to a non deformable medium and incompressible fluids (ϕ, ρ_w and ρ_a are constant). Under these assumptions the total flux density, i.e. $Q(t) = q_w + q_a$, is constant throughout the column. Then, a combination of Eqs. (15), (18), (16) and (20), together with relation $S_w = S_w(p_c)$, provides the differential equation governing function $S_w(z, t)$

$$\phi \frac{\partial S_w}{\partial t} = \frac{\partial^2}{\partial z^2} \varphi(S_w) + \frac{\partial}{\partial z} f(S_w, t) \qquad (24)$$

in which the two functions φ and f are defined as follows

$$\varphi(S_w) = \int_1^{S_w} \kappa(s)ds \qquad\qquad \kappa(S_w) = -\frac{k_w k_a}{k_w + k_a}\frac{dp_c}{dS_w}$$

$$f(S_w) = \frac{k_w k_a}{k_w k_a}(\rho_w - \rho_a)g - \frac{k_a}{k_w k_a}Q(t)$$

The equation (24) can be ranked as a non-linear degenerate parabolic equations in the sense that there is a physical value of S_w, $S_w = 1$, for which the diffusion vanishes, $\kappa(1) = 0$. This equation is mathematically similar to equation $\dfrac{\partial u}{\partial t} = \dfrac{\partial^2 u^m}{\partial z^2}$ whose properties have been studied extensively (Peletier, 1981). Let us point out the properties of the solution of equation (24).

Applying the jump operator to (24) and using (22) and (23) yields

$$\lim_{S_w \to 1}\left(\phi V \frac{\partial S_w}{\partial z} + \frac{\partial^2 \varphi}{\partial z^2} + \frac{\partial f}{\partial z}\right) = 0 \qquad (25)$$

Mass conservation entails continuity of water and air fluxes at any point within the column and therefore continuity of $\partial\varphi/\partial z$. This can be shown from a mathematical analysis of (24). Using (22), it entails

$$\lim_{S_w \to 1}\left(\frac{\partial\varphi}{\partial z}\right) = 0 \qquad (26)$$

In the vicinity of the water table the permeability to air can be matched by the function $b(1-S_w)^\beta$ (with $\beta=3$ and $b=7.96 \ 10^{-8} \ m^2/Pa \ s$), and the capillary pressure by $a \ (1-S_w)^\alpha$ (with $\alpha=0.41$ and $a=2.56 \ 10^4 \ Pa$). Thus, in this region, the function $\kappa(S_w)$ is given by $ab\alpha \ (1-S_w)^n$ (with $n=\alpha+\beta-1$), while the order of magnitude of $f(S_w,t)$ is $(1-S_w)^\beta$. Using (26) and retaining the leading terms in (25), some calculations show that the gradient of S_w has to satisfy the relation

$$\frac{\partial S_w}{\partial z} = \phi V \frac{(1-S_w)}{\kappa(S_w)} + o(1) \qquad (27)$$

As $\kappa(S_w)=O((1-S_w)^n)$ with $n>1$ (27) stipulates that, at any time, the gradient of S_w is infinite at the water table location. This agrees with numerical computations (figure 4a). Equation (27) can be integrated with respect to z, yielding the velocity of the water table location in the form

$$V = \left(\frac{\partial F}{\partial z}\right)_\zeta \qquad F(S_w) = \frac{1}{\phi}\int_0^{\varphi(S_w)} \frac{dx}{1-\varphi^{-1}(x)} \qquad (28)$$

The function $-F(S_w)$ is equivalent to $c(1-S_w)^n$ around $S_w=1$. The constants $n=\alpha+\beta-1$ and $c=ab\alpha/\phi n$ take the values 2.41 and $1.167 \ 10^{-3} \ m^2/s$ respectively.

Relation (28) shows that the propagation of the water table is only governed by the local properties of the material through function $F(S_w)$. Relation (28) can be numerically checked. To this aim let us define the dimensionless functions $\tau(\bar{z},\bar{t})=\bar{t}+\bar{c}(1-S_w(z,t))^n$ and $\bar{\zeta}(\bar{t})=\zeta(t)/H$, where $\bar{z}=z/H$, $\bar{t}=t/T$ and $\bar{c}=cT/H^2$. T is an arbitrary period of time conveniently chosen in order to represent the whole computed solutions, namely $T=3$ days. The profiles $\tau(\bar{z},\bar{t})$ have been plotted for different times in figure 6, as also the curve $\bar{\zeta}(\bar{t})$. They intersect at $\bar{z}=\bar{\zeta}(\bar{t})$ where formula (27) implies that $\dfrac{d\bar{\zeta}}{d\bar{t}}=-\dfrac{\partial\tau}{\partial\bar{z}}$. Consequently, the profiles $\tau(\bar{z},\bar{t})$ have to be perpendicular to the curve $\bar{\zeta}(\bar{t})$ at the intersection point as shown in figure 6.

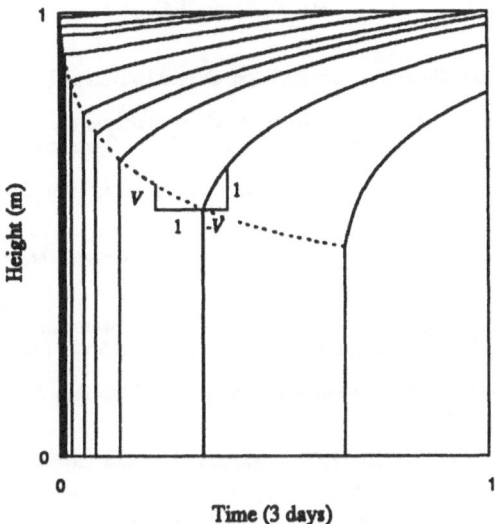

Figure 6 The formula (28) is checked through the orthogonality of the two following curves: in the horizontal axis $\tau(\bar{z},\bar{t}) = \bar{t} + \bar{c}(1 - S_w(z,t))^n$ is plotted as a function of $\bar{z}$ and for several time $\bar{t}$ (solid line); in the vertical axis $\bar{\zeta}(\bar{t})$ is plotted as a function of $\bar{t}$ (dotted line).

6 Conclusion

A non-linear poroelastic model for partially saturated deformable porous materials has been presented. Isothermal sorption experiments on ordinary concrete allowed to identify some parameters and to validate the model by accurately predicting the observed capillary shrinkage. The derivation of the poro-elasticity relations was based on the thermodynamics of open continua. It offers a consistent and relevant framework to formulate, directly at the macroscopic level, a series of behaviours as sophisticated as chemico-mechanical couplings (Ulm, 1995), non-ideal saturating mixtures (Dormieux, 1995) or poro-plasticity (Coussy, 1995). In a second part of this paper the analysis of the problem of the drainage of a column has highlighted a particular behaviour around the water table, which is essentially governed by a degenerate parabolic equation. The theoretical analysis of this equation has revealed that the gradient of the water degree of saturation is infinite at the water table. The water table has been shown to propagate with a velocity V governed by a local non linear equation. Regarding the numerical solutions proposed in the literature, the asymptotic behaviour, revealed by the theoretical analysis, could be a key indicator to demonstrate the quality of numerical methods. The equation mentioned above, namely $V=(\partial F/\partial z)_\zeta$, only involves the properties of the material through the function of the degree of saturation $F(S_w)$.

REFERENCES

Atkins P. W. (1990), *Physical Chemistry*, Fourth edition, Oxford University Press.

Baroghel-Bouny V. (1994), Caractérisation microscruturale et hydrique des pâtes de ciments et des bétons ordinaires et à très hautes performances, PhD. Thesis of l'École des Ponts et Chaussées.

Bedford A. and Drumheller D.S. (1980), Theories of immiscible and structured mixtures, *Int. J. Engng Sci.*, vol. 21, No 8, 863-960.

Biot M. (1972), Theory of deformations of porous solids, *Indiana University Mathematical Journal*, 21, 597-620.

Biot M. (1977), Variational Lagrangian-thermodynamics of non isothermal finite strain. Mechanics of porous solids and thermomolecular diffusion, *International Journal of Solids and Structures*, 13, 579-597.

Brooks R. N. and Corey A. T (1966)., Properties of porous media affecting fluid flow, *J. Irrig. Drain. Div. Am. Soc. Civ. Eng.*, 92(IR2), 61-68.

Brooks R. N. and Corey A. T (1964), Hydraulic properties of porous media, *Colorado State Univ. Hydrol. Paper*, No. 3, 27 pp., Mar.

Burland J. B. (1965), Some aspects of the mechanical behaviour of partially saturated soils. *Moisture Equilibria and Moisture Changes Beneath Covered Areas*: 270-278. Butterworths. Sydney.

Coussy O. (1995), *Mechanics of porous continua*, J. Wiley & Sons.

Dormieux L., P. Barboux, O. Coussy and P. Dangla (1995), A macroscopic model of the swelling phenomena of a saturated clay, *European Journal of Mechanics A/Solids*, vol. 14, N°6.

Gardner W. R. (1958), Some steady state solutions of the unsaturated moisture flow equation with application to evaporation from a water table, *Soil Sci.* 85(4), 228-232.

Hadamard J. (1903), *Leçons sur la propagation des ondes et les equations de l'hydrodynamique*, Hermann, Paris.

Lassabatère T. (1994), Couplages hydromécaniques en milieu poreux non saturé avec changement de phase: application au retrait de dessication. PHD thesis, Ecole Nationale des Ponts et Chausées, Paris.

Liakopoulos A. C. (1965), Transient flow through unsaturated porous media, D. Eng. dissertation, Univ. of Calif., Berkeley.

Matyas E. L. (1967), Air and water permeability of compacted soils. *Permeability and capillarity of soils*, ASTM ASTP 417 Amer. Soc. Testing and Materials, pp. 160-175.

Peletier L. A. (1981), The porous media equation. In *Application of Nonlinear Analysis in the Physical Sciences (Ch. 11)*, H. Amann, Bazley, K. Kirchgassner, Pitman (ed.), 229-241, London.

Schrefler B. A. and Zhan Xiaoyong (1993), A fully coupled model for water flow and airflow in deformable porous media, *Water Resources Research*, 29, 155-167.

Terzaghi K. (1925), Principles of soil mechanics, A summary of expeimental results of clay and sand, *Eng. News Rec.*, 3-98.

Truesdell C. (1984), *Rational thermodynamics*, second edition, Springer-Verlag.

Ulm F.J. and Coussy O. (1995), Modelling of thermochemomechanical couplings of concrete at early ages, *Journal of Engineering Mechanics*, pp. 785-794, July.

Yoshimi V. and Osterberg J. O. (1963), Compression of partially saturated cohesive soils. *J. Soil Mech. Found. Eng. Div.* ASCE 89(SM4), pp. 1-24.

MECHANICS OF SAND PRODUCTION FOR ACTING WELLS

S. B. GRAFUTKO, V. N. NIKOLAEVSKIY
United Institute of Physics of the Earth, Russian Acad. Sci.
B.Gruzinskaya 10, Moscow 123810, Russia.

1. Introduction.

Sand production from a poorly consolidated formation during gas, oil or water recovery is a traditional, but unsolved problem of petroleum and geomechanical engineering. In practice sand production can be avoided by simply limiting production rates of the wells by reducing the pore pressure gradients. Correspondingly, it is necessary to determine the maximum possible filter velocity, above which sand grains (fragments of totally destroyed formation matrix) or fines (damaged products of the matrix) begin to move in pore channels. Moreover, their accumulation in the close vicinity of the well can lead to "thrombosis" of pore channels, because of convergent underground flows in the formation. The value of the critical velocity for sand production beginning (Lapuk, 1948) depends on the mechanical properties of the matrix - its composition, cementage, strength, etc.

Presently, sand production is a serious threat to efficient gas recovery from such extremely rich gas reservoir as Urengoy (Western Siberia) where sand is flowing in thin layers of loose fractions of senoman sandstones (Maslennikov & Remizov,1993).

Earlier this century problems of sand production were important for oil reservoirs of the Emba region (West Kazakhstan) and of Baku (Azerbaijan). Nowdays the same problem is connected mainly with production of heavy oils in Canada (Selby & Fouruk Ali, 1993).

Previous publications on this topic were devoted to determining the size and nature of the plastic zones around the acting wells, but the movement of the matrix material to the well in the form of plastic flow was not considered. Moreover, sand flow in narrow layers can be generated by localization effect of filter flow into a finite number of channels. In this case the rock stress anisotropy (Detournay & Fairhurst, 1987) is essential. It was also found that transport of fines (dust) essentially changes phase permeabilities of the formation.

In this paper the solution of stationary joint motion of fluid and of fragmented matrix (of sand) is suggested basing on the assumption that the theory of dilatant plasticity can be employed to describe fragmentation and sand flow during pore pressure gradient driven filtration flow. In this study we use the elasto-plastic dilatancy theory with the nonasssociated flow rule (Nikolaevskiy, 1990;1996), developed earlier in connection with designing underground explosions, generalized for the case of saturated porous rocks and soils, but the effects of inertia, naturally, are neglected.

439

N. A. Fleck and A. C. F. Cocks (eds.),
IUTAM Symposium on Mechanics of Granular and Porous Materials, 439–450.
© 1997 *Kluwer Academic Publishers.*

The dilatancy effect (volume change due to shear) was found to be essential in the case considered here: porosity has to change, although, for simplicity, we shall neglect interconnected changes of permeability of the deforming matrix. As a result we obtain a simple analytical solution, although, in principle, it is possible to take into account both porosity and permeability changes as well as strength parameters of the layer in the close vicinity of the borehole.

2. Basic system of equations.

We begin from the traditional balances of masses of fluid and solid phase (matrix), which in the axisymmetric case have the following form :

$$\frac{\partial}{\partial t}(m\rho_f) + \frac{1}{r}\frac{\partial}{\partial r}(r\rho_f mw) = 0 \tag{2.1}$$

$$\frac{\partial}{\partial t}(1-m)\rho_s + \frac{1}{r}\frac{\partial}{\partial r}(r\rho_s(1-m)v) = 0 \tag{2.2}$$

where m - porosity, w, v - "true" velocities (nontrivial radial components) of fluid and solid phases, ρ_f, ρ_s - density of fluid and the material composing the matrix.

Darcy's law is formulated in a coordinate system, moving with the solid matrix:

$$m(w - v) = -\frac{k}{\mu}\frac{\partial p}{\partial r} \tag{2.3}$$

where k - permeability, μ - viscosity of fluid. Neglecting solid phase motion in (2.3) is inadmissible for the problem considered here. (2.3) is equivalent to the momentum balance for the fluid phase (Nikolaevskiy, 1990, 1996).

The equilibrium condition for the matrix is defined by the following equation :

$$\frac{\partial \sigma_r}{\partial r} + \frac{\sigma_r - \sigma_\theta}{r} - \frac{\partial p}{\partial r} = 0 \tag{2.4}$$

where the effective principal (radial and tangential) stresses appear as well as pore pressure p.

We assume that damaging of the matrix in the close vicinity of the borehole occurs if a critical plasticity condition is reached. In the rupture zone there is a sharp decrease of cohesion Y. However, in the calculation of the stationary solution the parameters of rupture zone α (solid friction coefficient) and Y appear only :

$$(\sigma_r - \sigma_\theta)\theta_\sigma = -\alpha(\sigma_r + \sigma_\theta) + Y \tag{2.5}$$

We now transform condition (2.5) to its effective form:

$$\sigma_\theta = N\sigma_r - Y/(\theta_\sigma - \alpha), \qquad N = \frac{\theta_\sigma + \alpha}{\theta_\sigma - \alpha} \qquad (2.6)$$

Here the sign of action of tangent stresses is incorporated: $\theta_\sigma = \text{sgn}(\sigma_r - \sigma_\theta)$.

The dilatancy relationship between increments of volume and shear deformation of the matrix is given in the terms of velocities and has the following form :

$$\frac{\partial v}{\partial r} + \frac{v}{r} = \Lambda \theta \left(\frac{\partial v}{\partial r} - \frac{v}{r} \right) \qquad (2.7)$$

where Λ- dilatancy rate, and the sign associated with the direction of shear is used, which coincides with the sign of tangent stresses: $\theta = \text{sgn}(\partial v / \partial r - v/r) = \theta_\sigma$ (also, this equivalence follows from the coaxiality of the stress tensor and the tensor of plastic deformation rates or, equivalently, from the condition of positive dissipation of work during plastic deforming).

The system considered here is closed if laws governing changes of density of the phases (with pressure), permeability (with porosity and pressure), solid friction coefficient, cohesion and dilatancy rate changes (with porosity) are known (Nikolaevskiy, 1990, 1996).

3. Stationary inflow to the borehole under sand production.

The most simple stationary problem is formulated under the condition of coincidence of filter feeding contour and of outside plastic zone boundary. This is an over simplified idealized situation, but it can reveal all the main features of the relationship between the rates of fluid mass production Q_f and of sand production Q_s.

The fluid production rate is defined by the fluid velocity at the wall of the borehole, $(r=a, \ h$ is formation thickness) :

$$Q_f = -2\pi a h \, \rho_f(p) \, w(a) \, m_a \qquad (3.1)$$

and sand flow by the velocity of a solid phase in the same cross-section of the formation:

$$Q_s = -2\pi a h \, \rho_s \, v(a) \, (1 - m_a) \qquad (3.2)$$

At the feeding boundary R the constant pore pressure is given as well as the rock (total) pressure Γ :

$$p(R) = p_R \qquad (3.3)$$

$$\sigma_r(R) = \sigma_R = -(\Gamma_r - p_R) \qquad (3.4)$$

According to the condition of plasticity (2.6), the effective stress is defined by the lateral pressure factor N. One has $N = 3$ for rock strata [3], if tension is positive, as assumed in this work.

At the feeding boundary we will assume the initial value of formation porosity:

$$m(R) = m_R \qquad (3.5)$$

and at the borehole wall the pore pressure is given by :

$$p(a) = p_a \qquad (3.6)$$

We further assume that the radial effective stress, at the open wall of the uncased is a zero:

$$\sigma(a) = \sigma_a = 0 \qquad (3.7)$$

Due to the condition of stationary the equations of mass balance give the integrals:

$$Q_f = -2\pi h r m \rho_f w(r) \qquad (3.8)$$

$$Q_s = -2\pi h \rho_s r(1-m)v(r) \qquad (3.9)$$

with Q_f, Q_s (flow rates of fluid and sand masses) constant within the zone of the underground flow.

In the plastic flow zone the dilatancy condition (2.7) gives :

$$v = \frac{C}{r^n} \qquad (3.10)$$

where

$$n = \frac{1 + \theta \Lambda}{1 - \theta \Lambda} \qquad (3.11)$$

Since the density of the matrix material is constant (grains of sand are incompressible), (3.9) and (3.10) define a constant of integration in terms of the mass "production rate" of sand Q_s and the porosity m_R :

$$1 - m = -\frac{Q_s}{2\pi h \rho_s} \frac{r^{n-1}}{C} = (1 - m_R)\frac{r^{n-1}}{R^{n-1}} \qquad (3.12)$$

One can see that for $n > 1$ loosening of the matrix occurs within the plastic zone $(r \cdot R)$, but for $n < 1$ the matrix will be compacted.

Hereinafter, integrals (3.8) - (3.10) allow us to find velocity fields for the fluid and sand, and, from (3.12), the porosity distribution can be obtained:

$$w(r) = -\frac{Q_f}{2\pi h r \rho_f m}, \qquad v(r) = -\frac{Q_s}{2\pi h R \rho_s (1 - m_R)}\left(\frac{R}{r}\right)^n \qquad (3.13)$$

$$m = 1 - (1 - m_R)\left(\frac{R}{r}\right)^{1-n} \qquad (3.14)$$

Substitution of (3.13) and (3.14) to Darcy's law (2.3) gives:

$$\frac{dp}{dr} = \frac{Q_f \mu}{2\pi k h \rho_f(p) r} - \frac{Q_s \mu}{2\pi k h \rho_s (1 - m_R) r}\left\{\left(\frac{r}{R}\right)^{1-n} - (1 - m_R)\right\} \qquad (3.15)$$

which can be integrated if the fluid is incompressible.

In order to account for changes inside the rupture zone (3.15) must be integrated numerically, together with the equilibrium equation (2.4) and the plasticity condition

(2.6) which lead to the following equation ($Y_0 = Y / (\theta_\sigma - \alpha)$) :

$$\frac{1}{N-1}\frac{d}{dr}\{(N-1)\sigma_r - Y_0\} - \frac{(N-1)\sigma_r - Y_0}{r} - \frac{dp}{dr} = 0 \tag{3.16}$$

4. Sand production under the stationary liquid flow conditions.

Consider now the most simple variant of sand production during recovery of liquid from the formation, which occurs when $\rho_f = const$.

First of all, note that in the case of a nondeformable porous space $(m = const)$ one can only define the relative phases influx to the borehole.

For $n = 1$ we have the filter equation:

$$\frac{dp}{dr} = \left(\frac{Q_f}{\rho_f} - \frac{Q_s}{\rho_s}\frac{m}{1-m}\right)\frac{\mu}{2\pi kh}\frac{1}{r} \tag{4.1}$$

and equation (3.16) which includes the rate of sand motion in the same combinations as in (4.1). Because of the simplification in the modeling (pore space incompressibility) the problem of sand production has not been solved .

Allowing for deformation of the porous space, integrating (3.15) gives:

$$p_R - p = \frac{(Q_f/\rho_f)+(Q_s/\rho_s)}{2\pi kh}\mu\ln\frac{R}{r} - \frac{Q_s/\rho_s}{1-m_R}\frac{\mu}{2\pi kh}\frac{1}{1-n}\left(1-\frac{r^{1-n}}{R^{1-n}}\right) \tag{4.2}$$

and integrating (3.16) gives consecutively:

$$(N-1)\sigma_r - Y_0 = \{(N-1)\sigma_R - Y_0\}\left(\frac{r}{R}\right)^{N-1} - (N-1)r^{N-1}\int_r^R \frac{dp}{d\rho}\frac{d\rho}{\rho^{N-1}} \tag{4.3}$$

$$(N-1)\sigma_r - Y_0 = [(N-1)\sigma_R - Y_0]\left(\frac{r}{R}\right)^{N-1} -$$

$$-\frac{\mu}{2\pi kh}\left(\frac{Q_f}{\rho_f}_f + \frac{Q_s}{\rho_s}\right)[1-\left(\frac{r}{R}\right)^{N-1}]+$$

$$+\frac{\mu Q_s}{2\pi kh\rho_s(1-m_R)}\frac{N-1}{N+n-2}[\left(\frac{r}{R}\right)^{1-n} - \left(\frac{r}{R}\right)^{N-1}] \tag{4.4}$$

Substitution of boundary conditions (3.6)-(3.7) at the borehole and at the feeding boundary (coinciding with the plastic boundary) into equations (4.2)-(4.4) gives :

$$\left(\frac{Q_f}{\rho_f} + \frac{Q_s}{\rho_s}\right) - \frac{Q_s}{\rho_s(1-m_R)(1-n)\ln(R/a)}[1-\left(\frac{a}{R}\right)^{1-n}] = \frac{2\pi kh}{\mu}\frac{p_R - p_a}{\ln(R/a)} \tag{4.5}$$

$$\frac{Q_s}{\rho_s(1-m_R)}\frac{N-1}{N+n-2}\left[\left(\frac{a}{R}\right)^{1-n}-\left(\frac{a}{R}\right)^{N-1}\right]-\left(\frac{Q_f}{\rho_f}+\frac{Q_s}{\rho_s}\right)\left[1-\left(\frac{a}{R}\right)^{N-1}\right]=$$

$$=\frac{2\pi kh}{\mu}\left\{(N-1)(\tilde{A}-p_R)\left(\frac{a}{R}\right)^{N-1}-(N-1)p_a-Y_0[1-\left(\frac{a}{R}\right)^{N-1}]\right\}\qquad(4.6)$$

and permits both the production rates of liquid and sand to be found.

Without the dilatancy effect $(n=1, m=const)$ the resulting formulae change their type, and they can only determine the relative velocities of fluid and sand, i.e. they cannot be defined separately (and this is the main difficulty of the problem)

$$m\left(\frac{Q_f}{\rho_f m}-\frac{Q_s}{\rho_s(1-m)}\right)=\frac{2\pi kh}{\mu}\frac{p_R-p_a}{\ln(R/a)}\qquad(4.7)$$

$$m\left(\frac{Q_f}{\rho_f m}-\frac{Q_s}{\rho_s(1-m)}\right)\left[1-\left(\frac{a}{R}\right)^{N-1}\right]=$$

$$=\frac{2\pi kh}{\mu}\left\{(N-1)(\tilde{A}-p_R)\left(\frac{a}{R}\right)^{N-1}-Y_0[1-\left(\frac{a}{R}\right)^{N-1}]\right\}\qquad(4.8)$$

The calculations can be simplified further by neglecting the terms with the ratio a/R compared with unit. Then for the general case $n\neq1$:

$$\left(\frac{Q_f}{\rho_f}+\frac{Q_s}{\rho_s}\right)-\frac{Q_s}{\rho_s(1-m)(1-n)\ln(R/a)}=\frac{2\pi kh}{\mu}\frac{p_R-p_a}{\ln(R/a)}\qquad(4.9)$$

$$\left(\frac{Q_f}{\rho_f}+\frac{Q_s}{\rho_s}\right)-\frac{N-1}{N+n-2}\frac{Q_s}{\rho_s(1-m_R)}=\frac{2\pi kh}{\mu}Y_0\qquad(4.10)$$

We can see that there is a critical pore pressure drop (or a critical fluid production rate) at which the sand production begins :

$$p_R - p_a = Y_0 \ln(R/a)\qquad(4.11)$$

Thus the drop of pressure mainly occurs to overcome the ruptured matrix cohesion Y_0. The rock pressure Γ only plays an important role for the case of low values of Y_0.

In Figure 1 the calculation results are given for the following parameters :
$N = 3$; $R = 10$ (m) ; $a = 0.1(m)$; $\sigma_R = -100$ $(10^5$ $Pa)$; $G = 4$ $(10^8 Pa)$;
$k = 10^{-13}$ (m^2); $\mu = 0.001(Pa$ $sec)$; $Y_0 = 3$ $(10^5$ $Pa)$; $Q_f = 300$ $(m^3$ $/day)$,
$\rho_f = 10^3$ (kg/m^3); $p_R = 50$ $(10^5$ $Pa)$; $h = 10$ (m).

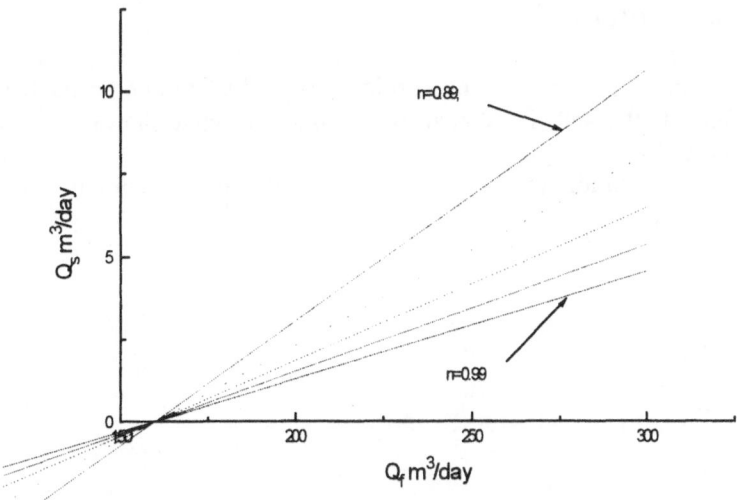

Figure. 1. Relationship between rates of oil (Q_f) and sand (Q_s) productions (stationary inflow).

5. Numerical calculations of stationary sand production from gas formation.

Consider now sand influx into the well bore by ideal gas flow when

$$\rho_f = \rho_0 \frac{p}{p_0} \tag{5.1}$$

In this case equation (3.15) must be solved numerically. The system of equations will be written now in the non-dimensional following form:

$$\frac{dp}{dr} = \frac{q_f}{r} \frac{p_k}{p} - \frac{q_s}{r(1 - m_0)(r^{(1-n)} - (1 - m_0)} \tag{5.2}$$

$$\frac{d\sigma_r}{dr} = \frac{(N-1)\sigma_r - 1}{r} + \frac{dp}{dr} \tag{5.3}$$

$$r' = \frac{r}{R}, \qquad p' = \frac{p}{Y_0}, \qquad u' = \frac{2G}{Y_0 R}u, \qquad \sigma'_r = \frac{\sigma'_r}{Y_0},$$

$$\tag{5.4}$$

$$q_f = \frac{\mu}{k} \frac{Q_f}{2\pi h \rho_0 Y_0}, \qquad q_s = \frac{\mu}{k} \frac{Q_s}{2\pi h \rho_s Y_0},$$

The data used for calculations given in Figure 2 are the following ones :

$N = 3;\ R = 10\ (m)\ ;\ a = 0.1(m);\ \sigma_R = -100\ (10^5\ Pa);\ G = 4\ (10^8\ Pa);$
$k = 10^{-14}\ (m^2\ );\ \mu = 10^{-5}\ (Pa\ sec);\ Y_0 = 10^3\ (Pa);\ Q_f = 300\ (m^3\ /day);$
$p_R = 50\ (10^5\ Pa);\ h = 10\ (m).$

We can see again the existence of a critical gas production rate for the beginning of sand movement to the well. The dotted straight line is given to show the curvature of lines corresponded to calculation.

However, the solution of these equations results in exceedingly large sand production rates (Figure 2).

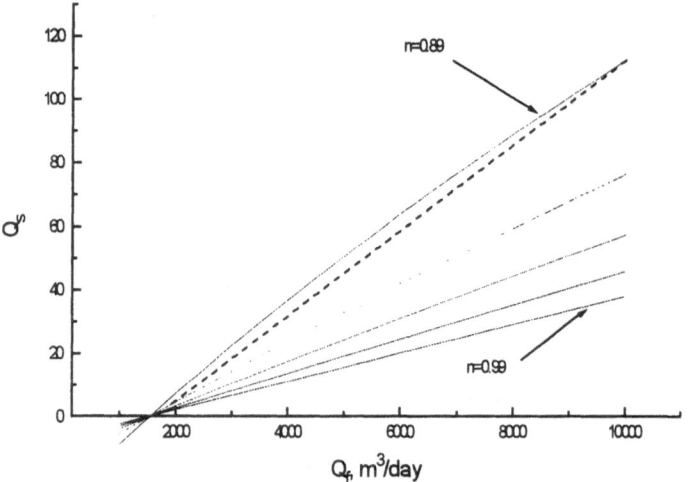

Figure 2. Relationship between the rates of gas production Q_f and of sand production Q_s (stationary inflow).

The first assumption is that plastic flow is realized only in the fine layer as this was confirmed in the practice of Urengoy gas reservoir [2]. Determination of such layer can be realized, for example, by radiometric measurements of loose sand states or by special seismic exploration. Here the existence of such a layers can be accounted for simply by diminishing of Q_s proportionally to the ratio of thicknesses of fine layer and all the formation.

The second assumption, which must be checked, is that that the plastic zone radius has to be less than the feeding contour. The apropriatenen of this assumption can be determined by employing a nonstationary analysis.

Nonstationarity due to the growing viscoplastic zone was introduced by Geilikman et al (1994) for the sand production zone. However, they used the associative flow rule (that is, $N = n$) and elasticity of outer filter zone was omitted.

6. Self-similar solution of sand production.

The situation when the plastic zone is localized within the filter zone needs in the nonstationary (for example, a self-similar) problem. Then the plastic zone corresponds to the close vicinity of acting well, and the main part of a reservoir is assumed to be under elastic deformation. In this case the system of equations has to be solved taking into account time dependency, because the elasto-plastic boundary is moving outwards. This is a consequence of continuity condition for displacements and velocities at the boundary.

Assume that the plastic state occurs in the form of plane flow inside an angular zone $a(t) \le r \le b(t)$, where $a(t)$ is the growing radius of the well and $b(t)$ is the elastoplastic boundary. The radius $a(t)$ is defined by the conditions of zero radial effective stress and of given well pressure p_a. We assume that $a < r_w$, where r_w is the real well radius. The elastoplastic boundary $b(t)$ will be found from the yield condition (2.6).

Additionally the conditions of constant sand and fluid productions are formulated for the same growing well - that is,

$$Q_f = const, \qquad Q_s = const, \qquad r = a(t) < r_w \qquad (6.1)$$

At the elastoplastic boundary the conditions arise from continuity of solid velocities and radial stresses:

$$[v] = 0 \quad , \qquad [\sigma_r] = 0 \quad , \quad r = b(t) \qquad (6.2)$$

At the infinity ($r/a \to \infty$) and at the zero time the same conditions of the rest are assumed to be valid :

$$m = m_0, \qquad p = p_0, \qquad \sigma_r = -(\Gamma - p_0) \qquad (6.3)$$

If the same limit condition is used as the criterion of geomaterial failure and for plastic flow in a fragmented state inside the plastic zone, the hoop stresses will be continuous. In or case condition (2.6) is used for simplicity at both the sides of the boundary $r = b(t)$, although it is a particular case.

We suggest that Darcy's law (2.3) has place in elastic and plastic zones.

For the elastic zone we have the equation of radial force balance for solid phase:

$$\frac{\partial \sigma_r}{\partial r} + \frac{2G}{r}\left(\frac{\partial u}{\partial r} - \frac{u}{r}\right) - \frac{\partial p}{\partial r} = 0 \qquad (6.4)$$

where the tangential stress is expressed by the radial displacement in accordance with the Hook law:

$$\sigma_r = 2G\left(\frac{\partial u}{\partial r} + \frac{v}{1-2v}\left(\frac{\partial u}{\partial r} + \frac{u}{r}\right)\right) - \varepsilon p \qquad (6.5)$$

$$\sigma_\theta = 2G\left(\frac{u}{r} + \frac{v}{1-2v}\left(\frac{\partial u}{\partial r} + \frac{u}{r}\right)\right) - \varepsilon p \qquad (6.6)$$

and ε is the ratio of the matrix and grain compressibilities [4,5]. For sands $\varepsilon \ll 1$ and the corresponding term is omitted in (6.5), (6.6) for simplicity.

The sand velocity is related to the sand displacement through the relationship

$$v = \frac{\partial u}{\partial t} \qquad (6.7)$$

For the plastic zone the equation of radial force balance in the form (3.16) and the kinematic dilatancy connection between volume and shear deformations (2.7) must be satisfied.

Now introduce the self-similar variable ξ such that

$$\xi = \frac{r}{\sqrt{t}}, \qquad \frac{\partial}{\partial t} = -\frac{1}{2t}\xi\frac{d}{d\xi}, \qquad \frac{\partial}{\partial r} = \frac{1}{\sqrt{t}}\frac{d}{d\xi}$$

$$w\sqrt{t} = \overline{w}, \qquad v\sqrt{t} = \overline{v},$$

(6.8)

Connection (6.7) has to be rewritten as

$$\frac{\overline{v}}{\sqrt{t}} = \frac{\partial}{\partial t}(\overline{u}\sqrt{t})$$

(6.9)

and then according to (6.8) we have

$$\overline{v} = \frac{\overline{u}}{2} - \frac{1}{2}\xi\frac{d\overline{u}}{d\xi}$$

(6.10)

Then all the boundary conditions, including (6.1), (6.2) and (6.3) (at $\xi = \infty$), become stationary, although, in real coordinates, these boundaries are moving outward at velocities inversely proportional to $\sqrt{t}$.

For computer calculation we assume that the condition at the infinity can be changed to the radius $R(t) \gg a(t)$, $b(t)$, analogous to the feeding contour in the stationary problem (for example, $R = \sqrt{t}$) and such that :

$$m_R = m_0, \qquad p_R = p_0, \qquad \sigma_R = -(\Gamma - p_0) \qquad \xi = \xi_R$$

(6.11)

The corresponding axisymmetric self-similar problem was solved numerically by a Kutta-Runge method. Six undefined functions p', u', v', σ'_r, m, w were determined as part of solution processs. At each step the Coulomb failure criterion (2.6) was checked. The displacement at the feeding boundary was chosen to ensure that $\sigma_r = 0$ and $p = p_a$ at the well radius $a = \xi_a\sqrt{t}$. The condition $a < r_w$ was checked also. The corresponding velocities permit us to find the rates of sand and fluid production. For simplicity we assume that porosity is constant in the elastic zone.

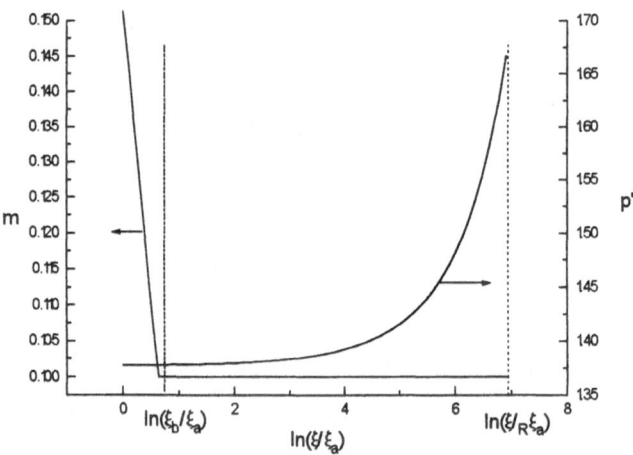

Figure 3. Porosity and pore pressure as functions of the self-similar variable.

In Figure 3,4,5 the results of calculation are given for the following data :

$n=1.1; N = 3; R=100 \ (m) \ ; \ \sigma_R =-100 \ (10^5 \ Pa); G = 4 \ (10^8 \ Pa); h = 10 \ (m);$
$k=10^{-13} \ (m^2); \ \mu = 0.001 \ (Pa \ sec); Y_0 = 30 \ (10^5 \ Pa); Q_f = 1.07 \ (m^3/day);$
$Q_s = 0.006 \ (m^3/day); \ \rho_f = 10^3 \ (kg/m^3); p_R = 50 \ (10^5 \ Pa).$

We use the dimensionless variables now:

$$p' = \frac{p}{Y_0}, \qquad \sigma'_r = \frac{\sigma_r}{Y_0}, \qquad \sigma'_\theta = \frac{\sigma_\theta}{Y_0}, \qquad (6.12)$$

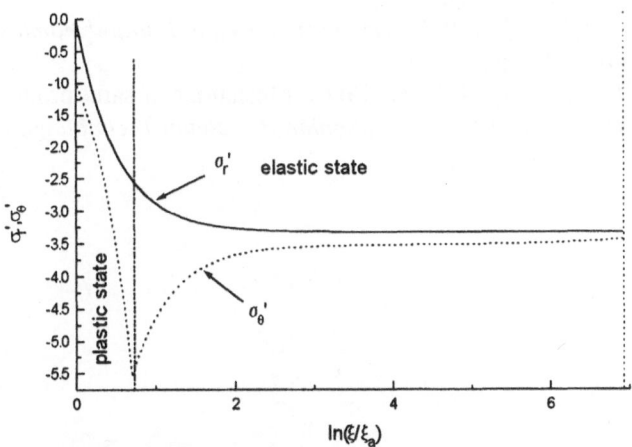

Figure 4. Stress distribution in the well vicinity.

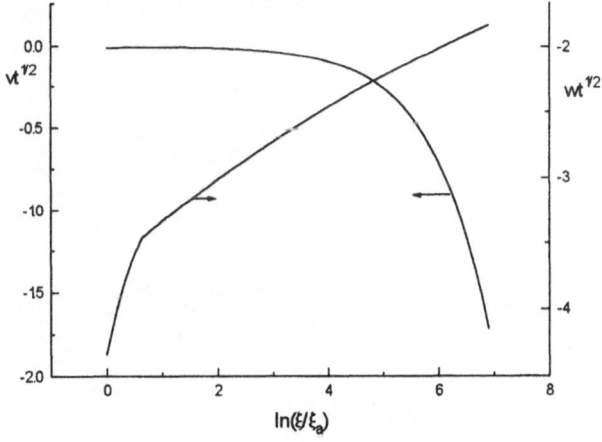

Figure 5. Sand and fluid velocities in the well vicinity.

7. References

Detournay E., Fairhurst C. (1987). Two-dimensional elastoplastic analysis of a long, cylindrical cavity under non-hydrostatic loading. *Int. J. Rock Mech. Min Sci. & Geomech. Abstr.* Vol. 24, No. 4, pp.197-211.

Geilikman M.B., Dusseault M.B., Dullien F.A. (1994) Sand production as a viscoplastic granular flow. *Soc. Petrol. Engrs.* SPE 27343, 41-50.

Lapuk B.B. (1948). Theoretical basement of development of natural gas reservoirs. Moscow : Gostoptekhizdat, 296 pp (in Russian).

Maslennikov V.V., Remizov V.V. (1993). *System Geophysical Control for Development of Giant Gas Reservoirs.* Moscow : Nedra, 304 pp (in Russian).

Nikolaevskiy V.N. (1990). *Mechanics of Porous and Fractured Media.* Singapore: World Scientific, 472 pp.

Nikolaevskiy V.N. (1996). *Geomechanics and Fluidodynamics.* Dordrecht, Kluwer Acad. Press, 348 pp.

Selby R.J., Farouq Ali S.M. (1986). Mechanics of sand production and the flow of fines in porous media. *J. of Canadian Petroleum Technology,* v. 27, # 3, 55 - 63.

Mechanics

SOLID MECHANICS AND ITS APPLICATIONS

Series Editor: G.M.L. Gladwell

Aims and Scope of the Series

The fundamental questions arising in mechanics are: *Why?, How?,* and *How much?* The aim of this series is to provide lucid accounts written by authoritative researchers giving vision and insight in answering these questions on the subject of mechanics as it relates to solids. The scope of the series covers the entire spectrum of solid mechanics. Thus it includes the foundation of mechanics; variational formulations; computational mechanics; statics, kinematics and dynamics of rigid and elastic bodies; vibrations of solids and structures; dynamical systems and chaos; the theories of elasticity, plasticity and viscoelasticity; composite materials; rods, beams, shells and membranes; structural control and stability; soils, rocks and geomechanics; fracture; tribology; experimental mechanics; biomechanics and machine design.

1. R.T. Haftka, Z. Gürdal and M.P. Kamat: *Elements of Structural Optimization.* 2nd rev.ed., 1990   ISBN 0-7923-0608-2
2. J.J. Kalker: *Three-Dimensional Elastic Bodies in Rolling Contact.* 1990
                                                                       ISBN 0-7923-0712-7
3. P. Karasudhi: *Foundations of Solid Mechanics.* 1991   ISBN 0-7923-0772-0
4. *Not published*
5. *Not published.*
6. J.F. Doyle: *Static and Dynamic Analysis of Structures.* With an Emphasis on Mechanics and Computer Matrix Methods. 1991   ISBN 0-7923-1124-8; Pb 0-7923-1208-2
7. O.O. Ochoa and J.N. Reddy: *Finite Element Analysis of Composite Laminates.*
                                                                       ISBN 0-7923-1125-6
8. M.H. Aliabadi and D.P. Rooke: *Numerical Fracture Mechanics.*   ISBN 0-7923-1175-2
9. J. Angeles and C.S. López-Cajún: *Optimization of Cam Mechanisms.* 1991
                                                                       ISBN 0-7923-1355-0
10. D.E. Grierson, A. Franchi and P. Riva (eds.): *Progress in Structural Engineering.* 1991
                                                                       ISBN 0-7923-1396-8
11. R.T. Haftka and Z. Gürdal: *Elements of Structural Optimization.* 3rd rev. and exp. ed. 1992
                                                       ISBN 0-7923-1504-9; Pb 0-7923-1505-7
12. J.R. Barber: *Elasticity.* 1992   ISBN 0-7923-1609-6; Pb 0-7923-1610-X
13. H.S. Tzou and G.L. Anderson (eds.): *Intelligent Structural Systems.* 1992
                                                                       ISBN 0-7923-1920-6
14. E.E. Gdoutos: *Fracture Mechanics.* An Introduction. 1993   ISBN 0-7923-1932-X
15. J.P. Ward: *Solid Mechanics.* An Introduction. 1992   ISBN 0-7923-1949-4
16. M. Farshad: *Design and Analysis of Shell Structures.* 1992   ISBN 0-7923-1950-8
17. H.S. Tzou and T. Fukuda (eds.): *Precision Sensors, Actuators and Systems.* 1992
                                                                       ISBN 0-7923-2015-8
18. J.R. Vinson: *The Behavior of Shells Composed of Isotropic and Composite Materials.* 1993
                                                                       ISBN 0-7923-2113-8
19. H.S. Tzou: *Piezoelectric Shells.* Distributed Sensing and Control of Continua. 1993
                                                                       ISBN 0-7923-2186 3

Kluwer Academic Publishers – Dordrecht / Boston / London

Mechanics

SOLID MECHANICS AND ITS APPLICATIONS

Series Editor: G.M.L. Gladwell

Kluwer Academic Publishers – Dordrecht / Boston / London